SOIL MAPPING AND PROCESS MODELING FOR SUSTAINABLE LAND USE MANAGEMENT

SOIL MAPPING AND PROCESS MODELING FOR SUSTAINABLE LAND USE MANAGEMENT

Edited by

PAULO PEREIRA
Mykolas Romeris University, Vilnius, Lithuania

ERIC C. BREVIK
Dickinson State University, Dickinson, ND, United States

MIRIAM MUÑOZ-ROJAS
The University of Western Australia, Crawley, WA, Australia; Kings Park and Botanic Garden, Perth, WA, Australia

BRADLEY A. MILLER
Leibniz Centre for Agricultural Landscape Research (ZALF), Müncheberg, Germany; Iowa State University, Ames, IA, United States

elsevier.com

Elsevier
Radarweg 29, PO Box 211, 1000 AE Amsterdam, Netherlands
The Boulevard, Langford Lane, Kidlington, Oxford OX5 1GB, United Kingdom
50 Hampshire Street, 5th Floor, Cambridge, MA 02139, United States

British Library Cataloguing-in-Publication Data
A catalogue record for this book is available from the British Library

Library of Congress Cataloging-in-Publication Data
A catalog record for this book is available from the Library of Congress

ISBN: 978-0-12-805200-6

For Information on all Elsevier publications visit our website at https://www.elsevier.com/books-and-journals

Publisher: Candice Janco
Acquisition Editor: Candice Janco
Editorial Project Manager: Emily Thomson
Production Project Manager: Mohanapriyan Rajendran
Designer: Victoria Pearson

Typeset by MPS Limited, Chennai, India

Contents

5. Geographic Information Systems and Spatial Statistics Applied for Soil Mapping: A Contribution to Land Use Management

BRADLEY A. MILLER

6. Soil Mapping and Processes Models for Sustainable Land Management Applied to Modern Challenges

MIRIAM MUÑOZ-ROJAS, PAULO PEREIRA, ERIC C. BREVIK, ARTEMI CERDÀ AND ANTONIO JORDÁN

III CASE STUDIES AND GUIDELINES

7. Modeling Agricultural Suitability Along Soil Transects Under Current Conditions and Improved Scenario of Soil Factors

SAMEH K. ABD-ELMABOD, ANTONIO JORDÁN, LUUK FLESKENS, JONATHAN D. PHILLIPS, MIRIAM MUÑOZ-ROJAS, MARTINE VAN DER PLOEG, MARÍA ANAYA-ROMERO, SOAD EL-ASHRY AND DIEGO DE LA ROSA

8. Soil and Land Use in the Alps—Challenges and Examples of Soil-Survey and Soil-Data Use to Support Sustainable Development

CLEMENS GEITNER, JASMIN BARUCK, MICHELE FREPPAZ, DANILO GODONE, SVEN GRASHEY-JANSEN, FABIAN E. GRUBER, KATI HEINRICH, ANDREAS PAPRITZ, ALOIS SIMON, SILVIA STANCHI, ROBERT TRAIDL, NINA VON ALBERTINI AND BORUT VRŠČAJ

9. Compilation of Functional Soil Maps for the Support of Spatial Planning and Land Management in Hungary

LÁSZLÓ PÁSZTOR, ANNAMÁRIA LABORCZI, KATALIN TAKÁCS, GÁBOR SZATMÁRI, NÁNDOR FODOR, GÁBOR ILLÉS, KINGA FARKAS-IVÁNYI, ZSÓFIA BAKACSI AND JÓZSEF SZABÓ

10. Mapping Ash $CaCO_3$, pH, and Extractable Elements Using Principal Component Analysis

PAULO PEREIRA, ERIC C. BREVIK, ARTEMI CERDÀ, XAVIER ÚBEDA, AGATA NOVARA, MARCOS FRANCOS, JESUS RODRIGO COMINO, IGOR BOGUNOVIC AND YONES KHALEDIAN

11. Human-Impacted Catenas in North-Central Iowa, United States: Ramifications for Soil Mapping

JENNY L. RICHTER AND C. LEE BURRAS

12. Mapping Soil Vulnerability to Floods Under Varying Land Use and Climate

ABDALLAH ALAOUI

List of Contributors

Sameh K. Abd-Elmabod National Research Centre, Cairo, Egypt; University of Seville, Seville, Spain

Abdallah Alaoui University of Bern, Bern, Switzerland

María Anaya-Romero Evenor-Tech, Seville, Spain

Zsófia Bakacsi Hungarian Academy of Sciences, Budapest, Hungary

Jasmin Baruck University of Innsbruck, Innsbruck, Austria

Igor Bogunovic The University of Zagreb, Zagreb, Croatia

Eric C. Brevik Dickinson State University, Dickinson, ND, United States

C. Lee Burras Iowa State University, Ames, IA, United States

Artemi Cerdà University of Valencia, Valencia, Spain

Sabine Chabrillat GFZ German Research Center for Geosciences, Potsdam, Germany

Jesus Rodrigo Comino Trier University, Trier, Germany; Málaga University, Málaga, Spain

Diego de la Rosa Earth Sciences Section, Royal Academy of Sciences, Seville, Spain

Daniel Depellegrin Mykolas Romeris University, Vilnius, Lithuania

Soad El-Ashry National Research Centre, Cairo, Egypt

Paula Escribano University of Almeria, Almería, Spain

Ferran Estebaranz Universitat de Barcelona, Barcelona, Spain

Kinga Farkas-Iványi Hungarian Academy of Sciences, Budapest, Hungary

Luuk Fleskens Wageningen University and Research Centre, Wageningen, The Netherlands

Nándor Fodor Hungarian Academy of Sciences, Martonvásár, Hungary

Marcos Francos University of Barcelona, Barcelona, Spain

Michele Freppaz University of Turin, Grugliasco, Italy

Mónica García Denmark Technical University (DTU), Kongens Lyngby, Denmark; Columbia University, New York City, NY, United States

Clemens Geitner University of Innsbruck, Innsbruck, Austria

Danilo Godone Geohazard Monitoring Group, CNR IRPI, Turin, Italy

Sven Grashey-Jansen University of Augsburg, Augsburg, Germany

Fabian E. Gruber University of Innsbruck, Innsbruck, Austria

Kati Heinrich Institute for Interdisciplinary Mountain Research, Austrian Academy of Sciences, Innsbruck, Austria

Gábor Illés National Agricultural Research and Innovation Centre, Sárvár, Hungary

Antonio Jordán University of Seville, Seville, Spain

Yones Khaledian Iowa State University, Ames, IA, United States

Annamária Laborczi Hungarian Academy of Sciences, Budapest, Hungary

Beatriz Lozano-García University of Córdoba, Cordoba, Spain

Oleksandr Menshov Taras Shevchenko National University of Kyiv, Kyiv, Ukraine

Bradley A. Miller Iowa State University, Ames, IA, United States; Leibniz Centre for Agricultural Landscape Research (ZALF), Müncheberg, Germany

Ieva Misiune Mykolas Romeris University, Vilnius, Lithuania

Miriam Muñoz-Rojas The University of Western Australia, Crawley, WA, Australia; Kings Park and Botanic Garden, Perth, WA, Australia

Agata Novara University of Palermo, Palermo, Italy

Marc Oliva University of Lisbon, Lisboa, Portugal

Andreas Papritz ETH Zurich, Zürich, Switzerland

Luis Parras-Alcántara University of Córdoba, Cordoba, Spain

László Pásztor Hungarian Academy of Sciences, Budapest, Hungary

Paulo Pereira Mykolas Romeris University, Vilnius, Lithuania

Jonathan D. Phillips University of Kentucky, Lexington, KY, United States

Jenny L. Richter Iowa State University, Ames, IA, United States

Emilio Rodríguez-Caballero Max Planck Institute for Chemistry, Mainz, Germany; University of Almeria, Almería, Spain

Thomas Schmid Center for Energy, Environment and Technology Research, Madrid, Spain

Alois Simon Provincial Government of Tyrol, Innsbruck, Austria

Anna Smetanova National Institute for Agricultural Research, Paris, France; Technical University Berlin, Berlin, Germany

Silvia Stanchi University of Turin, Grugliasco, Italy

József Szabó Hungarian Academy of Sciences, Budapest, Hungary

Gábor Szatmári Hungarian Academy of Sciences, Budapest, Hungary

Katalin Takács Hungarian Academy of Sciences, Budapest, Hungary

Robert Traidl Bavarian Environmental Agency, Marktredwitz, Germany

Xavier Úbeda University of Barcelona, Barcelona, Spain

Martine van der Ploeg Wageningen University and Research Centre, Wageningen, The Netherlands

Nina von Albertini Umwelt Boden Bau, Paspels, Switzerland

Borut Vrščaj Agricultural Institute of Slovenia, Ljubljana, Slovenia

Preface

Soils are the base of life on Earth. This thin layer of so-called "earth skin" provides an invaluable number of services that permit the planet to be habitable by life as it exists on Earth today. Soils are created at the interface of the lithosphere, atmosphere, biosphere, and hydrosphere. Their formation depends on parent material, topography, time, climate, and organisms, with other factors such as fire and humans gaining in importance. Soil formation is very slow and the soil itself is considered a nonrenewable resource over the human time scale.

The expansion of human activities is inducing tremendous soil degradation, without precedent in Earth's history. This uncontrolled expansion is leading to an important decrease in the services provided by soils at a global scale. Soil degradation is caused by climate change, conflict and wars, land use changes, deforestation, and other activities, threatening overall global food security, environmental sustainability and trigger famine, conflicts and wars.

Stopping this trend is a challenge for our time. Addressing this challenge is a duty and responsibility that we have to future generations to ensure them the provision of soil services that have existed in the past and that we have today. In this context, we scientists need to create knowledge, identify problems and offer solutions to invert this dynamic. It is essential that we provide sustainable measures to utilize soil resources without dilapidating or degrading them. Sustainable soil management is not an option, it is a necessity and a responsibility that scientists, stakeholders, decision makers and all the other agents involved in land management have to acknowledge and respond to out of respect for future generations and the health of planet Earth.

A key piece to understanding sustainable soil management is to recognize the unique characteristics of different soils as they are distributed across landscapes. Soil spatial variability can only be understood with modeling and maps. Maps are a simple, synthetic and clear representation of reality. Maps are spatial models that are tremendously useful for scientists to develop research and for land managers to intervene appropriately in the territory they control to protect and restore soil. Soil maps can identify and predict areas that are more vulnerable to degradation and thus promote sustainable use of the land to facilitate better and more customized management, contributing to the optimal allocation of resources for continued long-term use of the soil resource.

Soil Mapping and Process Modeling for Sustainable Land Use Management is an original book and the first published on this topic. The intent is to transfer knowledge of the current state of the art to students, scientists, land managers, and stakeholders to facilitate sustainable use of land resources. The chapters of this book were written by leading scientists who have several years of experience in this field.

The book is organized in two parts. The first is composed of six chapters focused on the theoretical aspects of soil mapping and process modeling, where historical and current aspects of soil mapping and sustainable land use management are analyzed. The importance of the integration of soil mapping and traditional know-how for sustainable use of the land, use of remote sensing for mapping and monitoring, application of GIS tools to soil mapping, analysis, and land use management, and the use of soil mapping and process modeling to address modern challenges are also discussed. The second part of the book has a practical orientation, where the methods discussed in the first part have been applied to several areas in Europe, the United States, and Africa.

Soil Mapping and Process Modeling for Sustainable Land Use Management is a product of several years of research and collaboration between the editors and authors of the book. The idea to create this book was discussed prior to and during the European Geoscience Union Assembly in Vienna in 2015, during the organization and execution of a short course titled "*Short course on soil mapping methods.*" Some of the authors of this book have collaborated for a decade and we joined our knowledge and efforts to provide what we hope will be an important contribution about *Soil Mapping and Process Modeling for Sustainable Land Use Management*. We truly believe this topic represents a crucial challenge in the present that will significantly impact future generations.

We would like to express our appreciation for the enormous support provided by Marisa LaFleur, Emily Thomson, and Rajesh Manohar, for their incredible editorial and technical support that was fundamental for the compilation of this monograph. We would also like to thank all the contributing authors that helped make it possible to bring this book to light. It was only with their commitment and enthusiasm that this project became a reality.

The Editors
Paulo Pereira
Eric C. Brevik
Miriam Muñoz-Rojas
Bradley A. Miller

PART I

THEORY

CHAPTER

1

Historical Perspectives on Soil Mapping and Process Modeling for Sustainable Land Use Management

Eric C. Brevik[1], *Paulo Pereira*[2], *Miriam Muñoz-Rojas*[3], *Bradley A. Miller*[4,5], *Artemi Cerdà*[6], *Luis Parras-Alcántara*[7] *and Beatriz Lozano-García*[7]

[1]Dickinson State University, Dickinson, ND, United States [2]Mykolas Romeris University, Vilnius, Lithuania [3]The University of Western Australia, Crawley, WA, Australia [4]Iowa State University, Ames, IA, United States [5]Leibniz Centre for Agricultural Landscape Research (ZALF), Müncheberg, Germany [6]University of Valencia, Valencia, Spain [7]University of Córdoba, Cordoba, Spain

INTRODUCTION

Basic soil management goes back to the earliest days of agricultural practices, approximately 9000 BCE. Through time humans developed soil management techniques of ever increasing complexity, including plows, contour tillage, terracing, and irrigation. Spatial soil patterns were being recognized as early as 3000 BCE, but the first soil maps did not appear until the 1700s and the first soil models finally arrived in the 1880s. The beginning of the 20th century saw an increase in standardization in many soil science methods and wide-spread soil mapping in many parts of the world, particularly in developed countries. However, the classification systems used, mapping scale, and national coverage varied considerably from country to country. Major advances were made in pedologic modeling starting in the 1940s, and in erosion modeling starting in the 1950s. In the 1970s and 1980s advances in computing power, remote and proximal sensing, geographic information systems (GIS), global positioning systems (GPS), and statistics and spatial statistics among other numerical techniques significantly enhanced our ability to map and model soils. These types of advances positioned soil science to make meaningful contributions to sustainable land use management as we moved into the 21st century.

Soil Mapping and Process Modeling for Sustainable Land Use Management.
DOI: http://dx.doi.org/10.1016/B978-0-12-805200-6.00001-3

BRIEF REVIEW OF DEVELOPMENTS PRIOR TO THE 20TH CENTURY

In many respects we can say that soil science has a long prehistory and a brief history (De la Rosa, 2008, 2013). Soil science has long standing ties to agriculture. The earliest evidence of agricultural practices comes from an area near Jarmo, Iraq dating to 9000 BCE, and there is evidence of irrigation from southern Iraq dating to 7500 BCE (Troeh et al., 2004). Between 6000 and 500 BCE soil management techniques including early plows, terracing, drainage, and contour tillage were developed in various parts of Europe (Fig. 1.1) (Brevik and Hartemink, 2010) and by the Maya and pre-Inca in Central and South America, who also engineered soils (Hillel, 1991; Jensen et al., 2007). Along with advances in production, various forms of land degradation, including soil erosion and salinization, became a problem very early in the history of agriculture (Hillel, 1991; Troeh et al., 2004). It is likely that early humans used a trial and error approach to determine which sites would work well for agricultural production, but by 3000–2000 BCE there is good evidence that humans were recognizing spatial patterns in soil and utilizing the more desirable soils for cropping (Krupenikov, 1992; Miller and Schaetzl, 2014). During the Sumerian and Babylonian civilizations, until 1000 BCE, agriculture continued to be developed. Soils were distinguished by their natural fertility and aptitude to support irrigation. From 2000 BCE the Greeks improved numerous treatises in which they explained their knowledge about different soil properties. Soil erosion was a serious problem in Ancient Greece; therefore it was thoroughly studied. Likewise, by about 500 BCE settlement patterns in many parts of the world were correlated to the kinds of soils present (Miller and Schaetzl, 2014). The Romans continued the Greek's studies. From 200 BCE, Catón, Varrón, Plinio, and later (in the first century AC) Columela proclaimed agriculture as a science, and considered soil as one of the most important components.

Knowledge about a subject must be accumulated before that subject can be classified (Marbut, 1922), and classification of soils began thousands of years ago. Early examples include the Chinese classification from 2000 BCE (Gong et al., 2003) and that of the Greek philosopher Theophrastus from c. 300 BCE (Brevik and

FIGURE 1.1 Terraces, such as these in Spain, have been used for thousands of years to make steep slopes suitable for agricultrual production. *Source: Photograph by Artemi Cerdà.*

Hartemink, 2010). In addition, the Romans developed a soil classification system for the soils of Italy and improved previous knowledge about soil fertility and ways to maintain and restore it. There are very important and interesting written works, such as *Res Rustica* (Columela, 42 CE) where the author describes soils in detail. In the Western Hemisphere the Maya civilization in Central America created a detailed soil classification that they used to guide their agricultural decisions long before Europeans arrived (Wells and Mihok, 2010). Therefore humans have sought to describe and manage soils based on their properties and have recognized a spatial distribution to those properties for thousands of years. However, while this was a precursor to soil mapping and modeling, recognizing the existence of spatial distribution of soil properties is different than actually mapping and modeling those properties.

The first recordings of spatial soil information were written accounts linking soil properties and attributes to land ownership documents. These were utilized in China as early as 300 CE, Arabia as early as 500 CE, and Europe as early as 800 CE (Miller and Schaetzl, 2014). Soil properties and attributes were first mapped in Europe beginning in the 1700s (Brevik and Hartemink, 2010), something that was made possible by improved base maps (Miller and Schaetzl, 2014). The 1800s saw increasing interest in soil mapping in Europe and the United States; much of the mapping in the United States was done by state geological surveys in an attempt to justify their budgets to state legislatures that were looking for a return on their investment (Aldrich, 1979).

In parallel with these advances in the first recordings of spatial soil information, it is important to point out the treatise "Agricultura General" (de Herrera, 1513), based on the previous studies of Columela, where the author introduced highlighted points about soil quality. After that, during the 19th century, advances were made in many areas that would ultimately prove to be important to understanding soil science for the purpose of sustainable management. The "Mineral Theory" of plant nutrition was first proposed by C. Sprengel in the late 1820s (Feller et al., 2003a) and became widely accepted after von Liebig's (1840) publication of *Chemistry as a Supplement to Farming and Plant Physiology*, which was a major improvement for both soil fertility and soil chemistry (Sparks, 2006). Many advances were made in soil mapping and cartography in both Europe and the United States, and the soil profile concept was developed (Brevik and Hartemink, 2010). Through his work on the influence of earthworms on soil development, Charles Darwin became a pioneer in soil biology (Feller et al., 2003b).

A major breakthrough in soil mapping and modeling occurred in Russia with the publication of Dokuchaev's (1883) classic work "Russian Chernozem." This work included a map showing the distribution of Chernozems in European Russia, but more importantly it introduced the concept of soil forming factors that ultimately led to the recognition of soil science as a stand-alone scientific discipline (Muir, 1962; Krupenikov, 1992; Krasilnikov et al., 2009). Dokuchaev's functional–factoral model was one of the first developed to explain soil formation (Brevik et al., 2016) and introduced the five soil forming factors: climate, parent material, organisms, topography, and time (Brevik and Hartemink, 2010). These five factors would eventually be cast into a state-factor model by Jenny, one of the most influential models in the history of soil science. Therefore Dokuchaev's work remains highly influential to the current day. Eugene Hilgard published ideas about soil formation quite similar to Dokuchaev's in 1860 (Hilgard, 1860), and for these ideas Jenny (1961) felt that Hilgard should be regarded as a cofounder of modern soil science along with Dokuchaev.

Unfortunately, Hilgard's advanced ideas did not catch on in the United States at the time he presented them and were left instead to be discovered by other soil scientists decades after they were originally published (Brevik et al., 2015).

Through most of human history pursuit of soil knowledge was motivated by and linked to agriculture. At the end of the 19th century, soil mapping was only about 100 years old and soil modeling had just begun. However, the slow and steady accumulation of knowledge about soils as well as advances in several related fields (biology, chemistry, geography, geology, and physics) meant that by the end of the 19th century soil mapping and modeling was positioned to make major strides in the 20th century. Those strides would vastly improve the ability of soil scientists to utilize soil information for agricultural management and would also take soils beyond agriculture and into areas like human health, urban planning, and environmental quality. Soil knowledge was poised to become a major player in sustainable land use management.

DEVELOPMENTS IN THE 20TH CENTURY

Much of the 20th century was, in many ways, a golden era of soil science, particularly from the 1930s through the 1970s. During this time, budgets for soil work were relatively strong, including funding for international work in developing countries (Brevik et al., 2015). Soil ideas were exchanged internationally through the development of meetings like the World Congresses of Soil Science and conservation tillage techniques were developed (Brevik and Hartemink, 2010). A number of methods and standards that would be important through the last half of the 20th century were established, including the use of aerial photographs as a base for soil mapping, standards for describing soil structure, use of the Munsell color charts (Hudson, 1999), and publication of standard soil survey laboratory and sample collection methods in the United States (Nettleton and Lynn, 2008). These conditions and advances were major milestones in soil survey and set the stage for the creation of much of the information available in the soil maps of today. Mapping products generated during the 20th century included everything from detailed maps to those produced at national scales (Fig. 1.2).

The latter part of the 20th century also saw increasing interest in sustainable land use. However, as pointed out by Blum (1998), there were considerable differences in the interpretation of "sustainable land use" at the end of the 20th century, and much of the discussion focused on agricultural land use without considering other kinds of land use. Blum (1998) proposed the following definition for sustainable land use: "The spatial (local or regional) and temporal harmonization of all six soil functions [1. agricultural and forest production, 2. source of raw materials, 3. geogenic and cultural heritage forming landscapes, 4. gene reserve and protection, 5. filtering, buffering, and transformation, and 6. Infrastructure] through minimizing irreversible uses, e.g., sealing, excavation, sedimentation, acidification, contamination or pollution, salinization and others." However, there are many challenges to defining sustainable land use, and well into the 2000s there still was not a globally accepted comprehensive definition (Kaphengst, 2014). Both Blum (1998) and Kaphengst (2014) agree that sustainable land use extends beyond the natural sciences to encompass social aspects such as political and economic considerations, making sustainable land use a truly transdisciplinary topic. Unfortunately, sustainable land use and management was also rare at the end of the 20th century. For example, Eswaran et al. (2001) estimated that only 10% of land in Asia was used sustainably.

There was a general global economic downturn in the 1980s (Garrett, 1998) that was

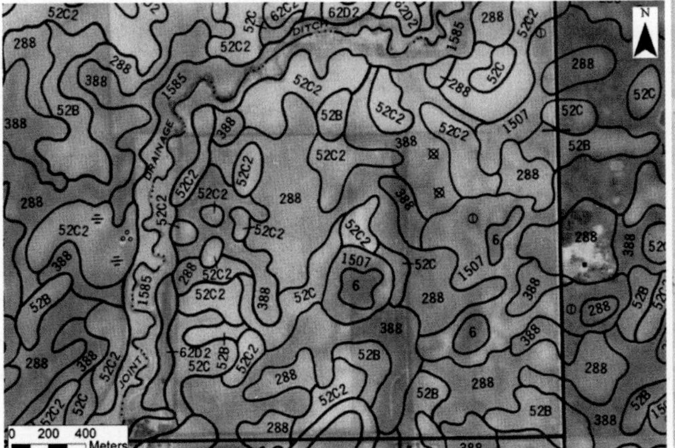

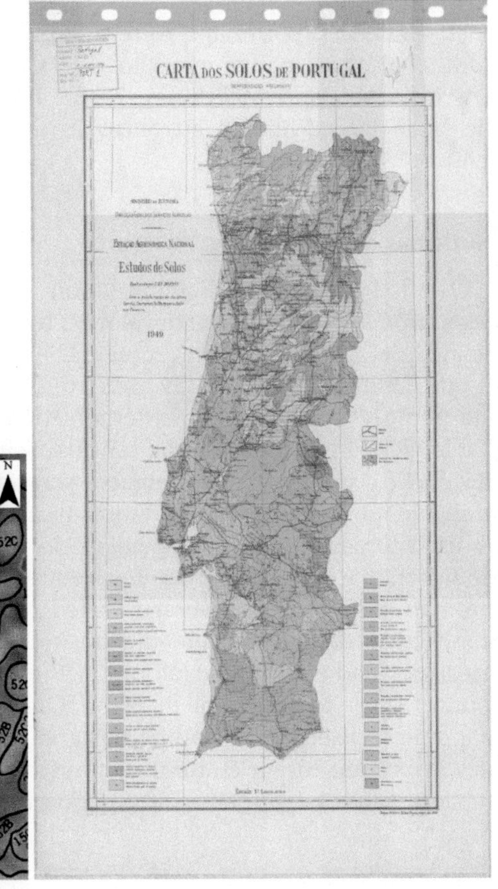

FIGURE 1.2 Examples of soil maps created by national soil survey programs during the 20th century include detailed maps such as the 1:15,840 map from the United States (left; Jones, 1997) and less detailed maps such as the national map of Portugal at the scale of 1:1,000,000 produced in 1949 (right) (http://esdac.jrc.ec.europa.eu/images/Eudasm/PT/port_x21.jpg).

accompanied by corresponding declines in soil science budgets (Hartemink and McBratney, 2008). However, new tools and technologies such as GIS, GPS, remote and proximal sensing techniques, and the emergence of more robust statistical methods and spatial statistics helped to overcome some of the obstacles created by reduced financing. The availability of inexpensive, increasingly powerful computers allowed for the storing and rapid processing of large amounts of data. The ability to collect environmental covariates with proximal and remote sensing coupled with spatial statistics and other numerical techniques allowed greater detail in the mapping of soil properties as well as better quantification of those properties (Brevik et al., 2016). While there is still much left to accomplish to improve soil mapping products to support the types of models that are essential for sustainable land management (Sanchez et al., 2009), by the end of the 20th century soil surveys had recognized the need to provide more quantitative data (Indorante et al., 1996) and soil maps were moving from traditional static paper maps to digital products (Minasny and McBratney, 2016). This set the stage for additional advances in the 21st century.

National Soil Mapping Programs

Detailed nationally organized soil survey began in many parts of the developed world in the first decades of the 20th century. This began in the United States in 1899 (Marbut, 1928) and rapidly spread to many other countries (Table 1.1). By the end of the century, several developed countries had detailed soil maps available for portions of the country that could be used to assist with management decisions. However, the amount of land surveyed and the map scale of that coverage varied considerably between countries, as did the soil characteristics and depth of exploration that each country chose to base their maps on. Often the mapping focused on soil properties and attributes important to agricultural or forestry production. Table 1.2 presents information on the mapping status of several developed countries at or near the end of the 20th century based on the most detailed maps produced by their national soil mapping program. It shows that mapping coverage ranged from essentially complete (100%) to barely mapped (0.25%) and that map scales for national mapping programs ranged from 1:2000 to 1:126,720, with the most common mapping scales being about 1:25,000–1:50,000 (Fig. 1.3). Some countries (e.g., Austria, Greece, Portugal, Sweden, and the United Kingdom) focused on agricultural areas, while a few countries (e.g., Bulgaria, Croatia) produced maps at a larger scale (1:1000–1:10,000) than their typical national soil mapping scales (1:25,000–1:50,000 for Bulgaria and Croatia, respectively) as part of their national surveys to address selected areas with special problems or needs including irrigation, drainage, contamination, and remediation (Jones et al., 2005). A number of different soil taxonomic systems were also employed in undertaking the mapping, often developed to address problems or needs that were very specific to each individual country (Krasilnikov et al., 2009). The combination of highly variable mapping coverage and scale between countries and the lack of a common nomenclature to communicate soil information led to nonuniform coverage that, along with a lack of quantitative soil information in most soil mapping, impeded the inclusion of soil information in modeling to support land management decisions (Sanchez et al., 2009).

Soil mapping and soil classification are mutually dependent activities (McCracken and Helms, 1994); therefore the quality of soil classification systems is closely related to the quality of soil mapping and vice versa (Cline, 1977). For this reason, it is important that soil mapping and soil classification be studied jointly when evaluating our understanding of soils. Ideas about soil classification changed considerably over the 20th century in several countries, and dozens of countries have their own classification systems. These systems

TABLE 1.1 The Beginning Date for Detailed Nationally Organized Soil Survey for Select Countries

Country	Date	Country	Date
United States of America	1899	Sri Lanka	1930
Russia	1908	China	1931
Canada	1914	Poland	1935
Australia	1920s	The Netherlands	1945
Great Britain	1920s	Ghana	1946
Mexico	1926	Malaysia	1955

Source: Brevik, E.C., Calzolari, C., Miller, B.A., Pereira, P., Kabala, C., Baumgarten, A., et al., 2016. Soil mapping, classification, and modeling: history and future directions. Geoderma 264, 256–274. http://dx.doi.org/10.1016/j.geoderma.2015.05.017.

TABLE 1.2 Percent Country Mapped at a Detailed Scale by the End of the 20th Century for Several Countries, Showing the Range in Mapping Coverage and Scale Even in Developed Countries

Country	%Mapped	Scale	Additional Notes	Reference
Bulgaria	100	1:25,000	1:10,000 scale mapping underway, and selected problem areas at scales of 1:1000–1:5000	Kolchakov et al. (2005)
Croatia	100	1:50,000	Some 1:5000–1:10,000 scale maps available for areas with special needs	Bašić (2005)
Czech Republic	100	1:5000–1:50,000	All but urban areas mapped	Němeček and Kozák (2005)
Hungary	100	1:25,000	70% of agricultural areas mapped at 1:10,000	Várallyay (2005)
The Netherlands	100	1:50,000		van der Pouw and Finke (2005)
Slovenia	100	1:25,000		Vrščaj et al. (2005)
Belgium	85	1:20,000		Dudal et al. (2005)
USA	>85	1:15,840–1:24,000		Indorante et al. (1996)
Romania	80	1:50,000–1:100,000		Munteanu et al. (2005)
Portugal	55	1:50,000		Gonçalves et al. (2005)
Ireland	44	1:126,720		Lee and Coulter (2005)
Austria	38	1:25,000	Larger scale soil taxation survey maps (1:2000) are also available. All land under agricultural use mapped	Haslmayr et al. (2016)
Finland	33	1:20,000–1:50,000		Sippola and Yli-Halla (2005)
United Kingdom	~24	1:25,000–1:63,360	About 24% of England and Wales, most of the arable land in Scotland, all of Northern Ireland at 1:50,000	Thompson et al. (2005)
Germany	13+	1:25,000	Some state soil quality maps are available for about 48% of Germany at 1:5000 and 1:10,000	Zitzmann (1994), Eckelmann (2005)
France	~12	1:100,000		King et al. (2005)
Switzerland	7	1:25,000		Bonnard (2005)
Greece	6	1:5000–1:20,000	About 39% of the high-quality agricultural land mapped	Yassoglou (2005)
Sweden	0.25	1:20,000	About 3% of the arable land mapped	Olsson (2005)

were often developed to address soil properties or management needs that were specific to the country in which they were developed, and it can be difficult to correlate the system of one country to the soil classification systems of other countries (Krasilnikov et al., 2009). By the early 2000s two classification systems had become the most widely utilized

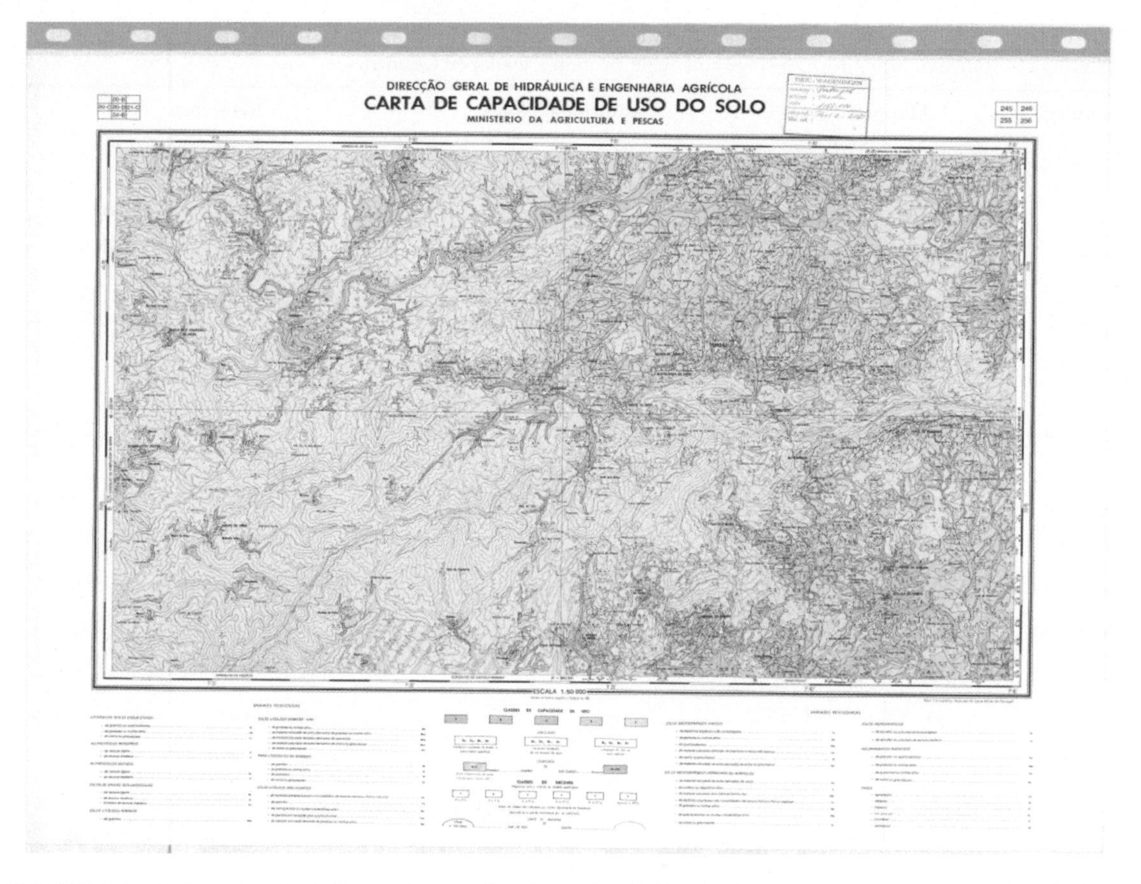

FIGURE 1.3 Soil use capacity in Portugal mapped at the scale of 1:50,000. Map produced in 1980. *Source: http://esdac.jrc.ec.europa.eu/images/Eudasm/PT/port2_20d.jpg.*

in the world, US Soil Taxonomy and the World Reference Base (WRB) (Brevik et al., 2016). However, a soil classification system that established an international standard had not been agreed on by the end of the 20th century. A uniform international system of soil classification that communicates a wide range of information about the soils classified and mapped would facilitate international communication (Sanchez et al., 2009; Hempel et al., 2013). Such standardization would support the compilation of national mapping efforts at a variety of scales and thus the use of spatial soil information for modeling in support of sustainable land management over large areas.

Models in Support of Soil Mapping and Land Use Management

Several models have been developed to explain soil formation, and many of these models have also been used in support of soil mapping. One of the most influential models of soil formation is that of Jenny (1941), who considered soil as a dynamic system and cast the soil forming factors that had been discussed by Hilgard (1860) and Dokuchaev (1883) into a state-factor equation:

$$s = f(cl, o, r, p, t, \ldots) \quad (1.1)$$

This equation can be quantitatively solved in theory, but a number of obstacles to successfully doing so still exist despite many attempts to solve it (Yaalon, 1975; Phillips, 1989). Rather, Jenny's model has been influential because it changed the way that soil studies were approached, leading to studies where one factor was allowed to vary while the others were held constant, thereby investigating the influence of the varying factor on soil properties and processes. This approach is also important for sustainable management planning, in that it views the soil as a part of the overall environment (Jenny, 1941) and thus can be used to investigate how a given change in the overall environment, including changes due to human management, influence the soil system (Yaalon and Yaron, 1966). Finally, from a mapping perspective, Jenny's model has been important in that it helps explain and predict the geographic distribution of soils (Holliday, 2006), a fundamental aspect of mapping.

Another pedogenic model that has been important in understanding how soil changes was the process-systems model developed by Simonson (1959). While Jenny focused on external factors that influenced the final soil created at a given location, Simonson focused on processes that occur within a soil. Also, unlike Jenny's model, Simonson's model was not cast into potentially quantifiable terms. It was a qualitative model meant to help the user understand soil processes, but that was not designed to be mathematically solved. Simonson's model is particularly useful in the study of soil individuals (Schaetzl and Anderson, 2005), which makes its concepts useful to understand human impacts on the soil resource at very large scales. The process-systems approach is also more useful than the functional–factoral approach to understand movement in a soil–landscape (Wysocki et al., 2000), which brings soil–landscape modeling closer to a mass balance approach.

A large number of the legacy soil maps available today, which still serve as the single largest source of accessible soil mapping data (Brevik et al., 2016), were created using soil–landscape relationship models. Once the relationship between the soils in a given area and the landscape were understood, soil–landscape models allowed a soil surveyor to map the soils in a given area with reasonable speed and accuracy using a minimal number of soil samples. To define reasonable accuracy the USA National Cooperative Soil Survey (NCSS) expected soil maps based on soil–landform relationships to have 50% or greater purity in soil map units. The understanding of soil–landform relationships was advanced by a number of studies beginning in the 1930s. Soil geomorphology studies in the United States from the 1930s through the 1970s made major contributions to this understanding (Brevik et al., 2015), as did work in Africa (Milne, 1935), Europe (Gerrard, 1992), and Australia (Butler, 1950). In the modern world, soil–landscape models have had a great influence on mapping and sustainable management through their impact on legacy maps.

Models are increasingly being used as decision support systems (DSSs), which combine available soil, climate, and land use and management data from different sources. DSS can evaluate information under different scenarios helping to support complex decision-making and problems. Among DSS the MicroLEIS DSS has been widely used in land evaluation (De la Rosa et al., 2004) to assist decision-makers with specific agro-ecological problems. MicroLEIS was designed as a knowledge-based approach, incorporating a set of information tools, linked to each other. Thus custom applications can be performed on a wide variety of problems related to land productivity and land degradation.

A major area of interest as we neared the end of the 20th century involved the role of soils in the carbon cycle. Some of the main challenges with soil carbon monitoring include the large amount of work needed to collect the necessary data and the consequently high costs compounded by the lack of consistency between different methods of data collection. To overcome these difficulties, several soil carbon models have been developed in the last few decades with different features and limitations, e.g., CENTURY (Parton et al., 1987), RothC (Coleman and Jenkinson, 1996), and CarboSOIL (Muñoz-Rojas et al., 2013). These models can be linked to spatial data sets (soil, land use, climate, etc.) to assess soil organic C dynamics and to determine current and future estimates of regional soil C stocks and sequestration (Falloon et al., 1998).

Recognizing Erosion as a Problem

Soil erosion is one of the major issues that threatens the sustainable use of the world's soil resources (Pimentel et al., 1995). Soil erosion problems have led to major problems for civilizations worldwide dating back thousands of years (Diamond, 2005). With the exception of some selected individuals who sought to bring attention to the problem, erosion was not widely recognized as a serious issue until about 100 years ago (Brevik and Hartemink, 2010). In the early 1900s in the United States, Milton Whitney, the head of the Bureau of Soils, hired William John McGee and Edward Elway Free to lead studies in soil erosion by water and wind, respectively (Brevik et al., 2015). McGee (1911) and Free (1911) both produced influential publications that provided in-depth reviews of the status of soil erosion knowledge to that time and presented the results of new studies that investigated erosion processes as well as ways to prevent erosion. Free's work has been particularly praised from a soil science perspective because it may be the first work to recognize the impact of windblown materials on soil genesis rather than just investigating wind and windblown materials as a geomorphic process and deposit.

Despite these advances, soil erosion was not recognized as a problem by many in the United States until the great environmental disaster known as the Dust Bowl, which lasted through the drought stricken 1930s in the Great Plains of the United States. The Dust Bowl was marked by extreme water and wind erosion of exposed production agriculture soils; by 1938 it was estimated that 4,047,000 ha of land had lost the top 12.5 cm of its topsoil and another 5,463,000 ha had lost at least 5 cm of topsoil, representing an average loss of 1,076,000 kg of soil ha^{-1} (Hansen and Libecap, 2004). In response to this soil loss the Soil Erosion Service (SES) was formed in 1933 under the direction of Hugh Hammond Bennett as part of President Franklin D. Roosevelt's public works legislation. The SES later became the Soil Conservation Service (SCS) by an act of Congress in 1935 (Helms, 2008). The SES rapidly established several erosion projects that tested and demonstrated soil conservation measures (Helms, 2010) and conservation tillage techniques were developed (Holland, 2004). When similar drought conditions occurred in the Great Plains again in the 1950s and 1970s, erosion on the scale of the Dust Bowl did not occur thanks to conservation measures that had been implemented during and following the 1930s (Hansen and Libecap, 2004).

Still, soil erosion continued to be a major problem. In a study conducted near the end of the 20th century, Pimentel et al. (1995) estimated that approximately one-third of the world's agricultural lands had been lost to erosion in the previous 50 years, with about 1.0×10^6 ha of additional agricultural land lost annually as a consequence of accelerated soil erosion. Soil losses to erosion were estimated as $17\,Mg\,ha^{-1}\,year^{-1}$ in the United States and Europe and $35\,Mg\,ha^{-1}\,year^{-1}$ in Asia, Africa, and South America (Pimentel et al., 1995). It was estimated that soil erosion cost the United States $27 billion annually in onsite costs and $17 billion annually in offsite costs, for a total of $44 billion annually, or about $100 annually ha^{-1} of cropland and pasture. The cost of preventing that erosion was estimated to be $8.4 billion annually. These values would be approximately $68.5 billion annually, $156 annually ha^{-1}, and $13 billion annually, respectively, in 2015 dollars (US BLS, 2016). In all respects these numbers indicated a serious environmental problem that needed to be solved to attain sustainability.

By the end of the 20th century the United States was probably the only country that had long-term soil erosion data collected using standardized methods; other countries had more sporadic (Cerdan et al., 2010) and/or shorter term (Dregne, 1995) erosion data coverage. In fact, Morgan and Rickson (1990) state that as we neared the end of the 20th century, the annual extent of erosion was not known for a single country in Europe. What was known of erosion rates in countries other than the United States was assessed primarily through models of large areas (Yang et al., 2003; Cerdan et al., 2010). That being said, erosion issues were being recognized and documented in other parts of the world during the 20th century (Morgan et al., 1998a), even if the overall effort did not have the same level of national coordination as seen in the United States. While agriculture has been practiced for millennia in Europe, there was not wide-spread concern about the effects of erosion and other agriculturally related environmental problems until the second half of the 20th century (Morgan and Rickson, 1990; Stoate et al., 2001). Strong interest in soil erosion began in New Zealand in the 1930s, but the first systematic national assessment of soil erosion did not occur until the 1970s (Dregne, 1995). Within Australia, where soil conservation efforts are primarily the responsibility of the individual States and Territories, New South Wales established a SCS in 1938, but the first national assessment of land degradation, including soil erosion, did not occur until 1975 (Dregne, 1995). Likewise, wide-spread concern over soil erosion did not take hold in Africa or India until later in the 20th century (Pretty and Shah, 1997). Pimentel et al. (1995) estimated that soil erosion cost $400 billion annually worldwide, or about $70\,person^{-1}\,year^{-1}$. This translates into about $623 billion annually in 2015 dollars (US BLS, 2016), which is about $85\,person^{-1}\,year^{-1}$ at the world's present population of approximately 7.3 billion (US Census Bureau, 2016). Panagos et al. (2015) estimated that early 21st century soil losses to erosion averaged $2.46\,Mg\,ha^{-1}\,year^{-1}$ in Europe while Verheijen et al. (2009) estimated that soil formation in Europe only averaged $1.4\,Mg\,ha^{-1}\,year^{-1}$, indicating that soil in Europe was still being lost to erosion much more rapidly than it was being replaced by pedogenesis as the 20th century ended.

In response to soil erosion issues, many countries or other governmental agencies developed programs that provided incentives and/or requirements for farmers to conserve soil (Morgan and Rickson, 1990; Dregne, 1995; Pretty and Shah, 1997), although in many countries there was still a need to develop soil conservation programs even late into the 20th century and beyond (Morgan and Rickson, 1990; Fullen, 2003). While the details of these programs differ considerably in terms of conservation techniques promoted and the approach to motivate farmers to participate, they shared the general theme that soil conservation provides a public benefit that is deserving of public investment (Fullen, 2003; Troeh et al., 2004). However, farmer perception of the erosion problem and how to best address it, or even if it needs to be addressed, has often been different than that of scientists. In a study in the United States, farmers tended to disagree with government assessment of what constituted highly erodible land and did not accurately perceive the severity of erosion occurring in their fields. The farmers were concerned about potential economic losses through reduced crop yields but did not see erosion as a problem in and of itself (Osterman and Hicks, 1988). In addition, there is debate over the best way to administer conservation programs, with some contending that the conservation programs developed in the 20th century failed to conserve soil, failed to spend program funding wisely, and in some cases actually increased erosion (Pretty and Shah, 1997; Boardman et al., 2003).

Erosion Modeling

To truly understand and address a problem such as soil erosion at the landscape scale, it is necessary to be able to model it. It is also important to note that soil mapping is an important part of modeling soil erosion (Fullen, 2003), because the map provides many key model variables. To that end, several soil erosion models were developed during the 20th century. In many respects the United States led the way in erosion modeling, beginning with the US Department of Agriculture's (USDA) development of the Universal Soil Loss Equation (USLE) in the 1950s. The USLE was developed to predict annual losses due to rill and interill erosion in the eastern half of the United States (Troeh et al., 2004). It was widely used and its use was rapidly extended beyond the area it was developed for, but it did not work well outside the eastern United States. To address this issue the modified USLE was released in 1978 followed by the revised USLE (RUSLE) in 1992 (Troeh et al., 2004). The RUSLE and its improved versions have become one of the most utilized soil erosion models worldwide to estimate annual soil loss to water erosion (Fig. 1.4) (Pal and Al-Tabbaa, 2009; Boni et al., 2015). In recent years, RUSLE has been adopted for use with computer systems, but it was originally developed to be solved in the field using paper tables and graphs (Troeh et al., 2004). RUSLE2, a 21st century improvement on RUSLE, now provides calculations at daily time steps, but still does not include gully erosion and has not been tested at the watershed scale.

Another commonly used water erosion model available from USDA is the Water Erosion Prediction Project (WEPP). The development of WEPP began in 1985 with initial model delivery in 1995. The WEPP was created to simulate physical processes that influence water erosion such as infiltration, runoff, raindrop and flow detachment, sediment transport and deposition, plant growth, and residue decomposition to replace empirically based erosion prediction models (Flanagan et al., 2007). The most widely used wind erosion model developed by USDA is the Wind Erosion Prediction System (WEPS), which was developed beginning in 1985 (Wagner, 2013). The WEPS simulates weather and field conditions to estimate wind erosion losses (Troeh et al., 2004). A weakness in the soil erosion models available from USDA at the end of the 20th century was that water and wind erosion could not be estimated within a single model, and therefore had to be modeled separately when estimates of both were desired (Langdale et al., 1991; Cooper et al., 2010). There have been efforts to combine WEPP and WEPS to create a single water and wind erosion model platform (Flanagan et al., 2007). Soil phases as mapped on National Cooperative Soil Survey maps were also used to estimate total erosion in the later part of the 20th century (Olson et al., 1994).

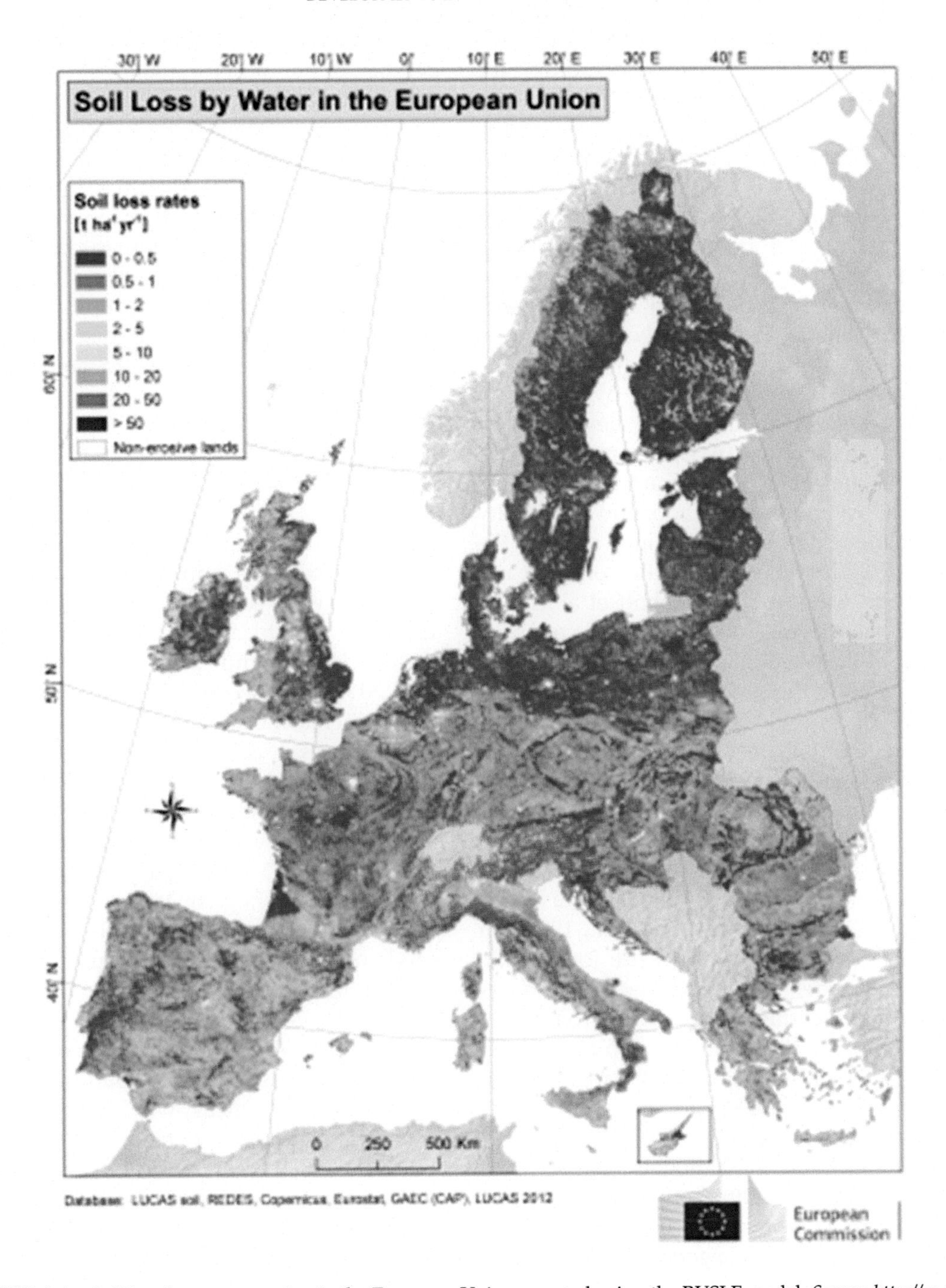

FIGURE 1.4 Soil loss by water erosion in the European Union mapped using the RUSLE model. *Source: http://ec.europa.eu/eurostat/statistics-explained/index.php/Agri-environmental_indicator_-_soil_erosion.*

Other soil erosion models were also developed in the 20th century, including the Système Hydrologique Européen (SHE) model (Abbott et al., 1986), the European Soil Erosion Model (EUROSEM) (Morgan et al., 1998a), the Limburg Soil Erosion Model (LISEM) (de Roo et al., 1996), and the soil erosion model for Mediterranean regions (SEMMED) (de Jong et al., 1999). Rose et al. (1983) developed an early mathematical model in Australia that described runoff on a plane assuming kinematic flow. All of the models discussed here were created to model erosion by water. In many cases these models were developed to address shortcomings in USDA models such as RUSLE and WEPP. For example, SHE was developed to address limitations in the ability of other models to evaluate things such as the impact of anthropogenic activities on land use change and water quality (Abbott et al., 1986). Some of the driving forces behind developing EUROSEM included that RUSLE could not predict deposition, the pathways taken by eroded material, or provide erosion information for individual rainfall events. Also, WEPP could not model peak sediment discharge or the pattern of sediment discharge over time (Morgan et al., 1998a). LISEM was incorporated into a raster-based GIS, which allowed the inclusion of remotely sensed data and was seen as being user friendly (de Roo et al., 1996). In other cases, such as SEMMED (de Jong et al., 1999), the model was developed to address the conditions within a specific environmental setting. Some of these models also saw widespread use; Morgan et al. (1998b) reported on the growing use of EUROSEM beyond Europe. Based on citation numbers in Google Scholar the SHE and EUROSEM models appear to be the most used of the 20th century water erosion models developed outside of the United States, with LISEM also getting a good amount of use.

Soil erosion models can tell how rapidly soil is lost given a set of conditions, but to determine if the rate of soil loss is a problem it is also important to know how rapidly pedogenesis might replace that lost soil. Several studies that investigated rates of soil formation were conducted during the 20th century; a number of those studies are summarized in Brevik (2013). These studies indicated that soil formation rates are often only fractions of a mm year^{-1}. However, the studies available are also heavily slanted to the United States and Europe. More studies covering wider geographic ranges are needed, especially in areas that are highly vulnerable to soil and land degradation.

Concept of Soil Quality/Health

The terms soil quality and soil health are generally used interchangeably within the scientific literature and are functionally synonymous, with scientists often preferring the term soil quality and farmers preferring soil health (Harris and Bezdicek, 1994; Karlen et al., 1997). However, the scientific community is increasingly using the term soil health as it implies a connection with soil biology, which is becoming a larger focal point in soils studies. Western culture has often viewed soil in a negative way, with terms such as "dirt-poor," "soiled," and "dirty minded" being common in the English language (Henry and Cring, 2013). Erosion of soils (Lieskovský and Kenderessy, 2014) and land management practices commonly used during the 20th century (Miao et al., 2015) often led to large-scale land degradation. The overall cultural underappreciation of soil and degradation caused by management practices was a driving force behind development of the soil quality/health concept (Karlen et al., 1997; Schjønning et al., 2004). Accurate soil maps and the information they contain are critical to fully understanding soil quality/health issues (Norfleet et al., 2003; Melakeberhan and Avendaño, 2008; Sanchez et al., 2009). However, existing soil maps are rarely detailed enough to adequately inform such decisions at the field or finer scale. The availability of larger scale maps may be useful to aid in tackling these problems.

The soil quality/health concept is closely tied to studies on the influence of soils on human health (Karlen et al., 1997; Schjønning et al., 2004). The relationship between soils and human health is another area that received increasing attention during the 20th century. Healthy soils influence human health by producing food products to support a balanced diet, providing a balanced supply of essential nutrients, filtering contaminants from water supplies, and as a source of medicines. However, unhealthy soils may act as possible points of contact with a variety of chemicals and pathogens that can negatively influence human health (Brevik, 2009). There are several ways that soil mapping can assist in understanding threats and improving human health. Some of these are quite traditional, for example, soil maps have long been used to provide information in support of agronomic management decisions related to crop production (Rust and Hanson, 1975; Karlen et al., 1990; Reynolds et al., 2000). Soil maps have also been an important component of water quality (Zhang et al., 1997; Chaplot, 2005) and soil contamination (Wu et al., 2002) assessment. Other uses of soil maps to support human health are less traditional. Some soil organisms are human pathogens, and a knowledge of soil properties and their distribution can help to create models to determine populations that are at risk of exposure to certain diseases (Tabor et al., 2011). Appropriate zoning policies that promote appropriate land uses based on information available in soil maps can also support public health (Neff et al., 2013). Therefore soil maps have had a role in supporting human health for many years and have the potential to have an enhanced role in the future as those maps become more quantitative and informative, while our understanding of some of these more complex environmental relationships improves.

Global Positioning Systems and Geographic Information Systems

Advances such as remote and proximal sensing were of limited practical use in support of soil mapping until ways were developed to precisely locate, manage, and manipulate the information contained within large data sets. GPS provided the means to precisely locate where the data were observed, and GIS programs run on rapidly improving computer technology provided the means to manage, manipulate, model, and analyze ever increasing amounts of spatial data.

The first publically available GPS was developed by the US military in the 1970s, however, signal accuracy was degraded so that inaccuracies of up to 500 m would occur (Hannay, 2009). That meant early GPS systems were of limited use to soil scientists. Signal degradation was reduced to 100 m in 1983 and was removed in 2000 (Hannay, 2009). As signal degradation was reduced the applicability of GPS for use in soil studies increased. The ability to precisely locate the position that data points were collected from revolutionized soil mapping and modeling, as sample sites could be accurately revisited to track trends over time, spatial relationships could be accurately intersected and investigated, and spatial statistical techniques could be used more effectively to model soil properties in-between sampling points. GPS was able to rapidly and inexpensively provide location information for data that could then be fed into a GIS.

The idea of laying multiple maps on top of one another to investigate the spatial relationships between related objects is not new; soil scientists have done so since the second half of the 19th century (Marbut, 1951). However, overlying multiple maps on top of one another could rapidly create an abundance of information that was difficult to effectively analyze visually and understand (Aguirre, 2014). The desire to be able to analyze the relationship between multiple spatial variables was a motivating factor behind proposing the first GIS in the 1960s (Tomlinson, 1962). The development of both commercially available and open source GIS programs through the latter part of the 20th century greatly enhanced the ability to quantitatively analyze spatial relationships between the items depicted in various maps of separate, but possibility related, natural features.

GIS also altered the concepts of map scale. Prior to the advent of GIS the level of detail that could be shown on a map was essentially determined by the size of the paper the map would be printed on and the amount of area the map would cover. In other words a map of an entire country printed on a small piece of paper could not show much detail for the item (e.g., soil) being mapped (Fig. 1.2), while a map on the same sized piece of paper that only covered a few square km could show much more detail for the same mapped item (Fig. 1.2). However, digital maps created with GIS can show multiple levels of detail, as the same GIS-based map can be zoomed out to show an entire country or zoomed in to show just a few square km within that country, all using the same data-base but with different levels of mapping detail displayed based on the level of zoom. By the end of the 20th century the combination of GPS and GIS allowed spatial analyses of soil properties and attributes and modeling of soil relationships and processes rapidly and inexpensively at a level of detail that had never before been possible.

Remote and Proximal Sensing

Remote sensing refers to a wide range of technologies used to detect Earth's surface, usually using aerial or satellite platforms. The earliest use of remote sensing in soil science was the development of aerial photographs as base maps for soil survey in the United States in the 1920s and 1930s (Bushnell, 1929), which represented a major advance over creating base maps using plane tables and odometers (Worthen, 1909) or using topographic maps when they were available as was common prior to the use of aerial photography (Miller and Schaetzl, 2014).

Digital remote sensing information was made widely available in the 1970s when the United States launched the Landsat program, one of the most popular sources of data for digital soil mapping. Seven Landsat satellites were launched during the 20th century with progressively increasing resolution and capabilities (Table 1.3). Another remote sensing technique developed in the 20th century that is seeing increasing use in modern soil science is LiDAR (McBratney et al., 2003; Brubaker et al., 2013). Aerial laser profiling systems date back to the 1970s, but it took advances in GPS, inertial measurement units, and inertial navigation systems to make LiDAR practical, something that did not occur until the mid-1990s (Carson et al., 2004). LiDAR represented an increase in data density and resolution of more than two orders of magnitude over traditional topographic information, significantly enhancing the ability of scientists to study landscapes, improving preplanning for field work and sampling (Roering et al., 2013), and making LiDAR an invaluable information layer in GIS-based analyses (Fisher et al., 2005). Satellite- and airplane-based radar technologies and airborne gamma-ray spectrometry are additional remote sensing techniques that were available in the late 20th century that have been used to aid in soil mapping (McBratney et al., 2003). Because remote sensing data are collected from aerial or satellite platforms, the sensors can quickly collect information over large areas.

One limitation of remote sensing is that it is largely confined to sensing conditions at the Earth's surface, with limited depth of penetration. Proximal sensing techniques have the ability to probe deeper into the soil profile, but are not able to cover large areas as quickly

TABLE 1.3 History of the Landsat Satellites Launched Prior to 2000

Satellite	Operational Dates	Notes
Landsat 1	July 1972–January 1978	Two sensors with 80 m ground resolution. Sensor 1—Return Beam Vidicon (RBV) with three bands: 1—visible blue-green (475–575 nm), 2—visible orange-red (580–680 nm), and 3—visible red to near-infrared (690–830 nm). Sensor 2—multispectral scanner (MSS) with four bands: 4—visible green (0.5–0.6 μm), 5—visible red (0.6–0.7 μm), 6—near-infrared (0.7–0.8 μm), and 7—near-infrared (0.8–1.1 μm). Ground sampling interval (pixel size): 57 × 79 m. Scene size: 170 km × 185 km
Landsat 2	January 1975–July 1983	Two sensors with 80 m ground resolution. Sensor 1—RBV with three bands. Sensor 2—MSS with four bands. Ground sampling interval (pixel size): 57 × 79 m. Scene size: 170 km × 185 km
Landsat 3	March 1978–September 1983	Two sensors with 40 m ground resolution. Sensor 1—RBV with three bands. Sensor 2—MSS with five bands: 4—visible green (0.5–0.6 μm), 5—visible red (0.6–0.7 μm), 6—near-infrared (0.7–0.8 μm), 7—near-infrared (0.8–1.1 μm), 8—thermal (10.4–12.6 μm). Ground sampling interval (pixel size): 57 × 79 m. Scene size: 170 km × 185 km
Landsat 4	July 1982–December 1993	Two sensors. Sensor 1—MSS with four bands: 4—visible green (0.5–0.6 μm), 5—visible red (0.6–0.7 μm), 6—near-infrared (0.7–0.8 μm), 7—near-infrared (0.8–1.1 μm). Ground sampling interval (pixel size): 57 × 79 m. Sensor 2—thematic mapper (TM) with seven bands: 1—visible (0.45–0.52 μm), 2—visible (0.52–0.60 μm), 3—visible (0.63–0.69 μm), 4—near-infrared (0.76–0.90 μm), 5—near-infrared (1.55–1.75 μm), 6—thermal (10.40–12.50 μm), 7—mid-infrared (IR) (2.08–2.35 μm). Ground sampling interval (pixel size): 30 m reflective, 120 m thermal. Scene size: 170 km × 185 km
Landsat 5	March 1984–January 2013	Two sensors. Sensor 1—MSS with four bands: 4—visible green (0.5–0.6 μm), 5—visible red (0.6–0.7 μm), 6—near-infrared (0.7–0.8 μm), 7—near-infrared (0.8–1.1 μm). Ground sampling interval (pixel size): 57 × 79 m. Sensor 2—thematic mapper (TM) with seven bands: 1—visible (0.45–0.52 μm), 2—visible (0.52–0.60 μm), 3—visible (0.63–0.69 μm), 4—near-infrared (0.76–0.90 μm), 5—near-infrared (1.55–1.75 μm), 6—thermal (10.40–12.50 μm), 7—mid-infrared (IR) (2.08–2.35 μm). Ground sampling interval (pixel size): 30 m reflective, 120 m thermal. Scene size: 170 km × 185 km
Landsat 6	October 1993	Failed to achieve orbit
Landsat 7	April 1999–present	One sensor, Enhanced Thematic Mapper Plus (ETM+) with eight bands: 1—visible (0.45–0.52 μm), 2—visible (0.52–0.60 μm), 3—visible (0.63–0.69 μm), 4—near-infrared (0.77–0.90 μm), 5—near-infrared (1.55–1.75 μm), 6—thermal (10.40–12.50 μm), low gain/high gain, 7—mid-infrared (2.08–2.35 μm), 8—panchromatic (PAN) (0.52–0.90 μm). Ground sampling interval (pixel size): 30 m reflective, 60 m thermal, 15 m panchromatic. Scene size: 170 km × 185 km

Source: USGS, 2015. Landsat missions: imaging the Earth since 1972. <http://landsat.usgs.gov/about_mission_history.php> (accessed 19.01.16).

as remote sensing. Several different proximal sensing technologies were experimented within the 20th century to investigate their potential application to soil work (Adamchuk et al., 2015), but the two that received the most attention were electromagnetic induction (EMI) and ground-penetrating radar (GPR) (Allred et al., 2008, 2010).

EMI was originally used to assess soil salinity (de Jong et al., 1979; Rhoades and Corwin, 1981; van der Lelij, 1983; Williams and Baker, 1982), but uses rapidly spread to other areas including measuring soil water content (Kachanoski et al., 1988; Khakural et al., 1998; Sheets and Hendrickx, 1995), clay content (Williams and Hoey, 1987), compaction (Brevik and Fenton, 2004), and exchangeable Ca and Mg (McBride et al., 1990). Each of these soil properties or attributes could be mapped with a great deal of spatial resolution using a georeferenced EMI survey if strong relationships could be found between the property or attribute of interest and the apparent electrical conductivity (EC_a) readings provided by the EMI instrument. Because of its ability to be linked to a GPS receiver and be correlated to a wide range of soil properties and attributes, EMI also attracted attention as a soil mapping tool starting in the 1990s (Jaynes et al., 1993, Doolittle et al., 1994; 1996; Fenton and Lauterbach, 1999). However, drawbacks to EMI surveys include that the EC_a-soil property/attribute relationships had to be established for each location, they were not universal, and changes in transient soil properties like soil water content and temperature change the absolute values (Brevik et al., 2004; Brevik et al., 2006) and, in some cases, the relative values (Brevik et al., 2006) of EMI readings over time even at a given location.

GPR was also used for the first time in soil studies in the 1970s (Benson and Glaccum, 1979; Johnson et al., 1979). GPR was successfully used to investigate several soil properties and attributes, including lateral extent of soil horizons and pans, depth to bedrock and water tables, and determine soil texture, organic matter content, and degree of cementation. However, many soils were found to be unsuitable for GPR investigations, including those with high soluble salt, clay, and water contents (Doolittle et al., 2007). Therefore use of GPR was limited to soils with favorable properties (Fig. 1.5) (Annan, 2002).

Remote and proximal sensing have both became important ways to rapidly collect large amounts of spatial data that can be related to soil properties and attributes. Analyzing and mapping the data collected with such techniques provided considerable information about the spatial distribution of soil properties and attributes that could then be entered into models (Brevik et al., 2016). In addition, the data could be collected at a much lower cost than with traditional field soil survey techniques (McBratney et al., 2000).

Spatial Statistics and Other Numerical Techniques

Research into the application of mathematical methods to study soil mapping and genesis issues, an approach that came to be called pedometrics, began in the 1980s (Minasny and McBratney, 2016). A number of different spatial statistics and other numerical techniques were being utilized to analyze and model the spatial variation of soil properties and attributes by the end of the 20th century (McBratney et al., 2000). While many of these techniques, such as kriging (Krige, 1951) and indices and models of diversity (e.g., Simpson, 1949; Margalef, 1958) have been around for decades, they were developed to address issues in other disciplines. Kriging was originally applied to the evaluation of ores and their distribution by the mining industry (Krige, 1951) and diversity approaches were widely used in ecological studies (Ibáñez et al., 2005). These techniques were applied to soil science

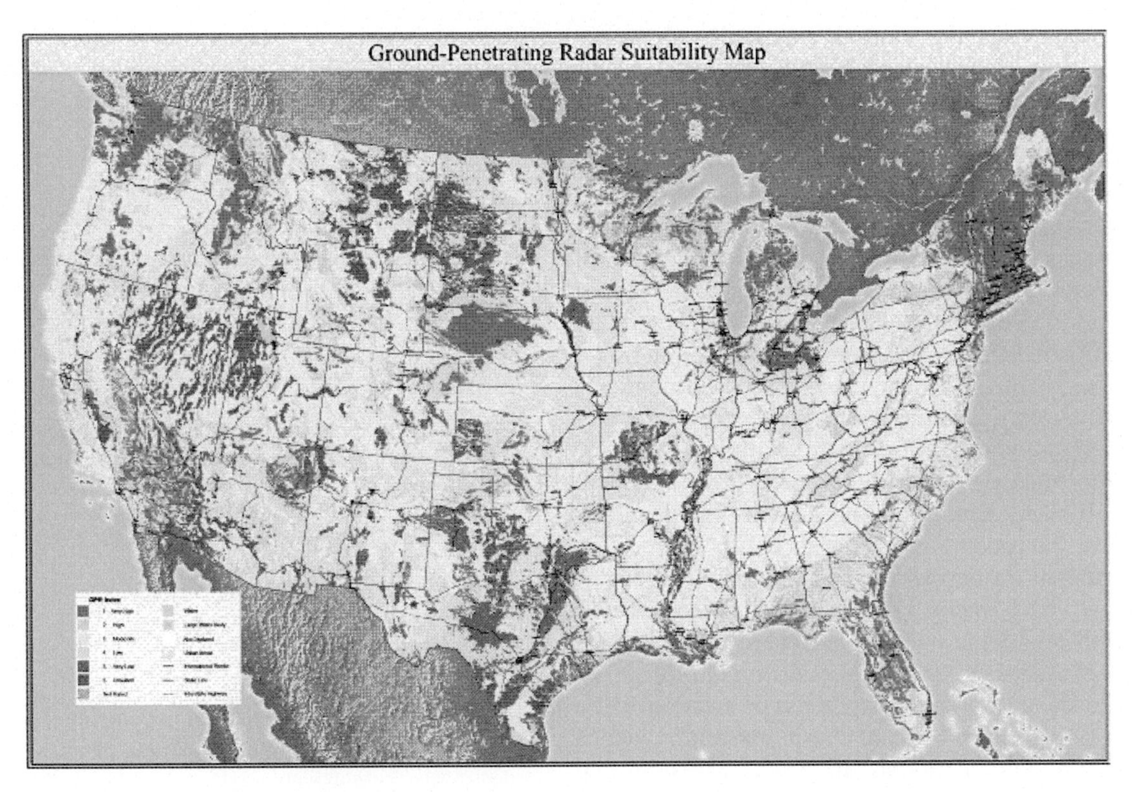

FIGURE 1.5 The GPR soil suitability map for the conterminous United States. Areas in dark green have soils most suitable to exploration using GPR, while areas in purple are least suitable (Soil Survey Staff, 2009). *Source: Figure courtesy of USDA-NRCS.*

questions in the final 20 years of the 20th century and were proven to be useful to soil scientists to model spatial distribution of soil properties, attributes, and pedodiversity, with several different variations of both techniques available (McBratney et al., 2000; Ibáñez et al., 2005). Cokriging, where the covariance with more readily observed variables were used to inform spatial predictions, proved particularly useful to soil scientists (McBratney et al., 2000; Minasny and McBratney, 2016) because it increased the accuracy of predictions. Another mathematical innovation was fuzzy sets and fuzzy logic, which were applied to soil classification (De Gruijter and McBratney, 1988) and soil survey (McBratney et al., 2000). Fuzzy applications are well suited to soil science because they allow continuous determination of the degree of soil class membership, much as occurs in a natural soil system. Increased computing power and the ability to precisely locate and manipulate the data in large data sets together with new mathematical techniques allowed for a revolution in the analysis of spatial data, and soil scientists took advantage of these new opportunities.

CONCLUDING COMMENTS

Soil science had come a long way by the end of the 20th century. The trial and error approaches the earliest agricultural societies used to determine which soils would best support their crops had been replaced by georeferenced soil data and predictive interpretations that were being analyzed and modeled in high-powered computer systems using a variety of mathematical and statistical techniques. Despite that, there were still significant needs to move soil survey forward and allow the information collected and displayed on maps to become more useful to a wider range of end users. There was a continued need for increased quantification of soil survey information, standardization in the communication of information, and ready access to up-to-date soil survey information from practically any location (Beaudette and O'Geen, 2010). There were still soil properties and processes that were not well understood and not well incorporated into pedologic models. For example, the influence of aspect and vegetation type, altitudinal gradient, and soil sampling type needed to be better understood, and these limitations meant that pedologic models still needed considerable additional work. There was a trend towards less field work in soil science at the end of the 20th century, with more reliance on remote and proximal sensing techniques. Remote and proximal sensing provides a great abundance of very valuable data at less expense than traditional field work, but field work is still essential to calibrate remote and proximal sensing data. Therefore it is important that funding continue to be provided for such work.

Furthermore, there are many end users of the products created by modern soil mapping and modeling. It is critical that soil scientists work with other scientists and with other stakeholders, such as land managers, policy makers, and the general public, to ensure that the final mapping and modeling products are useful, usable, and understandable to a wide range of end users (Bouma, 2015). Soil maps and models have been used to assist in making a number of management decisions, including agricultural, forestry, urban, and environmental decisions, often made by nonscientists. Accurate soil maps and models are critical to sustainable management of Earth's resources as we move into the future.

Acknowledgements

E.C. Brevik was partially supported by the National Science Foundation under Grant Number IIA-1355466 during this project.

References

Abbott, M.B., Bathurst, J.C., Cunge, J.A., O'Connell, P.E., Rasmussen, J., 1986. An introduction to the European Hydrological System-Systeme Hydrologique Europeen, "SHE", 1: History and philosophy of a physically-based, distributed modelling system. J. Hydrol. 87 (1–2), 45–59.

Adamchuk, V.I., Allred, B., Doolittle, J., Grote, K., Viscarra Rossel, R.A., 2015. Tools for proximal soil sensing. In: Ditzler, C., West, L. (Eds.), Soil Survey Manual, Soil Survey Staff. Natural Resources Conservation Service, Washington, DC. U.S. Department of Agriculture Handbook 18. Available from: http://www.nrcs.usda.gov/wps/portal/nrcs/detail/soils/scientists/?cid=nrcseprd329418 (accessed 20.01.16).

Aguirre, J.C., 2014. The unlikely history of the origins of modern maps. Smithsonian Magazine. http://www.smithsonianmag.com/history/unlikely-history-origins-modern-maps-180951617/?no-ist (accessed 24.01.16).

Aldrich, M.L., 1979. American state geological surveys, 1820–1845. In: Schneer, C.J. (Ed.), Two Hundred Years of Geology in America. Univ. Press of New England, Hanover, NH, pp. 133–143.

Allred, B.J., Ehsani, M.R., Daniels, J.J., 2008. General considerations for geophysical methods applied to agriculture. In: Allred, B.J., Daniels, J.J., Ehsani, M.R. (Eds.), Handbook of Agricultural Geophysics. CRC Press, Taylor & Francis, Boca Raton, FL, pp. 3–16.

Allred, B.J., Freeland, R.S., Farahani, H.J., Collins, M.E., 2010. Agricultural geophysics: past, present, and future. In: Proceedings of the Symposium on the Application of Geophysics to Engineering and Environmental Problems, SAGEEP 2010, pp. 190–202.

Annan, A.P., 2002. GPR-history, trends, and future developments. Subsurf. Sens. Technol. Appl. 3 (4), 253–270.

Bašić, F., 2005. Soil resources of Croatia. In: Jones, R.J.A., Houšková, B., Bullock, P., Montanarella, L. (Eds.), Soil Resources of Europe, second ed. European Soil Bureau, Institute for Environment & Sustainability, JRC Ispra, Italy, pp. 89–96.

Beaudette, D.E., O'Geen, A.T., 2010. An iPhone application for on-demand access to digital soil survey information. Soil Sci. Soc. Am. J. 74 (5), 1682–1684.

Benson, R., Glaccum, R., 1979. The Application of Ground Penetrating Radar to Soil Surveying. Final Report. NASA, Cape Kennedy Space Center, FL.

Blum, W.E.H., 1998. Agriculture in a sustainable environment—a holistic approach. Int. Agrophys. 12, 13–24.

Boardman, J., Poesen, J., Evans, R., 2003. Socio-economic factors in soil erosion and conservation. Environ. Sci. Policy 6, 1–6.

Boni, I., Giovannozzi, M., Martalò, P.F., Mensio, F., 2015. Soil erosion assessment in Piedmont: a territorial approach under the rural development program. In: Proceedings 8th European Congress on Regional Geoscientic Cartography and Information Systems. Barcelona Spain, 15–17 June, pp. 183–184.

Bonnard, L.F., 2005. Soil survey in Switzerland. In: Jones, R.J.A., Houšková, B., Bullock, P., Montanarella, L. (Eds.), Soil Resources of Europe, second ed. European Soil Bureau, Institute for Environment & Sustainability, JRC Ispra, Italy, pp. 365–370.

Bouma, J., 2015. Engaging soil science in transdisciplinary research facing "wicked" problems in the information society. Soil Sci. Soc. Am. J. 79, 454–458.

Brevik, E.C., 2009. Soil, food security, and human health. In: Verheye, W. (Ed.), Soils, Plant Growth and Crop Production. EOLSS Publishers, Oxford, UK. Encyclopedia of Life Support Systems (EOLSS), Developed under the Auspices of the UNESCO. http://www.eolss.net. (accessed 21.01.16).

Brevik, E.C., 2013. Forty years of soil formation in a South Georgia, USA borrow pit. Soil Horizons 54 (1), 20–29. http://dx.doi.org/10.2136/sh12-08-0025.

Brevik, E.C., Calzolari, C., Miller, B.A., Pereira, P., Kabala, C., Baumgarten, A., et al., 2016. Soil mapping, classification, and modeling: history and future directions. Geoderma 264, 256–274. http://dx.doi.org/10.1016/j.geoderma.2015.05.017.

Brevik, E.C., Fenton, T.E., 2004. The effect of changes in bulk density on soil electrical conductivity as measured with the Geonics® EM-38. Soil Surv. Horizons 45 (3), 96–102.

Brevik, E.C., Fenton, T.E., Homburg, J.A., 2015. Historical highlights in American soil science—prehistory to the 1970s. Catena. http://dx.doi.org/10.1016/j.catena.2015.10.003.

Brevik, E.C., Fenton, T.E., Horton, R., 2004. Effect of daily soil temperature fluctuations on soil electrical conductivity as measured with the Geonics® EM-38. Precis. Agric. 5, 143–150.

Brevik, E.C., Fenton, T.E., Lazari, A., 2006. Soil electrical conductivity as a function of soil water content and implications for soil mapping. Precis. Agric. 7, 393–404.

Brevik, E.C., Hartemink, A.E., 2010. Early soil knowledge and the birth and development of soil science. Catena 83, 23–33.

Brubaker, K.M., Myers, W.L., Drohan, P.J., Miller, D.A., Boyer, E.W., 2013. The use of LiDAR terrain data in characterizing surface roughness and microtopography. Appl. Environ. Soil Sci. http://dx.doi.org/10.1155/2013/891534.

Bushnell, T.M., 1929. Aerial photography and soil survey. Am. Soil Survey. Assoc. Bull. B10, 23–28. http://dx.doi.org/10.2136/sssaj1929.036159950B1020010004x.

Butler, B.W., 1950. A theory of prior streams as a casual factor of soil occurrence in the Riverine Plain of southeastern Australia. Crop Pasture Sci. 1 (3), 231–252.

Carson, W.W., Andersen, H.E., Reutebuch, S.E., McGaughey, R.J., 2004. LiDAR applications in forestry—an overview. ASPRS Annual Conference Proceedings, pp. 1–9.

Cerdan, O., Govers, G., Le Bissonnais, Y., Van Oost, K., Poesen, J., Saby, N., et al., 2010. Rates and spatial variations of soil erosion in Europe: a study based on erosion plot data. Geomorphology 122, 167–177.

Chaplot, V., 2005. Impact of DEM mesh size and soil map scale on SWAT runoff, sediment, and NO_3–N loads predictions. J. Hydrol. 312, 207–222.

Cline, M.G., 1977. The soils we classify and the soils we map. In: New York Soil Survey Conference, Bergamo East. USDA-SCS, U.S. Gov. Print Office, Washington, DC, pp. 5–19.

Coleman, K., Jenkinson, D.S., 1996. RothC-26.3-A model for the turnover of carbon in soil Evaluation of Soil Organic Matter Models. Springer, Berlin. 237–246.

Columela, L.J.M., 42 CE. De Res Rustica. Siglo I dC Los doce libros de Agricultura. Barcelona, Spain: Editorial Iberia (reprinted in 1979).

Cooper, D., Foster, C., Goodey, R., Hallett, P., Hobbs, P., Irvine, B., et al., 2010. Use of 'UKCIP08 Scenarios' to Determine the Potential Impact of Climate Change on the Pressures/Threats to Soils in England and Wales. Final Report to Defra, Project SP0571.

De Gruijter, J.J., McBratney, A.B., 1988. A modified fuzzy k-means method for predictive classification. In: Bock, H.H. (Ed.), Classification and Related Methods of Data Analysis. Elsevier, Amsterdam, pp. 97–104.

de Herrera, A., 1513. Agricultura General. Ministerio de Agricultura, Pesca y Alimentación, Madrid, (reprinted in 1981).

de Jong, E., Ballantyne, A.K., Cameron, D.R., Read, D.L., 1979. Measurement of apparent electrical conductivity of soils by an electromagnetic induction probe to aid salinity surveys. Soil Sci. Soc. Am. J. 43, 810–812.

de Jong, S.M., Paracchini, M.L., Bertolo, F., Folving, S., Megier, J., de Roo, A.P.J., 1999. Regional assessment of soil erosion using the distributed model SEMMED and remotely sensed data. Catena 37, 291–308.

De la Rosa, D., 2008. Evaluacion agro-ecologica de suelos para el densarrollo rural sostenible. Eds. Mundi-Prensa, Madrid. ISBN:9788484763611.

De la Rosa, D., 2013. Una agricultura a la medida de cada suelo: desde el conocimiento científico y la experiencia práctica a los sistemas de ayuda a la decisión. Inaugural speech at the Academy of Sciences in Sevilla, Spain. http://digital.csic.es/bitstream/10261/77729/1/Una%20agricultura%20a%20la%20medida%20de%20cada%20suelo.pdf (accessed 29.02.16).

De la Rosa, D., Mayol, F., Diaz-Pereira, E., Fernández, M., De la Rosa Jr, D., 2004. A land evaluation decision support system (MicroLEIS DSS) for agricultural soil protection: with special reference to the Mediterranean region. Environ. Modell. Softw. 19, 929–942.

de Roo, A.P.J., Wesseling, C.G., Ritsema, C.J., 1996. LISEM: a single-event physically based hydrological and soil erosion model for drainage basins. I: theory, input and output. Hydrol. Process. 10 (8), 1107–1117.

Diamond, J., 2005. Collapse: How Societies Choose to Fail or Succeed. Penguin Books, New York.

Dokuchaev, V.V., 1883. Russian Chernozem: Selected Works of V.V. Dokuchaev, vol. I. Israel Program for Scientific Translations, Jerusalem (translated in 1967).

Doolittle, J.A., Minzenmayer, F.E., Waltman, S.W., Benham, E.C., Tuttle, J.W., Peaslee, S.D., 2007. Ground-penetrating radar soil suitability map of the conterminous United States. Geoderma 141, 416–421.

Doolittle, J., Murphy, R., Parks, G., Warner, J., 1996. Electromagnetic induction investigations of a soil delineation in Reno County, Kansas. Soil Surv. Horizons 37, 11–20.

Doolittle, J.A., Sudduth, K.A., Kitchen, N.R., Indorante, S.J., 1994. Estimating depth to claypans using electromagnetic inductive methods. J. Soil Water Conserv. 49 (6), 552–555.

Dregne, H.E., 1995. Erosion and soil productivity in Australia and New Zealand. Land Degrad. Rehabil. 6, 71–78.

Dudal, R., Deckers, J., Van Orshoven, J., Van Ranst, E., 2005. Soil survey in Belgium and its applications. In: Jones, R.J.A., Houšková, B., Bullock, P., Montanarella, L. (Eds.), Soil Resources of Europe, second ed. European Soil Bureau, Institute for Environment & Sustainability, JRC Ispra, Italy, pp. 63–71.

Eckelmann, W., 2005. Soil information for Germany: the 2004 position. In: Jones, R.J.A., Houšková, B., Bullock, P., Montanarella, L. (Eds.), Soil Resources of Europe, second ed. European Soil Bureau, Institute for Environment & Sustainability, JRC Ispra, Italy, pp. 147–157.

Eswaran, H., Reich, P.F., Padmanabhan, E., 2001. Challenges of managing the land resources of Asia. In: Issues in the Management of Agricultural Resources. Proceedings of a Seminar in Commemoration of FFTC's 30th Anniversary, National Taiwan University, Taipei, Taiwan ROC, 6–8 September 2000. Food and Fertilizer Technology Center for the Asian and Pacific Region, pp. 85–104.

Falloon, P.D., Smith, P., Smith, J.U., Szabó, J., Coleman, K., Marshall, S., 1998. Regional estimates of carbon sequestration potential: linking the Rothamsted carbon model to GIS databases. Biol. Fertil. Soils 27, 236–241.

Feller, C., Brown, G.G., Blanchart, E., Deleporte, P., Chernyanskii, S.S., 2003b. Charles Darwin, earthworms and the natural sciences: various lessons from past to future. Agric. Ecosyst. Environ. 99, 29–49.

Feller, C., Thuriès, L.J.-M., Manlay, R.J., Robin, P., Frossard, E., 2003a. "The principles of rational agriculture" by Albrecht Daniel Thaer (1752–1828). An approach to the sustainability of cropping systems at the beginning of the 19th century. J. Plant Nutr. Soil Sci. 166, 687–698.

Fenton, T.E., Lauterbach, M.A., 1999. Soil map unit composition and scale of mapping related to interpretations for precision soil and crop management in Iowa Proceeding of the 4th International Conference on Precision Agriculture. American Society of Agronomy, Madison, WI, pp. 239–251.

Fisher, R.F., Fox, T.R., Harrison, R.B., Terry, T., 2005. Forest soils education and research: trends, needs, and wild ideas. Forest Ecol. Manage. 220, 1–16.

Flanagan, D.C., Gilley, J.E., Franti, T.G., 2007. Water erosion prediction project (WEPP): development history, model capabilities, and future enhancements. Trans. ASABE 50 (5), 1603–1612.

Free, E.E., 1911. The Movement of Soil Material by the Wind. U.S. Government Printing Office, Washington D.C., USDA Bureau of Soils Bulletin No. 68.

Fullen, M.A., 2003. Soil erosion and conservation in northern Europe. Prog. Phys. Geogr. 27 (3), 331–358.

Garrett, G., 1998. Partisan Politics in the Global Economy. Cambridge University Press, New York.

Gerrard, A.J., 1992. Soil Geomorphology. Springer Science & Business Media, Berlin.

Gonçalves, M.C., Reis, L.C.L., Pereira, M.V., 2005. Progress of soil survey in Portugal. In: Jones, R.J.A., Houšková, B., Bullock, P., Montanarella, L. (Eds.), Soil Resources of Europe, second ed. European Soil Bureau, Institute for Environment & Sustainability, JRC Ispra, Italy, pp. 275–279.

Gong, Z., Zhang, X., Chen, J., Zhang, G., 2003. Origin and development of soil science in ancient China. Geoderma 115, 3–13.

Hannay, P., 2009. Satellite navigation forensics techniques. Proceedings of the 7th Australian Digital Forensics Conference, 14–18.

Hansen, Z.K., Libecap, G.D., 2004. Small farms, externalities, and the Dust Bowl of the 1930s. J. Polit. Econ. 112 (3), 665–694.

Harris, R.F., Bezdicek, D.F., 1994. Descriptive aspects of soil quality/health. In: Defining Soil Quality for a Sustainable Environment, SSSA Special Publication No. 35. SSSA, Madison, WI, pp. 23–35.

Hartemink, A.E., McBratney, A.B., 2008. A soil science renaissance. Geoderma 148, 123–129.

Haslmayr, H.P., Geitner, C., Sutor, G., Knoll, A., Baumgarten, A., 2016. Soil function evaluation in Austria-development, concepts and examples. Geoderma 264, 379–387.

Helms, D., 2008. Hugh Hammond Bennett and the Creation of the Soil Erosion Service. USDA-NRCS Historical Insights Number 8, pp. 1–13.

Helms, D., 2010. Hugh Hammond Bennett and the creation of the Soil Conservation Service. J. Soil Water Conserv. 65 (2), 37A–47A.

Hempel, J., Micheli, E., Owens, P., McBratney, A., 2013. Universal soil classification system report from the International Union of Soil Sciences Working Group. Soil Horizons 54 (2), 1–6.

Henry, J.M., Cring, F.D., 2013. Geophagy: an anthropological perspective. In: Brevik, E.C., Burgess, L.C. (Eds.), Soils and Human Health. CRC Press, Boca Raton, pp. 179–198.

Hilgard, E.W., 1860. Report on the Geology and Agriculture of the State of Mississippi. State Printer, Jackson, MS.

Hillel, D., 1991. Out of the Earth: Civilization and the Life of the Soil. University of California Press, Berkeley, CA.

Holland, J.M., 2004. The environmental consequences of adopting conservation tillage in Europe: reviewing the evidence. Agric. Ecosyst. Environ. 103, 1–25.

Holliday, V.T., 2006. In: Warkentin, B.P. (Ed.), A history of soil geomorphology in the United States. Elsevier, Amsterdam, pp. 187–254.

Hudson, B.D., 1999. Some interesting historical facts about the American soil survey. Soil Surv. Horizons 40, 21–26.

Ibáñez, J.J., De-Alba, S., Bermúdez, F.F., García-Álvarez, A., 1995. Pedodiversity: concepts and measures. Catena 24, 215–232.

Indorante, S.J., McLeese, R.L., Hammer, R.D., Thompson, B.W., Alexander, D.L., 1996. Positioning soil survey for the 21st century. J. Soil Water Conserv. 51, 21–28.

Jaynes, D.B., Colvin, T.S., Ambuel, J., 1993. Soil type and crop yields determination from ground conductivity surveys. In: 1993 International Meeting of American Society of Agricultural Engineers. Paper No. 933552. American Society of Agricultural Engineers, St. Joseph, Michigan.

Jenny, H., 1941. Factors of Soil Formation: A System of Quantitative Pedology. Dover Publications, Mineola, NY, (reprinted in 1994).

Jenny, H., 1961. E.W. Hilgard and the Birth of Modern Soil Science. Collana della Revisa Agrochemical, Pisa.

Jensen, C.T., Moriarty, M.D., Johnson, K.D., Terry, R.E., Emery, K.F., Nelson, S.D., 2007. Soil resources of the Motul De San José Maya: correlating Soil Taxonomy and modern Itzá Maya soil classification within a classic Maya archaeological zone. Geoarchaeology 22 (3), 337–357.

Johnson, R.W., Glaccum, R., Wojtasinski, R., 1979. Application of ground penetrating radar to soil survey. Proc. Soil Crop Soc. Florida 39, 68–72.

Jones, R., 1997. Soil Survey of Emmet County, Iowa. Natural Resources Conservation Service. U.S. Department of Agriculture.

Jones, R.J.A., Houšková, B., Bullock, P., Montanarella, L. (Eds.), 2005. Soil Resources of Europe, second ed. European Soil Bureau, Institute for Environment & Sustainability, JRC Ispra, Italy.

Kachanoski, R.G., Gregorich, E.G., vanWesenbeeck, I.J., 1988. Estimating spatial variations of soil water content using noncontacting electromagnetic inductive methods. Can. J. Soil Sci. 68, 715–722.

Kaphengst, T. 2014. Towards a definition of global sustainable land use? A discussion on theory, concepts and implications for governance. Discussion paper produced within the research project "GLOBALANDS – Global Land Use and Sustainability". http://www.ecologic.eu/globalands/about. (accessed 14.08.16).

Karlen, D.L., Mausbach, M.J., Doran, J.W., Cline, R.G., Harris, R.F., Schuman, G.E., 1997. Soil quality: a concept, definition, and framework for evaluation. Soil Sci. Soc. Am. J. 61, 4–10.

Karlen, D.L., Sadler, E.J., Busscher, W.J., 1990. Crop yield variation associated with Coastal Plain soil map units. Soil Sci. Soc. Am. J. 54, 859–865.

Khakural, B.R., Robert, P.C., Hugins, D.R., 1998. Use of noncontacting electromagnetic inductive method for estimating soil moisture across a landscape. Commun. Soil Sci. Plant Anal. 29 (11–14), 2055–2065.

King, D., Stengel, P., Jamagne, M., Le Bas, C., Arrouays, D., 2005. Soil mapping and soil monitoring: state of progress and use in France. In: Jones, R.J.A., Houšková, B., Bullock, P., Montanarella, L. (Eds.), Soil Resources of Europe, second ed. European Soil Bureau, Institute for Environment & Sustainability, JRC Ispra, Italy, pp. 139–146.

Kolchakov, I., Rousseva, S., Georgiev, B., Stoychev, D., 2005. Soil survey and soil mapping in Bulgaria. In: Jones, R.J.A., Houšková, B., Bullock, P., Montanarella, L. (Eds.), Soil Resources of Europe, second ed. European Soil Bureau, Institute for Environment & Sustainability, JRC Ispra, Italy, pp. 83–87.

Krasilnikov, P., Ibáñez-Martí, J.J., Arnold, R., Shoba, S., 2009. A Handbook of Soil Terminology, Correlation, and Classification. Earthscan, London, UK.

Krige, D.G., 1951. A statistical approach to some basic mine valuation problems on the Witwatersrand. J. Chem. Metallur. Mining Soc. S. Afr. 52, 119–139.

Krupenikov, I.A., 1992. History of Soil Science from its Inception to the Present. Oxonian Press, New Delhi.

Langdale, G.W., Blevins, R.L., Karlen, D.L., McCool, D.K., Nearing, M.A., Skidmore, E.L., et al., 1991. Cover crop effects on soil erosion by wind and water Cover Crops for Clean Water. Soil and Water Conservation Society, Ankeny, IA, pp. 15–22.

Lee, J., Coulter, B., 2005. Application of soils data to land use and environmental problems in Ireland. In: Jones, R.J.A., Houšková, B., Bullock, P., Montanarella, L. (Eds.), Soil Resources of Europe, second ed. European Soil Bureau, Institute for Environment & Sustainability, JRC Ispra, Italy, pp. 187–191.

Lieskovský, J., Kenderessy, P., 2014. Modelling the effect of vegetation cover and different tillage practices on soil erosion in vineyards: a case study in Vráble (Slovakia) using WATEM/SEDEM. Land Degrad. Develop. 25, 288–296.

Marbut, C.F., 1922. Soil classification. Am. Soil Surv. Assoc. Bull. B3, 24–33.

Marbut, C.F., 1928. History of soil survey ideas. In: Weber, G.A. (Ed.), The Bureau of Chemistry and Soils: Its History, Activities, and Organization. Brookings Institution Institute for Government Research, Washington, DC, pp. 91–98.

Marbut, C.F., 1951. Soils: their genesis and classification A Memorial Volume of Lectures Given in the Graduate School of the United States Department of Agriculture in 1928. Soil Science Society of America, Madison, WI.

Margalef, R., 1958. Information theory in ecology. Gen. Syst. Bull. 3, 36–71.

McBratney, A.B., Mendonça Santos, M.L., Minasny, B., 2003. On digital soil mapping. Geoderma 117, 3–52.

McBratney, A.B., Odeh, I.O.A., Bishop, T.F.A., Dunbar, M.S., Shatar, T.M., 2000. An overview of pedometric techniques for use in soil survey. Geoderma 97, 293–327.

McBride, R.A., Gordon, A.M., Shrive, S.C., 1990. Estimating forest soil quality from terrain measurements of apparent electrical conductivity. Soil Sci. Soc. Am. J. 54 (1), 290–293.

McCracken, R.J., Helms, D., 1994. Soil surveys and maps. In: McDonald, P. (Ed.), The Literature of Soil Science. Cornell Univ. Press, Ithaca, pp. 275–311.

McGee, W.J., 1911. Soil Erosion. Bureau of Soils Bulletin No. 71. U.S. Government Printing Office, Washington, DC.

Melakeberhan, H., Avendaño, F., 2008. Spatio-temporal consideration of soil conditions and site-specific management of nematodes. Precis. Agric. 9, 341–354.

Miao, L., Moore, J.C., Zeng, F., Lei, J., Ding, J., He, B., et al., 2015. Footprint of research in desertification management in China. Land Degrad. Develop. 26, 450–457.

Miller, B.A., Schaetzl, R.J., 2014. The historical role of base maps in soil geography. Geoderma 230–231, 329–339.

Milne, G., 1935. Some suggested units of classification and mapping particularly for East African soils. Soil Res. 4 (3), 183–198.

Minasny, B., McBratney, A.B., 2016. Digital soil mapping: a brief history and some lessons. Geoderma 264, 301–311.

Morgan, R.P.C., Quinton, J.N., Smith, R.E., Govers, G., Poesen, J.W.A., Auerswald, K., et al., 1998a. The European soil erosion model (EUROSEM): a dynamic approach for predicting sediment transport from fields and small catchments. Earth Surf. Process. Landforms 23, 527–544.

Morgan, R.P.C., Quinton, J.N., Smith, R.E., Govers, G., Poesen, J.W.A., Auerswald, K., et al., 1998b. The European Soil Erosion Model (EUROSEM): Documentation and User Guide, Version 3.6. Silsoe College, Cranfield University.

Morgan, R.P.C., Rickson, R.J., 1990. Issues on soil erosion in Europe: the need for a soil conservation policy. In: Boardman, J., Foster, I.D.L., Dearing, J.A. (Eds.), Soil Erosion on Agricultural Land. John Wiley and Sons Ltd., Hoboken, pp. 591–603.

Muir, J.W., 1962. The general principles of classification with reference to soils. J. Soil Sci. 13 (1), 22–30.

Muñoz-Rojas, M., Jordán, A., Zavala, L.M., González-Peñaloza, F.A., De la Rosa, D., Pino-Mejias, R., et al., 2013. Modelling soil organic carbon stocks in global change scenarios: a CarboSOIL application. Biogeosciences 10, 8253–8268.

Munteanu, I., Dumitru, M., Florea, N., Canarache, A., Lacatusu, R., Vlad, V., et al., 2005. Status of soil mapping, monitoring, and database compilation in Romania at the beginning of the 21st century. In: Jones, R.J.A., Houšková, B., Bullock, P., Montanarella, L. (Eds.), Soil Resources of Europe, second ed. European Soil Bureau, Institute for Environment & Sustainability, JRC Ispra, Italy, pp. 281–296.

Neff, R.A., Carmona, C., Kanter, R., 2013. Addressing soil impacts on public health: issues and recommendations. In: Brevik, E.C., Burgess, L.C. (Eds.), Soils and Human Health. CRC Press, Boca Raton, pp. 259–295.

Němeček, J., Kozák, J., 2005. Status of soil surveys, inventory and soil monitoring in the Czech Republic. In: Jones, R.J.A., Houšková, B., Bullock, P., Montanarella, L. (Eds.), Soil Resources of Europe, second ed. European Soil Bureau, Institute for Environment & Sustainability, JRC Ispra, Italy, pp. 103–109.

Nettleton, W.D., Lynn, W.C., 2008. Development in soil survey: soil survey investigations laboratories' support of taxonomic concepts. Soil Surv. Horizons 49, 48–51.

Norfleet, M.L., Ditzler, C.A., Puckett, W.E., Grossman, R.B., Shaw, J.N., 2003. Soil quality and its relationship to pedology. Soil Sci. 168 (3), 149–155.

Olson, K.R., Norton, L.D., Fenton, T.E., Lal, R., 1994. Quantification of soil loss from eroded soil phases. J. Soil Water Conserv. 49 (6), 591–596.

Olsson, M., 2005. Soil survey in Sweden. In: Jones, R.J.A., Houšková, B., Bullock, P., Montanarella, L. (Eds.), Soil Resources of Europe, second ed. European Soil Bureau, Institute for Environment & Sustainability, JRC Ispra, Italy, pp. 357–363.

Osterman, D.A., Hicks, T.L., 1988. Highly erodible land: farmer perceptions versus actual measurements. J. Soil Water Conserv. 43 (2), 177–182.

Pal, I., Al-Tabbaa, A., 2009. Suitability of different erosivity models used in RUSLE2 for the south west Indian region. Environmentalist 29, 405–410.

Panagos, P., Borrelli, P., Poesen, J., Ballabio, C., Lugato, E., Meusburger, K., et al., 2015. The new assessment of soil loss by water erosion in Europe. Environ. Sci. Policy 54, 438–447.

Parton, W.J., Schimel, D.S., Cole, C.V., Ojima, D.S., 1987. Analysis of factors controlling soil organic matter levels in Great Plains grasslands. Soil Sci. Soc. Am. J. 51, 1173–1179.

Phillips, J.D., 1989. An evaluation of the state factor model of soil ecosystems. Ecol. Model. 45, 165–177.

Pimentel, D., Harvey, C., Resosudarmo, P., Sinclair, K., Kurz, D., McNair, M., et al., 1995. Environmental and economic costs of soil erosion and conservation benefits. Science 267 (5201), 1117–1123.

Pretty, J.N., Shah, P., 1997. Making soil and water conservation sustainable: from coercion and control to partnerships and participation. Land Degrad. Develop. 8, 39–58.

Reynolds, C.A., Jackson, T.J., Rawls, W.J., 2000. Estimating soil water-holding capacities by linking the Food and Agriculture Organization soil map of the world with global pedon databases and continuous pedotransfer functions. Water Resour. Res. 36 (12), 3653–3662.

Rhoades, J.D., Corwin, D.L., 1981. Determining soil electrical conductivity-depth relations using an inductive electromagnetic soil conductivity meter. Soil Sci. Soc. Am. J. 45, 255–260.

Roering, J.J., Mackey, B.H., Marshall, J.A., Sweeney, K.E., Deligne, N.I., Booth, A.M., et al., 2013. 'You are HERE': connecting the dots with airborne lidar for geomorphic fieldwork. Geomorphology 200, 172–183.

Rose, C.W., Williams, J.R., Sander, G.C., Barry, D.A., 1983. A mathematical model of soil erosion and deposition processes: I. theory for a plane land element. Soil Sci. Soc. Am. J. 47, 991–995.

Rust, R.H., Hanson, L.D., 1975. Crop Equivalent Rating Guide for Soils of Minnesota. Agricultural Experiment Station, University of Minnesota. Miscellaneous Report 132-1975.

Sanchez, P.A., Ahamed, S., Carré, F., Hartemink, A.E., Hempel, J., Huising, J., et al., 2009. Digital soil map of the world. Science 325 (5941), 680–681. http://dx.doi.org/10.1126/science.1175084.

Schaetzl, R., Anderson, S., 2005. Soils: Genesis and Geomorphology. Cambridge University Press, New York.

Schjønning, P., Elmholt, S., Christensen, B.T., 2004. Soil quality management—concepts and terms. In: Schjønning, P., Elmholt, S., Christensen, B.T. (Eds.), Managing Soil Quality: Challenges in Modern Agriculture. CAB International, Wallingford, pp. 1–15.

Sheets, K.R., Hendrickx, J.M.H., 1995. Noninvasive soil water content measurements using electromagnetic induction. Water Resour. Res. 31, 2401–2409.

Simonson, R.W., 1959. Outline of a generalized theory of soil genesis. Soil Sci. Soc. Am. Proc. 23, 152–156.

Simpson, E.H., 1949. Measurement of diversity. Nature 163, 688.

Sippola, J., Yli-Halla, M., 2005. Status of soil mapping in Finland. In: Jones, R.J.A., Houšková, B., Bullock, P., Montanarella, L. (Eds.), Soil Resources of Europe, second ed. European Soil Bureau, Institute for Environment & Sustainability, JRC Ispra, Italy, pp. 133–138.

Soil Survey Staff, 2009. Ground-penetrating radar soil suitability maps. <http://www.nrcs.usda.gov/wps/portal/nrcs/detail/soils/survey/geo/?cid=nrcs142p2_053622> (accessed 21.01.16).

Sparks, D.L., 2006. Historical aspects of soil chemistry. In: Warkentin, B.P. (Ed.), Footprints in the Soil: People and Ideas in Soil History. Elsevier, Amsterdam, pp. 307–337.

Stoate, C., Boatman, N.D., Borralho, R.J., Rio Carvalho, C., de Snoo, G.R., Eden, P., 2001. Ecological impacts of arable intensification in Europe. J. Environ. Manage. 63, 337–365.

Tabor, J.A., O'Rourke, M.K., Lebowitz, M.D., Harris, R.B., 2011. Landscape-epidemiological study design to investigate an environmentally based disease. J. Exposure Sci. Environ. Epidemiol. 21, 197–211.

Thompson, T.R.E., Bradley, R.I., Hollis, J.M., Bellamy, P.H., Dufour, M.J.D., 2005. Soil information and its application in the United Kingdom: an update. In: Jones, R.J.A., Houšková, B., Bullock, P., Montanarella, L. (Eds.), Soil Resources of Europe, second ed. European Soil Bureau, Institute for Environment & Sustainability, JRC Ispra, Italy, pp. 377–393.

Tomlinson, R.F., 1962. An introduction to the use of electronic computers in the storage, compilation and assessment of natural and economic data for the evaluation of marginal lands. National Land Capability Inventory Seminar, Agricultural Rehabilitation and Development Administration of the Canada Department of Agriculture, Ottawa, Canada. <https://gisandscience.files.wordpress.com/2012/08/4-computermapping.pdf> (accessed 24.01.16).

Troeh, F.R., Hobbs, J.A., Donahue, R.L., 2004. Soil and Water Conservation for Productivity and Environmental Protection, fourth ed. Prentice Hall, Upper Saddle River, NJ.

US BLS, 2016. CPI inflation calculator. <http://www.bls.gov/data/inflation_calculator.htm> (accessed 17.01.16).

US Census Bureau, 2016. U.S. and world population clock. <http://www.census.gov/popclock/> (accessed 25.01.16).

USGS, 2015. Landsat missions: imaging the Earth since 1972. <http://landsat.usgs.gov/about_mission_history.php> (accessed 19.01.16).

van der Lelij, A., 1983. Use of Electromagnetic Induction Instrument (type EM-38) for Mapping Soil Salinity. Water Resources Commission, Murrumbidgee Division, New South Wales, Australia.

van der Pouw, B.J.A., Finke, P.A., 2005. Development and perspective of soil survey in the Netherlands. In: Jones, R.J.A., Houšková, B., Bullock, P., Montanarella, L. (Eds.), Soil Resources of Europe, second ed. European Soil Bureau, Institute for Environment & Sustainability, JRC Ispra, Italy, pp. 245–255.

Várallyay, G., 2005. Soil survey and soil monitoring in Hungary. In: Jones, R.J.A., Houšková, B., Bullock, P., Montanarella, L. (Eds.), Soil Resources of Europe, second ed. European Soil Bureau, Institute for Environment & Sustainability, JRC Ispra, Italy, pp. 169–179.

Verheijen, F.G.A., Jones, R.J.A., Rickson, R.J., Smith, C.J., 2009. Tolerable versus actual soil erosion rates in Europe. Earth-Sci. Rev. 94, 23–38.

von Liebig, J., 1840. Die organische Chemie in ihrer Anwendung auf Agrikultur und Physiologie.

Vrščaj, B., Prus, T., Lobnik, F., 2005. Soil information and soil data use in Slovenia. In: Jones, R.J.A., Houšková, B., Bullock, P., Montanarella, L. (Eds.), Soil Resources of Europe, second ed. European Soil Bureau, Institute for Environment & Sustainability, JRC Ispra, Italy, pp. 331–344.

Wagner, L.E., 2013. A history of wind erosion prediction models in the United States Department of Agriculture: the wind erosion prediction system (WEPS). Aeolian Res. 10, 9–24.

Wells, E.C., Mihok, L.D., 2010. Ancient Maya perceptions of soil, land, and Earth. In: Landa, E.R., Feller, C. (Eds.), Soil and Culture. Springer Science + Business Media, Berlin, pp. 311–327.

Williams, B.G., Baker, G.C., 1982. An electromagnetic induction technique for reconnaissance surveys of soil salinity hazards. Aust. J. Soil Res. 20, 107–118.

Williams, B.G., Hoey, D., 1987. The use of electromagnetic induction to detect the spatial variability of the salt and clay contents of soils. Aust. J. Soil Res. 25, 21–27.

Worthen, E.L., 1909. Methods of soil surveying. Agron. J. 1 (1), 185–191.

Wu, J., Norvell, W.A., Hopkins, D.G., Welch, R.M., 2002. Spatial variability of grain cadmium and soil characteristics in a durum wheat field. Soil Sci. Soc. Am. J. 66, 268–275.

Wysocki, D.A., Schoeneberger, P.J., LaGarry, H.E., 2000. Geomorphology of soil landscapes. In: Sumner, M.E. (Ed.), Handbook of Soil Science. CRC Press, Boca Raton, FL, pp. E-5–E-39.

Yaalon, D.H., 1975. Conceptual models in pedogenesis: can the soil-forming functions be solved? Geoderma 13, 189–205.

Yaalon, D.H., Yaron, B., 1966. Framework for man-made soil changes-an outline of metapedogenesis. Soil Sci. 102 (4), 272–277.

Yang, D., Kanae, S., Oki, T., Koike, T., Musiake, K., 2003. Global potential soil erosion with reference to land use and climate changes. Hydrol. Process. 17, 2913–2928.

Yassoglou, N., 2005. Soil survey in Greece. In: Jones, R.J.A., Houšková, B., Bullock, P., Montanarella, L. (Eds.), Soil Resources of Europe, second ed. European Soil Bureau, Institute for Environment & Sustainability, JRC Ispra, Italy, pp. 159–168.

Zhang, M., Geng, S., Ustin, S.L., Tanji, K.K., 1997. Pesticide occurrence in groundwater in Tulare County. Calif. Environ. Monit. Assess. 45, 101–127.

Zitzmann, A., 1994. Geowissenschaftliche Karten in der Bundesrepublik Deutschland. Zeitschrift der Deutschen Geologischen Gesellschaft Band 145, 38–87.

CHAPTER

2

Soil Mapping and Processes Modeling for Sustainable Land Management

Paulo Pereira[1], *Eric C. Brevik*[2], *Miriam Muñoz-Rojas*[3,4], *Bradley A. Miller*[5,6], *Anna Smetanova*[7,8], *Daniel Depellegrin*[1], *Ieva Misiune*[1], *Agata Novara*[9] *and Artemi Cerdà*[10]

[1]Mykolas Romeris University, Vilnius, Lithuania [2]Dickinson State University, Dickinson, ND, United States [3]The University of Western Australia, Crawley, WA, Australia [4]Kings Park and Botanic Garden, Perth, WA, Australia [5]Iowa State University, Ames, IA, United States [6]Leibniz Centre for Agricultural Landscape Research (ZALF), Müncheberg, Germany [7]National Institute for Agricultural Research, Paris, France [8]Technical University Berlin, Berlin, Germany [9]University of Palermo, Palermo, Italy [10]University of Valencia, Valencia, Spain

INTRODUCTION

Soil is the basis of life and a major supplier of ecosystem services. It is a nonrenewable resource at the human time scale and a medium of interaction among several spheres: the atmosphere, biosphere, hydrosphere, and lithosphere, and recently with the antroposphere as a consequence of the tremendous impact humans now have on soil properties through agriculture, urbanization, landfills, pollution, and other activities (Yaalon and Yaron, 1966; Richter and Yaalon, 2012; Brevik et al., in press). Soil degradation is a worldwide problem, and it is understood as "*a change in the soil health status resulting in a diminished capacity of the ecosystem to provide goods and services for its beneficiaries. Degraded soils have a health status such, that they do not provide the normal goods and services of the particular soil in its ecosystem*[1]." Soil degradation is not an exclusive problem of arid and semiarid environments as a consequence of farming activities.

Soil degradation is a consequence of intensive land use management, which is assumed to be caused by human impact, poverty, and a response to economic opportunities at the global level (Lambin et al., 2001). There are several examples of human-induced soil

[1] http://www.fao.org/soils-portal/soil-degradation-restoration/en/ (consulted on 21.01.16).

Soil Mapping and Process Modeling for Sustainable Land Use Management.
DOI: http://dx.doi.org/10.1016/B978-0-12-805200-6.00002-5

degradation in arctic (Jefferies and Rockwell, 2002), humid (Graves et al., 2015; Varallya, 1989), tropical (Ali, 2006), and alpine environments (Upadhyay et al., 2005; Wu and Tiessen, 2002) in addition to arid and semi-arid environments (García-Orenes et al., 2009). Soil degradation poses several threats, such as loss of ecosystem services delivery, biodiversity protection, climate change, energy sustainability, food and water security, and productivity stagnation. All of these aspects are important obstacles to sustainability (Bouma and McBratney, 2013). Soil degradation is attributed to erosion, sealing, compaction, nutrient depletion, pollution, salinization, and other indirect actions, such as creating unfavorable conditions for soil formation and productivity (Bindraban et al., 2012). In Europe, mean soil losses are estimated to be 2.46 t $year^{-1}$ and 0.032 t $ha^{-1} MJ^{-1} mm^{-1}$ (Panagos et al., 2014b, 2015).

Soils are the base of economic activity and the costs of degradation are extremely high (Görlach et al., 2004; Pimentel et al., 1995). Soil degradation has been estimated to cost England and Wales between £0.9 billion and £1.4 billion per year, which are especially attributed to the loss of organic matter, erosion, and compaction (Graves et al., 2015). The economic and environmental costs of the use of pesticides is estimated to be $8 billion per year (Pimentel et al., 1992) and soil erosion $44 billion per year in the United States and $400 billion per year worldwide (Pimentel et al., 1995). Soil cadmium remediation by replacement of contaminated soil is estimated to be United States $3 million ha^{-1} (Chaney et al., 2004). The remediation cost of soil contaminants through stabilization/stagnation technology in situ varies from US$80 for shallow applications to US$330 for deeper applications per cubic meter (Khan et al., 2004). Looking at the values above, soil degradation and pollution is extremely expensive. In this context, soil degradation is of major importance from an environmental, social, and economic point of view.

Maps are widely used to gain a better understanding of human impacts on the landscape. Degradation processes can be studied and evaluated using remote sensing techniques (Raina et al., 1993; Vagen et al., 2016), soil erosion models (Prashun et al., 2013), geostatistical models (Diodato and Ceccarelli, 2004), and expert analysis and satellite images (Kheir et al., 2006) in urban and rural environments at diverse scales. The maps produced by these works are important in understanding our impact on the landscape and are an important contribution to develop better territorial planning.

Soil maps and soil models are important to plan sustainable use of a given territory and to help identify areas that are vulnerable to human activities, creating a high probability of degradation. Good spatial information and planning can reduce exposure to environmental hazards and risks, the impact of human activities on soil and land degradation, adverse effects on human health, and economic losses and loss of lives (Anaya-Romero et al., 2011). Good planning can contribute to a better environment (e.g., pollution reduction) and a general correct use of the land.

SOIL AND SUSTAINABLE DEVELOPMENT INTERDEPENDENCE

Sustainable development cannot be understood without considering soils. Soils are a natural capital and are the source of a number of regulating, provisioning, cultural, waste processing, and supporting ecosystem services (Adhikari and Hartemink, 2016; Calzolari et al., 2016; Robinson et al., 2013) that are indispensable for our existence (Fig. 2.1). These services can be divided into agricultural and nonagricultural (Fig. 2.2) (Pulleman et al., 2012). According to Powlson et al. (2011), soils provide a wide variety of services to society that are of high environmental significance, such as

(1) influence water quality and regulate nutrient runoff and percolation, (2) serve as the basis for soil biodiversity, (3) water retention for vegetation use and transfer to water bodies, (4) influence atmospheric chemistry and act as a sink for greenhouse gases, (5) serve as the base for vegetation development and support for all the living elements of this world, and (6) are the basis for several human and natural activities.

The unsustainable use of soil ecosystem services will lead to soil degradation and the emergence of problems with food production and security (Gregory, 2012; Montanarella and Vargas, 2012), one of the most important factors for human social and economic development. Studies in the Midwestern United States showed that moderate soil erosion led to yield reductions of 16%–23% and severe erosion led to yield reductions of 25%–36% as compared to crops grown in fields with only slight erosion (Troeh et al., 2004). The unstainable use of soil services is an issue transversal to the three spheres of sustainable development (Fig. 2.3). The correct or incorrect management of the

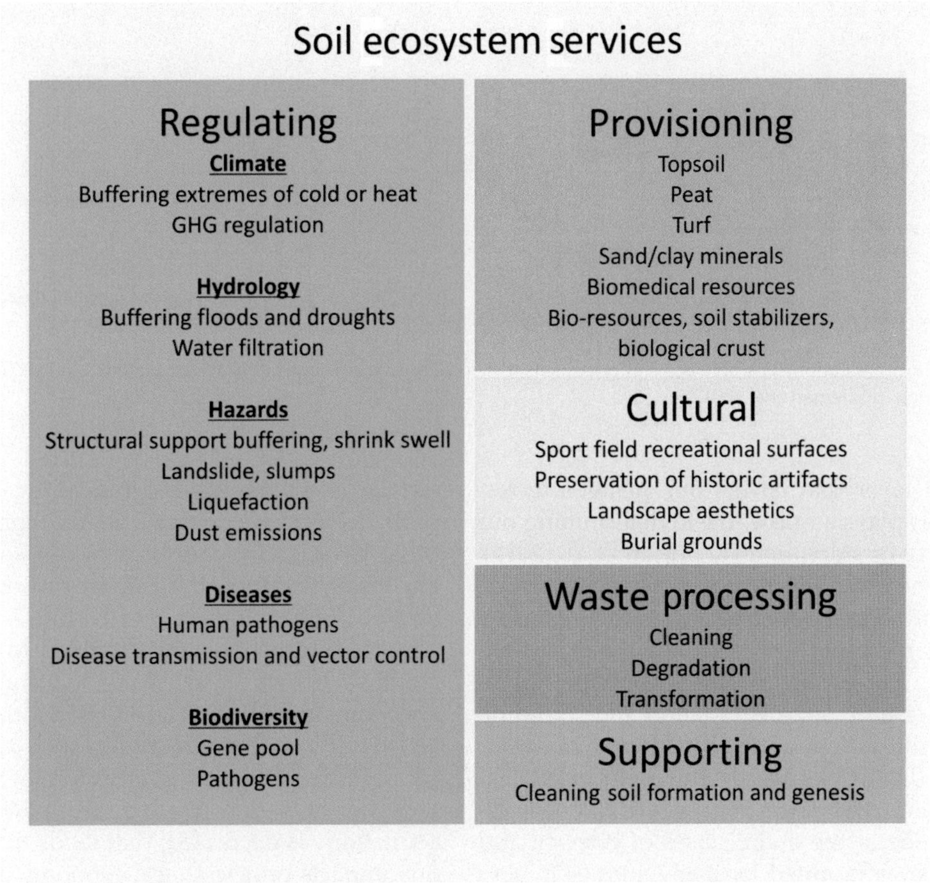

Soil ecosystem services

Regulating
Climate
Buffering extremes of cold or heat
GHG regulation
Hydrology
Buffering floods and droughts
Water filtration
Hazards
Structural support buffering, shrink swell
Landslide, slumps
Liquefaction
Dust emissions
Diseases
Human pathogens
Disease transmission and vector control
Biodiversity
Gene pool
Pathogens

Provisioning
Topsoil
Peat
Turf
Sand/clay minerals
Biomedical resources
Bio-resources, soil stabilizers, biological crust

Cultural
Sport field recreational surfaces
Preservation of historic artifacts
Landscape aesthetics
Burial grounds

Waste processing
Cleaning
Degradation
Transformation

Supporting
Cleaning soil formation and genesis

FIGURE 2.1 Soil ecosystem services. *Adapted from Robinson, D.A., Hockley, N., Cooper, D.M., Emmett, B.A., Keith, A.M., Lebron, I., et al., 2013. Natural capital and ecosystem services, developing an appropriate soils framework as a basis for valuation. Soil Boil. Biochem. 57, 1023–1033. http://dx.doi.org/10.1016/j.soilbio.2012.09.008.*

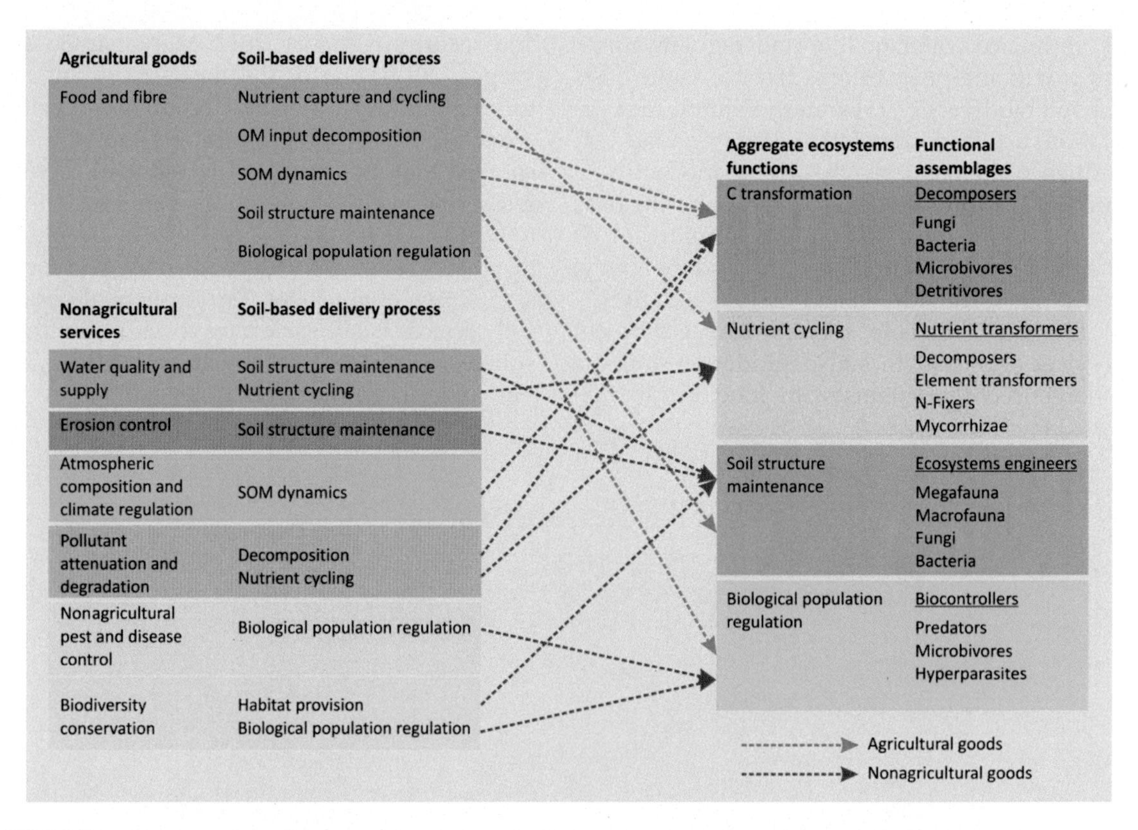

FIGURE 2.2 Relationships between soil organisms, their ecosystem functions and the ecosystem services that they provide to society (Pulleman et al., 2012).

thin soil layer that covers our planet's terrestrial areas plays a major role in determining our prosperity or starvation (Robinson et al., 2012). 52% of the areas used for agriculture are moderately or severely affected by soil degradation. At the same time, 4–6 million ha of cultivated soils are lost each year as a consequence of human-induced soil degradation and 75 billion tons of soil is lost annually to wind or water erosion (UNCCD, 2009). Human-induced soil degradation and corresponding loss of soil services is one of the main causes of poverty and starvation as reported by many studies in several environments (Barbier, 2000; Bindraban et al., 2012; Burras et al., 2013; Ludeke et al., 1999; Scherr, 2000). Soil nutrition status in Africa is statistically significantly correlated with the rate of poverty on the continent; in other words, in countries where soil nutrient losses are high the rate of poverty is high as well (ELD Initiative and UNEP, 2015). Food security and production is related to wars and conflicts (Lynch et al., 2013), natural hazards, and climate change related effects that reduce soil quality and productivity, such as extreme droughts and floods (Vermulen et al., 2012; Wheeler and Von Braun, 2013). When food availability is decreased, that tends to have serious impacts on social and economic aspects of households and individuals, problems related to the reduced capacity to work, vulnerability to diseases, and negative impacts on the mental

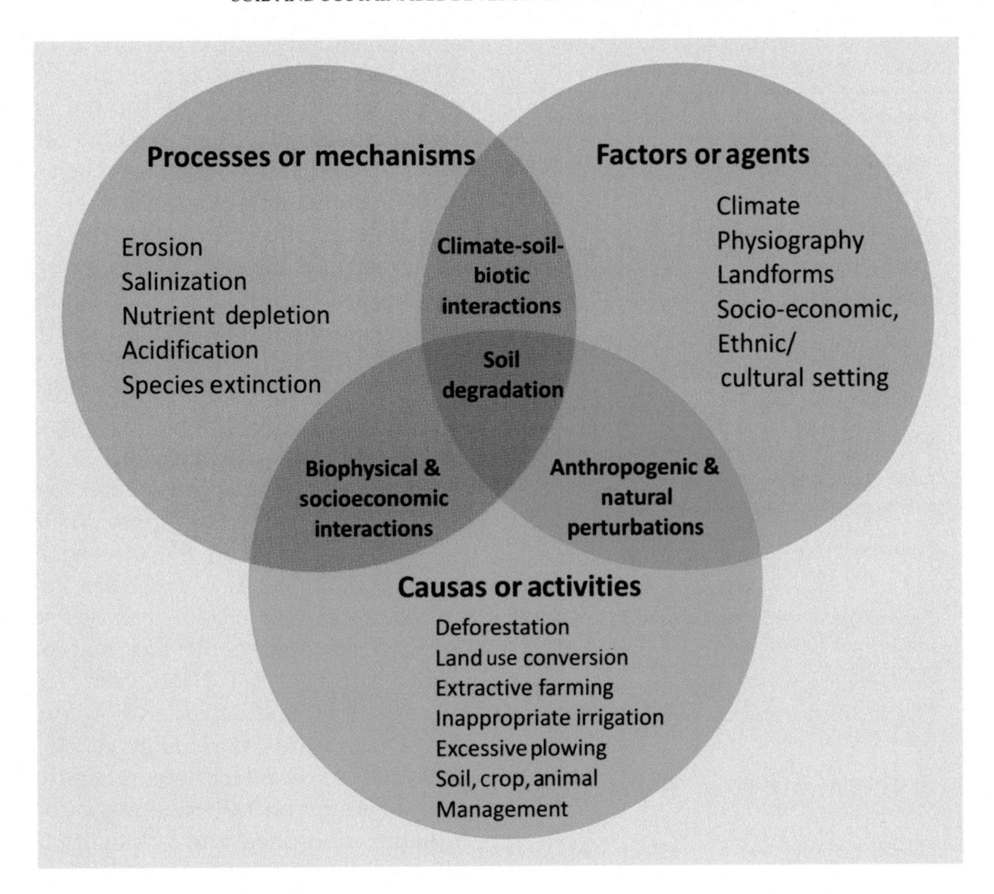

FIGURE 2.3 Soil degradation causes and drivers (Lal, 2015).

and educational development of children (FAO, 2002; Arndt et al., 2012; Wheeler and Von Braun, 2013).

In 2015 the world population was 7.3 billion and is estimated to reach approximately 9.5 billion in 2050. From 2005 to 2050, population growth will increase the demand for agricultural production by approximately 70% (Lal, 2015). In 2013, 38% of the Earth's soil had been converted into agricultural land, while only 11% of Earth's soils are considered suitable for farming (FAO, 2002; World Bank, 2008).[2]

[2]http://wdi.worldbank.org/table/3.2# (consulted on 02.02.16).

This shows that we are greatly exceeding the capacity of our soils due to population growth and demand for food. According to the World Bank, from the 1960s until 2014, there was an increase of more than 100% in crop and food production, livestock production, and cereal yield. A high increase in the use of agricultural machinery and land for agricultural production was identified. On the other hand a decrease of arable hectares per person and in the rural population was observed (Table 2.1). These activities are normally related to an unsustainable use of soil and land degradation. Feeding a growing population in the future will be a major challenge (Godfray et al., 2010), but the

TABLE 2.1 Percent of Variation of Some Agriculture and Rural Development Variables for the World

Acronym	Variable	% of Variation
AMT	Agricultural machinery, tractors	+60.34 (1962–2001)
ALK	Agricultural land (km^2)	+21.05 (1962–2014)
AGL%	Agricultural land (% of land area)	+4.48 (1962–2014)
ALP	Arable land (hectares per person)	−88.42 (1962–2014)
AL%	Arable land (% of land area)	+10.67 (1962–2014)
LCP	Land under cereal production (ha)	+27.74 (1962–2014)
PC%	Permanent cropland (% of land area)	+38.31 (1962–2014)
AMTSQ	Agricultural machinery, tractors per 100 km^2 of arable land	+ 50.01 (1962–1999)
CPI	Crop production index (2004–2006 = 100)	+142.43 (1962–2014)
FPI	Food production index (2004–2006 = 100)	+138.85 (1962–2014)
LPI	Livestock production index (2004–2006 = 100)	+129.88 (1962–2014)
CY	Cereal yield ($kg\,ha^{-1}$)	+125.82 (1962–2014)
RP	Rural population (% of total population)	−42.53 (1962–2014)

Source: World Bank Database.[a]

[a]*http://data.worldbank.org/topic/agriculture-and-rural-development?display=default (accessed 02.06.16).*

challenge is not limited to this. The intensification of agriculture, overexploitation of soil resources and degradation of soil services are one of the main causes of poverty and is a real threat to food security (Bommarco et al., 2013; Das Gupta, 2016). Agriculture practices also contribute significantly to greenhouse emissions. It is estimated that between 2001 and 2011, greenhouse gas emissions increased 14% (EEA, 2015). Intensive agriculture and livestock production is responsible for the emission of great amounts of carbon dioxide (Lal, 2004a) as well as other greenhouse gases such as nitrogen oxide and methane (Linquist et al., 2012). This is mainly attributed to increasing population, consumer demands and changing of food habits, which contributed to unsustainable farming practices and soil degradation (De Boer et al., 2013). A shift in human consumption patterns, especially in regards to meat, is a key to reduce agricultural contributions to greenhouse gas emissions (Bouwman et al., 2013).

Soils are the largest active reservoir of carbon (±1500 PgC), containing approximately double the carbon present in the atmosphere (Smith, 2012). Soil degradation processes influence the carbon cycle. Soil erosion releases soil organic carbon, and despite the fact that part of this eroded carbon (0.06–0.27 $PgC\,year^{-1}$) is deposited and stored in landscapes, erosion leads to a net global lateral flux of 0.61 $PgC\,year^{-1}$ (Van Oost et al., 2007). Soil–plant systems contribute to carbon sequestration by removing carbon dioxide from the atmosphere and locking it up in the soil as organic matter, thereby contributing to climate change mitigation. Nevertheless, this capacity to sequester carbon depends on soil texture, structure, rainfall, temperature, farming system, and soil management. No-till management has been widely reported to release less carbon dioxide into the atmosphere compared to intensively tilled systems (Lal, 2004c), although this has been questioned by several researchers (Baker et al., 2007; Blanco-Canqui and Lal, 2008; Christopher et al., 2009; Khan et al., 2007) and carbon sequestration benefits may be limited to locations with an appropriate climate (Carr et al., 2015; Van den Bygaart et al., 2003). Including cover crops in agricultural management is another technique that holds great promise for sequestration of carbon in soils (Olson et al., 2014; Poeplau and Don, 2015), and even the effects of management and land use on carbon sequestration in urban soils has been studied and influences found (Bae and Ryu, 2015; Beesley, 2012; Weissert et al., 2016). Thus the way we use any given soil will influence our contribution to or mitigation of global climate change. In the present soil landscape, carbon pools are much reduced as compared to before human intervention. It is estimated that soils have lost between 40 and 90 PgC due to cultivation and other disturbances. The correct management of soil, including no-tilling practices, cover crops, and other management techniques that reduce soil degradation, e.g., afforestation, natural rehabilitation, terracing, and organic farming will contribute to a decrease in carbon dioxide emissions and increase soil carbon sequestration (Lal, 2004b).

Managing soil carbon is extremely important since soil organic matter has an important impact on several soil ecosystem functions. Small changes in soil carbon can have large impacts on soil physical properties (Powlson et al., 2011). In addition, soil carbon sequestration is an extremely valuable regulating ecosystem service and a relatively low-cost option to reduce emissions that is very attractive to governments. In this context, for sustainable soil use, it is important to encourage management practices that promote the preservation and restoration of carbon to soils (Lal, 2004b; Powlson et al., 2011). Several studies have pointed out that carbon farming is one of the most cost-effective alternatives to offset carbon emissions and to deliver biodiversity benefits via ecosystems restoration and other economic and social benefits dependent on atmospheric carbon reduction (Evans et al., 2015; Funk et al., 2014) that also increase soil carbon (Becker et al., 2013; Cowie et al., 2013). A study carried out in Australia by Evans et al. (2015) observed that assisted natural regeneration sequestered 1.6–2.2 times more carbon than plantations. In addition, the costs for natural regeneration were 60% lower than the plantations. Natural processes are much less expensive than engineering solutions, such as the transformation of carbon dioxide into carbonates (Lal, 2009).

There is much discussion about the economic value of soil ecosystem services. Although establishing exact financial values for any given service is difficult, the ecosystem services provided by soils can have considerable value. In New Zealand, Dominati et al. (2014) estimated that the soils they studied provided ecosystem services valued at NZ$16,390 $ha^{-1} year^{-1}$ (approximately US$13,110 $ha^{-1} year^{-1}$). Services included in the Dominati et al. (2014) evaluation were food quantity and quality, support for human infrastructure, support for animals, flood mitigation, filtering of nitrogen, phosphorus, and other contaminants, recycling of wastes, N_2O regulation, CH_4 oxidation, and regulation of pest and disease populations. Many of the services provided by soils discussed earlier in this chapter can be seen in the economic evaluation completed by Dominati et al. (2014). However, demonstrating the difficulty of generating these values and the variability of soils, other researchers have reached very different values for ecosystem services. In another New Zealand study, Sandhu et al. (2008) estimated the value of ecosystem services as being between US$1270 and 19,420 $ha^{-1} year^{-1}$, with management making a difference in the value of ecosystem services. Both the Dominati et al. (2014) and Sandhu et al. (2008) studies were done on agricultural soils, which should have a fairly high total ecosystem services value. McBratney et al. (2017) estimated that the ecosystem services for all lands globally, including nonagricultural lands, deserts, etc., were valued at about US$867 $ha^{-1} year^{-1}$, considerably less than the values typically calculated for agricultural lands. In all of these studies the use the land was put to, the ecosystem services considered (or left out), and the values assigned to each ecosystem service made a major difference in the final results.

SUSTAINABLE LAND MANAGEMENT AND SOIL MAPS

Definition and Principles

Sustainable land management aims to integrate water, biodiversity, land and environmental management aspects to meet increasing food, feed, fiber, and bioenergy demands while maintaining the sustainability of ecosystem services and livelihoods. Achieving this is a fundamental need, especially since intensive exploitation of soil and ecosystems can lead to land degradation and the loss of ecosystem services capacity, and undermines ecosystems' resilience and adaptability (Schwilch et al., 2010; World Bank, 2008). The Earth Summit (1992) defined sustainable land management as "*the use of land resources, including soils, water, animals, and plants, for the production of goods to meet changing human needs, while simultaneously ensuring the long-term productive potential of these resources and the maintenance of their environmental functions.*" According to the World Bank (2008) the goals of sustainable land management are

1. "*Preserving and enhancing the productive capabilities of cropland, forestland, and grazing land (such as upland areas, down-slope areas, flatlands, and bottomlands)*"
2. "*Sustaining productive forest areas and potentially commercial and non-commercial forest reserves*"
3. "*Maintaining the integrity of watersheds for water supply and hydropower generation needs and water conservation zones*"
4. "*Maintaining the ability of aquifers to serve the needs of farm and other productive activities*"

Management should be focused on reduced land degradation, increased productivity, and sustainable use of the soil resource. There should be a participative approach, involving all interested stakeholders in land use planning to arrive at the use of acceptable techniques and methods to avoid overexploitation of natural resources and inappropriate management. These goals should be achieved by empowering local communities and land managers, use of local resources in sustainable land management implementation, sharing information and experiences, and raising the importance of watershed management at the government level (UNDP, 2014).

Sustainable land management is divided into six components, (1) understanding the ecology of land use management, (2) maintaining or enhancing productivity, (3) maintenance of soil quality, (4) increasing diversity for high stability and resilience, (5) provision of economic and ecosystem service benefits for communities, and (6) social acceptability (Montavalli et al., 2013). According to FAO (1993), sustainable land management should meet four different criteria, (1) production levels should be maintained, (2) risk of production should not increase, (3) soil and water quality should be preserved, and (4) systems should be accepted by the society where they are being implemented and economically feasible. Finally, for TerraAfrica[3] sustainable land management principles are based on (1) increased land productivity, (2) improved livelihoods, and (3) improved ecosystems. Sustainable land management has a strong ecological, social, and economic component, dependent upon effectively combatting land degradation to ensure the sustainability of livelihoods and food security and ability to pay back the investments taken out by land user communities or governments (Liniger et al., 2011) (Fig. 2.4).

[3]http://www.terrafrica.org/sustainable-land-management-platform/what-does-slm-achieve (consulted on 02.02.16).

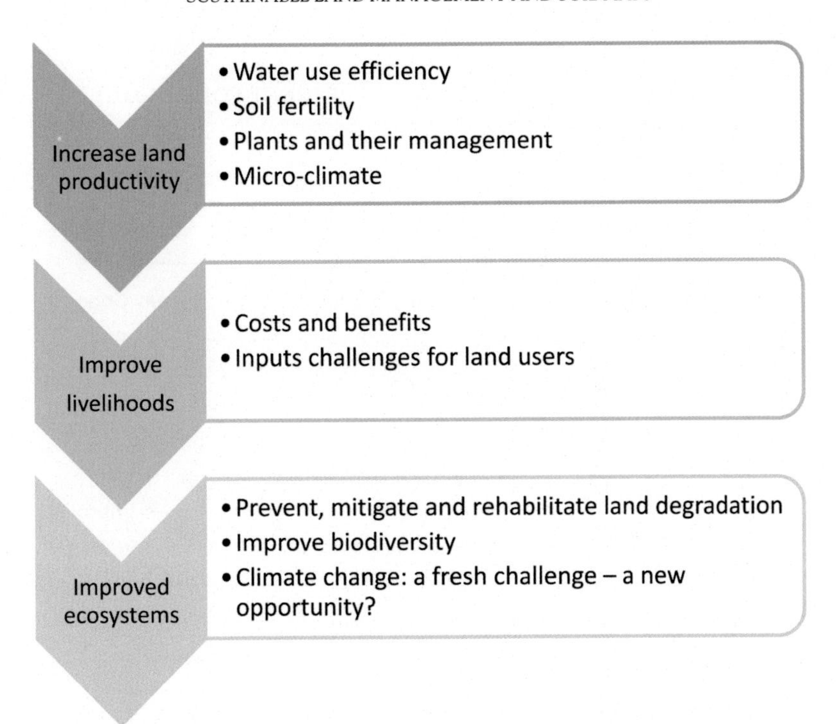

FIGURE 2.4 Principles for the best sustainable land management (Liniger et al., 2011).

Sustainable Land Management Need: The Water Question

Sustainable land management is fundamental for future generations. Human activities are indeed responsible for the transformation of Earth's surface and soil degradation, with humans now representing the single most defining geomorphic force of our time (Steffen et al., 2015; Zalasiewicz et al., 2015) and functioning as a soil forming factor (Yaalon and Yaron, 1966; Richter and Yaalon, 2012). According to a WWF (2014) report, we need 1.5 planets to meet our present demands on nature. We are consuming resources from the planet faster than they can be regenerated. Agriculture is having a huge impact on water consumption. Our unsustainable water demands and the increasing scarcity imposed by pollution and climate change are creating critical levels in water availability (Kresic, 2009; WWF, 2014). Globally, the intensive application of fertilizers and irrigation water to arable land is way too high (Aguilera et al., 2013), which can produce long-term loss of natural capital, including soil productivity and increased soil pollution with potential impacts on human health, especially if wastewater is used as a soil amendment (Khan et al., 2008; OECD, 2012; Wang et al., 2012). Irrigated systems are not well adapted to today's agriculture and the level of productivity is much reduced, representing a loss of resources, efficiency, and economic values. From 1961 to 2009 the irrigated cultivation area increased 117% and is expected to increase by 127%–129% by 2050 in relation to 1961 (FAO, 2011). This unsustainable growth leads to extremely high consumption of water resources. 10%–25% of rainfall is lost to runoff and evaporation, and as a consequence of these losses, only between

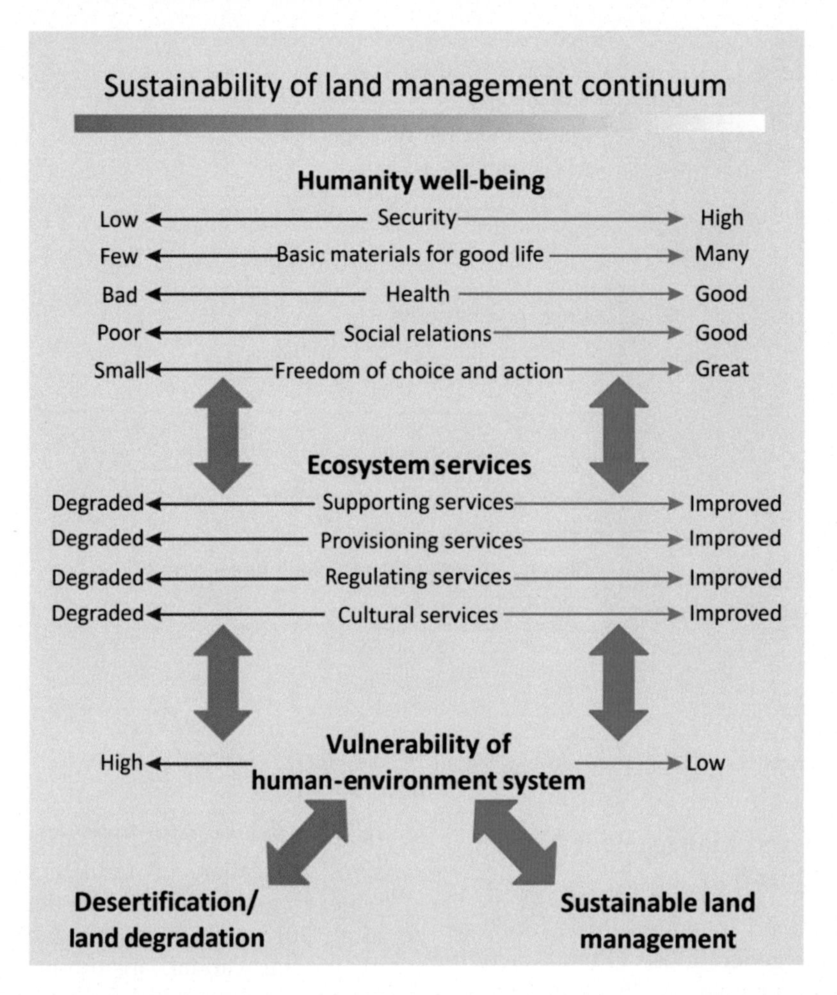

FIGURE 2.5 Interdependence between human well-being choices, ecosystems services, land use management, and the human–environment system (based on Buenemann et al., 2011).

15% and 30% of rain is typically used for plant development (FAO, 2011). Husbandry practices, intensive farming, and development of irrigation technologies are responsible for the increasing unstainable use of water resources (FAO, 2011; Liniger et al., 2011; World Bank, 2008). For these reasons, sustainable land management, which includes sustainable use of our water resources in the production of food, feed, fiber, and fuel, is extremely important to ensure sustainability for future generations.

Sustainable Land Management Practices and Indicators

Sustainable land management practices are fundamental for the preservation and quality of the soil. They are a key aspect of the delivery of regulating, supporting, providing and cultural ecosystem services, and are connected to our well-being as mentioned earlier in this chapter (Fig. 2.5). Several practices have been developed to ensure soil productivity. However, the

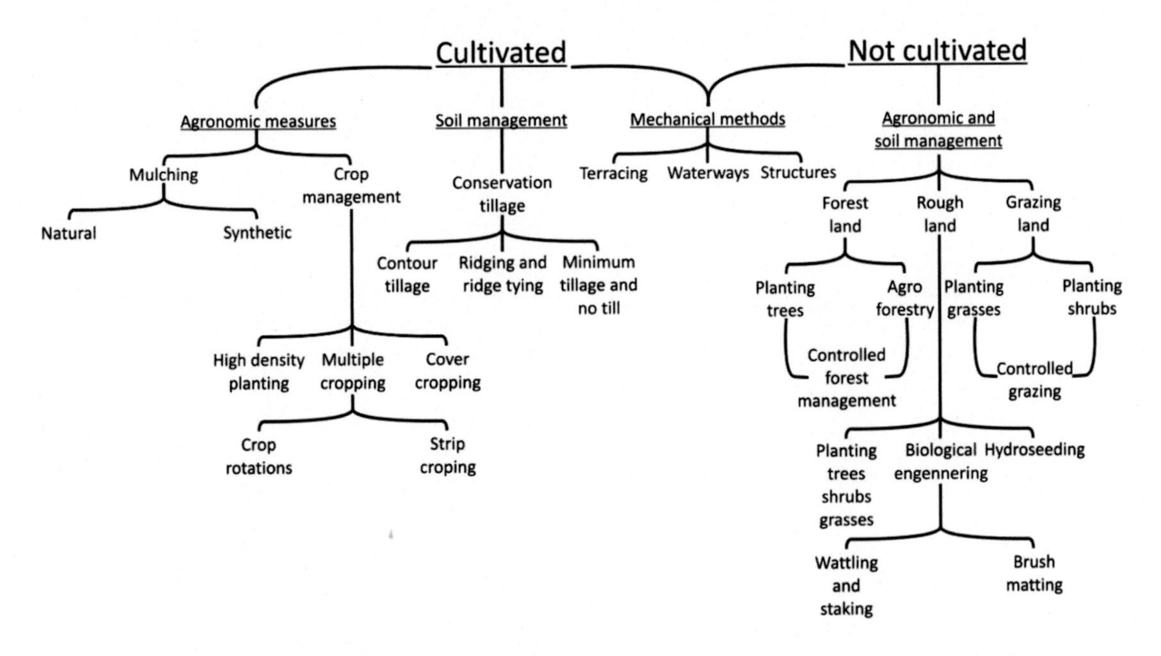

FIGURE 2.6 Sustainable land management practices in cultivated and noncultivated environments. *Adapted from UNDP, 2014. Sustainable Land Management Toolkit. Available from: http://www.ls.undp.org/content/lesotho/en/home/library/environment_energy/SLM-Toolkit.html (consulted on 15.03.16).*

application of these measures is often difficult to implement and adopt due to different interests of the stakeholders involved in land use management (World Bank, 2008). Sustainable land management is divided into cultivated and noncultivated techniques as shown in Fig. 2.6. Several methodologies have been developed to monitor and assess sustainable land management at local levels by applying the World Overview of Conservation Approaches and Technologies (WOCAT) guidelines, which have been used lately in the assessment of land degradation by the Land Degradation Assessment in Drylands (LADA) and EU-Desire projects. The main objective of monitoring and assessment procedures is to analyze and create solid information for decision and policy-makers at several levels (Schwilch et al., 2010).

Multiple attempts have been made to define the best indicators for assessing and monitoring sustainable land management. According to Cornforth (1999) the indicators should (1) be selected from the outputs of production, (2) influence the product value, and (3) have impacts on the production at local and other levels. The selected environmental indicators must also be (1) sensitive and responsive to changes in land management, (2) important in the assessed area, (3) related to ecosystem process, (4) scientifically valid, (5) use existing data, (6) easy and cheap to measure, (7) not complex, (8) accessible to land users, managers, scientists, and policy-makers, (9) internationally recognized, and (10) strong enough to support political decisions (Cornforth, 1999). Soil quality indicators, which are fundamental to assess sustainable land management are divided into three categories. These are (1) develop in the near term, (2) require longer term research, and (3) developed by other networks. Sustainable land management indicators, on the other hand, are based

TABLE 2.2 Common Indicators for Land Use Quality and Sustainable Land Management (Dumanski et al., 1998)

Land quality	Developed in the near term	Nutrient balance
		Yield gap
		Land use intensity
	Requiring longer term research	Soil quality
		Land degradation
		Agrobiodiversity
	Developed by other networks	Water quality
		Forest land quality
		Rangeland quality
		Soil pollution
Sustainable land management	Productivity	Crop yield
	Security	Soil cover
		Yield
		Variability
		Climate
	Protection	Soil and water quality/quantity
		Biological diversity
	Viability	Net farm profitability
		Input use efficiency
		Pesticides, fertilizers, nutrients
		Off-farm income
		Return to labor
	Acceptability	Use of conservation practices
		Farm decision-making criteria

on their productivity, security, protection, variability, and acceptability (Table 2.2) (Dumanski et al., 1998). More recently the KM: Land project developed five global indicators, measurable at the project level, in order to assess the complexities of land degradation processes and sustainable land management, which depend upon biophysical, social, political, economic, and cultural factors (UNU-INWEH, 2011). These indicators consider land use/cover aspects, productivity in different land use types and systems, water resources, and human well-being (Fig. 2.7). Despite the existence of these common and global indicators, it is important to develop indicators adapted to the local reality of the studied area. Several studies have pointed out the importance of integrating local with scientific knowledge in the development of effective sustainable land management plans and reducing land degradation on several continents, such as Africa (Reed et al., 2007), Asia and Oceania (Lefroy et al., 2000), and South and Central America (Barrera-Bassols and Toledo, 2005). In many cases, local knowledge is considered to be the core of the programs developed.

Sustainable Land Management Monitoring and Assessment

Monitoring and assessment studies have traditionally been more focused on land degradation rather than on the sustainable management of land. Studies focused on the social, economic, and environmental costs and benefits of sustainable land management are largely lacking. The available works show that sustainable land management is positively associated with land tenure security in middle and advanced economies. In countries with lower incomes, this association was not observed since secure land tenure is not related to unsustainable farming practices (Nkonya et al., 2008). Diao and Sarpong (2011) estimated that sustainable land management practices applied in Ghana between 2006 and 2015 increased total benefits by $6.4 billion, reducing poverty. If farmers perceive economic advantages from the adoption of sustainable land use practices, it will facilitate the implementation of these measures. Kassie et al. (2010) found that farmers

Land use/cover	Land productivity	Water availability	Human well-being
•Global land cover classes •National land use systems •Project defined land management practices	•Crop diversity •Production per unit of physical inputs •Annual a sivilcultural and agricultural production	•Water availability in the catchment •Extracted water volume by land use system across the catchment •Ratio between avaliable water and the extracted	•National maternal mortality ratio •National proportion of chornically undernourished children under the age of 5 in rural areas •Percentage of rural inhabitants bellow the national poverty line

FIGURE 2.7 Global sustainable land management impacts and measurable indicators at a project level (UNU-INWEH, 2011).

who used minimum tillage in areas with low agricultural potential had higher productivity compared to farmers who used commercial fertilizers. This facilitated the adoption of minimum tillage in the studied areas. Observed in a survey carried out in several parts of the world that the great majority of the farmers interviewed (97%) acknowledged the long-term benefits of the implementation of sustainable land use practices and technologies. Nevertheless, there are cases where such practices are not implemented because of lack of knowledge about these practices due the lack of communication between scientists and land managers and guidance on environmental questions. There are also cases where sustainable practices are not adapted for cultural reasons (Burras et al., 2013; Sandor et al., 2006). Therefore it is important to invest in communication of scientific results to land owners and managers to demonstrate the advantages of using sustainable land use practices, and it is also important to work with local communities to identify practices that are culturally acceptable. According to Mirzabaev et al. (2015), there are three reasons to promote more investments in sustainable land management (1) the social costs of land degradation are very high in the global community compared to private interests, (2) the private costs of land degradation in some cases are much higher than the costs of inaction; this may also partly be a consequence of lack of knowledge about sustainable management practices or barriers imposed by policy makers, and (3) despite the fact that land owners understand the direct costs imposed by land degradation, they are still resistant to invest in sustainable land management measures. The challenge is to show the advantage of long-term benefits to heads of households and decision makers that normally are not part of political agendas and to supply them with fiscal security during the transition period to new management practices. From the economic point of view, soil and land degradation do not need any intervention. This only happens when the market fails and the consequent results impact on the social sphere. At this level the costs of soil rehabilitation are very likely higher than the costs of sustainable land management practices (Mirzabaev et al., 2015; Shiferaw and Holden, 2000). In Africa it is estimated that the costs of inaction against land degradation are seven times higher than the costs of implementing sustainable land management practices (ELD Initiative and UNEP, 2015). It has been estimated that the global cost of soil erosion is about five times higher than the cost of prevention would be (Pimentel et al., 1995). Thus it is of major importance to continue to implement sustainable land management practices adapted to the different realities and get evidence that these practices are effective at decreasing soil and land degradation and that they improve soil productivity and the social and economic conditions of households that implement them.

Soil Spatial Analysis, Mapping, and Sustainable Land Management

Sustainable land management planning requires geospatial analyses and mapping that are integrative. Such planning needs to have the capacity to link quantitative and qualitative data that characterizes the natural and human environments. Spatial analysis enhances understanding of interactions occurring on the landscape and exactly where those interactions are likely to lead to soil degradation. Developing our abilities in these areas will contribute strongly to understanding the degree of the impacts of land use (Buenemann et al., 2011). A correct, effective, and integrative geospatial approach to monitoring and assessing sustainable land management needs to (1) provide spatial information about risks and vulnerabilities, (2) identify interrelations between human and environmental dimensions at micro, local, regional, and macroscales, (3) provide suggestions for alternative land management, (4) consider accuracy and uncertainty analysis, and (5) recognize the unique characteristics of local environments (Buenemann et al., 2011).

Sustainable land management needs to be done at a wide range of spatial scales. A large effort has been made to map land degradation using expert analysis within the framework of WOCAT, LADA, and DESIRE at the international scale (Bouma, 2002; Reed et al., 2011). The WOCAT–LADA–DESIRE mapping was based on land use systems at the national level, similar to CORINE land cover classification for the European Union. CORINE is the Coordination of Information on the Environment program promoted by the European Commission in 1985 for the assessment of environmental quality in Europe. The CORINE Land Cover project provides consistent information on land cover and land cover changes across Europe (Neumann et al., 2007). According to the different land uses, experts assessed the actual land degradation and the practices carried out for sustainable land use. The information obtained from the survey was georeferenced using geographic information systems (GIS) techniques, producing a map with the level of conservation practices and land degradation of the assessed area (LADA, 2013). Nowadays the use of land use classification is extensively used for expert evaluation of ecosystem services at national (Egoh et al., 2008), regional (Burkhard et al., 2009; Palomo et al., 2013), and catchment levels (Vrebos et al., 2015).

Despite the importance of the expert information, more reliable data are necessary to have a good assessment of land degradation and sustainable land management. One example of this is soil data. Soil maps are an extremely important source of information to assess these parameters at any scale. Soils are the basis for sustainable land management and in the identification of the first indicators of land degradation. In this context, it is of major relevance to have a high quality, quantitative soil database. Land use maps connected with soil, topographical, and climate maps allow us to create a spatial and temporal view of the areas that are most vulnerable to land degradation and that may need urgent implementation of sustainable land management practices. Several projects at the international level, such as the Global Assessment of Land Degradation and

Improvement (GLASOD), that aimed to map land degradation did not use real soil data but developed indices based on remote sensing techniques to estimate land degradation instead (GLASOD),[4] this is done when adequate soil information is not available. In the case of the GLASOD, net primary productivity was used as an indirect estimation of soil erosion, salinity, and nutrient depletion (Bai et al., 2010). Despite the large extent and the coarse resolution (8 km) used in this work, soil data would have been useful to validate the estimations made using net primary production, because expert evaluations in GLASOD were not very accurate nor reliable (Sonneveld and Dent, 2009). As in other cases such as the EU-project Pan-European Soil Erosion Risk Assessment (PESERA) (Kirby et al., 2004), the results produced have been criticized because of the lack of calibration and validation (Reed et al., 2011). At the European level, in the last few years there has been a big effort to create a better soil database in digital format and made freely available for public use[5] (Fig. 2.8), including the production of maps of different soil properties for policy making and public use (Panagos et al., 2012). The availability of this information is very important for scientists, but for land managers the use is quite limited because the resolution ($1 km^2$) is too coarse for land managers to utilize in a meaningful way.

Digital information at finer resolutions is needed to make the soil information useful for managers. This problem is also currently observed in other sources of information such as the Global Facility Soil Information[6] where the soil grid's information is at a resolution of $1 km^2$. Despite the evident value of this database, several concerns arise regarding their use as the accuracy of the predictions of soil properties and class values only exceeded 50% in a few cases. That modeling effort was not able to detect much of the spatial variability, and it is biased by an unequal distribution of the soil profiles used to create it. Areas of Canada, North Africa, Russia, and Central Asia are very poorly covered with quality soil data (Hengl et al., 2014). This creates problems regarding the validity of predictions made using these databases. In addition, many of the areas that are poorly represented are arid and semi-arid environments, among the most vulnerable areas to land degradation or sustainable land management issues. To tackle these questions a better spatial distribution of the soil information collected is needed. Efforts to address that issue compliment and support the need to use more robust statistical methods to improve the accuracy of soil property predictions (Brevik et al., 2016).

Recent work by Hengl et al. (2015) tried to tackle these problems by downscaling a $1 km^2$ resolution (global coverage) soil map (Hengl et al., 2014) to a $250 m^2$ resolution using the same database, but only applied to Africa. The resulting map had a finer resolution and predictions carried out at a 250 m resolution were better than the ones observed at $1 km^2$. The use of the random forests statistical techniques helped improve the accuracy of the spatial predictions from the $1 km^2$ resolution to the 250 m resolution applied to Africa. Soil data availability was extremely relevant for increasing the accuracy of the predictions (Hengl et al., 2015). The findings of this work are highly relevant to the question of land degradation assessment and the implementation of sustainable development practices in Africa, which is recognized as the continent with the most serious problems related to land degradation and most in need of sustainable land use practices. Each year, Africa loses approximately 280 million tons of cereal from 105 million hectares of croplands where soil erosion could be managed (ELD Initiative and UNEP, 2015).

[4] http://www.fao.org/nr/lada/gladis/gladis_db/.
[5] http://esdac.jrc.ec.europa.eu/resource-type/soil-data-maps.
[6] http://www.isric.org/projects/global-soil-information-facilities-gsif.

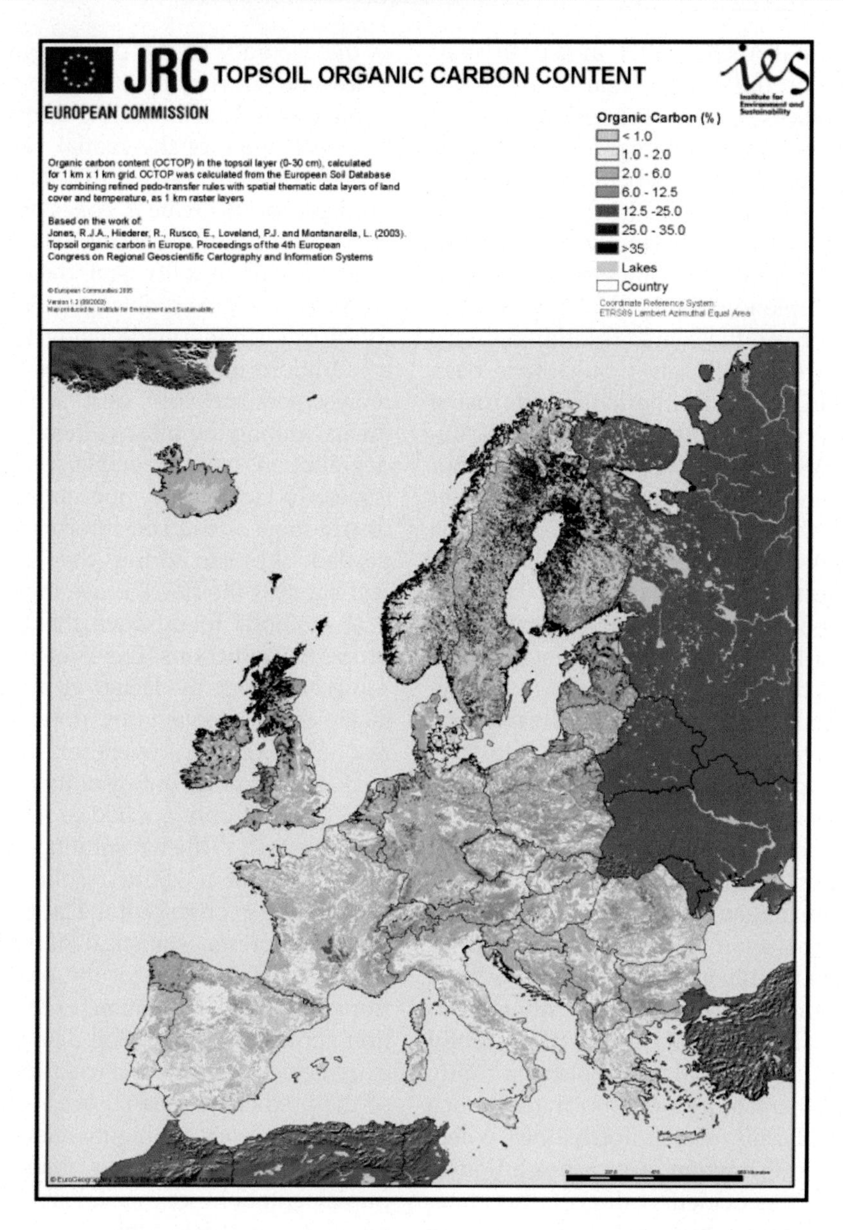

FIGURE 2.8 Topsoil carbon distribution in European Union countries. *Source: http://esdac.jrc.ec.europa.eu/public_path/OCTOP.png.*

Downscaling methods using remote sensing for mapping have been extensively applied to estimate soil hydraulic properties, especially water content (Crow et al., 2000; Djamai et al., 2015; Kim and Barros, 2002; Ray et al., 2010). Recently, several remote sensing methods have been applied to estimate and map other soil properties. A review of these methods can be found in Mulder et al. (2011) and Brevik et al. (2016). On the other side, upscaling soil properties has been frequently used for mapping soil properties at plot (Sundqvist et al., 2015), catchment (Crow et al., 2012; Taylor et al., 2013), regional (Horta et al., 2014; Xu et al., 2013), and country (Constantini and L'Abate, 2016) levels. Both upscaling and downscaling will continue to be relevant to soil mapping, particularly where high-resolution soil surveys are not a reality (Malone et al., 2013). More investments are needed to provide better spatial coverage of soil data.

Soil Models Contribution to Sustainable Land Management

Out of necessity, models simplify reality and are used to understand the complexity of environmental systems. There are three basic types of models used in environmental studies, process-based models, empirical models, and conceptual models. Empirical models are the most simple of the models, while conceptual models are considered to have a degree of complexity between empirical and process-based models (Letcher and Jakeman, 2010).

Process-based or mechanistic models are used to simulate past, present, and future changes, based on the representation of the components and their interactions in a determined environmental system. These models give a numerical solution over a determined time and space. In other words, they assume changes in the quantities of the studied variables (state variables). These variables are expressed by differential equations and driven by fluxes formulated as rates of processes, known as rate variables. Process-based models require large computing costs and are based on the existence of a large number of parameters distributed within the investigated space that can be, in theory, measured within the system analyzed; this is one of the limitations of their applicability. However, when a large number of parameters are involved and due to our inability to correctly measure the heterogeneity of the parameters involved, the errors of measurement can be important, increasing the uncertainty of the models. In practice, process-based models may include some empirical data and the correlation existents in empirical models can be useful to assume a link to a process. Process-based models are mainly applied in ocean and atmosphere models, climate modeling, and subsurface hydrological modeling (Adams et al., 2013; Letcher and Jakeman, 2010; Wali et al., 2010).

Empirical models (correlative or statistical models) are focused on the statistical correlation among the variables involved, but without describing the system behavior, rules, interactions, and structure in detail. Empirical modeling is divided into three stages, (1) selection of the predictor variables, (2) model calibration, and (3) validation. They are mainly designed to predict and depend on data to quantify the response of a determined system as a function of a small number causal variables. Several empirical models are based on data analysis using a stochastic approach, which is ideal to explore data patterns and identify hidden relations between the variables. These models do not require explanation of the processes or structures occurring in the studied system. In this type of model the uncertainty is reduced, however, some bias can be observed as a consequence of the exclusion of important variables or processes in the system. Empirical models are commonly applied in agricultural, ecological, and ecotoxicological studies (Adams et al., 2013; Bradford and Fierer, 2012; Koltermann and Gorelick, 1996; Wali et al., 2010). Some models use a hybrid approach and combine process-based methods and empirical representation of relationships (Adams et al., 2013; Korzukhin et al., 1996; Letcher and Jakeman, 2010; Makela et al., 2000; Perez-Cruzado et al., 2011).

Conceptual or mental models are based on simple representations of the system, allowing them to deal with nonlinear processes. Normally they combine processes at the scale where the outputs are simulated. Parameter values are determined by calibrating them against observed data. These models provide a general description of the system, but without a detailed explanation of the interactions and processes among them. Conceptual models give an indication of quantitative and qualitative changes in the system, and do not need a large amount of data. They are often graphical, can represent static and dynamic phenomena, and are very useful to understand environmental phenomena. Conceptual models are commonly applied in hydrology, especially in surface processes such as rainfall-runoff and water quality (Letcher and Jakeman, 2010; Rodila et al., 2015).

The choice between mechanistic or empirical models depends on whether we want to have a more accurate but more uncertain model (mechanistic approach) or a model that better predicts ecosystem processes but is less accurate (empirical approach). There is no correct answer to this question (Bradford and Fierer, 2012).

The models mentioned above have been extensively used in soil science in ways that are extremely relevant for sustainable land management. Process-based models have been applied to understand soil greenhouse gas emission (Giltrap et al., 2010), methane consumption (Ridgwell et al., 1999), earthworm distribution and abundance (Johnston et al., 2014), microbial activity (Manzoni et al., 2016), trace element solubility (Groenenberg et al., 2012), nutrients solubility (Messiga, et al., 2015), organic carbon sequestration (Yagasaki and Shirato, 2014), depth (Dietrich et al., 1995), texture (Groenendyk et al., 2015), erosion (Misra and Rose, 1996), spatial distribution (Park et al., 2001), and crop water use and productivity (Immerzeel et al., 2008).

Empirical models have been used to predict soil respiration and evapotranspiration (Bosch et al., 2016; Feng et al., 2016), infiltration (Pena et al., 2016), moisture (Egea et al., 2016), acidification (Caputo et al., 2016), heavy metals concentration (McGrath and Zhao, 2015), cation exchange capacity (Jafarzadeh et al., 2016), carbon formation and decomposition (Bradford et al., 2013; Sierra et al., 2015), erosion (Galdino et al., 2016; Gartner et al., 2014), compaction (Gao et al., 2016), thermal conductivity (Lu et al., 2014), bulk density (Tranter et al., 2007), moisture and roughness (Zribi and Dechambre, 2003), and nutrients transfer to plants (Ding et al., 2015).

Finally, conceptual models have been applied to study soil CO_2 and mercury fluxes (Briggs and Gustin, 2013; Hoffmann et al., 2015), carbon decomposition (Moore et al., 2015), organic matter stabilization (Castellano et al., 2016), carbon and nitrogen dynamics (Luo et al., 2014), organo-mineral interactions (Kleber et al., 2007), microbial production and processes (Liang and Balser, 2011), water retention (Assouline et al., 1998), runoff and soil erosion (Al-Hamdan et al., 2015), hydraulic conductivity (Assouline and Or, 2013), salinity (Giordano and Liersch, 2012), nutrient transport and translocation (Kaizer and Kalbitz, 2012), pedogenesis (Simonson, 1959), and soil spatial distribution.

Process-based empirical and conceptual models have been incorporated with GIS techniques, with the whole providing a better understanding of soil model results (Li et al., 2015; Ju et al., 2010). Two examples of this include the present efforts carried out by the Joint Research Centre to join data from several projects and sources to model soil erosion at the European level (Fig. 2.9) and the Global Land Degradation Information System (Fig. 2.10). Soil erosion and land susceptibility models are very important to assess soil and land degradation vulnerability, causes, and processes (Borrelli et al., 2016), and impact of land use management (Galdino et al., 2016). They are relevant for identifying the most important factors

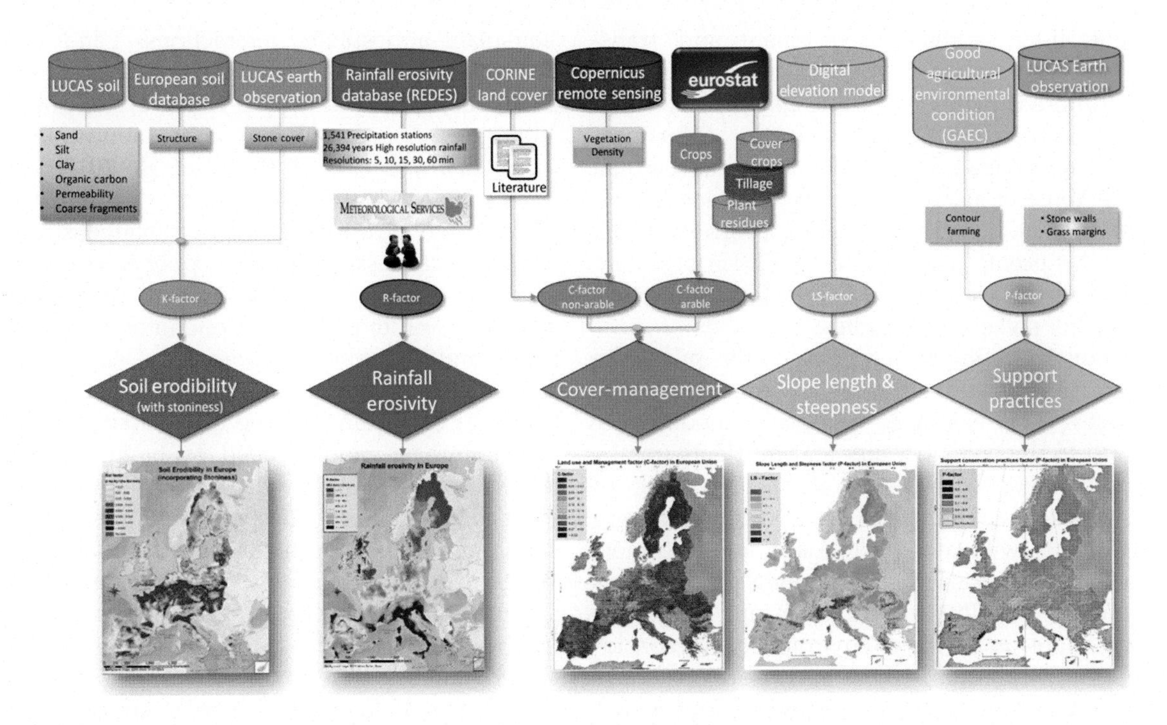

FIGURE 2.9 Structure of the soil erosion model for Europe using the RUSLE method. *Source: http://ec.europa.eu/eurostat/statistics-explained/index.php/File:RUSLE2015_soil_erosion_model_structure.png.*

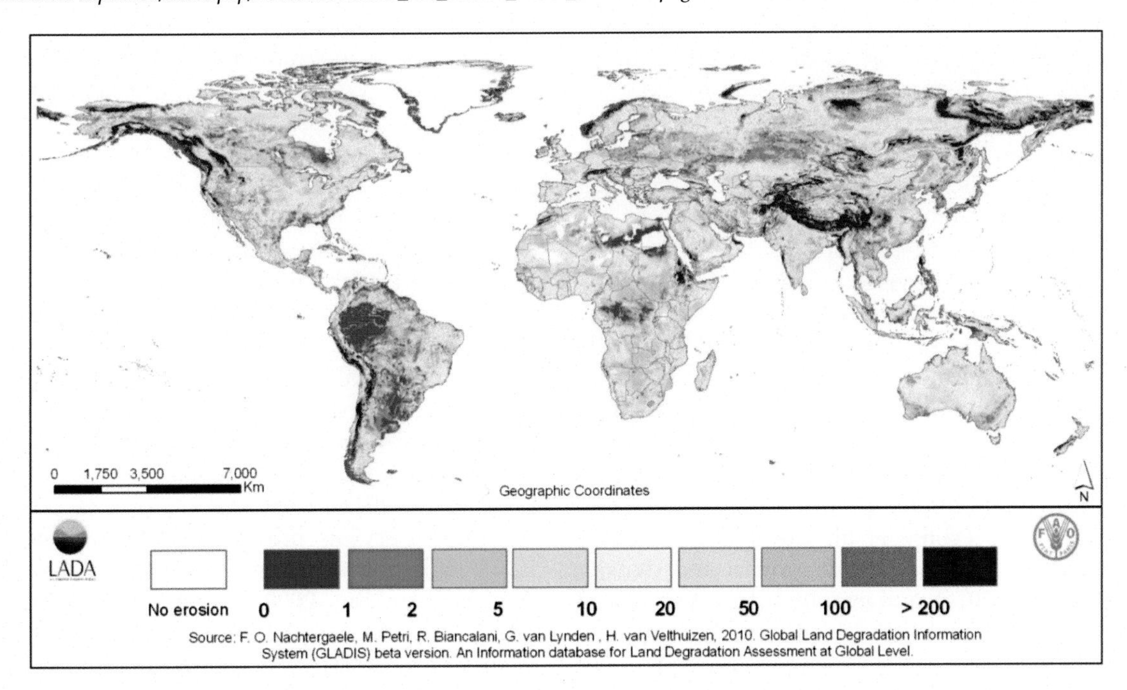

FIGURE 2.10 Estimated soil water erosion in tons $ha^{-1} year^{-1}$ (Nachtergaele et al., 2010).

controlling sediment detachment and transport. There are a number of soil models used at different scales, including hillslopes (Ziadat and Taimeh, 2013), subcatchment (Martinez-Casanovas et al., 2016), catchment and regional (De Vente et al., 2013), country (Panagos et al., 2014a), and continental levels (Borrelli et al., 2016). Recently De Vente et al. (2013) provided a detailed review of the most important models used for soil erosion studies.

Soil is an important piece of the puzzle in models for understanding environmental processes. For example, soil–plant models are used to understand water and nutrients uptake by plants (Rodrigues et al., 2012). Much of the water and nutrients transported from soil to plants will later be released to the atmosphere or be consumed by animals or humans (Lobet et al., 2014; Manzoni et al., 2013). The soil–plant–atmosphere environment is considered a continuum, where each develops interdependent and complex interactions of energy and mass, which is only possible to understand with the use of models. In this context, soil properties have direct or indirect implications for atmospheric chemistry and human health. This system is fundamental to the existence of life on Earth (Anderson et al., 2003). As in soil erosion models, soil–plant–atmosphere models are highly dependent on their spatial and temporal scales (Gharun et al., 2013). Soil–plant–atmosphere models have been applied in diverse fields with extreme relevance to sustainable land management, such as drought assessment (Anderson et al., 2013), groundwater reserves (Andreasen et al., 2013), groundwater impacts on plant water use and productivity (Soylu et al., 2014), rain and convective cloud formation (Manoli et al., 2016), water distribution in the rhizosphere, carbon uptake (Volpe et al., 2013) and emission (Zhang et al., 2016), plant growth and density (Ren et al., 2016), and evapotranspiration (Katul et al., 2012). Sustainable land use is only possible if we have a deep understanding about soil–plant–atmosphere interactions. Land use has strong impacts on soil–plant–atmosphere interactions, including the water cycle, pollution, greenhouse gas emissions, atmospheric chemistry, and climate (Beetz et al., 2013; Wu et al., 2012). Thus understanding the level of disturbance induced by land use change is fundamental to classify good and poor practices of land use and apply sustainable management practices. Soil models are a fundamental part of these multidisciplinary approaches, since they are relevant to assess soil quality, a fundamental indicator of sustainable land use (Herrick, 2000).

Biophysical models need to be integrated with socio-economic models to provide a complete picture of land use impacts on soil and land degradation and have better solutions and measures for sustainable land management implementation. Integration of biophysical and socio-economic models is an advantage, since it can identify potential problems and social risks related to project implementation, or adaptations to a new scenario, such as climate change (Fig. 2.11). Understanding and recognizing the biophysical and socio-economical causes of land degradation at local, regional, national, and continental levels is imperative to design and implement strategies to combat and reverse degradation processes (Vu et al., 2014; Salvati et al., 2015: Fleskens et al., 2014). Soil is an important part of biophysical and socio-economic models to evaluate agricultural production and adaptation under climate change conditions (Fraser et al., 2013; Weeb et al., 2013), agricultural sustainability (Nambiar et al., 2001), land use allocation (Du et al., 2013), land use intensity (Lambin et al., 2000), livestock distribution (Kerven et al., 2016), water resources and groundwater pollution (Baker et al., 2015; Lima et al., 2015), carbon and nutrient fluxes, ecosystem services provision (Swetnam et al., 2011), soil conservation and erosion hazard (Lu and Stocking, 2000), vegetation change in urban ecosystems (Luck et al., 2009), migration (Henry et al., 2003), food security (Schmidhuber

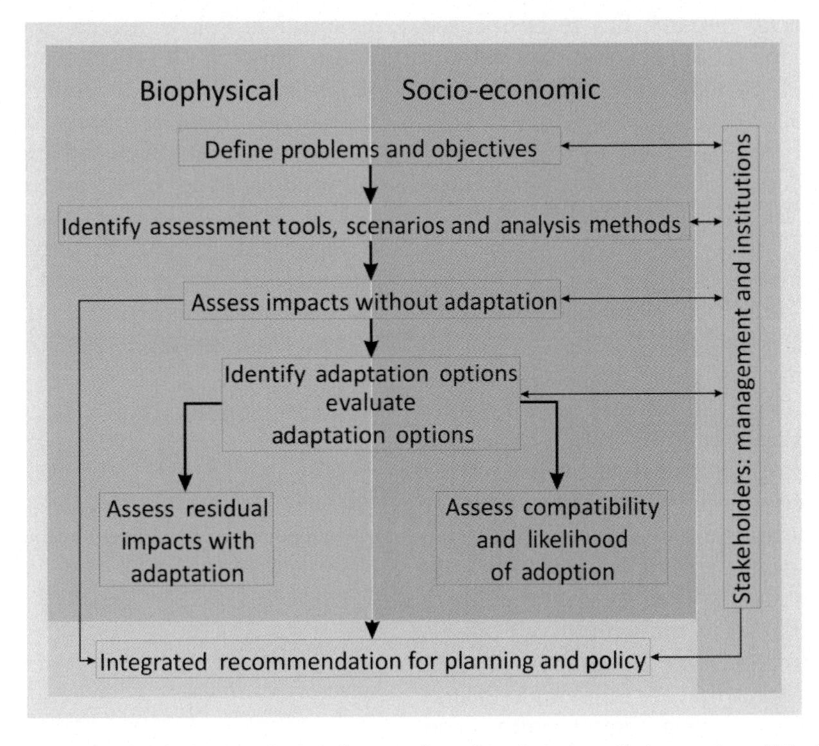

FIGURE 2.11 Framework for evaluating climate change adaptations, based on the integration of biophysical and socio-economic models (Weeb et al., 2013).

and Tubiello, 2007), and forest systems (Graves et al., 2007).

As mentioned previously in this chapter, soil provides a wide range of ecosystem services, and soil information integrated with other environmental, social and economic information is fundamental to quantify and assess these services, including supporting processes (soil formation, water and nutrient cycling, and biological activity), degrading processes (salinization, erosion, and compaction), regulating services (climate regulation, buffering, filtering and recycling of wastes), and provisioning services (biomass production for food, fiber and energy, physical support, and habitat) (Vereecken et al., 2016).

This all confirms the interdisciplinary nature of soil science, and how soil models alone or integrated with other disciplines contribute to a better understanding of spatial and temporal environmental changes. The interdisciplinary nature of soil science and the models developed are a key aspect of correct and sustainable land use management. Soils are where the major impacts of many human activities are reflected. Improved spatial soil information, increased quantification of soil information, and incorporation of soils into related models will help these models improve their results, and therefore facilitate their integration into territorial management and decision making. It is not possible to understand or to even imagine sustainable land use without the contribution of soil variables and models. Despite this, there is much work to do regarding soil models improvement, especially related to incorporating more covariates, model building methods, testing of accuracy, error reduction, and communicating uncertainty. Future research has to be focused on these issues and on the development of more quantitative and spatially relevant soil information (Brevik et al., 2016).

CONCLUDING REMARKS

Soil maps and models, alone or integrated with other sciences, are a fundamental piece of correct and sustainable land management. Correct land use is a global question because our impacts on the environment are no longer local, and land degradation is a problem at the planetary scale. The costs of unsustainable land use are enormous. Sustainable land use over the long term is always less expensive for the economy, society, and environment as compared to unsustainable use, which has contributed to the collapse of numerous civilizations throughout human history. Soil maps and models can identify areas that are vulnerable to land degradation, prevent degradation with good planning, reduce the costs of remediation when it is necessary, and contribute to issues related to climate change (e.g., reduction of greenhouse gas emissions) and human health (e.g., soil contamination).

In this context, it is clear that soil and sustainable land management are interdependent due to the services that soil provides to society, including food security, fuel security, water security, poverty reduction, and even prevention of war and other conflicts. This is especially relevant in the context of climate change and population increase, where overexploitation of soils by intensive nonsustainable agricultural practices are reaching a dangerous threshold. The awareness of sustainable land use issues by society and political leaders is one part of the solution to achieving a sustainable world. Sustainable land use implementation and the benefits derived from these practices are extremely relevant for our society and priceless when viewed from a long-term perspective. Policy makers should be informed about this in a socio-economical context, which is often easier to understand than when approached from a purely scientific context. The information produced by soil maps and soil models alone, or integrated with the information and models of other disciplines, is indispensable to present unbiased and robust information to managers and politicians about the urgency, necessity, and advantages of sustainable land use.

Acknowledgements

Paulo Pereira and Artemi Cerdà would like to gratefully acknowledge the support of COST action ES1306 (Connecting European connectivity research). E.C. Brevik was partially supported by the National Science Foundation under Grant Number IIA-1355466 during this project.

References

Adams, H.D., Williams, A.P., Xu, C., Rauscher, S.A., Jiang, X., McDowell, N.G., 2013. Empirical and process-based approaches to climate-induced forest mortality models. Front. Plant. Sci. 4, 1–5. http://dx.doi.org/10.3389/fpls.2013.00438.

Adhikari, K., Hartemink, A.E., 2016. Linking soils to ecosystem services. Geoderma 262, 101–111. http://dx.doi.org/10.1016/j.geoderma.2015.08.009.

Aguilera, E., Lassaletta, L., Sanz-Cobena, A., Garnier, J., Vallejo, A., 2013. The potential of organic fertilizers and water management to reduce N_2O emissions in Mediterranean climate cropping systems. A review. Agric. Ecosyst. Environ. 164, 32–52. http://dx.doi.org/10.1016/j.agee.2012.09.006.

Al-Hamdan, O.Z., Hernandez, M., Pierson, F.B., Nearing, M.A., Williams, C.J., Stone, J.J., et al., 2015. Rangeland hydrology and erosion model (RHEM) enhancements for applications on disturbed rangelands. Hydrol. Process. 29, 445–457. http://dx.doi.org/10.1002/hyp.10167.

Ali, A.M.S., 2006. Rice to shrimp: Land use/land cover changes on soil degradation in southwestern Bangladesh. Land Use Policy 23, 421–435. http://dx.doi.org/10.1016/j.landusepol.2005.02.001.

Anaya-Romero, M., Pino, R., Moreira, J.M., Muñoz-Rojas, M., De la Rosa, D., 2011. Analysis of soil capability versus land-use change by using CORINE Land Cover and MicroLEIS in Southern Spain. Int. Agrophys. 25, 395–398.

Anderson, M.C., Cammalleri, C., Hain, C., Otkin, J., Zhan, X., Kustas, W., 2013. Using a diagnostic soil–plant–atmosphere model for monitoring drought at field to Continental scales. Procedia Environ. Sci. 19, 47–56. http://dx.doi.org/10.1016/j.proenv.2013.06.006.

Anderson, M.C., Kustas, W.P., Norman, J.M., 2003. Upscaling and downscaling—a regional view of the soil–plant–atmosphere continuum. Agron. J 95, 1408–1423. http://dx.doi.org/10.2134/agronj2003.1408.

Andreasen, M., Andreasen, L.A., Jensen, K.H., Sonnenborg, T.O., Bircher, S., 2013. Estimation of regional groundwater recharge using data from a distributed soil moisture network. Vadose Zone J. 12 http://dx.doi.org/10.2136/vzj2013.01.0035.

Arndt, C., Farmer, W., Strzepek, K., Thurlow, J., 2012. Climate change, agriculture and food security in Tanzania. Rev. Dev. Econ. 16, 378–393. http://dx.doi.org/10.1111/j.1467-9361.2012.00669.x.

Assouline, S., Or, D., 2013. Conceptual and parametric representation of soil hydraulic properties. A review. Vadose Zone J 12. http://dx.doi.org/10.2136/vzj2013.07.0121.

Assouline, S., Tessier, A., Bruand, A., 1998. A conceptual model of soil water retention curve. Water Reour. Res 34, 223–231. http://dx.doi.org/10.1029/97WR03039.

Bae, J., Ryu, Y., 2015. Land use and land cover changes explain spatial and temporal variations of the soil organic carbon stocks in a constructed urban park. Landsc. Urban Plan 136, 57–67. http://dx.doi.org/10.1016/j.landurbplan.2014.11.015.

Bai, Z.G., Jong, R.D., van Lynden, G.W.J., 2010. An update of GLADA-Global assessment of land degradation and improvement. ISRIC report 2010/08, ISRIC–World Soil Information, Wageningen.

Baker, J.M., Ochsner, T.E., Venterea, R.T., Griffis, T.J., 2007. Tillage and soil carbon sequestration-what do we really know? Agric. Ecosyst. Environ. 118, 1–5. http://dx.doi.org/10.1016/j.agee.2006.05.014.

Baker, T.J., Cullen, B., Debevec, L., Abebe, Y., 2015. A socio-hydrological approach incorporating gender into pyo-physical models and implication for water resources management. Appl. Geogr. 62, 325–338. http://dx.doi.org/10.1016/j.apgeog.2015.05.008.

Barbier, E.B., 2000. The economic linkages between rural poverty and land degradation: some evidences from Africa. Agric. Ecosyst. Environ. 82, 355–370. http://dx.doi.org/10.1016/S0167-8809(00)00237-1.

Barrera-Bassols, N., Toledo, V.M., 2005. Ethnoecology of the Yucatec Maya: symbolism, knowledge and management of natural resources. J. Latin Am. Geogr. 4, 9–41.

Becker, K., Wulfmeyer, V., Berger, T., Gebel, J., Munch, W., 2013. Carbon farming in hot, dry coastal areas: an option for climate change mitigation. Earth Syst. Dyn. 4, 237–251. http://dx.doi.org/10.5194/esd-5-41-2014.

Beesley, L., 2012. Carbon storage and fluxes in existing and newly created urban soils. J. Environ. Manage. 104, 158–165. http://dx.doi.org/10.1016/j.jenvman.2012.03.024.

Beetz, S., Liebersbach, H., Glatzel, S., Jurasinski, G., Buczko, U., Hoper, H., 2013. Effects of land use intensity on the full greenhouse gas balance in an Atlantic peat bog. Biogeosciences 10, 1067–1082. http://dx.doi.org/10.5194/bg-10-1067-2013.

Bindraban, P.S., van der Velde, M., Ye, L., van den Berg, M., Materechera, S., Kiba, D.I., et al., 2012. Assessing the impact of soil degradation on food production. Curr. Opin. Environ. Sustain. 4, 478–488. http://dx.doi.org/10.1016/j.cosust.2012.09.015.

Blanco-Canqui, J., Lal, R., 2008. No-tillage and soil-profile carbon sequestration: an on-farm assessment. Soil Sci. Soc. Am. J. 72, 693–701. http://dx.doi.org/10.2136/sssaj2007.0233.

Bommarco, R., Kleijn, D., Potts, S.G., 2013. Ecological intensification: harnessing ecosystem services for food security. Trends Ecol. Evol. 28, 230–238. http://dx.doi.org/10.1016/j.tree.2012.10.012.

Borrelli, P., Panagos, P., Ballabio, C., Lugato, E., Weynants, M., Montanarella, L., 2016. Towards a pan-European assessment of land susceptibility to wind erosion. Land Degrad. Dev. http://dx.doi.org/10.1002/ldr.2318.

Bosch, A., Dorfer, C., He, J.S., Schmidt, K., Scholten, T., 2016. Predicting soil respiration for the Qinghai-Tibet Plateau: an empirical comparison of regression models. Pedobiologia 59, 41–49. http://dx.doi.org/10.1016/j.pedobi.2016.01.002.

Bouma, J., 2002. Land quality indicators of sustainable land management across scales. Agric. Ecosyst. Environ. 88, 129–136. http://dx.doi.org/10.1016/S0167-8809(01)00248-1.

Bouma, J., McBratney, A., 2013. Framing soils as an actor when dealing with wicked environmental problems. Geoderma 200-201, 130–139. http://dx.doi.org/10.1016/j.geoderma.2013.02.011.

Bouwman, L., Goldewijk, K.K., Van Der Hoek, K.W., Beusen, A.H.W., Van Vuuren, D.P., Willems, J., et al., 2013. Exploring global changes in nitrogen and phosphorous cycles in agriculture induced by livestock production over 1900–2050 period. Proc. Natl. Acad. Sci. U.S.A. 110, 20882–20887. http://dx.doi.org/10.1073/pnas.1012878108.

Bradford, M.A., Fierer, N., 2012. The biogeography of microbial communities and ecosystem processes: implications for soil and ecosystem models. In: Wall, D.H. (Ed.), Soil Ecology and Ecosystem Services. Oxford University Press, Oxford, UK, pp. 189–200.

Bradford, M.A., Keiser, A.D., Davies, C.A., Mersmann, C.A., Strickland, M.S., 2013. Empirical evidence that soil carbon formation from plant inputs is positively related to microbial growth. Biogeochemistry 113, 271–281. http://dx.doi.org/10.1007/s10533-012-9822-0.

Brevik, E., Pereira, P., Munoz-Rojas, M., Miller, B., Cerda, A., Parras-Alcantara, L., et al., 2006. Historical perspectives on soil mapping and process modelling for sustainable land use management. In: Pereira, P., Brevik, E., Munoz-Rojas, M., Miller, B. (Eds.), Soil Mapping and Process Modelling for Sustainable Land Use Management. Elsevier Publishing House ISBN: 978012805, in press.

Brevik, E.C., Calzolari, C., Miller, B.A., Pereira, P., Kabala, C., Baumgarten, A., et al., 2016. Soil mapping, classification, and pedologic modelling: history and future directions. Geoderma 264, 256–272. http://dx.doi.org/10.1016/j.geoderma.2015.05.017.

Brevik, E.C., Cerda, A., Mataix-Solera, J., Pereg, L., Quinton, J., Six, J., et al., 2015. The interdisciplinary nature of soil. Soil 1, 117–129. http://dx.doi.org/10.5194/soil-1-117-2015.

Briggs, C., Gustin, M.S., 2013. Building upon the conceptual model for soil mercury flux: evidence of a link between moisture evaporation and Hg evasion. Water Air Soil Pollut. 224, 1744–1755. http://dx.doi.org/10.1007/s11270-013-1744-5.

Buenemann, M., Martius, C., Jones, J.W., Herrmann, S.M., Klein, D., Mulligan, M., et al., 2011. Integrative geospatial approaches for the comprehensive monitoring and assessment of land management sustainability: rationale, potentials, and characteristics. Land Degrad. Dev 22, 226–239. http://dx.doi.org/10.1002/ldr.1074.

Burkhard, B., Kroll, F., Muller, F., Windhorst, W., 2009. Landscapes' capacities to provide ecosystem services—a concept for land-cover based assessments. Landsc. Online 15, 1–22. http://dx.doi.org/10.3097/LO.200915.

Burras, C.L., Nyasimi, M., Butler, L., 2013. Soils, human health, and wealth: a complicated relationship. In: Brevik, E.C., Burgess, L.C. (Eds.), Soils and Human Health. CRC Press, Boca Raton, pp. 215–226.

Calzolari, C., Ungaro, F., Filippi, N., Guermandi, M., Malucelli, F., Marchi, N., et al., 2016. A methodological framework to assess the multiple contributions of soils to ecosystem services delivery at regional scale. Geoderma 261, 190–203. http://dx.doi.org/10.1016/j.geoderma.2015.07.013.

Caputo, J., Beier, C.M., Sullivan, T.J., Lawrence, G.B., 2016. Modeled effects of soil acidification on long-term ecological and economic outcomes for managed forests in Adirondack region (USA). Sci. Total Environ. 565, 401–411. http://dx.doi.org/10.1016/j.scitotenv.2016.04.008.

Carr, P.M., Brevik, E.C., Horsley, R.D., Martin, G.B., 2015. Long-term no-tillage sequesters soil organic carbon in cool semi-arid regions. Soil Horizons 56 (6). http://dx.doi.org/10.2136/sh15-07-0016.

Castellano, M.J., Mueller, K.E., Olk, D.C., Sawyer, J.E., Six, J., 2016. Integrating plant litter quality, soil organic matter stabilization, and carbon saturated concept. Global Change Biol 21, 3200–3209. http://dx.doi.org/10.1111/gcb.12982.

Chaney, R., Reeves, P.G., Ryan, J.A., Simmons, R.W., Welch, R.M., Angle, J.S., 2004. An improved understanding of soil Cd risk to humans and low cost methods to phytoextract Cd from contaminated soils to prevent soil Cd risks. Biometals 17, 549–553. http://dx.doi.org/10.1023/B:BIOM.0000045737.85738.cf.

Christopher, S.F., Lal, R., Mishra, U., 2009. Regional study of no-till effects on carbon sequestration in the Midwestern United States. Soil Sci. Soc. Am. J. 73, 207–216.

Constantini, E.A.C., L'Abate, G., 2016. Beyond the concept of dominant soil: preserving pedodiversity in upscaling soil maps. Geoderma 271, 243–253. http://dx.doi.org/10.1016/j.geoderma.2015.11.024.

Cornforth, I.S., 1999. Selecting indicators for assessing sustainable land management. J. Environ. Manage. 56, 173–179. http://dx.doi.org/10.1006/jema.1999.0276.

Cowie, A.L., Lonergan, V.E., Rabbi, S.M.F., Fornasier, F., Macdonald, C., Harden, S., et al., 2013. Impact of carbon farming practices on soil carbon in northern New South Wales. Soil Res 51, 707–718. http://dx.doi.org/10.1071/SR13043.

Critchley, W.R.S., Reij, C., Willcocks, T.J., 1994. Indigenous soil and water conservation: a review of the state of knowledge and prospects for building on traditions. Land Degrad. Dev. 5, 293–314. http://dx.doi.org/10.1002/ldr.3400050406.

Crow, W., Wood, E., Dubayah, R., 2000. Potential for downscaling soil moisture maps derived from spaceborne imaging radar data. J. Geophys. Res-Atmos. 105, 2203–2212. http://dx.doi.org/10.1029/1999JD901010.

Crow, W.T., Berg, A.A., Cosh, M.H., Loew, A., Mohanty, B.P., Panciera, R., et al., 2012. Upscaling sparse ground based soil moisture observations for the validation of coarse-resolution satellite soil moisture products. Rev. Geophys. 50 RG2002. http://dx.doi.org/10.1029/2011RG000372.

Das Gupta, M., 2016. Population, poverty and climate change. World Bank Res. Obs. http://dx.doi.org/10.1093/wbro/lkt009.

De Boer, J., Schosler, H., Boeresma, J.J., 2013. Climate change and meat eating: an inconvenient couple? J. Environ. Psychol. 33, 1–8. http://dx.doi.org/10.1016/j.jenvp.2012.09.001.

De Vente, J., Poesen, J., Verstraeten, G., Govers, G., Vanmaercke, M., Van Rompaey, A., et al., 2013. Predicting soil erosion and sediment yield at regional scales: where do we stand. Earth-Sci. Rev. 127, 16–29. http://dx.doi.org/10.1016/j.earscirev.2013.08.014.

Diao, X., Sarpong, D.B., 2011. Poverty implications of agricultural land degradation in Ghana: an economy wide, Multimarked Model Assessment. Afr. Dev. Rev. 23, 263–275.

Dietrich, W.E., Reiss, R., Hsu, M.L., Montgomery, D.R., 1995. A process-based model for colluvial soil depth and shallow landsliding using digital elevation data. Hydrol. Process. 9, 383–400. http://dx.doi.org/10.1002/hyp.3360090311.

Ding, C., Zhou, F., Li, X., Zhang, T., Wang, X., 2015. Modelling the transfer of arsenic from soil to carrot (*Dactus carota* L.)—a greenhouse and field-based study. Environ. Sci. Poll. Res 22, 10627–10635. http://dx.doi.org/10.1007/s11356-015-4255-7.

Diodato, N., Ceccarelli, M., 2004. Multivariate indicator Kriging approach using a GIS to classify soil degradation for Mediterranean agricultural lands. Ecol. Ind. 4, 177–187. http://dx.doi.org/10.1016/j.ecolind.2004.03.002.

Djamai, N., Magagi, R., Goita, R., Goita, K., Merlin, O., Kerr, Y., et al., 2015. Disaggregation of SMOS soil moisture over the Canadian prairies. Remote Sens. Environ. 170, 255–268. http://dx.doi.org/10.1016/j.rse.2015.09.013.

Dominati, E., Mackay, A., Green, S., Patterson, M., 2014. A soil change-based methodology for the quantification and valuation of ecosystem services from agroecosystems: a case study of pastoral agriculture in New Zealand. Ecol. Econ. 100, 119–129. http://dx.doi.org/10.1016/j.ecolecon.2014.02.008.

Du, Y., Huffman, T., Toure, S., Feng, F., Gameda, S., Green, M., et al., 2013. Integrating socio-economic and biophysical assessments using land use allocation model. Soil Use Manage, 140–149. http://dx.doi.org/10.1111/sum.12018.

Dumanski, J., Gameds, S., Pieri, C., 1998. Indicators of Land Quality and Sustainable Land Management. World Bank, Washington.

EEA, 2015. Agriculture and climate change. Signals—living in a changing climate. Copenhagen. Available from: http://www.eea.europa.eu/downloads/a898650f58a641589e-b0ad2cd92b55be/1446559824/agriculture-and-climate-change.pdf (consulted on 02.02.16).

Egea, G., Diaz-Espejo, A., Fernandez, J.E., 2016. Soil moisture dynamics in a hedgerow olive orchard under well-watered and deficit irrigation regimens: assessment, prediction and scenario analysis. Agric. Water Manage. 164, 197–211. http://dx.doi.org/10.1016/j.agwat.2015.10.034.

Egoh, B., Reyers, B., Rouget, M., Richardson, D.M., Le Maitre, D.C., van Jaarsveld, A.S., 2008. Mapping ecosystem services for planning and management. Agric. Ecosyst. Environ. 127, 135–140. http://dx.doi.org/10.1016/j.agee.2008.03.013.

ELD Initiative & UNEP, 2015. The Economics of Land Degradation in Africa: Benefits of Action Outweigh the Costs. Available from: www.eld-initiative.org (consulted on 27.02.16).

Evans, M.C., Carwardine, J., Fensham, R.J., Butler, D.W., Wilson, K.A., Possingham, H.P., et al., 2015. Carbon farming via assisted natural regeneration as a cost-effective mechanism for restoring biodiversity in agricultural landscapes. Environ. Sci. Policy 50, 114–119. http://dx.doi.org/10.1016/j.envsci.2015.02.003.

FAO, 2002. World Agriculture: Towards 2015/2030. Summary Report. Food and Agriculture Organization of the United Nations., Available in: http://www.fao.org/docrep/004/y3557e/y3557e00.HTM (consulted on 02.02.16).

FAO, 2011. The State of World's Land and Water Resources for Food and Agriculture. Managing Systems at Risk, New York.

Feng, Y., Cui, N., Zhao, L., Hu, X., Gong, D., 2016. Comparison of ELM, GANN, WNN and empirical models for estimating reference evapotranspiration in humid region in southwest China. J. Hydrol. 536, 376–383. http://dx.doi.org/10.1016/j.jhydrol.2016.02.053.

Fleskens, L., Nainggolan, D., Stringer, L.C., 2014. An exploration of scenarios to support sustainable land management using integrated environmental socio-economic models. Environ. Manage. 54, 1005–1021. http://dx.doi.org/10.1007/s00267-013-02.

FAO, 1993. FESLM: An International Framework for Evaluating Sustainable Land Management. World Resources Report 73. FAO, Rome, Italy.

Fraser, E.D.G., Simelton, E., Termansen, M., Gosling, S.N., South, A., 2013. 'Vulnerability hotspots': integrating socio-economic and hydrological models to identify where cereal production may decline in the future due climate change induced drought. Agric. Forest Meteorol. 170, 195–205. http://dx.doi.org/10.1016/j.agrformet.2012.04.008.

Funk, J.M., Field, C.B., Kerr, S., Daigneault, A., 2014. Modelling the impact of carbon farming on land use in a New Zealand landscape. Environ. Sci. Policy 37, 1–10. http://dx.doi.org/10.1016/j.envsci.2013.08.008.

Galdino, S., Sano, E.E., Andrade, R., Grego, C.R., Nogueira, S.F., Bragantini, C., et al., 2016. Large-scale modelling of soil erosion with RUSLE for conservational planning of degraded cultivated Brazilian pastures. Land Degrad. Dev 27, 773–787. http://dx.doi.org/10.1002/ldr.2414.

Gao, W., Whalley, R., Tian, Z., Liu, J., Ren, T., 2016. A simple model to predict soil penetrometer resistance as function of density, drying and depth in the field. Soil Tillage Res. 155, 190–198. http://dx.doi.org/10.1016/j.still.2015.08.004.

García-Orenes, F., Cerdà, A., Mataix-Solera, J., Guerrero, C., Bodí, M.B., Arcenegui, V., et al., 2009. Effects of agricultural management on surface soil properties and soil–water losses in eastern Spain. Soil Tillage Res. 106, 117–123. http://dx.doi.org/10.1016/j.still.2009.06.002.

Gartner, J.E., Cannon, S.H., Santi, P.M., 2014. Empirical models for predicting volumes of sediment deposited in debris flows and sediment-laden floods in a transverse ranges of south California. Eng. Geol. 176, 45–56. http://dx.doi.org/10.1016/j.enggeo.2014.04.008.

Gharun, M., Trunbull, T.L., Adams, M.A., 2013. Validation of canopy transpiration in a mixed-species foothill eucalypt forest using a soil–plant–atmosphere model. J. Hydrol. 492, 219–277. http://dx.doi.org/10.1016/j.jhydrol.2013.03.051.

Giltrap, D.L., Li, C., Saggar, S., 2010. DNDC: a process-based model for greenhouse gas fluxes from agricultural soils. Agric. Ecosyst. Environ. 136, 292–300. http://dx.doi.org/10.1016/j.agee.2009.06.014.

Giordano, R., Liersch, S., 2012. A fussy GIS-based system to integrate local and technical knowledge in soil salinity monitoring. Environ. Model. Softw. 36, 49–63. http://dx.doi.org/10.1016/j.envsoft.2011.09.004.

Godfray, H.C., Beddington, J.R., Crute, I.R., Haddad, L., Lawrence, D., Muir, J.F., et al., 2010. Food security: the challenge of feeding 9 billion people. Science 327, 812–818. http://dx.doi.org/10.1126/science.1185383.

Görlach, B., Landgrebe-Trinkunaite, R., Interwies, E., 2004. Assessing the Economic Impacts of Soil Degradation. Volume I: Literature Review. Ecologic, Berlin, Study Commissioned by the European Commission, DG Environment, Study Contract ENV.B.1/ETU/2003/0024.

Graves, A.R., Burgess, P.J., Palma, J.H.N., Herzog, F., Moreno, G., Bertomeu, M., et al., 2007. Development and application of bio-economic to compare silvoarable, arable, and forestry systems in three European countries. Ecol. Eng. 29, 434–449. http://dx.doi.org/10.1016/j.ecoleng.2006.09.018.

Graves, A.R., Morris, J., Deeks, L.K., Rickson, R.J., Kibblewhite, M.G., Harris, J.A., et al., 2015. The total costs of soil degradation in England and Wales. Ecol. Econ. 119, 399–413. http://dx.doi.org/10.1016/j.ecolecon.2015.07.026.

Gregory, P.J., 2012. Soils and food security: challenges and opportunities. In: Hester, R.E., Harrison, R.M. (Eds.), Issues in Environmental Science and Technology. Royal Society of Chemistry, Cambridge, pp. 2–30.

Groenenberg, J.E., Dijkstra, J.J., Bonten, L.T.C., de Vries, W., Comans, R.N.J., 2012. Evaluation of the performance and limitations of empirical-partition relations and process based multisurface models to predict element solubility in soils. Environ. Pollut. 166, 98–107. http://dx.doi.org/10.1016/j.envpol.2012.03.011.

Groenendyk, D.G., Ferre, T.P.A., Thorp, K.R., Rice, A.K., 2015. Hydrologic-process based soil texture classifications for imporved visualization of landscape function. PLoS One, e0131299. http://dx.doi.org/10.1371/journal.pone.0131299.

Hengl, T., de Jesus, J.M., MacMillan, R.A., Batjes, N.H., Heuvelink, G.B.M., Ribeiro, E., et al., 2014. SoilGrids1Km—global soil information based on automated mapping. PLoS One 9, e114788. http://dx.doi.org/10.1371/journal.pone.0105992.

Hengl, T., Heuvelink, G.B.M., Kempen, B., Leenaars, J.G.B., Walsh, M.G., Shepherd, K.D., et al., 2015. Mapping soil properties of Africa at 250 m resolution: random forests significantly improve current predictions. PLoS One 10, e0125814. http://dx.doi.org/10.1371/journal.pone.0125814.

Henry, S., Boyle, P., Lambin, E.F., 2003. Modelling inter-provincial migration in Burkina Faso, West Africa: the role of socio-demographic and environmental factors. Appl. Geogr. 23, 115–136. http://dx.doi.org/10.1016/j.apgeog.2002.08.001.

Herrick, J.E., 2000. Soil quality: an indicator of sustainable land management. Appl. Soil. Ecol. 15, 75–83. http://dx.doi.org/10.1016/S0929-1393(00)00073-1.

Hoffmann, M., Jurisch, N., Borraz, E.A., Hagemann, U., Drosler, M., Sommer, M., et al., 2015. Automated modelling of ecosystem CO_2 fluxes based on periodical closed chamber measurments: a standardized conceptual and practical approach. Agric. Forest Meterol. 200, 30–45. http://dx.doi.org/10.1016/j.agrformet.2014.09.005.

Horta, A., Pereira, M.J.P., Goncalves, M., Ramos, T., Soares, A., 2014. Spatial modelling of hydraulic soil properties integrating different supports. J. Hydrol. 511, 1–9. http://dx.doi.org/10.1016/j.jhydrol.2014.01.027.

Immerzeel, W.W., Gaur, A., Zwart, S.J., 2008. Integrating remote sensing and a process-based hydrological model to evaluate water use and productivity in a south Indian catchment. Agric. Water Manage. 95, 11–24. http://dx.doi.org/10.1016/j.agwat.2007.08.006.

Jafarzadeh, A.A., Pal, M., Servati, M., Fazelifard, M.H., Ghorbani, M.A., 2016. Comparative analysis of support vector machine and artificial neural network models for soil cation exchange capacity. Int. J. Environ. Sci. Technol. 13, 87–96. http://dx.doi.org/10.1007/s13762-015-0856-4.

Jefferies, R.L., Rockwell, R.F., 2002. Foraging geese, vegetation loss and soil degradation in an Artic saltmarsh. Appl. Veg. Sci 5, 7–16.

Johnston, A.S.A., Holmstrup, M., Hodson, M.E., Thorbek, P., Alvarez, T., Silby, R.M., 2014. Earthworm distribution and abundance predicted by a process-based model. Appl. Soil Ecol. 84, 112–123. http://dx.doi.org/10.1016/j.apsoil.2014.06.001.

Ju, W., Gao, P., Wang, J., Zhou, Y., Zhang, X., 2010. Combining an ecological with remote sensing and GIS techniques to monitor soil water content of croplands with monsoon climate. Agric. Water Manage. 97, 1221–1231. http://dx.doi.org/10.1016/j.agwat.2009.12.007.

Kaizer, K., Kalbitz, K., 2012. Cycling downwards—dissolved organic matter in soils. Soil Biol. Biochem. 52, 29–32. http://dx.doi.org/10.1016/j.soilbio.2012.04.002.

Kassie, M., Zikhali, P., Pender, J., Kohlin, G., 2010. The economics of sustainable land management practices in Ethiopian Highlands. J. Agric. Econ 61, 605–627. http://dx.doi.org/10.1111/j.1477-9552.2010.00263.x.

Katul, G.G., Oren, R., Manzoni, S., Higgings, C., Parlange, M.B., 2012. Evapotranspiration: a process driving mass transport and energy exchange in soil–plant–atmosphere climate system. Rev. Geophys. 50 RG3002, http://dx.doi.org/10.1029/2011RG000366.

Kerven, C., Robinson, S., Kushenov, K., Milner-Gulland, E.J., 2016. Horseflies, wolves and wells: biophysical and socio-economic factors influencing livestock distribution in Kazakhstan's rangelands. Land Use Policy 52, 392–409. http://dx.doi.org/10.1016/j.landusepol.2015.12.030.

Khan, F.I., Husain, T., Hejazi, R., 2004. An overview and analysis of site remediation technologies. J. Environ. Manage. 71, 95–122. http://dx.doi.org/10.1016/j.jenvman.2004.02.003.

Khan, S., Cao, Q., Zheng, Y.M., Huang, Y.Z., Zhu, Y.G., 2008. Health risks of heavy metals in contaminated soils and food crops irrigated with wastewater in Beijing, China. Environ. Pollut. 152, 686–692. http://dx.doi.org/10.1016/j.envpol.2007.06.056.

Khan, S.A., Muvaney, R.L., Ellsworth, T.R., Boast, C.W., 2007. The myth of nitrogen fertilization for soil carbon sequestration. J. Environ. Qual 36, 1821–1832.

Kheir, R.B., Cerdan, O., Abdallah, C., 2006. Regional soil erosion risk mapping in Lebanon. Geomorphology 82, 34–359. http://dx.doi.org/10.1016/j.geomorph.2006.05.012.

Kim, G., Barros, A., 2002. Downscaling of remotely sensed soil moisture with a modified fractal interpolation method using a contraction mapping ancillary data. Remote Sens. Environ. 83, 400–413. http://dx.doi.org/10.1016/S0034-4257(02)00044-5.

Kirby, M.J., Jones, R.J.A., Irvine, B., Gobin, A, Govers, G., Cerdan, O., et al., 2004. European Soil Bureau Research Report No. 16, EUR 21176, 18 pp. and 1 map in ISO B1 format. Office for Official Publications of the European Communities, Luxembourg.

Kleber, M., Sollins, P., Sutton, R., 2007. A conceptual model of organo-mineral interactions in soils: self-assembly of organic molecular fragments into zonal structures on mineral surfaces. Biogeochemistry 85, 9–24. http://dx.doi.org/10.1007/s10533-007-9103-5.

Koltermann, C.E., Gorelick, S.M., 1996. Heterogeneity in sedimentary deposits: a review of structure-imitating, process-imitating, and descriptive approaches. Water Resour. Res. 32, 2617–2658. http://dx.doi.org/10.1029/96WR00025.

Korzukhin, M.D., Ter-Mikaelian, M.T., Wagner, R.G., 1996. Processes versus empirical models: which approach for forest ecosystem management? Can. J. Forest Res. 26, 879–887. http://dx.doi.org/10.1139/x26-096.

Kresic, N., 2009. Groundwater Resources: Sustainability, Management, and Restoration. McGraw Hill, New York.

LADA, 2013. Questionnaire for mapping land degradation and sustainable land management. Version 2. FAO, Rome.

Lal, R., 2004a. Carbon emission from farm operations. Environ. Int. 30, 981–990. http://dx.doi.org/10.1016/j.envint.2004.03.005.

Lal, R., 2004b. Soil carbon sequestration to mitigate climate change. Geoderma 123, 1–22. http://dx.doi.org/10.1016/j.geoderma.2004.01.032.

Lal, R., 2004c. Soil carbon sequestration impacts on global climate change and food security. Science 304, 1623–1627. http://dx.doi.org/10.1126/science.1097396.

Lal, R., 2009. Soil Carbon Sequestration. SOLAW Background Thematic Report–TRO4B, Land and Water Use Options for Climate Change Adaptation and Mitigation in Agriculture. Available from: http://www.fao.org/fileadmin/templates/solaw/files/thematic_reports/TR_04b_web.pdf (consulted on 02.02.16).

Lal, R., 2015. Restoring soil quality to mitigate soil degradation. Sustainability 7, 5875–5895. http://dx.doi.org/10.3390/su7055875.

Lambin, E.F., Rounsevell, M.D.A., Geist, H.J., 2000. Are agricultural land use models able to predict changes in land-use intensity? Agric. Ecosyst. Environ. 82, 321–331. http://dx.doi.org/10.1016/S0167-8809(00)00235-8.

Lambin, E.F., Turner, B.L., Geist, H., Agbola, S.B., Angelsen, A., Bruce, J.W., et al., 2001. The causes of land-use and land-cover change moving beyond the myths. Glob. Environ. Change 11, 261–269. http://dx.doi.org/10.1016/S0959-3780(01)00007-3.

Lefroy, R.D.B., Bechstedt, H.D., Rais, M., 2000. Indicators for sustainable land management based on farmer surveys in Vietnam, Indonesia, and Thailand. Agric. Ecosyst. Environ. 81, 137–146. http://dx.doi.org/10.1016/S0167-8809(00)00187-0.

Letcher, R.A., Jakeman, A.J., 2010. Types of environmental models. In: Sydow, A. (Ed.), Ecyclopedia of Life Support Systems, Volume II, Environmental Systems. Eolss Publishers Co., Ltd, Oxford, UK, pp. 193–215.

Li, H.Y., Webster, R., Shi, Z., 2015. Mapping soil salinity in the Yangtze delta: REML and universal kriging (E-BLUP) revisited. Geoderma 237–238, 71–77. http://dx.doi.org/10.1016/j.geoderma.2014.08.008.

Liang, C., Balser, T.C., 2011. Microbial production of recalcitrant organic matter in global soils. Implications for productivity and climate policy. Nat. Rev. Microbiol 9, 75. http://dx.doi.org/10.1038/nrmicro2386-c1.

Lima, M.L., Romanelli, A., Massone, H.E., 2015. Assessing groundwater pollution hazard changes under different socio-economic and environmental scenarios in an agricultural watershed. Sci. Total Environ 530-531, 333–346. http://dx.doi.org/10.1016/j.scitotenv.2015.05.026.

Liniger, H.P., Mekdaschi Studer, R., Hauert, C., Gurter, M., 2011. Sustainable land use management in practice—guidelines for Sub-Saharan Africa. TerrAfrica. World Overview of Conservation Approaches and Technologies (WOCAT) and Food and Agriculture Organization of the United Nations (FAO).

Linquist, B., Van Groeningen, K.J., Adviento-Borbe, M.A., Pittelkow, C., Kessel, C.V., 2012. An agronomic assessment of greenhouse gas emissions from major cereal crops. Global Change Biol. 18, 194–209. http://dx.doi.org/10.1111/j.1365-2486.2011.02502.x.

Lobet, G., Pages, L., Draye, X., 2014. A modelling approach to determine the importance of dynamic regulation of plant-hydraulic conductivities on the water uptake dynamics in the soil–plant–atmosphere system. Ecol. Model. 290, 65–75. http://dx.doi.org/10.1016/j.ecolmodel.2013.11.025.

Lu, Y., Lu, S., Horton, R., Ren, T., 2014. An empirical model for estimating soil thermal conductivity from texture, water content and bulk density. Soil Sci. Soc. Am. J. 78, 1859–1868. http://dx.doi.org/10.2136/sssaj2014.05.0218.

Lu, Y., Stocking, M., 2000. Integrating biophysical and socio-economic aspects of soil conservation on Loess Plateau. Land Degrad. Dev. 11, 153–165. http://dx.doi.org/10.1002/(SICI)1099-145X(200003/04)11:2<153::AID-LDR374>3.0.CO;2-#.

Luck, G.W., Smallbone, L.T., O'Brian, R., 2009. Socio-economic and vegetation change in urban ecosystems: patterns in space and time. Ecosystems 12, 604–620. http://dx.doi.org/10.1007/s10021-009-9244-6.

Ludeke, M.K.B.D., Moldenhaure, O., Petschel-Held, G., 1999. Rural poverty driven by soil degradation under climate change: the sensitivity of the disposition towards the Sahel syndrome with respect to climate change. Environ. Model. Assess 4, 315–326. http://dx.doi.org/10.1023/A:1019032821703.

Luo, Z., Wang, E., Fillery, I.R.P., Macdonald, L.M., Huth, N., Baldock, J., 2014. Modeling soil carbon and nitrogen dynamics using measurable and conceptual matter pools in APSIM. Agric. Ecosyst. Environ. 186, 94–104. http://dx.doi.org/10.1016/j.agee.2014.01.019.

Lynch, K., Maconachie, R., Binns, T., Tengbe, P., Bangura, K., 2013. Meeting the urban challenge? Urban agriculture and food security in post-conflict Freetown, Sierra Leone. Appl. Geogr. 36, 31–39. http://dx.doi.org/10.1016/j.apgeog.2012.06.007.

Makela, A., Landsberg, J., Ek, A.R., Burk, T.E., Ter-Mikaelian, M., Agren, G.I., et al., 2000. Process-based models for forest ecosystem management: currents state of the art and challenges for practical implementation. Tree Physiol. 20, 289–298.

Malone, B.P., McBratney, A.B., Minasny, B., 2013. Spatial scaling for digital soil mapping. Soil Sci. Soc. Am. J. 77, 890–902. http://dx.doi.org/10.2136/sssaj2012.0419.

Manoli, G., Domec, J.C., Novick, K., Oishi, A.C., Noormets, A., Marani, M., et al., 2016. Soil–plant–atmosphere conditions regulating convective cloud formation above south eastern pine plantations. US Global Change Biol. http://dx.doi.org/10.1111/gcb.13221.

Manzoni, S., Moyano, F., Katterer, T., Schimel, J., 2016. Modeling coupled enzymatic and solute transport controls on decomposition in drying soils. Soil Biol. Biochem. 95, 275–287. http://dx.doi.org/10.1016/j.soilbio.2016.01.006.

Manzoni, S., Vico, G., Porporato, A., Katul, G., 2013. Biological constrains on water transport in the soil–plant–atmosphere system. Adv. Water Resour. 51, 292–304. http://dx.doi.org/10.1016/j.advwatres.2012.03.016.

Martinez-Casanovas, J.A., Ramos, M.C., Benites, G., 2016. Soil and water assessment tool soil loss simulation at the sub-basin scale at Alt-Penedes Vineyard region (NE Spain) in the 2000s. Land Degrad. Dev. 27, 160–170. http://dx.doi.org/10.1002/ldr.2355.

McBratney, A.B., Morgan, C.L.S., Jarrett, L.E. 2017. The Value of Soil's Contributions to Ecosystem Services. In: Field, D.J., McBratney, A.B., Morgan, C.L.S. (Eds.), Global Soil Security. Progress in Soil Science Series, Springer. http://link.springer.com/chapter/10.1007/978-3-319-43394-3_20.

McGrath, S.P., Zhao, F.J., 2015. Concentrations of metals and metalloids in soil have the potential to exceedance of maximum limit concentration of contaminants in food and feed. Soil Use Manage 31, 34–45. http://dx.doi.org/10.1111/sum.12080.

Messiga, A.J., Ziadi, N., Parent, L.E., Schenider, A., Morel, C., 2015. Process-based mass-balance modelling of soil phosphorous availability: testing different scenarios in a long-term maize monoculture. Geoderma 243-244, 41–49. http://dx.doi.org/10.1016/j.geoderma.2014.12.009.

Mirzabaev, A., Nkonya, E., von Braun, J., 2015. Economics of sustainable land management. Curr. Opin. Environ. Sust. 15, 9–19. http://dx.doi.org/10.1016/j.cosust.2015.07.004.

Misra, R.K., Rose, C.W., 1996. Application and sensivity analysis of process-based erosion model GUEST. Eur. J. Soil Sci 47, 593–604. http://dx.doi.org/10.1111/j.1365-2389.1996.tb01858.x.

Montanarella, L., Vargas, R., 2012. Global governance of soil resources as a necessary condition for sustainable development. Curr. Opin. Environ. Sust. 4, 559–564. http://dx.doi.org/10.1016/j.cosust.2012.06.007.

Montavalli, P., Nelson, K., Udawatta, R., Jose, S., Bardhan, S., 2013. Global achievements in sustainable land management. Int. Soil Water Conserv. Res. 1, 1–10. http://dx.doi.org/10.1016/S2095-6339(15)30044-7.

Moore, J.A.M., Jiang, J., Post, W.M., Classen, A.T., 2015. Decomposition by ectomycorrhizal fungi alters carbon storage in a simulation model. Ecosphere 6, 1–16. http://dx.doi.org/10.1890/ES14-00301.1.

Mulder, V.L., de Bruin, S., Schaepman, M.E., Mayr, T.R., 2011. The use of remote sensing in soil and terrain mapping—a review. Geoderma 162, 1–19. http://dx.doi.org/10.1016/j.geoderma.2010.12.018.

Nachtergaele, F.O., Petri, M., Biancalani, R., van Lynden, G., van Velthuzien, H., 2010. Global Degradation Information System (GLADIS) Beta Version. Available from: http://www.fao.org/nr/lada/index.php?option=com_content&view=article&id=161&Itemid=1 (consulted on 28.02.16).

Nambiar, K.K.M., Gupta, A.P., Fu, Q., Li, S., 2001. Biophysical, chemical and socio-economic indicators for assessing agricultural sustainability in Chinese coastal zone. Agric. Ecosyst. Environ. 87, 209–214. http://dx.doi.org/10.1016/S0167-8809(01)00279-1.

Neumann, K., Herold, M., Hartley, A., Schmullius, C., 2007. Comparative assessment of CORINE2000 and GLC2000: spatial analysis of land cover data for Europe. Int. J. Appl. Earth Obs. Geoinf. 9, 425–437. http://dx.doi.org/10.1016/j.jag.2007.02.004.

Nkonya, E., Pender, J., Kaizzi, E., Moguarura, S., Ssali, H., Muwonge, J., 2008. Linkages between land management, land degradation, and poverty in sub-Saharan Africa: the case of Uganda. IFPRI Research Report 159, Washington.

OECD, 2012. Agriculture and Water Quality: Monetary Costs and Benefits Across OECD Members. Available from: https://www.oecd.org/tad/sustainable-agriculture/49841343.pdf (consulted on 18.03.16).

Olson, K., Ebelhar, S.A., Lang, J.M., 2014. Long-term effects of cover crops on crop yields, soil organic carbon stocks and sequestration. Open J. Soil Sci. 4 Article ID:48993. http://dx.doi.org/10.4236/ojss.2014.48030.

Palomo, I., Martin-Lopez, B., Potschin, M., Haines-Young, R., Montes, C., 2013. National parks, buffer zones and surrounding lands: mapping ecosystem services flows. Ecosyst. Serv. 4, 104–116. http://dx.doi.org/10.1016/j.ecoser.2012.09.001.

Panagos, P., Borrelli, P., Poesen, P., Ballabio, C., Lugato, E., Meusburger, K., et al., 2015. The new assessment of soil loss by water erosion in Europe. Environ. Sci. Policy 54, 438–447. http://dx.doi.org/10.1016/j.envsci.2015.08.012.

Panagos, P., Christos, K., Cristiano, B., Ioannis, G., 2014a. Seasonal monitoring of soil erosion at regional scale: an application of the G2 model in Crete focusing on agricultural land uses. Int. J. Appl. Earth Obs. Geoinf. 27, 147–155. http://dx.doi.org/10.1016/j.jag.2013.09.012.

Panagos, P., Meusburger, K., Ballabio, B., Borrelli, P., Alewell, C., 2014b. Soil erodibility in Europe: a high resolution dataset based on LUCAS. Sci. Total Environ. 479-480, 189–200. http://dx.doi.org/10.1016/j.scitotenv.2014.02.010.

Panagos, P., Van Liedekerke, M., Jones, A., Montanarella, L., 2012. European soil data centre: response to European policy support and data requirements. Land Use Policy 29, 329–338. http://dx.doi.org/10.1016/j.landusepol.2011.07.003.

Park, S.J., McSweeney, K., Lowery, B., 2001. Identification of the spatial distribution of soils using a process based terrain characterization. Geoderma 103, 249–272. http://dx.doi.org/10.1016/S0016-7061(01)00042-8.

Pena, S.B., Abreu, M.M., Magalhaes, M.R., 2016. Planning landscape with water infiltration. Empirical model to assess maximum infiltration areas in Mediterranean landscapes. Water Resour. Manage. 30, 2343–2360. http://dx.doi.org/10.1007/s11269-016-1291-0.

Perez-Cruzado, C., Munoz-Saez, F., Basurco, F., Riesco, G., Rodriguez-Soalleiro, R., 2011. Combining empirical models and process-based model 3-PG to predict Eucalyptus nitens plantations growth in Spain. Forest Ecol. Manage. 262, 1067–1077. http://dx.doi.org/10.1016/j.foreco.2011.05.045.

Pimentel, D., Acquay, H., Biltonen, M., Rice, P., Silva, M., Nelson, J., et al., 1992. Environmental and economic cost of pesticide use. Bioscience 42, 750–760.

Pimentel, D., Harvey, C., Resosudarmo, P., Sinclair, K., Kurz, D., McNair, M., et al., 1995. Environmental and economic costs of soil erosion and conservation benefits. Science 267, 1117–1123.

Poeplau, C., Don, A., 2015. Carbon sequestration in agricultural soils via cultivation of cover crops—a meta-analysis. Agric. Ecosyst. Environ. 200, 33–41. http://dx.doi.org/10.1016/j.agee.2014.10.024.

Powlson, D.S., Gregory, P.J., Whalley, W.R., Quinton, J.N., Hopkins, D.W., Whitmore, A.P., et al., 2011. Soil management in relation to sustainable agriculture and ecosystem services. Food Policy 36, S72–S84. http://dx.doi.org/10.1016/j.foodpol.2010.11.025.

Prashun, V., Liniger, H., Gisler, S., Herweg, K., Candinas, A., Clement, J.P., 2013. A high-resolution soil erosion risk map of Switzerland and strategic policy support system. Land Use Policy 32, 281–291. http://dx.doi.org/10.1016/j.landusepol.2012.11.006.

Pulleman, M., Creamer, R., Hamer, U., Helder, J., Pelosi, C., Peres, G., et al., 2012. Soil biodiversity, biological indicators and ecosystem services—an overview of European approaches. Curr. Opin. Environ. Sust 4, 529–538. http://dx.doi.org/10.1016/j.cosust.2012.10.009.

Raina, P., Joshi, D.C., Kolarkar, A.S., 1993. Mapping soil degradation by using remote sensing on alluvial plain, Rajasthan, India. Arid Soil Res. Rehabil 7, 145–161.

Ray, R.L., Jacobs, J.M., Cosh, M.H., 2010. Landslide susceptibility mapping using downscaled AMSRE-soil moisture: a case study from Cleveland Corral, California. Remote Sens. Environ. 114, 2624–2636. http://dx.doi.org/10.1016/j.rse.2010.05.033.

Reed, M.S., Buenemann, M., Atlhopheng, J., Akhtar-Schuster, M., Bachmann, F., Bastin, G., et al., 2011. Cross-scale monitoring and assessment of land degradation and sustainable land management: a methodological framework for knowledge management. Land Degrad. Dev. 22, 261–271. http://dx.doi.org/10.1002/ldr.1087.

Reed, M.S., Dougill, A.J., Taylor, M.J., 2007. Integrating local and scientific knowledge for adaptation to land degradation: Kalahari rangeland management options. Land Degrad. Dev 17, 249–268. http://dx.doi.org/10.1002/ldr.777.

Ren, X., Sun, D., Wang, Q., 2016. Modelling the effects of plant density on maize productivity and water balance in the Loess Plateau of China. Agric. Forest Meteorol. 171, 40–48. http://dx.doi.org/10.1016/j.agwat.2016.03.014.

Richter, D., Yaalon, D.H., 2012. "The changing model of soil" revisited. Soil Sci. Soc. Am. J. 76, 766–778. http://dx.doi.org/10.2136/sssaj2011.0407.

Ridgwell, A.J., Marshall, S.J., Gregson, K., 1999. Consumption of atmospheric methane by soils. A process based model. Global Biogeochem. Cycles 13, 59–70. http://dx.doi.org/10.1029/1998GB900004.

Robinson, A., Emmett, B.A., Reynolds, B., Rowe, E.C., Spurgeon, D., Keith, A.M., et al., 2012. Soil natural capital and ecosystem service delivery in a world of global soil change. In: Hester, R.E., Harrison, R.M. (Eds.), Issues in Environmental Science and Technology. Royal Society of Chemistry, Cambridge, pp. 41–68.

Robinson, D.A., Hockley, N., Cooper, D.M., Emmett, B.A., Keith, A.M., Lebron, I., et al., 2013. Natural capital and ecosystem services, developing an appropriate soils framework as a basis for valuation. Soil Boil. Biochem 57, 1023–1033. http://dx.doi.org/10.1016/j.soilbio.2012.09.008.

Rodila, D., Ray, N., Gorgan, D., 2015. Conceptual model for environmental science applications on parallel and distributed infrastructures. Environ. Syst. Res. 4, 23. http://dx.doi.org/10.1186/s40068-015-0050-1.

Rodrigues, S.M., Pereira, M.E., Duarte, A.C., Romkens, P.F.A. M., 2012. Soil–plant–animal transfer models to improve soil protection guidelines: a case study in Portugal. Environ. Int. 39, 27–37. http://dx.doi.org/10.1016/j.envint.2011.09.005.

Salvati, L., Kosmas, C., Kairis, O., Karvitis, C., Acikalin, S., Belgacem, A., et al., 2015. Unveiling soil degradation and desertification risk in Mediterranean basin: a data mining analysis of the relationships between biophysical and socioeconomic factors in agro-forest landscapes. J. Environ. Plan. Manage 58, 1789–1803. http://dx.doi.org/10.1080/09640568.2014.958609.

Sandhu, H.S., Wratten, S.D., Cullen, R., Case, B., 2008. The future of farming: the value of ecosystem services in conventional and organic arable land. An experimental approach. Ecol. Econ. 64, 835–848. http://dx.doi.org/10.1016/j.ecolecon.2007.05.007.

Sandor, J.A., WinklerPrins, A.M.G.A., Barrera-Bassols, N., Zinck, J.A., 2006. The heritage of soil knowledge among the world's cultures. In: Warkentin, B. (Ed.), Footprints in the Soil: People and Ideas in Soil History. Elsevier, Amsterdam, pp. 43–84.

Scherr, S.J., 2000. A downward spiral? Research evidence on relationship between poverty and natural resource degradation. Food Policy 25, 479–498. http://dx.doi.org/10.1016/S0306-9192(00)00022-1.

Schmidhuber, J., Tubiello, F.N., 2007. Global food security under climate change. Proc. Natl. Acad. Sci. U.S.A 104, 19703–19708. http://dx.doi.org/10.1073/pnas.0701976104.

Shiferaw, S.T., Holden, B., 2000. Policy instruments for sustainable land management: the case of highland smallholders in Ethiopia. Agric. Econ. 22, 217–232. http://dx.doi.org/10.1016/S0169-5150(00)00046-3.

Sierra, C.A., Trumbore, S.E., Davidson, E.A., Vicca, S., Janssens, I., 2015. Sensivity of decomposition rates of soil organic matter with respect to simultaneous changes in temperature and moisture. J. Adv. Model. Earth Syst. 7, 335–356. http://dx.doi.org/10.1002/2014MS000358.

Simonson, R.W., 1959. Outline of a generalized theory of soil genesis. Soil Sci. Soc. Am. Proc. 23, 152–156.

Smith, P., 2012. Soils and climate change. Curr. Opin. Environ. Sust 4, 539–544. http://dx.doi.org/10.1016/j.cosust.2012.06.005.

Sonneveld, B.G.J.S., Dent, D.L., 2009. How good is GLASOD. J. Environ. Manage. 90, 274–283. http://dx.doi.org/10.1016/j.jenvman.2007.09.008.

Soylu, M.E., Kucharik, C.J., Loheide II, S.P., 2014. Influence of groundwater on plant water use and productivity: development of an integrated ecosystem. Variable saturated soil water flow model. Agric. Forest Meteorol. 189-190, 198–210. http://dx.doi.org/10.1016/j.agrformet.2014.01.019.

Steffen, W., Broadgate, W., Deutsch, L., Gaffney, O., Ludwig, C., 2015. The trajectory of the Anthropocene: the great acceleration. Anthropocene Rev. 2, 81–98. http://dx.doi.org/10.1177/2053019614564785.

Sundqvist, E., Persson, A., Kljun, N., Vestin, P., Chasmer, L., Hopkinson, C., et al., 2015. Upscaling of methane exchange in a boreal forest using soil chamber measurements and high-resolution LiDAR elevation data. Agric. Forest Meteorol. 214-215, 393–401. http://dx.doi.org/10.1016/j.agrformet.2015.09.003.

Swetnam, R.D., Fisher, B., Mbilinyi, B.P., Munishi, P.K.T., Willcock, S., Ricketts, T., et al., 2011. Mapping socio-economic scenarios of land cover change: a GIS method to enable ecosystem service modelling. J. Environ. Manage. 92, 563–574. http://dx.doi.org/10.1016/j.jenvman.2010.09.007.

Taylor, J.A., Jacob, F., Galleguillos, M., Prevot, L., Guix, N., Lagacherie, P., 2013. The utility of remotely-sensed vegetative and terrain covariates at different spatial resolutions in modelling soil and watertable depth (for digital soil mapping). Geoderma 193-194, 83–93. http://dx.doi.org/10.1016/j.geoderma.2012.09.009.

Tranter, G., Minasny, B., McBratney, A.B., Murphy, B., Mckenzie, N.J., Grundy, M., et al., 2007. Building and testing conceptual and empirical models for predicting soil bulk density. Soil Use Manage 43, 437–443. http://dx.doi.org/10.1111/j.1475-2743.2007.00092.x.

Troeh, F.R., Hobbs, J.A., Donahue, R.L., 2004. Soil and Water Conservation for Productivity and Environmental Protection, fourth ed. Prentice Hall, Upper Saddle River, NJ.

UNCCD, 2009. Benefits of Sustainable Land Management. WOCAT. Available from: https://www.wocat.net/en/news-events/global-news/newsdetail/article/benefits-of-sustainable-land-management.html (consulted on 15.03.16).

UNDP, 2014. Sustainable Land Management Toolkit. Available from: http://www.ls.undp.org/content/lesotho/en/home/library/environment_energy/SLM-Toolkit.html (consulted on 15.03.16).

UNU-INWEH, 2011. Guidelines for the Preparation and Reporting on Globally-Relevant SLM Impact Indicators for Project-Level Monitoring. United Nations University, Hamilton, Canada.

Upadhyay, T.P., Sankhayan, P.L., Solberg, B., 2005. A review of soil sequestration dynamics in the Himalayan region as a function of land-use change and forest/soil degradation with special reference to Nepal. Agric. Ecosyst. Environ. 105, 449–465. http://dx.doi.org/10.1016/j.agee.2004.09.007.

Vagen, T.G., Winowiecki, L.A., Tondoh, J.E., Desta, L.T., Gumbricht, T., 2016. Mapping of soil properties and land degradation risk in Africa using Modis reflectance. Geoderma 263, 216–225. http://dx.doi.org/10.1016/j.geoderma.2015.06.023.

Van den Bygaart, A.J., Gregorich, E.G., Angers, D.A., 2003. Influence of agricultural management on soil organic carbon: a compendium and assessment of Canadian studies. Can. J. Soil Sci 83, 363–380. http://dx.doi.org/10.3334/CDIAC/tcm.001.

Van Oost, K., Quine, T.A., Govers, G., De Gryze, S., Six, J., Harden, J.W., et al., 2007. The impact of agricultural soil erosion on the global carbon cycle. Science 318, 626–629.

Varallya, G., 1989. Soil degradation processes and their control in Hungary. Land Degrad. Dev 3, 171–188. http://dx.doi.org/10.1126/science.1145724.

Vereecken, H., Schnepf, A., Hopmans, J.W., Javaux, M., Roose, D.O.T., Vanderborght, J., et al., 2016. Modelling soil processes: review, key challenges, and new perspectives. Vadose Zone J 15, 1–57. http://dx.doi.org/10.2136/vzj2015.09.0131.

Vermulen, S.J., Aggarwal, P.K., Ainslie, A., Angelone, C., Campbell, B.M., Challinor, A.J., et al., 2012. Options for support to agriculture and food security under climate change. Environ. Sci. Policy 15, 136–144. http://dx.doi.org/10.1016/j.envsci.2011.09.003.

Volpe, V., Marani, M., Albertson, J.D., Katul, G., 2013. Root controls on water redistribution and carbon uptake in the soil–plant system under current and future climate change. Adv. Water Resour. 60, 110–120. http://dx.doi.org/10.1016/j.advwatres.2013.07.008.

Vrebos, D., Staes, J., Vandenbroucke, T., D'Haeyer, T., Johnston, R., Muhumuza, M., et al., 2015. Mapping ecosystem services flows with land cover scoring maps for data scarce regions. Ecosyst. Serv. 13, 28–40. http://dx.doi.org/10.1016/j.ecoser.2014.11.005.

Vu, Q.M., Le, Q.B., Frossard, E., Vlek, P.L.G., 2014. Socio-economic and biophysical determinates of land degradation in Vietnam: an integrated causal analysis at national level. Land Use Policy 36, 605–617. http://dx.doi.org/10.1016/j.landusepol.2013.10.012.

Wali, M.K., Evrendilek, F., Fennessy, M.S., 2010. The Environment. Science, Issues and Solutions. CRS Boca Raton.

Wang, Y., Qiao, M., Liu, Y., Zhu, Y., 2012. Health risk assessment of heavy metals in soils and vegetables from wastewater irrigated area, Beijing-Tianjin cluster, China. J. Environ. Sci. 24, 690–698. http://dx.doi.org/10.1016/S1001-0742(11)60833-4.

Weeb, N.P., Stokes, C.J., Marshall, N.A., 2013. Integrating biophysical and socio-economic evaluations to improve the efficacy of adaptation assessments for agriculture. Global Environ. Change 23, 1164–1177. http://dx.doi.org/10.1016/j.gloenvcha.2013.04.007.

Weissert, L.F., Salmond, J.A., Schwendenmann, L., 2016. Variability of soil organic carbon stocks and soil CO_2 efflux across urban land use and soil cover types. Geoderma 271, 80–90. http://dx.doi.org/10.1016/j.geoderma.2016.02.014.

Wheeler, T., von Braun, J., 2013. Climate change impacts on global food security. Science 341, 508–513. http://dx.doi.org/10.1126/science.1239402.

World Bank, 2008. Sustainable Land Management Sourcebook. Word Bank, Washington.

Wu, R., Tiessen, H., 2002. Effect of land use on soil degradation in alpine grassland soil, China. Soil Sci. Soc. Am. J. 66, 1648–1655. http://dx.doi.org/10.2136/sssaj2002.1648.

Wu, S., Mickley, L.J., Kaplan, J.O., Jacob, D.J., 2012. Impacts of changes in land use and land cover on atmospheric chemistry and air quality over the 21st century. At. Chem. Phys. 12, 1597–1609. http://dx.doi.org/10.5194/acp-12-1597-2012.

WWF, 2014. Living Planet Report, Sciences and Species, Gland, pp. 176.

Xu, S., Zhao, Y., Shi, X., Yu, D., Li, C., Wang, S., et al., 2013. Map scale effects of soil databases on modelling organic carbon dynamics for paddy soils in China. Catena 104, 67–76. http://dx.doi.org/10.1016/j.catena.2012.10.017.

Yaalon, D.H., Yaron, B., 1966. Framework for man-made soil changes-an outline of metapedogenesis. Soil Sci. 102, 272–277.

Yagasaki, Y., Shirato, Y., 2014. Assessment on the rates and potentials of soil organic carbon sequestration in agricultural lands in Japan using a process-based model and a spatially explicit land-use change inventories—Part 1: Historical trend and validation based on nation-wide soil monitoring. Biogeosciences 11, 4429–4442. http://dx.doi.org/10.5194/bg-11-4429-2014.

Zalasiewicz, J., Waters, C.N., Williams, M., Barnosky, A.D., Cearreta, A., Crutzen, P., et al., 2015. When did the Anthropocene begin? A mid-twentieth century boundary level is stratigraphically optimal. Quat. Int. 383, 196–203. http://dx.doi.org/10.1016/j.quaint.2014.11.045.

Zhang, X., Xu, M., Liu, J., Sun, N., Wang, B., Wu, L., 2016. Greenhouse gas emissions and stocks of soil carbon and nitrogen from a 20-year fertilized wheat-maize intercropping system: a model approach. J. Environ. Manage. 167, 105–114. http://dx.doi.org/10.1016/j.jenvman.2015.11.014.

Ziadat, F.M., Taimeh, A.Y., 2013. Effect of rainfall intensity, slope and land use and antecedent soil moisture on soil erosion in an arid environment. Land Degrad. Dev 24, 582–590. http://dx.doi.org/10.1002/ldr.2239.

Zribi, M., Dechambre, M., 2003. A new empirical model to retrieve soil moisture and roughness. Remote Sens. Environ. 84, 42–52. http://dx.doi.org/10.1016/S0034-4257(02)00069-X.

CHAPTER

3

Goal Oriented Soil Mapping: Applying Modern Methods Supported by Local Knowledge

Paulo Pereira[1], Eric C. Brevik[2], Marc Oliva[3], Ferran Estebaranz[4], Daniel Depellegrin[1], Agata Novara[5], Artemi Cerdà[6] and Oleksandr Menshov[7]

[1]Mykolas Romeris University, Vilnius, Lithuania [2]Dickinson State University, Dickinson, ND, United States [3]University of Lisbon, Lisboa, Portugal [4]Universitat de Barcelona, Barcelona, Spain [5]University of Palermo, Palermo, Italy [6]University of Valencia, Valencia, Spain [7]Taras Shevchenko National University of Kyiv, Kyiv, Ukraine

INTRODUCTION

Soil mapping results are extremely relevant for sustainable land use management. Soil maps need to be useful for land managers and farmers to help increase soil productivity, reduce soil degradation, and facilitate soil restoration. For these reasons, recent advances in spatial statistical methods and soil mapping techniques are a great step forward in improving mapping accuracy, important to allow managers and decision makers to implement the results of scientific investigations. The majority of current studies use geostatistical methods to interpolate soil data (Lu et al., 2012; Fernandez-Calvino et al., 2013; Zhang et al., 2016). There are a great number of studies investigating several interpolation methods with the objective to find the best technique to map soil properties in forest (Hoffmann et al., 2014) and agricultural areas (e.g., Gumiere et al., 2014; Miller et al., 2015; Chen et al., 2016) at diverse scales and using different auxiliary variables. Test the best methods of interpolation are extremely important for producing accurate maps for land managers' use for territorial planning (Pereira et al., 2010). Nowadays, remote and proximal sensing have increased the amount of data available for mapping and the accuracy of the maps produced (Escribano et al., 2017).

Traditional knowledge about the environment in general and soil in particular has

Soil Mapping and Process Modeling for Sustainable Land Use Management.
DOI: http://dx.doi.org/10.1016/B978-0-12-805200-6.00003-7

long been acknowledged by science in several parts of the world, such as Africa (Ajayi, 2007; Rushemuka et al., 2014; Tesfahunegn et al., 2016), Asia (Colen et al., 2016; Jyoti Nath et al., 2015a,b), Central America (Pauli et al., 2012; Falkowski et al., 2016), Central Europe (Burgi et al., 2013; Wahlhutter et al., 2016), the Mediterranean (Marques et al., 2015; Capra et al., 2016), and North America (Brevik et al., 2016c). Soil management practices have been transmitted among generations through both written documentation and oral tradition. In fact, traditional know-how is the fundamental basis of sustainable land management and the success of indigenous farmers (Venkateswarlu et al., 2013; Kangalawe et al., 2014). The integration of scientific and traditional knowledge is an important step for an effective sustainable land management (Barrera-Bassols and Toledo, 2005; Sandor et al., 2006; Mairura et al., 2008; Brevik et al., in press). In fact, failure to do so has led to failures when scientifically developed agricultural systems have been introduced into developing countries (Critchley et al., 1994; Hellin and Haigh, 2002). In this context, the knowledge and information produced by soil maps combined with traditional soil knowledge has the potential to increase the efficiency of problem solving and create solutions for better land management.

RECENT METHODS USED FOR SOIL MAPPING

Maps have been fundamental in the advancement for our scientific knowledge about soils and to communicate information in an understandable manner (Brevik and Hartemink, 2013). Despite the great advances observed since the end of the 19th century, the emergence of new technologies (e.g., geographic information systems (GIS), global positioning systems (GPS), remote and proximal sensing, data loggers, and geophysical instrumentation) and the development of statistical and geostatistical methods increased our capacity to collect, analyze, and predict soil information with a better accuracy substantially (Pereira et al., 2017; Brevik et al., 2016a). There has been a great effort to create soil information that is available for mapping. However, some of these data have been collected using different methods and standards, since they were collected at the national level using different scales (Mulder et al., 2011) and soil properties were determined using different analyses; in fact, many soil analyses were not standardized until the second half of the 20th century (Brevik et al., 2016c). Despite the existence of large databases, soil data are unequally distributed and this may create problems for modeling since some areas have more observations than necessary while others do not have enough (see also Brevik et al., 2017; Pereira et al., 2017).

Remote and Proximal Sensing

Remote and proximal sensing methods are cost-effective and, when combined with appropriate modeling, reduce the number of arduous and time-consuming laboratory analyses needed for mapping soil properties (Heilig et al., 2011). In addition, they are rapid and nondestructive techniques that can collect large amounts of data without inducing soil disturbance. Because of this, when coupled with modern GPS, soil properties can also be investigated at the same locations multiple times, allowing better temporal investigations into soil heterogeneity than traditional sampling. Traditional sampling can be destructive as it removes samples from the field, allowing to measure properties at that exact location only once. Despite these advantages, several concerns have been raised concerning the validity of these techniques, particularly concerns about lack of ground truthing of remote and proximally sensed data (Brevik and Hartemink, 2010). Some works have been carried out to validate the use of remote or proximal sensing methods with laboratory analysis with good results (Zhu et al., 2011; Nocita et al., 2013; Anne et al., 2014; Sharma et al., 2016).

In soil science the application of remote sensing techniques works best on bare surfaces since vegetated areas do not allow direct measurement of soil reflectance. Several indices have been created to tackle this problem using variations of the Normalized Difference Vegetation Index (NDVI), such as the Soil Adjusted Vegetation Index (SAVI), generalized SAVI, transformed SAVI, modified SAVI, and the Global Environmental Monitoring Index. In urban areas paved surfaces do not provide important information about soil proprieties and the information collected with remote sensing is incomplete. Other problems may also occur in the estimation of soil properties using remote sensing such as atmospheric influences, geometric distortions, spectral mixture of the features, lower spectral and spatial resolution, structural effects and the fact that the information is limited to the surface layer (Gilabert et al., 2002; Mulder et al., 2011). Nevertheless, in spite of all these limitations, several works have been published mapping soil properties using remote sensing methods, including studies to determine taxonomy (Brungard et al., 2015), organic matter (Poggio et al., 2013), organic carbon (Gomez et al., 2008), texture (Anne et al., 2014), pH, phosphorous, and potassium (Lopez-Granados et al., 2005), moisture (Kim and Barros, 2002), electrical conductivity (Ben-Dor et al., 2002), erosion (Vrieling, 2006), salinity (Allbed and Kumar, 2013), carbonate content, mineralogy, iron and nitrogen content (Mulder et al., 2011), and hydrological roughness (Kaiser et al., 2015).

Proximal sensing for mapping purposes is applied at a smaller scale and the accuracy of the predictions is better than with remote sensing (Gooley et al., 2014; Aldabaa et al., 2015; Muñoz-Rojas et al., 2017). Several methods are used in the field or in the laboratory (Brevik et al., 2016a). Here, for mapping proposes, we will focus only on proximal sensing methods used in the field. Proximal sensing is very useful for rapid assessment of soil proprieties in the field. The majority of these "on-the-go" sensors are based on electromagnetic or electrical sensing. These methods cannot measure individual soil properties, but they are often most useful in situations where one or two soil properties of interest dominate the readings obtained from the sensor.

Despite the fact that proximal sensing methods are less accurate than laboratory methods, the spatial resolution of the maps produced is much better (Kodaira and Shibusawa, 2013). The new devices are equipped with GPS, facilitating mapping tasks, and are very cost-effective. For example, just a few hours in the field with an electromagnetic induction (EMI) sensor can result in thousands of electrical conductivity readings over tens or hundreds of ha that can be related back to certain soil properties. The same amount of time would allow for the collection of only a limited number of soil samples and would not allow for laboratory analyses. The use of proximal sensors also collects a much denser data set much more rapidly than traditional soil mapping (Brevik et al., 2003).

There are a large number of methods used for proximal sensing (please see Mulder et al., 2011; Brevik et al., 2016a). For example, Portable X-ray fluorescence spectroscopy (PXRF) has been used to map soil heavy metals contamination in urban soils and old mines (Carr et al., 2008; Mamindy-Pajany et al., 2014; Weindorf et al., 2013). EMI is frequently used to map soil salinity, texture, compaction, bulk density, water content, organic matter, pH, cation exchange capacity, surface soil horizons, and apparent electrical conductivity (Doolittle and Brevik, 2014; Huang et al., 2016). Visible near infrared has been applied in the field to map soil moisture content, organic matter, pH, electrical conductivity, total carbon, cation exchange capacity, hot water extractable nitrogen, ammonium nitrogen, nitrate nitrogen, total nitrogen, phosphorous active coefficient, and available phosphorous (Kodaira and Shibusawa, 2013). Electrical resistance tomography, ground-penetrating radar and time domain reflectometry are widely used to map soil water content in the field (Huisman et al., 2002; Minet et al., 2012; Andre et al., 2012; Beff et al., 2013; Tran et al., 2014). Ground-penetrating radar has also been used to map peat thickness (Comas et al., 2015), soil organic horizons (Laamrani et al., 2013) and permafrost tables (Pan et al., 2014). The gamma radiometer has been used to map soil depth, clay content, organic carbon, pH, iron content, potassium, uranium, and thorium (Castrignano et al., 2012; Stockmann et al., 2015).

The use of one type of sensor can be an advantage in some studies but disadvantageous in others because each proximal sensing method has strengths and weakness. Comparing different sensors is fundamental to identify the most accurate to estimate a particular soil property given local soil conditions (Piikki et al., 2013). For example, EMI sensors cannot differentiate between sand and gravel soils and the presence of soluble salts reduces the capacity of EMI to distinguish clay soils from saline sandy soils because both can have high bulk electrical conductivities (Castrignano et al., 2012). Likewise, soils that have high soluble salt, clay, or water contents are not suitable for investigation by ground-penetrating radar (GPR) because they limit the penetration depth and resolution of GPR (Brevik et al., 2016b). The information given by a single sensor is often considered to be of limited use. To tackle this problem, several studies have combined different sensors in order to collect more accurate data for mapping soil proprieties (De Benedetto et al., 2012; Rodrigues Jr et al., 2015) or, in the case of EMI studies, have simultaneously collected data from multiple depths (Doolittle and Brevik, 2014). These types of approaches are expected to become standard because several sources of information are often needed to make correct decisions about land use management (De Benedetto et al., 2013).

There are some works that combine both remote and proximal sensing techniques to increase the accuracy of soil properties prediction. Buchanan et al. (2012) combined EMI with Landsat thematic mapper data to estimate soil texture. Gomez et al. (2008) combined hyperspectral data with vis-NIR spectroscopy to estimate soil organic carbon. Aldabaa et al. (2015) used remote sensing data, PXRF, and VisNIR to estimate soil salinity. Triantafilis et al. (2009) also estimated soil salinity by combining EMI and remotely sensed data. The combination of proximal and remotely sensed data for soil mapping is advantageous since it increases the amount of information that can be used as covariates and can identify the most precise method to estimate a determined soil property. The increase in the number of ancillaries contributes to error reduction and reduced uncertainty of the predictions. However, on the other hand, the use of multiple sources of information can increase the complexity of the analysis, which may require robust statistical methods.

Traditional and Spatial Statistics

Due to the exponential increase in information derived from proximal and remote sensing as compared to traditional field sampling approaches, the use of simple statistical methods is no longer enough to understand the complex relationships among covariates (Brevik et al., 2016a). In recent years, several quantitative methods have been applied to study the spatial distribution of soil properties. This came about due to the need to rationalizing and quantify the knowledge of soil surveyors, since they often have difficulties explaining how they developed their soil classifications, which makes it difficult for users to apply those classifications (Bourennane et al., 2014). Most of the new statistical methods are derived from modern statistics or geostatistics (Miller, 2017). The accuracy of a given pedometric technique depends on the objective of the survey and the quality of the work (McBratney et al., 2000). According to Hengl et al. (2007), there are five main groups of soil classification techniques using modern statistical methods (1) pure classification techniques, using remote sensing, (2) pure regression techniques, based on data mining techniques and regression techniques, such as supporting vector machines, neural networks, generalized linear models random forests, factor discriminant analysis, logistic regressions, partial least squares, general additive models, and regression trees), (3) pure geostatistical methods, (4) hybrid techniques, a combination between kriging techniques and univariate or multivariate methods, and (5) expert systems, using qualitative and auxiliary data. Another type of statistical approach that is currently used in soil classifications is fuzzy sets and pedodiversity (Bourennane et al., 2014; Brevik et al., 2016a).

For mapping proposes, geostatistical (univariate) and hybrid (multivariate) methods are the most important because they are able to detect the spatial autocorrelation of soil proprieties (McBratney et al., 2000). Geostatistical techniques are based on the theory of the regionalized variables and provide some descriptive tools such as the variogram to describe soil spatial distribution. One of the important contributions of geostatistical techniques is the assessment of the uncertainty in relation to the unsampled points. These analyses of probability are fundamental to assess the probability of exceeding a critical threshold of a soil attribute (Goovaerts, 1999; Miller, 2017). Thus geostatistical techniques are fundamental for sustainable land management proposes since it is possible to identify areas where restoration is needed and areas where it is not needed.

Kriging is the interpolation method used by geostatistical methods. Kriging methods can be univariate (modeling one variable) and multivariate (using one or more covariates to predict a primary variable). The most used univariate methods are ordinary kriging, simple kriging, and universal kriging. There are a large number of multivariate kriging methods used for soil data modeling such as regression kriging (Hengl et al., 2007), cokriging (Chen et al., 2016), kriging with external drift (Shi et al., 2016), factorial kriging (Bevington et al., 2016), and empirical bayesian kriging (Peng et al., 2015). Other univariate and multivariate kriging methods are also used in soil science. A review of the univariate and multivariate methods used in soil and other environmental sciences is provided by Li and Heap (2014). Normally, since multivariate methods include ancillary variables, the prediction of the primary variable is more accurate than univariate variables, as reported by several works (Zhang et al., 2012, 2014; Bilgli, 2013; Dai et al. 2014; Chen et al., 2016).

Other nongeostatistical or deterministic methods used for soil data interpolation include nearest neighbor, inverse distance to a weight, regression models, natural neighbors, trend surface analysis, splines, thin plate splines, global polynomial, local polynomial, regression tree, classification, and radial basis function methods, to mention some (Gumiere et al., 2014; Pereira et al., 2015). Some hybrid methods were created using nongeostatistical techniques, such as the combination of support vector machine, random forests methods, and boosted decision tree with inverse distance to a weight (Li and Heap, 2014). However, to our knowledge they have not been applied in soil science. In some cases, deterministic methods can be more accurate than geostatistical methods as identified in previous works (Wu et al., 2013; Gumiere et al., 2014; Emadi and Baghernejad, 2014; Souza et al., 2016) and this is attributed to the data's spatial distribution and the lack of spatial autocorrelation. When data does not have a spatial pattern, deterministic methods can be more accurate to predict soil variables.

Several indices have been developed to calculate the spatial autocorrelation of environmental variables. Global Moran's *I* index can be used to measure the strength and identify the positive or negative spatial correlation of a determined variable. This index is similar to the Pearson correlation coefficient and varies between −1 and 1. A higher negative spatial correlation shows that variables have a dispersed pattern while a high positive correlation shows that the variables are clustered. If the index is close to 0, it shows that the variable in the study has a random pattern. The Global Moran's *I* calculates the *Z*-score. If the *Z*-score value is less than −1.96 the variable is significantly dispersed at a $P < 0.05$. If the *Z*-score is greater than 1.96, the variable is significantly clustered at a $P < 0.05$. This index just gives a global assessment of the spatial correlation (Huo et al., 2011).

A variant of the Global Moran's *I* method was recently developed. The incremental spatial autocorrelation index measures the autocorrelation at different distances using the Moran's *I* index. This method can identify the distance at which the spatial correlation is high (clustered) or low (dispersed), and if they are significant or not. This can be observed through the peaks identified by this index. For more information about this index please consult.[1] To identify map clusters and spatial outliers we used the Local Moran's *I* index. The principles applied to identify clusters or outliers are the same as mentioned above. High–high clusters show high values in the neighborhood, while low–low clusters show the opposite (Fig. 3.1A). Local Moran's results are important to identify areas with significantly higher or lower concentrations of a specific soil property (Huo et al., 2011; Fu et al., 2014). In the example in Fig. 3.1A, it can be observed that soil clay contents were significantly autocorrelated (Moran's Index $I = 0.67$, $P < 0.000$). High clay values were typically located in the western part of Lithuania, while the low values were observed in the eastern part. Overall, Global Moran's *I* is important to identify the overall clustering/dispersed pattern, incremental Moran's *I* to detect at what distance this occurs, and Local Moran's *I* to detect in the map where the clustered/dispersed areas are located.

Other spatial statistical models, similar to Global Moran's *I*, have been developed. Geary's *c* method assesses the spatial autocorrelation of a specific variable. In this method the values observed are in the range between 0 and 2. Zero shows that the variable has a cluster pattern, while 2 indicates that the pattern is dispersed (Erdogan, 2009). The Getis-Ord General *G* is an index similar to Global Moran's *I* and the Getis-Ord Gi is similar to Local Moran's *I* (Getis and Ord, 1992). The Getis-Ord General *G* identifies the overall high/low clustering similar to Global Moran's *I* and Getis-Ord Gi similar to Local Moran's *I*. However, they measure different things. Moran's *I* measures the spatial patterns of the data, which is used to identify the clustering of high (hot spots) and low (cold spots) index values (Truong and Somenahally, 2011). Moran's *I* method shows the presence of high/low clustering values, but it cannot differentiate between situations. Getis-Ord methods can identify if the cluster is composed of high values (hot spot) or low values (cold spot) (Fig. 3.1B) (Erdogan, 2009). The results of the Getis-Ord General *G* in Fig. 3.1B showed that soil clay content was clustered (*Z*-score = 973.26, $P < 0.000$). Hot and cold spots showed a similar pattern to that identified in the Local Moran's *I*. However, in this case, we have a quantification of the standard deviations and the areas' levels of significance of each cluster (Fig. 3.1B).

[1] http://desktop.arcgis.com/en/arcmap/10.3/tools/spatial-statistics-toolbox/incremental-spatial-autocorrelation.htm.

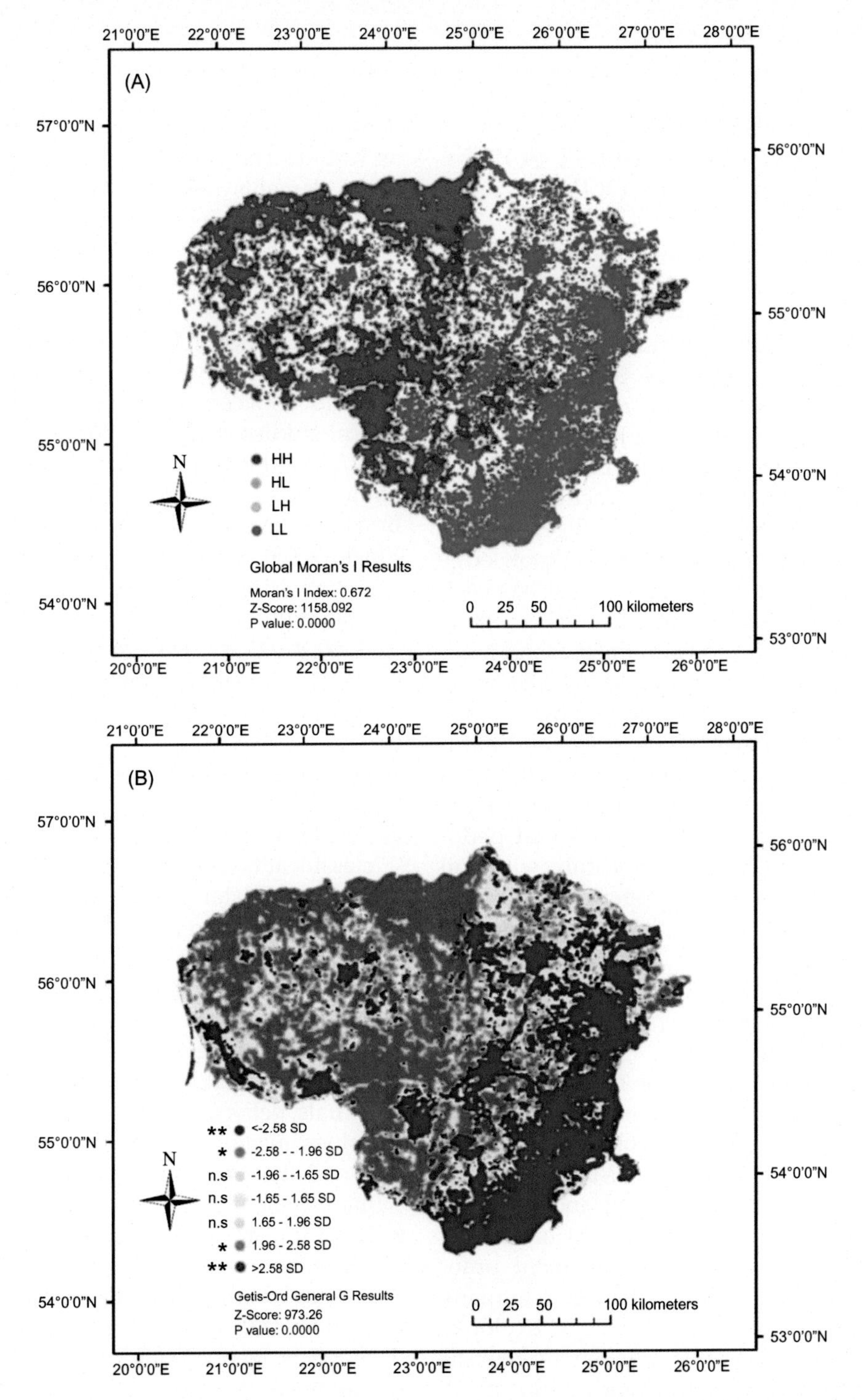

FIGURE 3.1 Topsoil clay content in percentage Moran's Local spatial correlation index (A), and hot spot Getis-Ord Gi Index (B) in Lithuania. Both figures show the global results of Morans *I* and Getis-Ord indexes. *HH*, significantly high–high clusters; *LL* (significant low–low clusters; *HL*, high value outlier surrounded by low values; *LH*, low value outlier surrounded by high values. *SD*, standard deviation. ** and *, significant hot/cold spot at $P < 0.01$ (red/dark blue) and $P < 0.05$ (blue/orange). Soil clay content data are from the LUCAS project (Toth et al., 2013).

TRADITIONAL PERCEPTIONS AND KNOW-HOW ABOUT SOIL: ETHNOPEDOLOGY

The study of traditional soil knowledge, including knowledge of soil properties, management, and taxonomy, is known as ethnopedology. This is defined by the understanding and documentation of soil appraisal, classification, perception, use, and management (Krasilnikov and Tabor, 2003; Jyoti Nath et al., 2015a). Traditional knowledge is considered a mix between knowledge and practice and it is very difficult to separate them. Indigenous cultures have a tremendous and useful knowledge about their soil and environment that has been transferred through generations. This capital is invaluable for a long and sustainable land management (WinklerPrins and Sandor, 2003).

Ethnopedology is considered a hybrid discipline supported by both the social and natural sciences, and includes all soil and land information from the most traditional to the most modern (Fig. 3.2). It assumes that pedodiversity is strongly linked not only with soil physical and chemical properties and soil classification, but also with human management, cultural practices, and historic land uses. Ethnopedology holistically considers the local knowledge about a specific environment (Capra et al., 2015).

Ethnopedology studies the role of land and soil in the management of natural resources and as a part of the local ecological and economic processes. It encompasses both technical and farmers' knowledge of soil and analyzes the role of soil in the management of natural resources. Traditional soil management practices are an important link between humans and the environment. The soil is explored as a polysemic cognitive domain, a natural resource with multiple uses and as an object with symbolic values and meanings. Local people understand soil as a resource, through the interaction of symbolism (Kosmos), knowledge (Corpus), and management practices, known as K–C–P. The complex relationships between these elements result of the fusion of knowledge and experience, facts and values, matter and mind, and sacred and secular features (Barrera-Bassols and Zinck, 2003; Adderley et al., 2004).

The ethnopedological richness is correlated with biological diversity. There is a strong relationship between the numbers of ethnopedological studies and plant domestication centers such as Indonesia, Thailand, the Philippines, Papua and New Guinea, Bolivia, Peru, Mexico, Brazil, India, and Nepal (Barrera-Bassols and Zinck, 2003), showing the importance of local soil knowledge for agriculture existence and development.

According to Barrera-Bassols and Zinck (2003) the most important fields of ethnopedology are:

1. Local soil and land uses and management practices;
2. Local soil and land taxonomy and classification nomenclatures;
3. Local understanding of the relationship between soil and land with biophysical variables, process, and elements;
4. Perception of local communities about soil and land resources and the explanation of the proprieties, processes, structure, dynamics, and distribution of soil;
5. Rituals, beliefs, myths, as other symbolic values, meanings, and practices connected to soil quality evaluation and land management;
6. Local transformation, renewal, and adaptation;
7. Integration of ethnopedology with modern soil science, agricultural and rural practices, geopedological survey, and agroecological strategies in order to stimulate participation in land use planning and land evaluation practices for local sustainable development.

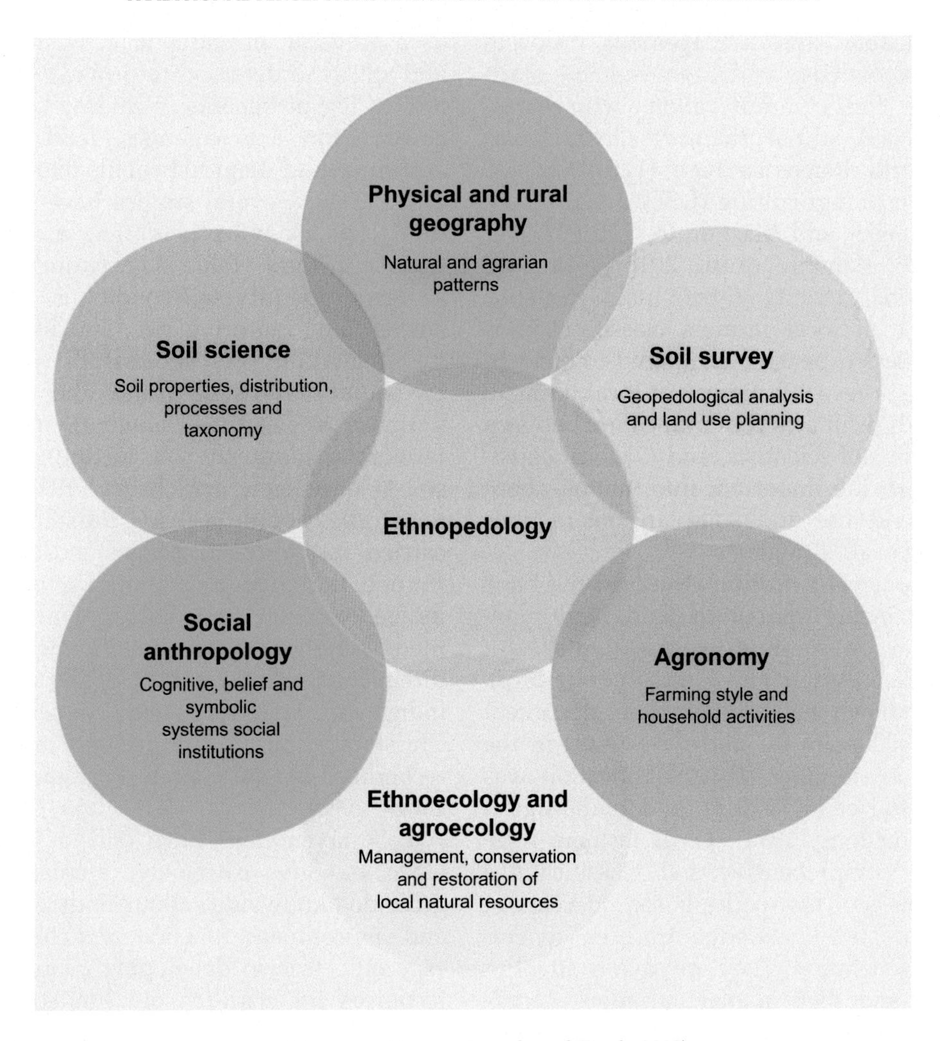

FIGURE 3.2 Ethnopedology as a hybrid science (Barrera-Bassols and Zinck, 2003).

To development more fluent and understandable communications between farmers and scientists, many of the ethnopedology studies that were carried out focused on the poor results obtained from technical studies and top-down interventions in developing territories, something that was normally disastrous from a sustainable land management perspective (Barrera-Bassols and Zinck, 2003; Niemeijer and Mazzucato, 2003; Barrios et al., 2006; Zuniga et al., 2013; Capra et al., 2015).

Soil Classification

Indigenous cultures have a long tradition of soil classification that reflects the way local communities managed, valued, and understood soil resources (Krasilnikov and Tabor, 2003). This capacity and knowledge are extremely valuable and productive as regards practical application (Beyene, 2015; Nyssen et al., 2015a,b; Coomes and Miltner, 2016; Grum et al., 2016; Zoumides et al., 2016).

For sustainable land management proposes soil local knowledge is a more suitable starting point for communication with farmers than Food and Agriculture Organization (FAO), World Reference Base (WRB), or US Department of Agriculture (USDA) soil taxonomy (Niemeijer and Mazzucato, 2003; Hillyer et al., 2006; Kambire et al., 2016). Soil local classification presents three main benefits: (1) the use of local farmers classification is faster and less expensive compared to conventional soil surveys, (2) the use of local terms to classify soils facilitates communication between farmers and soil scientists, and (3) local classifications provide important information about farmers' land use and soil-plant interactions (Jyoti Nath et al., 2015a,b).

Indigenous classification is indeed the basis for a sustainable agriculture and land management in developing regions (Jyoti Nath et al., 2015a,b; Assefa and Hans-Rudolf, 2016). The first known soil classification document dates to 2500 years BC and was found in the Chinese book Yogong. This classification was based on soil color, texture, and hydrological properties (Zitong, 1994). Local farmers have developed comprehensive soil classifications in many parts of the world, based on years of observation and knowledge transfer generation after generation. They are able to identify soil types and their spatial variation according to topography, land use, and mesoclimate. This knowledge was refined with time and local communities developed an extremely high understanding of the environment that surrounded them (Barrera-Bassols et al., 2006; Pauli et al., 2012).

Local communities understand soil as one of the multiple components of nature. For soil type, indigenous people typically understand soil as a three-dimensional object. Regarding farming activities, farmers typically understand soil as a two-dimensional body, manageable according to its bioclimatic characteristics. Indigenous farmers understand soil as a valuable provider of ecosystem services and soil is understood to provide society with a long list of benefits, including materials for construction and ceramics, food, medicines, and ritual and magical benefits (Barrera-Bassols et al., 2006). Several studies have investigated local soil classifications that are based on diverse criteria (Table 3.1). Farmers use their senses to identify soil conditions, such as the smells (e.g., maturing rice, and indicator that the plantation was successful), sounds (e.g., the sound of birds, bees, and other insects is an indicator of clean environment), and feel (e.g., farmers distinguish soil texture) of a given soil (Concepcion and Batjes, 1997). Most of the studies about local soil classifications were carried out in Asia, Africa, and America. In Europe, little documentation is available about indigenous soil knowledge, which is attributed to the European scientific method, which diminished traditional knowledge, and because indigenous knowledge was connected to pre-Christian agrarian cultures, not approved or recognized by the church (Barrera-Bassols and Zinck, 2003).

Folk taxonomies have only a local importance, but give extremely valuable and well-grounded knowledge about landscape structure and environment, function, and change. This is especially true in developing countries, where resources for research are limited. However, due to the current globalization this knowledge is being lost as a consequence of the introduction of scientific knowledge, migrations, and economic evolution. It is essential that traditional soil classifications be studied because they give us great insights into land use and management (Krasilnikov and Tabor, 2003).

Soil and Land Degradation

Soil and land degradation are one of the major causes of hunger and poverty in developing countries (Pereira et al., 2017; Kassa et al., 2016). Local know-how and local farming

TABLE 3.1 Farmers Criteria for Soil Classification in Different Environments

Farmers Criteria for Classification	Country, Continent	Reference
Gravel, texture, color, topographical position, and water infiltration	Ivory Coast, Africa	Birmingham (2003)
Gravel, texture, topography, texture, and color	Burkina Faso, Africa	Niemeijer and Mazzucato (2003)
Color, water retention, topographic position, surface rock fragments, and water retention	Mexico, Central America	Estrada-Medina et al. (2013)
Consistence, texture, color, and stoniness	Mexico, Central America	Barrera-Bassols et al. (2006)
Color, texture, and stoniness	North East India, Asia	Jyoti Nath et al. (2015a,b)
Texture, color, and fertility	Niger, Africa	Osbahar and Allan (2003)
Color, texture, erosion potential, water retention, vegetation, crops suitability, and soil depth	Tanzania, Africa	Oudwater and Martin (2003)
Texture, color, and gravel content	Republica Dominicana, Central America	Ryder (2003)
Earthworms, color, crop productivity, humus layer, surface compaction, erosion, slope, surface water, and moisture	Northern Vietnam, Asia	Trung et al. (2008)
Color, depth, color, slope, and stoniness	Honduras, Central America	Barrios and Trejo (2003)
Color, fertility, moisture, and plant growth	Philippines, Asia	Price (2007)
Consistence, texture, hardness, granularity, color, and fertility	East Africa and Bangladesh, Asia	Payton et al. (2003)
Texture, color, presence of gravel, presence of blocks of laterite, hardpan, and water growth	Southwestern Burkina Faso, Africa	Gray and Morant (2003)
Color, slope gradient, and slope position	Michoacan, Mexico	Pulido and Bocco (2003)
Texture, depth, mineral nodules, color, soil surface features, and landscape physiography	Ghana, Africa	Asiamah et al. (1997)
Drainage, texture, and flooding	Philippines, Asia	Concepcion and Batjes (1997)
Geographic location, texture, color, water content, and hardpan	Peru, South America	Kauffman and Ramos (1998)
Texture, organic matter, drainage, salinity, color, consistence, and fertility of the soils	Bangladesh, Asia	Shajaat Ali (2003)
Texture, water holding capacity, color, hardness, and consistency	Burkina Faso, Africa	Bonsu (2004)
Color, texture, fertility, water retention, vegetative indicators, and possible inundation[a]	North of Mono Province of southwest Benin, South Africa	Brouwers (1993)
Texture, fertility, consistency, geographical location, color, and drainage[a]	Burkina Faso, Africa	Dialla (1993)
Color, rock presence, and texture[a]	Central Java, Indonesia, Asia	Grobben (1992)
Texture, slope, vegetation, and color[a]	Northern Zambia, Africa	Kerven et al. (1995)
Color, texture, crop suitability, fertility, and permeability	Semiarid tropics in India, Asia	Talawar (1991)

[a]*This information was taken from the work of Talawar and Rhoades (1998).*

practices are fundamental to tackle this problem (Norton et al., 1998); however, the application of remediation measures is highly dependent on the economic and educational capacities of each community (Adimassu et al., 2013). Farmers are among the first to identify landscape features related to degradation and are the most directly affected by this process. In Africa, some studies have pointed out that soil and land degradation are mainly observed using these indicators: decreasing or disappearing of some plant species, deforestation, water erosion, gully formation, and loss of soil fertility. Farmers associate top soil color with soil erosion and fertility. Black soil is an indicator of a healthy soil, while white and yellow colors represent an eroded soil (Wezel and Haigis, 2000; Chizana et al., 2007; Davies et al., 2010; Adimassu et al., 2013; Teshome et al., 2016). In India, farmers' perceived soil and land degradation as occurring when rills developed on the soil surface, lowlands became waterlogged, and top soil was lost (Kumar Shit et al., 2015). In Pakistan Cochard and Dar (2014) found that mountain farmers associated environmental degradation with the reduction of medicinal plant availability. More recently, Tarrason et al. (2016) observed that farmers identified soil and land degradation in Nicaragua using more indicators, such as brush encroachment, presence of unpalatable plants, reduction of native grasses and overall biodiversity, decreases in plant and root shoots, reduction of nutrients supplied by grazing, reduction of soil depth and organic matter, soil crack formation, temporal changes in flowering, seed production, and germination. As a consequence of this, farmers developed a wide number of techniques to reduce soil and land degradation (Assefa and Hans-Rudolf, 2016; Pereira et al., 2017).

Many studies have investigated the implementation of restoration of degraded lands in diverse environments (Muñoz-Rojas et al., 2017; Asefa et al., 2003; Gisladottir and Stocking, 2005; Hu et al., 2016; Khaledian et al., 2016; Wolfgramm et al., 2016; Muñoz-Rojas et al., 2016; Panagos et al., 2016). Local knowledge is fundamental for the success of these land restoration programs. Measures considering only scientific knowledge and/or experiences from another environment may be unsuccessful (Reed et al., 2007). Scientific knowledge about agriculture is much generalized and may be is not applicable to local conditions. On the other hand, farmers' understanding is based on the experience and local ecological knowledge. Sometimes farmers can be extremely resistant to change because of their short-term objectives do not match with the long-term vision defended by researchers. Before recommending something like the implementation of conservation agriculture and/or land restoration measures, it is essential to consider the farmer's ecological knowledge, priorities, and concerns. Often there is the feeling that when the soil is more fertile, farmers' more intensively exploit the land, increasing the potential for soil and land degradation. In that case, some information from scientists may be needed to raise farmers' awareness regarding this problem (Anaya-Romero et al., 2011; Halbrendt et al., 2014; Tesfaye et al. 2014). Tesfaye et al. (2014) observed that in Ethiopia, smallholders will to implement sustainable land management measures depended on the investment in cash, the size of the property, and hand-labor availability in the family. The immediate profit is linked to the amount of money invested by the farmers for land restoration. Adimassu et al. (2013) also found that in Ethiopia farmers invested five times more money in soil fertility measures than in erosion control. Soil fertility measures were also easier to implement, and the farmers applied soil fertility amendments regardless of terrain slope and erosion vulnerability. In this case, farmers did not invest too much in land degradation measures, despite the fact that they perceived degradation as a problem, due the lack of funds.

Soil Quality

Good soil quality is essential for the implementation of long-term sustainable land management. It is also a great concern of farmers, especially those with fewer resources since their livelihoods rely on soil productivity (Pauli et al. (2012); Lancriet et al., 2015; Giger et al., 2016). Their long relationship with nature allowed these farmers to develop very important knowledge about their local soils, and therefore they developed a number of soil quality indicators, important for farmers to evaluate appropriate land use and implement sustainable land management (WinklerPrins and Sandor, 2003; Trung et al., 2008). Farmers used to link soil quality with agricultural production. They can determine if a soil is of high quality using several indicators, including dominating grasses, soil texture, color, water holding capacity, presence of invertebrate's, workability, infiltration, organic matter, depth, landform-type, crop yield and vigor, leaf color and size, time to flowering, abundance of local species, abundance or diversity of weed species, slope, fallowing history, previous crops, and weather and climate conditions (Ericksen and Ardon, 2003; Gray and Morant, 2003; Barrios et al., 2006; Maiura et al., 2007; Dawoe et al., 2012). According to Ericksen and Ardon (2003) perceptions about soil quality were based on:

1. Texture: Silty soils were easy to work, and clay soils were valued due their high fertility. Heavy rains decreased the workability of clayed soils.
2. Slope: Flat areas were preferred because they are easier to plow;
3. Perceived fertility: Soil texture and vegetation type were indicators of soil fertility conditions;
4. Drainage: Depended on the soil texture and hydrologic patterns within the catchment. Farmer's evaluation of this depended on the season and the crop;
5. Disease history: The capacity of the farmers to eradicate it was limited and they agreed that some crops could not be planted;
6. Irrigation: Water availability was important to plot value and determined where the intensive plots were located.

However, not all the variables had the same value for any given farmer's assessment of soil quality, and the variables used and their values are different in different regions of the world. Murage et al. (2000) observed that in Kenya crop yield was the most important indicator, followed by soil tilth, water retention, color, and indicator species. Niemeijer and Mazzucato (2003) observed that farmers in Burkina Faso considered that soil quality depended primarily on the amount of rainfall and the micro and macrotopography. In this area, farmers understood that soil quality depended on weather patterns. Some soils were high quality in wet years (rich in gravels) and others with an impermeable layer were considered good in dry years. In Vietnam, Trung et al. (2008) found that farmers considered the presence of certain soil fauna to be the most important criteria for soil quality, with the presence of earthworms providing evidence of a fertile soil.

INTEGRATION OF SCIENTIFIC AND LOCAL KNOWLEDGE FOR SUSTAINABLE LAND MANAGEMENT

Local soil knowledge is a tremendously valuable resource that can allow citizens and scientists to learn how humans have responded to environmental changes throughout history (WinklerPrins and Sandor, 2003). Therefore local soil knowledge should be used to support the scientific knowledge produced by modern science to improve the accuracy of classifications (Tabor, 1992). Using farmers' and scientific knowledge to complement each other can help

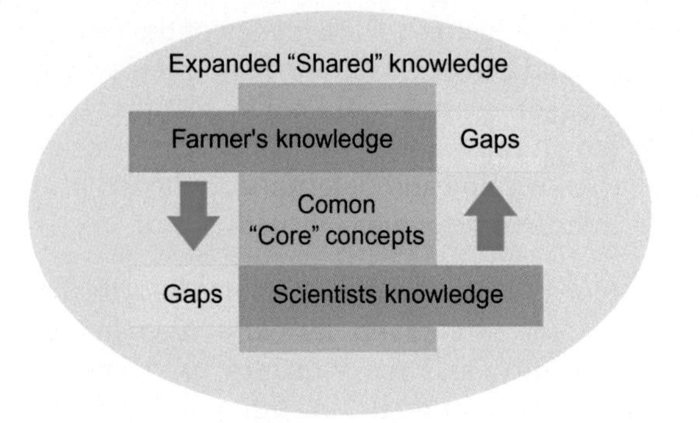

FIGURE 3.3 Integrating farmer's and scientific knowledge into a expanded shared knowledge (Barrios et al., 2006).

solve the gaps existent in each domain (Fig. 3.3) (Barrios et al., 2006).

One of the main goals of soil scientists is to integrate scientific with local knowledge, especially regarding soil classification (Barrera-Bassols et al., 2006; Barrios et al., 2006; Estrada-Medina, 2013; Capra et al., 2015) and soil quality (Ericksen and Ardon, 2003; Maiura et al., 2007; Trung et al., 2008; Pauli et al., 2012; Jyoti Nath et al., 2015a). In some cases, there is a good agreement between folk and scientific taxonomies (Gray and Morant, 2003; Barrera-Bassols et al., 2006; Trung et al., 2008; Jyoti Nath et al., 2015b) while in others, the classifications do not match up (Niemeijer and Mazzucato, 2003). Several criticisms have been raised regarding the comparison of local and scientific classifications. Farmers' soil classification is different from those carried out by scientists. Local smallholders are mainly interested in the top-soil layer because it is most responsible for soil fertility. They are not concerned about the pedogenesis, and they are typically concerned about relatively small, local areas. On the other hand, scientists have also focused on the deep soil layers (Birmingham, 2003; Zuniga et al., 2013; Juilleret et al., 2016) and are typically interested in classification over much larger areas (Brevik and Hartemink, 2013). In some cases these comparisons are absolutely impossible since scientific taxonomies are based on soil genesis (French classification), laboratory criteria, and in several cases considered artificial as is the case with the FAO, WRB, and USDA classifications, since not all natural units are classified and similar types of soils can be classified incorrectly (Niemeijer and Mazzucato, 2003). For example, Estrada-Medina et al. (2013) observed that some soils identified by farmers could be in one or more WRB classes and vice versa. In addition, soil local classification is based on environmental characteristics, while technical classification is based on soil properties that may or may not always reflect present environmental conditions. Another important aspect is that farmers' knowledge is related to their experience growing crops in their fields, meaning they have better knowledge about the soils of their own proprieties than of soils located in distant areas (Payton et al., 2003). Scientific classifications are useful for homogenization of soil nomenclature at national or international levels and for multiple potential land uses, however, for agriculture proposes, these taxonomies have been strongly criticized (Niemeijer and Mazzucato, 2003). These issues impose important problems on attempts to establish relationships between scientific and local classifications.

Mapping local soil classifications has been an objective for some soil scientists to try to better understand how they may match with technical classifications (Asiamah et al., 1997). They try to find the spatial correlations between folk and scientific soil classifications utilizing GIS methods. Despite the fact that maps can contribute to understanding farmers' classifications, the results should be interpreted carefully. Participatory mapping often encourages the delineation of land use areas rather than the spatial variability of soils. GIS facilitates data organization and management integrating local and scientific knowledge. However, the use of GIS in local knowledge mapping and the delineation of map boundaries into maps are often inadequate for spatial analysis. In this context the integration of farmers' mental map and scientific maps produced by GIS may be problematic (Oudwater and Martin, 2003; Payton et al., 2003).

The studies that are available show contradictory results. For example, Barrera-Bassols et al. (2006) found that the spatial correlation of local and scientific classifications based on the classification of the Purhépecha community as compared to USDA in central Mexico, depended on the type of soil analyzed. Despite this, there was a good agreement between local and scientific soil map units in 25% and 28% of the studied areas, respectively. Payton et al. (2003) observed that the map units developed by farmers in Tanzania, matched poorly with the WRB. However, research carried out in Bangladesh by Payton et al. (2003) found that the WRB and local classifications were similar. Schuler et al. (2006) observed that a local soil map in northern Thailand was very dissimilar from the WRB soil map and closer to the petrographic map. The Thai farmers classified their soils according to the color, which was closely related to the parent materials.

There are several reasons why in some cases the local and scientific classifications closely matched and in others they did not. One of is the landscape structure. Topography typically has a great influence on the farmers' classifications and they normally divide landscapes according to the relief characteristics. Therefore their soil classifications reproduce landscape features. If the landscape is diverse, the folk and scientific classifications are similar as many scientific classifications also include a strong landscape component (Bockheim et al., 2014; Brevik and Miller, 2015). However, the opposite occurs if there are no clear geomorphic units. In several cases, perceptions of local community about soil are not related to agriculture or biophysical characteristics. Some communities are tied to cultural aspects such as belonging to a place, local identity, history of land tenure, and family inheritance, and they connect these aspects to their soil classification. In these circumstances the relationships between local and scientific taxonomies are expected to be low (Barrera-Bassols et al., 2009).

More advanced techniques are needed to incorporate aspects of both local and scientific knowledge. The incorporation of local knowledge in spatial predictions is essential for the accuracy of predictions and to reduce survey costs. Fleming et al. (2004) observed that through soil color analysis, farmers' identification of soil productivity was very similar to a scientific assessment using an EMI sensor. Oliver et al. (2010) compared farmers' spatial extent of poor performance soils with NDVI and yield maps. They observed that the farmers were able to correctly map their properties and clearly identify areas that performed above and below average. Oberthur et al. (1999) used geostatistical models to show that the incorporation of farmers' local knowledge increased the accuracy of soil texture prediction. Local knowledge is an important information source to be considered in soil models and can be incorporated into advanced statistical (e.g., fuzzy) or spatial (e.g., regression or cokriging) models.

Sustainable land management can only be achieved with the active participation of local communities. It is clear that local knowledge is an important source of information that can lead to the production of more accurate maps. Some problems may arise in the comparison and blending of folk and scientific taxonomies. Nevertheless, in many circumstances, the knowledge of specific soil and environmental aspects by a local culture may provide the basis for improvement in overall soil classification, mapping, and modeling.

CONCLUDING REMARKS

Soil mapping techniques have developed exponentially, which has allowed us to collect an enormous quantity of data. Remote and proximal sensor technologies have been largely responsible for this growth due the amount of data collected, significantly reducing the need for laborious and expensive traditional field sampling and laboratory analyses. The combination of remote and proximal sensing increased the amount of explanatory variables that can be modeled by spatial statistical techniques and increased the accuracy of models. However, despite the tremendous evolution of technology, local soil knowledge can still provide important information for mapping, classification, and modeling. Local soil knowledge is highly useful to indigenous farmers and may prove critical to the widespread implementation of sustainable land management practices in developing parts of the world. Farmers are the heart of sustainable land management and understanding and incorporating their knowledge is critical to get their cooperation in implementing scientific management plans based on mapping and modeling results. Without this cooperation, investments in technology-driven plans will be useless. Several problems have been identified in correlating folk and scientific soil classifications due to the many cultural variables that influence the interactions between local communities and the environment. Additional research that seeks to bridge these gaps is important. Such research also represents an opportunity to better understand the investigated cultures and learn from indigenous people the knowledge they have accumulated over millennia. Mapping and modeling should start from knowing the natural and cultural peculiarities of each study area and involve as soon as possible farmers and other stakeholders in the mapping process because they will be the end-users of this work. Independent of the method used, the final goal should be the sustainable management of local soils and prosperity of the communities.

Acknowledgments

Paulo Pereira and Artemi Cerdà would like to gratefully acknowledge the support of COST action ES1306 (Connecting European connectivity research). E.C. Brevik was partially supported by the National Science Foundation under Grant Number IIA-1355466 during this project.

References

Adderley, W.P., Simpson, I.A., Kirscht, H., Adam, M., Spencer, J.Q., Sanderson, D.C.W., 2004. Enhancing ethno-pedology: integrated approaches to Kanuri and Shuwa Arab definitions in Kala-Balge region, northeast Nigeria. Catena 58, 41–64.

Adimassu, Z., Kessler, A., Yirga, C., Stroosnijder, L., 2013. Farmer's perception of land degradation and their investments in Land management: a case study in the Central Rift Valley of Ethiopia. Environ. Manage. 51, 989–998.

Ajayi, O.C., 2007. User acceptability of sustainable soil fertility technologies: lessons from farmers' knowledge, attitude and practice in Southern Africa. J. Sust. Dev. 30, 21–40.

Aldabaa, A.A.A., Weindorf, D.C., Chakraborty, S., Sharma, A., Li, B., 2015. Combination of proximal and remote sensing methods for rapid salinity quantification. Geoderma 239-240, 34–46.

Allbed, A., Kumar, L., 2013. Soil salinity mapping and monitoring in arid and semi-arid regions using remote sensing technology: a review. Adv. Rem. Sens. 2, 373–385.

Anaya-Romero, M., Pino, R., Moreira, J.M., Muñoz-Rojas, M., De la Rosa, D., 2011. Analysis of soil capability versus land-use change by using CORINE Land Cover and MicroLEIS in Southern Spain. Int. Agrophys. 25, 395–398.

Andre, F., van Leeuwen, C., Saussez, S., van Durmen, R., Bogaert, P., Moghadas, D., et al., 2012. High-resolution imaging of a vineyard in south of France using ground-penetrating radar, electromagnetic induction and electrical resistance tomography. J. Appl. Phys. 78, 113–122.

Anne, N.J.P., Abd-Elrahaman, A.H., Lewis, D.B., Hewitt, N.A., 2014. Modeling soil parameters using hyperspectral image reflectance in subtropical coastal wetlands. Int. J. Appl. Earth Obs. Geoinf. 33, 47–56.

Asefa, D.T., Oba, G., Weladji, R.B., Colman, J.E., 2003. An assessment of restoration of biodiversity in high mountain grazing lands in Northern Ethiopia. Land Degard. Dev. 14, 25–38.

Asiamah, R.D., Senayah, J.K., Adjei-Gyapong, T., Spaargaren, O.C., 1997. Ethno-Pedology Surveys in the Semi-arid Savanna Zone of Northern Ghana. Soil

Research Institute, Kwadaso-Kumasi and International Soil Reference and Information Centre. Kumasi, and ISRIC, Wageningen, The Netherlands.

Assefa, E., Hans-Rudolf, B., 2016. Farmer's perception of land degradation and traditional knowledge in Southern Ethiopia—resilience and stability. Land Degard. Dev. http://dx.doi.org/10.1002/ldr.2364.

Barrera-Bassols, N., Toledo, V.M., 2005. Ethnoecology of the Yucatec Maya: symbolism, knowledge and management of natural resources. J. Latin Am. Geogr. 4 (1), 9–41.

Barrera-Bassols, N., Zinck, J.A., 2003. Ethnopedology: a worldwide view on the soil knowledge of local people. Geoderma 111, 171–195.

Barrera-Bassols, N., Zinck, J.A., Van Ranst, E., 2006. Local soil classification and comparison of indigenous and technical soil maps in a Mesoamerican community using spatial analysis. Geoderma 135, 140–162.

Barrera-Bassols, N., Zinck, J.A., van Ranst, E., 2009. Participatory soil survey: experience in working with a Mesoamerican indigenous community. Soil Use Manage. 25, 43–56.

Barrios, E., Delve, R.J., Bekunda, M., Mowo, J., Agunda, J., Ramisch, J., et al., 2006. Indicators of soil quality: a South-South development of a methodological guide for linking local and technical knowledge. Geoderma 135, 248–259.

Barrios, E., Trejo, M.T., 2003. Implications of local soil knowledge for integrated soil management in Latin America. Geoderma 111, 217–231.

Beff, L., Gunther, T., Vandoorne, B., Couvreur, V., Javaux, M., 2013. Three-dimensional monitoring of soil water content in a maize field using Electrical Restivity Tomography. Hydrol. Earth Syst. Sci. 17, 595–609.

Ben-Dor, E., Patkin, K., Banin, A., Karnieli, A., 2002. Mapping of several soil properties using DAIS-7915 hyperspectral scanner data—a case study over clayey soils in Israel. Int. J. Remote Sens. 23, 1043–1062.

Bevington, J., Piragnolo, D., Teatini, P., Vellidis, G., Morari, F., 2016. On the spatial variability of soil hydraulic properties in a Holocene coastal farmland. Geoderma 262, 294–305.

Beyene, F., 2015. Incentives and challenges in community-based rangeland management: evidence from Eastern Ethiopia. Land Degrad. Dev. http://dx.doi.org/10.1002/ldr.2340.

Bilgli, A.V., 2013. Spatial assessment of soil salinity in the Harran Plain using multiple kriging techniques. Environ. Monit. Assess. 185, 777–795.

Birmingham, D.M., 2003. Local Knowledge of soils. The case of contrast in Cote d'Ivoire. Geoderma 111, 481–502.

Bockheim, J.G., Gennadiyev, A.N., Hartemink, A.E., Brevik, E.C., 2014. Soil-forming factors and soil taxonomy. Geoderma 226, 231–237.

Bonsu, M., 2004. Indigenous knowledge: the basis of survival of the peasant farmer in Africa. J. Philos. Culture 1, 49–61.

Bourennane et al., 2014. http://www.sciencedirect.com/science/article/pii/S0016706114000081.

Brevik, E., Pereira, P., Munoz-Rojas, M., Miller, B., Cerda, A., Parras-Alcantara, L., et al., 2017. Historical perspectives on soil mapping and process modelling for sustainable land use management. In: Pereira, P., Brevik, E., Munoz-Rojas, M., Miller, B. (Eds.), Soil Mapping and Process Modelling for Sustainable Land Use Management. Elsevier, pp. 3–30.

Brevik, E.C., Calzolari, C., Miller, B.A., Pereira, P., Kabala, C., Baumgarten, A., et al., 2016a. Soil mapping, classification, and pedologic modelling: history and future directions. Geoderma 264, 256–274.

Brevik, E.C., Fenton, T.E., Homburg, J.A., 2016c. Historical highlights in American soil science—prehistory to the 1970s. Catena 146, 111–127.

Brevik, E.C., Fenton, T.E., Jaynes, D.B., 2003. Evaluation of the accuracy of a central Iowa soil survey and implications for precision soil management. Precis. Agric. 4, 323–334.

Brevik, E.C., Hartemink, A.E., 2010. Early soil knowledge and the birth and development of soil science. Catena 83, 23–33.

Brevik, E.C., Hartemink, A.E., 2013. Soil maps of the United States of America. Soil Sci. Soc. Am. J. 77, 1117–1132.

Brevik, E.C., Homburg, J.A., Miller, B.A., Fenton, T.E., Doolittle, J.A., Indorante, S.J., 2016b. Selected highlights in American soil science history from the 1980s to the mid-2010s. Catena 146, 128–146.

Brevik, E.C., Homburg, J.A., Sandor, J.A., Soils, climate, and ancient civilizations. In: William Horwath and Yakov Kuzyakov (Eds.), Changing Soil Processes and Ecosystem Properties in the Anthropocene. Developments in Soil Science Series. Elsevier, Amsterdam, in press. https://www.elsevier.com/books/climate-change-impacts-on-soil-processes-and-ecosystem-properties/horwath/978-0-444-63865-6.

Brevik, E.C., Miller, B.A., 2015. The use of soil surveys to aid in geologic mapping with an emphasis on the eastern and Midwestern United States. Soil Horizons, 56. http://dx.doi.org/10.2136/sh15-01-0001.

Brouwers, J.H.A.M., 1993. Rural people's response to soil fertility decline. The Adja case (Benin). Wageningen Agricultural University Papers, 93–94, Wageningen Agricultural University.

Brungard, C.W., Boettinger, J.L., Duniway, M.C., Wills, S.A., Edwards Jr, T.C., 2015. Machine learning for predicting soil classes in three semi-arid landscapes. Geoderma 239-240, 68–93.

Buchanan, S., Triantafilis, J., Odeh, I.O.A., Subansinghe, R., 2012. Digital soil mapping of compositional particle size-fractions using proximal and remotely sensed anciliary data. Geophysics 77, 201–211.

Burgi, M., Gimmi, U., Stuber, M., 2013. Assessing traditional knowledge on forest uses to understand forest ecosystem dynamics. Forest Ecol. Manage. 289, 115–122.

Capra, G.F., Ganga, A., Buondonno, A., Grilli, E., Gaviano, C., Vacca, S., 2015. Ethnopedology in the study of toponyms connected to the indigenous knowledge of soil resource. Plos One 10, e0120240.

Capra, G.F., Ganga, A., Filzmoser, P., Gaviano, C., Vacca, S., 2016. Combining places names and scientific knowledge on soil resources through an integrated ethnopedological approach. Catena 142, 89–101.

Carr, R., Zhang, C., Moles, N., Harder, M., 2008. Identification and mapping of heavy metals pollution in soils of a sports ground in Galway City, Ireland, using a portable XRF analyser and GIS. Environ. Geochem. Health 30, 45–52.

Castrignano, A., Wong, M.T.F., Stelluti, M., De Benedetto, D., Sollitto, D., 2012. Use of EMI, gama-ray emission and GPS height as multisensory data for soil characterization. Geoderma 175-176, 78–89.

Chen, T., Chang, Q., Liu, J., Clevers, J.G.P.W., Kooistra, L., 2016. Identification of heavy metal sources and improvement in spatial mapping based on soil spectra information: a case study in northwest of China. Sci. Total Environ. 565, 155–164.

Chizana, C.T., Mapfumo, P., Albrecht, A., Van Wijk, M., Giller, K., 2007. Smallholder famer's perception and land degradation and soil erosion in Zimbabwe. Afr. Crop Sci. Conf. Proc. 8, 1485–1490.

Cochard, R., Dar, M.E.U.I., 2014. Mountain farmers' livelihoods and perceptions of forest resource degradation at Machiara National Park, Pakistan-administered Kashmir. Environ. Dev. 10, 84–103.

Colen, L., Turkelboom, F., Van Seenwinkel, S., Al Ahmed, K., Deckers, J., Poesen, J., 2016. How the soil moves upward in the olive orchards of NW Syria: sustainable analysis of local innovation. Land Degrad. Dev. 27, 416–426.

Comas, X., Terry, N., Slater, L., Warren, M., Kolka, R., Kristiyono, A., et al., 2015. Imaging tropical peatlands in Indonesia using ground-penetrating radar (GPR) and electrical resistivity imaging (ERI): implications for carbon stock estimates and peat soil characterization. Biogeosciences 12, 2995–3007.

Concepcion, R.N., Batjes, N.H., 1997. A farmers guided classification system for the Phillippines: a case study for barangays Triala and Santa Rosa Ecija, Central Luzon. Report 97/03, Bureau of Soil and Water Management (BSWM), Manila and ISRIC, Wageningen, The Netherlands.

Coomes, O.T., Miltner, B.C., 2016. Indigenous charcoal and biochar production: potential for soil improvement under shifting cultivation systems. Land Degrad. Dev. http://dx.doi.org/10.1002/ldr.2500.

Critchley, W.R.S., Reij, C., Willcocks, T.J., 1994. Indigenous soil and water conservation: a review of the state of knowledge and prospects for building on traditions. Land Degrad. Dev. 5, 293–314.

Dai, F., Zhou, Q., Lv, Z., Wang, X., Liu, G., 2014. Spatial prediction of soil organic matter content integrating artificial neural network and ordinary kriging in Tibetean Plateau. Ecol. Indic. 45, 184–194.

Davies, G.M., Pollard, L., Mwenda, M.D., 2010. Perceptions of land-degradation, forest restoration and fire management. Land Degrad. Dev. 21, 546–556.

Dawoe, E.K., Quashie-Sam, J., Isaac, M.E., Oppong, S.K., 2012. Exploring farmers' local knowledge and perceptions of soil fertility and management in the Ashanti Region of Ghana. Geoderma 179–180, 96–103.

De Benedetto, D., Castrigiano, A., Sollitto, D., Modugno, F., Buttafuoco, G., lo Papa, G., 2012. Integrating geophysical and geostatistical analysis techniques to map spatial variation of clay. Geoderma 171-172, 53–63.

De Benedetto, D., Castrinano, A., Rinaldi, M., Ruggieri, S., Santoro, F., Figorito, B., et al., 2013. An approach for delineating homogeneous zones by using multy-sensor data. Geoderma 199, 117–127.

Dialla, B.E., 1993. The Mossi indigenous soil classification in Burkina Faso. Indigenous Knowledge Dev. Monit. 1, 17–18.

Doolittle, J.A., Brevik, E.C., 2014. The use of electromagnetic induction techniques in soil studies. Geoderma 223-225, 33–45.

Emadi, M., Baghernejad, M., 2014. Comparison of spatial interpolation techniques for mapping pH and salinity in agricultural coastal areas, Northern Iran. Arch. Agron. Soil Sci. 60, 1315–1327.

Erdogan, S., 2009. Explorative spatial analysis of traffic accident statistics and road mortality among provinces in Turkey. J. Safety Res. 40, 341–351.

Ericksen, P.J., Ardon, M., 2003. Similarities and differences between farmer and scientist views on soil quality issues in central Honduras. Geoderma 111, 233–248.

Escribano, P., Schmidt, T., Chabrillat, S., Rodriguez-Caballero, E., Garcia, M., 2017. Optical remote sensing for soil mapping and monitoring. In: Pereira, P., Brevik, E., Munoz-Rojas, M., Miller, B. (Eds.), Soil Mapping and Process Modelling for Sustainable Land Use Management. Elsevier, pp. 91–128.

Estrada-Medina, H., Bautista, F., Jimenez-Osorio, J.J.J., Gonzalez-Iturbe, J.A., Cordero, W.J.A., 2013. Maya and WRB soil classification in Yucatan, Mexico: differences and similarities. ISRN Soil Sci. Article ID 634260, http://dx.doi.org/10.1155/2013/634260.

Falkowski, T.B., Diemont, S.A.W., Chankin, A., Douterlungne, D., 2016. Lacandon Maya traditional ecological knowledge and rainforest restoration: soil fertility beneath six agroforestry system trees. Ecol. Eng. 92, 210–217.

Fernandez-Calvino, D., Garrido-Rodriguez, B., Lopez-Periago, J.E., Paradelo, M., Arias-Estevez, M., 2013. Spatial distribution of copper fractions in a vineyard soil. Land Degrad. Dev. 24, 556–563.

Fleming, K.L., Heermann, D.F., Westfall, D.G., 2004. Evaluating soil color with farmer input and apparent soil electrical conductivity for management zone delineation. Agron. J 96, 1581–1587.

Fu, W.J., Jiang, P.K., Zhou, G.M., Zhao, K.L., 2014. Using Moran's *I* and GIS to study the spatial pattern of forest litter carbon density in a subtropical region of southeastern China. Biogeosciences 11, 2401–2409.

Getis, A., Ord, J.K., 1992. The analysis of spatial association by use of distance statistics. Geogr. Anal. 24, 189–206.

Giger, M., Liniger, H., Sauter, C., Schwilch, G., 2016. Economic benefits and costs of sustainable land management technologies: an analysis of WOCAT'S Global Data. Land Degrad. Dev. http://dx.doi.org/10.1002/ldr.2429.

Gilabert, M.A., Gonzalez-Piqueras, J., Garcia-Haro, F.J., Melia, J., 2002. A generalized soil-adjusted vegetation index. Remote Sens. Environ. 82, 303–310.

Gisladottir, G., Stocking, M., 2005. Land degradation control and its global environmental benefits. Land Degrad. Dev. 16, 99–112.

Gomez, C., Viscarra-Rossell, R.A., McBratney, A.B., 2008. Soil organic carbon prediction by hyperspectral remote sensing and field vis-NIR spectroscopy: an Australian study case. Geoderma 146, 403–411.

Gooley, L., Huang, J., Page, D., Triantafilis, J., 2014. Digital soil mapping of available water content using proximal and remotely sensed data. Soil Use Manage 30, 139–151.

Goovaerts, P., 1999. Geostatistics in soil science: state-of-the-art and perspectives. Geoderma 89, 1–45.

Gray, L.C., Morant, P., 2003. Reconciling indigenous knowledge with scientific assessment of soil fertility changes in Southwestern Burkina Faso. Geoderma 111, 425–437.

Grobben, P., 1992. Using farmers knowledge about soil classification. A case study of farming system. AT Source: Q. Dev. 20, 6–9.

Grum, B., Assefa, D., Hessel, R., Wolderegay, K., Kessler, A., Ritsema, C., et al., 2016. Effect of *in situ* water harvesting techniques on soil and nutrient losses in semi-arid Northern Ethiopia. Land Degrad. Dev. http://dx.doi.org/10.1002/ldr.2603.

Gumiere, S.J., Lafond, J.A., Hallema, D.W., Periard, Y., Caron, J., Gallichand, J., 2014. Mapping soil hydraulic conductivity and matrix potential for water management of a cranberry: characterization and spatial interpolation methods. Byosyst. Eng. 128, 29–40.

Halbrendt, J., Gray, S.A., Crow, S., Radovich, T., Kimura, A.H., Tamang, B.B., 2014. Differences in farmer and expert beliefs and perceived impacts of conservation agriculture. Glob. Environ. Chang. 28, 50–62.

Heilig, J., Kempenich, J., Doolittle, J., Brevik, E.C., Ulmer, M., 2011. Evaluation of electromagnetic induction to characterize and map sodium-affected soils in the Northern Great Plains. Soil Surv. Horizons, 77–88.

Hellin, J., Haigh, M.J., 2002. Better land husbandry in Honduras: towards the new paradigm in conserving soil, water and productivity. Land Degrad. Dev. 13, 233–250.

Hengl, T., Toomanian, N., Reuter, H.I., Malakouti, M., 2007. Methods to interpolate soil categorical variables from profile observations: lessons from Iran. Geoderma 140, 417–427.

Hillyer, A.E.M., McDonagh, J.F., Verlinden, A., 2006. Land-use and legumes in northern Namibia—the value of a local classification system. Agric. Ecosyst. Environ. 117, 251–265.

Hoffmann, U., Hoffmann, T., Jurasinski, G., Glatzel, S., Kuhn, N.J., 2014. Assessing the spatial variability of soil organic carbon stocks in an alpine setting (Grindelwald, Swiss Alps). Geoderma 232–234, 270–283.

Hu, G., Liu, H., Yin, Y., Song, Z., 2016. The role of legumes in plant community succession of degraded grasslands. Land Degrad. Dev. 27, 366–372.

Huang, J., Scudiero, E., Choo, H., Corwin, D.L., Triantafilis, J., 2016. Mapping soil moisture across an inrrigated field using electromagnetic conductivity imaging. Agric. Water Manage. 163, 285–294.

Huisman, J.A., Snepvangers, J.J.J.C., Bouten, W., Heuvelink, G.B.M., 2002. Mapping spatial variation in surface water content: comparison of ground-penetrating radar and time domain reflectometry. J. Hydrol. 269, 194–207.

Huo, X.N., Zhang, W.W., Sun, D., Li, H., Zhou, L.D., Bao-Guo, L., 2011. Spatial pattern analysis of heavy metals in Beijing agricultural soils based on Spatial Autocorrelation Statistics. Int. J. Environ. Res. Public Health 8, 2074–2089.

Juilleret, J., Dondeyne, S., Vancampenhout, K., Deckers, J., Hissler, C., 2016. Mind the gap: a classification system for integrating the subsolum into soil surveys. Geoderma 264, 332–339.

Jyoti Nath, A., Lal, R., Kumar Das, A., 2015a. Ethnopedology and soil quality of bamboo (*Bambusa* Sp.). Sci. Total Environ. 521-522, 372–379.

Jyoti Nath, A., Lal, R., Kumar Das, A., 2015b. Ethnopedology and soil properties in Bamboo (Bambusa sp.) based on agroforestry system in North East India. Catena 135, 92–99.

Kaiser, A., Neugirg, F., Haas, F., Schmidt, J., Becht, M., Schindewolf, M., 2015. Determination of hydrological roughness by means of close range remote sensing. Soil 1, 613–620.

Kambire, F., Bielders, C.L., Kestemont, M.P., 2016. Optimizing indigenous soil fertility assessments. A case study in cotton-based systems in Burkina-Faso. Land Degrad. Dev. http://dx.doi.org/10.1002/ldr.2603.

Kangalawe, R.Y.M., Noe, C., Tungaraza, F.S.K., Naimani, G., Mlele, M., 2014. Understanding of traditional knowledge and Indegenous institutions on sustainable land management in Kilimanjaro Region, Tanzania. Open J. Soil Sci. 4, 469–493.

Kassa, H., Dondeyne, S., Poesen, J., Frankl, A., Nyssen, J., 2016. Transition from Forest-based to cereal-based agricultural systems: a review of the drivers of land use and degradation in Southwest of Ethiopia. Land Degrad. Dev. http://dx.doi.org/10.1002/ldr.2575.

Kauffman, S., Valencia Ramos, M. 1998. A farmer-assisted soil survey and land evaluation in the Peruvian Andes: a case study for the San Marcos-Cajamarca and Quilcas-Huancayo ILEIA project areas. Report 98/03, Soil Department of the Universidad Nacional de la Molina, Lima, and ISRIC, Wageningen.

Kerven, C., Dolva, H., Renna, R., 1995. Indigenous soil classification systems in Northern Zambia. In: Warren, D.M., Slikkerveer, L.J., Brokensha, D. (Eds.), The Cultural Dimension of Development. Indigenous Knowledge Systems. Intermediate Technology Publications, London, pp. 82–87.

Khaledian, Y., Kiani, F., Ebrahimi, S., Brevik, E.C., Aitkenhead-Peterson, J., 2016. Assessment and monitoring of soil degradation during land use change using multivariate analysis. Land Degrad. Dev. http://dx.doi.org/10.1002/ldr.2541.

Kim, G., Barros, A.P., 2002. Downscaling of remotely sensed soil moisture with a modified fractal interpolation method using contracting mapping and ancillary data. Remote Sens. Environ. 83, 400–413.

Kodaira, M., Shibusawa, S., 2013. Using a mobile real-time soil visible-near infrared sensor for high resolution soil property mapping. Geoderma 199, 64–79.

Krasilnikov, P.V., Tabor, J.A., 2003. Perspectives on utilitarian ethnopedology. Geoderma 111, 197–215.

Kumar Shit, P., Sankar Bhunia, G., Maiti, R., 2015. Farmers' perceptions of soil erosion and management strategies in South Bengal, India. Eur. J. Geogr. 6, 85–100.

Laamrani, A., Valeria, O., Cheng, L.Z., Bergeron, Y., Camerlynck, C., 2013. The use of ground penetrating radar for remote sensing the organic layer—mineral soil interface in paludified boreal forests. Can. J. Remote Sens. 39, 74–88.

Lancriet, S., Derudder, B., Naudts, J., Bauer, H., Deckers, J., Haile, M., et al., 2015. A political ecology perspective of land degradation in the North Ethiopian highlands. Land Degrad. Dev. 26, 521–530.

Li, J., Heap, A.D., 2014. Spatial interpolation methods applied in the environmental sciences: a review. Environ. Model. Softw. 53, 173–189.

Lopez-Granados, F., Jurado-Exposito, M., Pena-Barragan, J.M., Garcia-Torres, L., 2005. Using geostatistical and remote sensing approaches for mapping soil properties. Eur. J. Agron. 23, 279–289.

Lu, A., Wang, J., Qin, X., Wang, K., Han, P., Zhang, S., 2012. Multivariate and geostatistical analysis of the spatial distribution and origin of heavy metals in agricultural soils in Shunyi, Beijing, China. Sci. Total Environ. 425, 66–74.

Mairura, F.S., Mugendi, D.N., Mwanje, J.I., Ramisch, J.J., Mbugua, P.K., Chianu, J.N., 2008. Scientific evaluation of smallholder land use knowledge in central Kenya. Land Degrad. Dev. 19, 77–90.

Maiura, F.S., Mugendi, D.N., Mwanje, J.I., Ramisch, J.J., Mbugua, P.K., Chianu, J.N., 2007. Integrating scientific and farmer's evaluation of soil quality indicators in Central Kenya. Geoderma 139, 134–143.

Mamindy-Pajany, Y., Sayent, S., Mosselmans, F.W., Guillon, E., 2014. Cooper, nickel and zinc speciation in a biosolid-amended soil: pH adsorption edge, μ-XRF and μ-XANES investigations. Environ. Sci. Technol. 48, 7237–7244.

Marques, M.J., Bienes, R., Cuadrado, J., Ruiz-Colmenero, M., Barbero-Sierra, C., Velasco, A., 2015. Analysing perceptions attitudes and responses of winegrowers about sustainable land management in Central Spain. Land Degrad. Dev. 26, 458–467.

McBratney, A.B., Odeh, I.O.A., Bishop, T.F.A., Dunbar, M.S., Shatar, T.M., 2000. An overview of pedometric techniques for use in soil survey. Geoderma 97, 293–327.

Miller, B.A., 2017. GIS and spatial statistics applied for soil mapping. In: Pereira, P., Brevik, E.C., Munoz-Rojas, M., Miller, B.A. (Eds.), Soil Mapping and Process Modeling for Sustainable Land Use Management. Elsevier, pp. 129–152.

Miller, B.A., Koszinski, S., Wehrhan, M., Sommer, M., 2015. Comparison of spatial association approaches for landscape mapping of soil organic carbon stocks. Soil 1, 217–233.

Minet, J., Bogaert, P., Vanclooster, M., Lambot, S., 2012. Validation of ground penetratinf radar full-wave form for field scale soil moisture mapping. J. Hydrol. 424-425, 112–123.

Mulder, V.L., de Bruin, S., Schaepman, M.E., Mayr, T.R., 2011. The use of remote sensing in soil and terrain mapping—a review. Geoderma 162, 1–19.

Muñoz-Rojas, M., Erickson, T., Dixon, K., Merritt, D., 2016. Soil quality indicators to assess functionality of restored soils in degraded semi-arid ecosystems. Restoration Ecol. http://dx.doi.org/10.1111/rec.12368.

Munoz-Rojas, M., Pereira, P., Brevik, E., Cerdo, A., Jordan, A., 2017. Soil mapping and processes models for sustainable land management applied to modern challenges. In: Pereira, P., Brevik, E.C., Munoz-Rojas, M., Miller, B.A. (Eds.), Soil Mapping and Process Modeling for Sustainable Land Use Management. Elsevier, pp. 153–194.

Murage, E.W., Karanja, N.K., Smithson, P.C., Woomer, P.L., 2000. Diagnostic indicators of soil quality in productive and non-productive smallholders' fields of Kenya's central highlands. Agric. Ecosyst. Environ. 79, 1–8.

Niemeijer, D., Mazzucato, V., 2003. Moving beyond indigenous soil taxonomies: local theories of soils for sustainable development. Geoderma 111, 403–424.

Nocita, M., Stevens, A., Noon, C., Weseamel, B., 2013. Prediction of soil organic carbon for different leves of soil moisture using Vis-NIR spectroscopy. Geoderma 199, 37–42.

Norton, J.B., Pawluk, R.R., Sandor, J.A., 1998. Observation and experience linking science and indigenous knowledge. J. Arid. Environ. 39, 331–340.

Nyssen, J., Frankl, A., Zenebe, A., Deckers, J., Poesen, J., 2015a. Land management in the Northern Ethiopian highlands: local and global perspective; past, present and future. Land Degrad. Dev. 26, 759–764.

Nyssen, J., Frankl, A., Zenebe, A., Poesen, J., Deckers, J., 2015b. Environmental conservation for food production and sustainable livelihood in Tropical Africa. Land Degrad. Dev. http://dx.doi.org/10.1002/ldr.2379.

Oberthur, T., Goovaerts, P., Dobermann, A., 1999. Mapping soil texture classes using field texturing, particle size distribution and local knowledge by both conventional and geostatistical methods. Eur. J. Soil Sci 50, 457–479.

Oliver, Y.M., Robertson, M.J., Wong, M.T.F., 2010. Integrating farmer knowledge, precision agriculture tools, and crop simulation modelling to evaluate management options for poor-performing patches in cropping fields. Eur. J. Agron. 32, 40–50.

Osbahar, H., Allan, C., 2003. Indigenous knowledge of soil fertility management in southwest Niger. Geoderma 111, 457–479.

Oudwater, N., Martin, A., 2003. Methods and issues in exploring local knowledge of soils. Geoderma 111, 387–401.

Pan, X., You, Y., Roth, K., Guo, L., Wang, X., Yu, Q., 2014. Mapping permafrost features that influence the hydrological processes of a Thermokarst Lake on the Qinghai-Tibet Plateau, China. Permafrost Periglac. Process. 25, 60–68.

Panagos, P., Imeson, A., Meusburger, K., Borrelli, P., Poesen, J., Alewell, C., 2016. Soil conservation in Europe: whish or reality. Land Degrad. Dev. 27, 1547–1551.

Pauli, N., Barrios, E., Conacher, A.J., Oberthur, T., 2012. Farmer knowledge of relationships among soil macrofauna, soil quality and tree species in a smallholder agroforestry system of western Honduras. Geoderma 189-190, 186–198.

Payton, R.W., Barr, J.F.F., Martin, A., Sillitoe, P., Deckers, J.F., Gowing, J.W., et al., 2003. Contrasting approaches to integrating indigenous knowledge about soils and scientific survey in East Africa and Bangladesh. Geoderma 111, 355–386.

Peng, Y., Xiong, X., Adhikari, K., Knadel, M., Geunwald, S., Greve, M.H., 2015. Modelling soil organic carbon at regional scale by combining multispectral images with laboratory spectra. Plos One 10, e0142295.

Pereira, P., Brevik, E., Munoz-Rojas, M., Miller, B., Smetanova, A., Depellegrin, D., et al., 2017. Soil mapping and process modelling for sustainable land management. In: Pereira, P., Brevik, E., Munoz-Rojas, M., Miller, B. (Eds.), Soil Mapping and Process Modelling for Sustainable Land Use Management. Elsevier, pp. 31–64.

Pereira, P., Oliva, M., Baltrenaité, E., 2010. Modelling extreme precipitation in mountain hazard areas. A contribution to landsc. planning and environmental management. J. Environ. Eng. Landsc. 18, 329–342.

Pereira, P., Pranskevicius, M., Bolutiene, V., Jordan, A., Zavala, L., Ubeda, X., et al., 2015. Short term spatio-temporal variability of soil water extractable Al and Zn after low severity grassland fire in Lithuania. Flamma 6, 50–57.

Piikki, K., Soderstrom, M., Stenberg, B., 2013. Sensor data fusion for topsoil clay mapping. Geoderma 199, 106–116.

Poggio, L., Gimona, A., Brewer, M.J., 2013. Regional scale mapping of soil properties and their uncertainity with a large number of satellite-derived covariates. Geoderma 209-210, 1–14.

Price, L.L., 2007. Locating farmer-based knowledge and vested interests in natural resources management: the interface of ethnopedology, land tenure and gender in soil erosion management in Manupali watershed, Philipines. J. Ethnobiol. Ethnomed. 3, 30.

Pulido, J.S., Bocco, G., 2003. The traditonal farming system of Mexican indigenous community: the case of Nuevo San Juan Parangaricutiro, Michoacan, Mexico. Geoderma 111, 249–265.

Reed, M.S., Dougill, A.J., Taylor, M.J., 2007. Integrating local and scientific knowledge for adaptation to land degradation: Kalahari rangeland management options. Land Degrad. Dev. 18, 249–268.

Rodrigues Jr, F.A., Bramley, R.G.V., Gobbett, D.L., 2015. Proximal soil sensing for Precision agriculture:

simultaneous use of electromagnetic induction and gamma radiometrics in contrasting soils. Geoderma 243-244, 183–195.

Rushemuka, N.P., Bizoza, R.A., Mowo, J.G., Bock, L., 2014. Farmers' soil knowledge for effective participatory integrated watershed management in Rwanda: towards soil-specific fertility management and farmers' judgment fertilizer use. Agric. Ecosyst. Environ. 183, 145–159.

Ryder, R., 2003. Local soil knowledge and site suitability evaluation in the Dominican Republic. Geoderma 111, 289–305.

Sandor, J.A., WinklerPrins, A.M.G.A., Barrera-Bassols, N., Zinck, J.A., 2006. The heritage of soil knowledge among the world's cultures. In: Warkentin, B. (Ed.), Footprints in the Soil: People and Ideas in Soil History. Elsevier, Amsterdam, pp. 43–84.

Schuler, U., Choocharoen, C., Elstner, P., Neef, A., Stahr, K., Zarei, M., et al., 2006. Soil mapping for land-use planning in a karst area of N Thailand with due consideration of local knowledge. J. Plant Nutr. Soil Sci. 169, 444–452.

Shajaat Ali, A.M., 2003. Farmers' knowledge of soils and the sustainability of agriculture in a saline water ecosystem in southwestern Bangladesh. Geoderma 111, 333–353.

Sharma, A., Weindorf, D.C., Wang, D., Chakraborty, S., 2016. Characterizing soils via portable X-ray fluorescence spectrometer: 4. Cation exchange capacity (CEC). Geoderma 239-240, 130–134.

Shi, W.J., Yue, T.X., Du, Z.P., Li, X.W., 2016. Surface modelling soil antibiotics. Sci. Total Environ. 543, 609–619.

Souza, E.G., Bazzi, C.L., Khosla, R., Uribe-Opazo, M.A., Reich, R.M., 2016. Interpolation type and data computation of crop yield is important for precision crop production. J. Plant Nutr. Soil Sci., 531–538.

Stockmann, U., Malone, B.P., McBratney, A.B., Minasny, B., 2015. Landscape-scale exploratory radiometric mapping using proximal soil sensing. Geoderma 239, 115–129.

Tabor, J.A., 1992. Ethnopedological surveys. Soil surveys that incorporate local systems of land classification. Soil Surv. Horizons 33, 1–5.

Talawar, S., 1991. Farm science of farmers in arid agriculture. Ph.D. Dissertation, Division of Agricultural Extension, Indian Agricultural Research Institute, New Delhi, India.

Talawar, S., Rhoades, R.E., 1998. Scientific and local classification and management of soils. Agric. Hum. Values 15, 3–14.

Tarrason, D., Ravera, F., Reed, M.S., Dougill, A.J., Gonzalez, L., 2016. Land degradation assessment through ecosystem services lens: integrating knowledge and methods appraisal in pastoral semi-arid systems. J. Arid Environ. 127, 205–213.

Tesfahunegn, G.B., Tamene, L., Vlek, P.L.G., 2016. Assessing farmers' knowledge of weed species, crop type and soil management practices in relation to soil quality status in Mai-Negus catchment, Northern Ethiopia. Land Degrad. Dev. 27, 120–133.

Tesfaye, A., Negatu, W., Brouwer, R., Van Der Zaag, P., 2014. Understanding soil conservation decision of farmers in the Gedeb watershed, Ethiopia. Land Degrad. Dev. 25, 71–79.

Teshome, A., de Graaff, J., Ritsema, C., Kassie, M., 2016. Farmers' perceptions about the influence of land quality, land fragmentation and tenure systems on sustainable land management in the North Western Ethiopian Highlands. Land Degrad. Dev. 27, 884–898.

Toth, G., Jones, A., Montanarella, L., 2013. The Lucas topsoil database and derived information on the regional variability of cropland topsoil properties in European Union. Environ. Monit. Assess. 185, 7409–7425.

Tran, A.P., Vanclooster, M., Zupanski, M., Lambot, S., 2014. Join estimation of moisture profile and hydraulic parameters by ground-penetration radar data assimilation with maximum likelihood ensemble filter. Water Resour. Res. 50, 3131–3146.

Triantafilis, J., Kerridge, B., Buchanan, S.M., 2009. Digital soil-class mapping from proximal and remotely sensed data at the field level. Agron. J 101, 841–853.

Trung, N.D., Verdoodt, A., Dusar, M., Tan Van, T., Van Ranst, E., 2008. Evaluating ethnopedological knowledge systems for classifying soil quality. A case study in Bo Hamlet with Muong People of Northern Vietnam. Geogr. Res. 46, 27–38.

Truong, L.T., Somenahally, V.C., 2011. Using GIS to identify pedestrian-vehicle crash hot spots and unsafe bus stops. J. Public Transport. 14, 99–114.

Venkateswarlu, B., Srinivasarao, C.H., Venkateswarlu, J., 2013. Traditional knowledge for sustainable management of soils. In: Lal, R., Stewart, B.A. (Eds.), Principles of Sustainable Soil Management in Agroecosystems. CRC Press, Boca Raton, pp. 303–336.

Vrieling, 2006. http://onlinelibrary.wiley.com/doi/10.1002/ldr.711/abstract.

Wahlhutter, S., Vogl, C.R., Eberhart, H., 2016. Soil as a key criteria in the construction of farmers' identities: the example of farming in Austrain province of Burgenland. Geoderma 269, 39–53.

Weindorf, D.C., Paulette, L., Man, T., 2013. In-situ assessment of metal contamination via portable X-ray fluorescence spectroscopy: Zlatna, Romenia. Environ. Pollut. 182, 92–100.

Wezel, A., Haigis, J., 2000. Farmer's perception of vegetation changes in semi-arid Niger. Land Degrad. Dev. 11, 523–534.

WinklerPrins, A.M.G.A., Sandor, J.A., 2003. Local soil knowledge: insights applications and challenges. Geoderma 111, 165–170.

Wolfgramm, B., Shigaeva, J., Dear, C., 2016. The research-action interface in sustainable land management in Kyrgyzstan and Tajikistan: challenges and recommendations. Land Degrad. Dev 26, 480–490.

Wu, Y.H., Hung, M.C., Patton, J., 2013. Assessment and visualization of spatial interpolation of soil pH values and farmland. Precision Agric. 14, 565–585.

Zhang, S., Kong, W., Huang, Y., Shen, C., Ye, H., 2014. Spatial prediction of topsoil texture in mountain-plan transition zone using univariate and multivariate methods based on symmetry logratio transformation. Intell. Autom. Soft. Co. 20, 115–129.

Zhang, S., Yan, L., Huang, J., Mu, L., Huang, Y., Zhang, X., et al., 2016. Spatial heterogeneity of soil C:N ratio in a Mollisol watershed on Northeast of China. Land Degrad. Dev. 27, 295–304.

Zhang, W., Wang, K., Chen, H., He, X., Zhang, J., 2012. Ancillary information improves kriging on soil organic carbon data for typical karst peak cluster depression landscape. J. Sci. Food Agric. 92, 1094–1102.

Zhu, Y., Weindorf, D.C., Zhang, W., 2011. Characterizing soils using a portable X-ray fluorescence spectrometer: 1. Soil texture. Geoderma 167-168, 167–177.

Zitong, G., 1994. Chinese soil taxonomy classification (first proposal). Institute of Soil Science, Academia Sinica, Nanjing. 93.

Zoumides, C., Bruggeman, A., Giannakis, E., Camera, C., Djuma, H., Eliades, M., et al., 2016. Community-based rehabilitation of mountain terraces in Cyprus. http://onlinelibrary.wiley.com/doi/10.1002/ldr.2586/abstract.

Zuniga, M.C., Feijoo, M.A., Quintero, H., Aldana, N.J., Carvajal, A.F., 2013. Farmers' perceptions of earthworms and their role in the soil. Appl. Soil Ecol. 69, 61–68.

Further Reading

Asamiah, R.D., Senayah, J.K., Adjei-Gyapong, T., Spaargaren, O.C., 1997. Ethno-Pedology Surveys in Semi-arid Savanna Zone of Northern Ghana. An ILEIA Initiated Report. Soil Research Institute, Kwadaso-Kumasi and International Soil Reference Information Centre (ISRIC), Wageningen, The Netherlands.

Barrera-Bassols, N., Zinck, A., 2002. Ethnopedologial research: a worldwide review. 17th WCSS, 14–21 August 2002, Thailand.

Blanes, S., 2014. Comparative performance of classification algorithms for the development of models of spatial distribution of landscape structures. Geoderma 219–220, 136–144.

PART II

INSTRUMENTATION AND SENSORS USED FOR GAINING INFORMATION ON SOIL RELATED PARAMETERS

CHAPTER

4

Optical Remote Sensing for Soil Mapping and Monitoring

Paula Escribano[1], Thomas Schmid[2], Sabine Chabrillat[3], Emilio Rodríguez-Caballero[4,5] and Mónica García[6,7]

[1]University of Almeria, Almería, Spain [2]Center for Energy, Environment and Technology Research, Madrid, Spain [3]GFZ German Research Center for Geosciences, Potsdam, Germany [4]Max Planck Institute for Chemistry, Mainz, Germany [5]University of Almeria, Almería, Spain [6]Denmark Technical University (DTU), Kongens Lyngby, Denmark [7]Columbia University, New York City, NY, United States

INTRODUCTION

Soils are considered a nonrenewable resource and soil preservation or restoration is a necessary global strategy to mitigate the effects of land degradation and soil erosion on the functions and services that they provide. It is difficult to name just one cause of soil degradation, rather, it is due to a combination of natural (biophysical or climatic) and human-induced (social-economic or political) factors (Stolte et al., 2016). All these factors interact at different spatial scales and therefore the issue of soil degradation and conservation it is difficult to tackle and an integrative and multiscale approach is needed. There exists a wide range of soil databases around the world (Mulder, 2011). Nevertheless there are huge discrepancies in data availability between regions, scales at which soil databases are retrieved and even the methods and protocols used for deriving such maps (Ballabio et al., 2016; Nachtergaele, 1999; Brevik et al., 2017). Different stakeholders, from users to policy makers, require easy access to soil data at different spatial scales for making rational use of natural resources and to prevent soil degradation (Manchanda et al., 2002; Brevik et al., 2016; Ballabio et al., 2016). Up-to-date soil properties maps (e.g., information on sand, silt, clay, or organic matter) are needed for modeling processes like soil erosion, hydrology, or carbon fluxes between soil and atmosphere (Muñoz-Rojas et al., 2017; Pereira et al., 2017). Countries like Canada the United States, Australia, or those in Europe typically have a wider range of soil maps at different spatial scales, while developing countries have fewer soil databases and mainly at small or medium

Soil Mapping and Process Modeling for Sustainable Land Use Management.
DOI: http://dx.doi.org/10.1016/B978-0-12-805200-6.00004-9

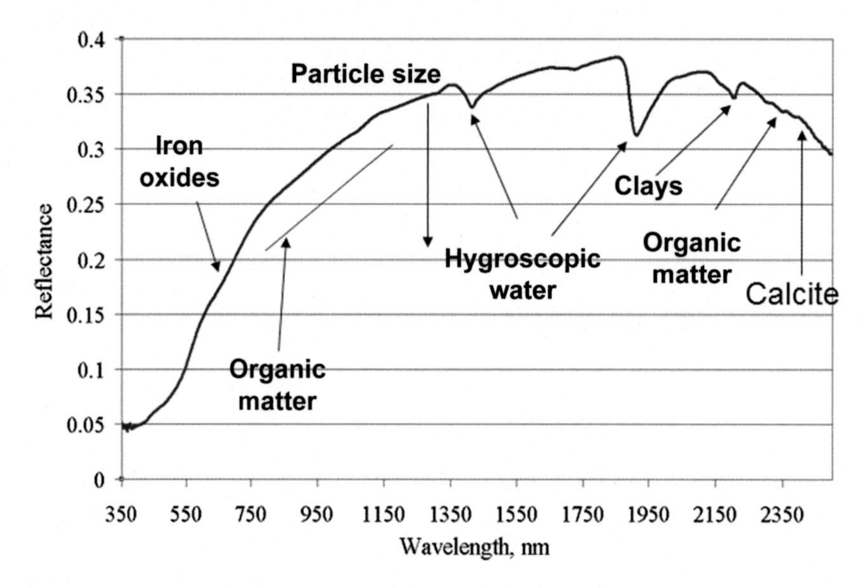

FIGURE 4.1 A soil spectrum (Haploxeralf) that represents the major chromophores in soils. *Source: After Ben-Dor, E., Taylor, G.R., Hill, J., Demattê, J.A.M., Whiting, M.L., Chabrillat, S., et al., 2008. Imaging spectrometry for soil applications. Adv. Agron. J. 97, 321–392.*

scales. Protocols and standards are also not precisely settled and important differences in soil maps between countries and even between regions within a country are obvious (Brevik et al., 2017).

Most of the conventional methods to map and to monitor soil properties give accurate information but can be expensive and time-consuming (Brevik et al., 2003; Viscarra Rossel et al., 2011, 2016). Sometimes the high spatial variability of some soil properties like organic content (Muñoz-Rojas et al., 2012) makes it unfeasible to produce regional or global maps based on traditional methods (e.g., Stevens et al., 2008). Remote sensing from satellite and proximal sensors can complement and, in some cases, even replace the use of traditional field and laboratory analyses. Most of these sensors are more time and cost efficient than traditional methods (Vaudour et al., 2016).

Optical remote sensing can only "view" the thin upper layer of soil, the first few centimeters, which limits its use for the study of the entire soil body. Subsurface soil properties can be inferred by surface soil properties but accurate knowledge of the entire soil body must come from measurements of traditional soil sampling or by the use of other proximal devices such ground penetrating radar or X-ray fluorescence meter (a good review of the synergic use of optical remote sensing and other proximal devices can be found in Hartemink and Minasny, 2014). Nevertheless, as pointed out by Fajardo et al. (2016), soil spectroscopy is becoming a promising tool for studies of the entire soil pedon alone or in combination with other methods.

Soil spectral reflectance is determined by both physical and chemical characteristics of soils (Ben-Dor et al., 2003; Shepherd and Walsh, 2002). Soil spectroscopy in the VISNIR (visible and near infrared, 400–1100 nm) and the SWIR (shortwave infrared, 1100–2500 nm) wavelengths has already become an established technique in the laboratory to detect and predict soil properties thanks to the presence of soil chromophores that can be used as predictors (Fig. 4.1). In these spectral regions, inorganic and organic components such as clay minerals, soil organic matter (SOM), iron oxides, or calcium carbonate ($CaCO_3$) interact with the electromagnetic radiation and produce characteristic absorption features in soil reflectance spectra that can be used to identify soil properties when soils are exposed at the surface and vegetation cover is low (Chabrillat et al., 2002; Nanni and Demattê, 2006; Stevens et al., 2013).

Recent reviews such as Viscarra Rossel et al. (2011) or Soriano-Disla et al. (2014) listed soil properties that could be determined by means of reflectance spectroscopy: soil water content, clay, sand, soil organic carbon (SOC), free iron oxides, salt, cation exchange capacity (CEC), exchangeable Ca and Mg, total N, pH, electrical conductivity (EC), and total concentration of potential pollutant metals/metalloids (As, Cd, Hg, and Pb), and showed that soil spectroscopy has the potential to revolutionize soil survey. More recently, Nocita et al. (2015) presented soil spectroscopy as an alternative to wet chemistry for soil monitoring.

Hyperspectral sensors, with more than a hundred contiguous bands and with high spatial resolution (from cm to a few meters), have a wide range of applications like vegetation cover and state (Xie et al., 2008), mapping soil properties (e.g., Gomez et al., 2012), soil erosion (e.g., Schmid et al., 2016), or soil salinity (e.g., Metternicht and Zinck, 2003) to name a few. The main advantage of using this type of data is that it gives the opportunity not only to identify but also to quantify several soil and vegetation properties. Multispectral data, on the contrary, give information on just few spectral bands but the temporal resolution of these sensors is higher allowing monitoring the temporal dynamics and changes in time, and the data archive dates back to the 1980s in the case of Landsat TM and therefore their use has been widely recognized. For mapping soil properties, their use has been more restricted though some examples of iron oxides identification or soil erosion are found in the literature (Castaldi et al., 2016; Manchanda et al., 2002; Mulder et al., 2011). Most of the current studies with multispectral data are based on vegetation indices like normalized difference vegetation index (NDVI) for a wide variety of processes related with land degradation, land use or land cover changes (Diouf and Lambin, 2001; Horion et al., 2016), desertification (Fensholt and Rasmussen, 2011; Karnieli and Dall'Olmo, 2003) or evapotranspiration (ET) indicators (Garcia et al., 2008), to name a few.

This chapter focuses on the use of optical remote sensing in the wavelength domain of VISNIR and SWIR. It gives an overview of the principal issues concerning the use of these techniques for soil mapping and monitoring, including different spectral (multispectral and hyperspectral data) and spatial scales (laboratory, field, and image). It also analyzes several aspects needed for the use of these types of data, like the information contained in a spectral signature, the main sources of uncertainty or the methods and approaches commonly used for soil mapping. Finally, it gives an overview of the common uses of soil spectroscopy showing in greater detail the state of the art of its use for mapping soil properties, soil erosion, biocrusts, and soil related processes like evaporation or land degradation. We also show how the synergic use of multispectral and thermal domains can be used for land degradation studies at regional scales as the energy reflected in the optical domain determines how much energy will be emitted in the thermal domain.

FROM LABORATORY TO IMAGE DATA

Several studies can be found in the literature using VISNIR spectroscopy for identification, prediction, or quantification of soil surface properties, but it is necessary to understand the type of environment and the spatial scale at which each of these properties have been

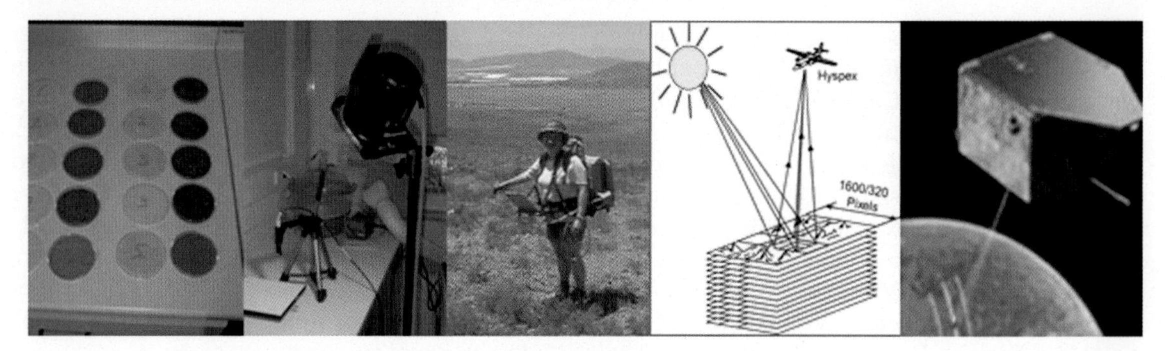

FIGURE 4.2 Soil spectroscopy measurements. Laboratory (left), field measurements, airborne imaging spectroscopy, and spaceborne missions (right) like the upcoming Environmental Mapping and Analysis Program (EnMAP) imaging spectrometer satellite that will be launched in 2019.

successfully quantified. This means whether the data came from laboratory spectroscopy (LS), field spectroscopy (FS), or image spectroscopy (IS) (Fig. 4.2). Several authors have contrasted the usefulness of LS, FS, and IS for different soil properties (Mulder et al., 2011; Lagacherie et al., 2008; Stevens et al., 2008).

Laboratory Spectroscopy

Laboratory spectral measurements are taken with a spectroradiometer under controlled illumination conditions with an artificial source of light. The soil samples must be prepared before measuring but, contrary to traditional laboratory measurements, this only involves drying, grinding, and sieving. Another advantage of this method is that measurements are taken in a few seconds and the spectral signatures can give information on several soil properties at once (Viscarra Rossel et al., 2016).

Several spectral properties have been properly determined by means of this method. For instance Stevens et al. (2008) reported accuracies of SOC content by means of LS close to those of standard methods. Other soil properties successfully measured are $CaCO_3$, extractable Fe, clay, silt, and CEC, less accurate results have been found for sand and pH (e.g., Viscarra Rossel et al., 2016).

It is necessary to develop soil spectral libraries (SSLs) to make the use of soil spectroscopy possible for local, regional, and continental estimations of soil properties by means of LS using standards and protocols to make it possible to share data from different users/regions (Ben-Dor et al., 2015; Kopačková and Ben-Dor, 2016). These databases contain soil spectra and laboratory data that can be used for calibration models for the prediction of soil properties exclusively by means of soil spectra (Ji et al., 2016). The first SSL date back to 1981 (Stoner and Baumgardner, 1981), since then a number of SSL have been developed at different scales. The most complete one to date is the Global VISNIR SSL (Viscarra Rossel et al., 2016) that includes more than 20,000 soil spectra distributed among 92 different countries.

Field Spectroscopy

FS refers to spectral measurements retrieved in the field, under natural conditions, with sensors that can be fixed or mounted on vehicles for on-the-go measurements. The use of on-the-go hyperspectral sensors is commonly used for applications like precision agriculture (Adamchuk et al., 2004). It is important to remark that besides FS, other proximal sensors commonly used for soil mapping include

X-ray fluorescence, ground penetrating radar, and electromagnetic induction (Viscarra Rossel et al., 2011; Brevik et al., 2016). While optical remote sensing is largely restricted to collecting information on conditions at the Earth's surface, proximal sensors such as ground penetrating radar and electromagnetic induction can provide information on conditions at various depths below the surface.

In general FS gives less accurate results than LS, mainly because of the uncontrolled natural conditions in the field, but it allows the capture of information over larger areas and can provide information on spatial variability of the soil properties. In general FS is less expensive and less laborious than traditional field collection of data and it can potentially give timely information. Several soil properties can be mapped with FS with relatively good accuracies, for example Ji et al. (2016) reported good accuracies when measuring simultaneously organic matter content, pH, and total nitrogen. Viscarra Rossel et al. (2016) also reported good accuracies for soil color, soil mineralogy, and clay content with field data. In addition, these data are an important and necessary source for image validation and calibration models because both are retrieved under natural conditions and therefore can be compared.

Soil moisture, roughness (especially important in agricultural areas), and mixing problems (presence of green or dry vegetation, litter, biological soil crusts, or pebbles in the field) are environmental factors that highly affect the spectral signature of a soil. These environmental factors are highly variable in space and time and the mixing effects that they have on a spectral signature are still not well understood. This is why making SSL based on FS including all the natural variability of soil and of the environmental conditions in which these soils can be found in nature is unfeasible. Studies by Viscarra Rossel et al. (2009), Mouazen et al. (2006), and Ji et al. (2015, 2016) are good examples of the effort made to minimize the spectral variability due to these environmental factors and going one step further toward the use of FS for up-to-date data that can be included in soil monitoring.

Imaging Spectroscopy

Soil spectroscopy methods have subsequently been transferred from the laboratory to the field and up to airborne and space-borne observations in order to test the capability to predict and map soil properties at local, regional, and global scales with multiscale and multisensor studies. LS and FS give accurate information on different soil properties but are based on point data, IS on the other hand offers the possibility to have spatially explicit data providing continuous information over large areas.

Nowadays IS is recognized as a robust approach to derive soil maps using operational methodologies that have a great potential for the determination of more quantitative soil products when ground soil data are available. Nevertheless the synergic use of optical IS data combined with thermal or active sensors (i.e., radar and lidar) or with *in situ* information from proximal sensors can highly improve our understanding of soil properties and related surface processes.

There are several sources of uncertainty when using IS data, where the atmosphere is one of the factors that influences the signal retrieved by the sensor (Ben-Dor et al., 2009). Several preprocessing techniques, including atmospheric correction, smoothing, and noise reduction have been developed and it is well recognized that these steps highly influence the final prediction model (Minu et al., 2016). The signal to noise ratio and the radiometric calibration of the sensor are other factors that affect the signal and thus predicted soil variables. Besides all this, environmental factors related to the characteristics of the surface at the moment of the image acquisition plays an important role. Soil moisture has a profound effect on

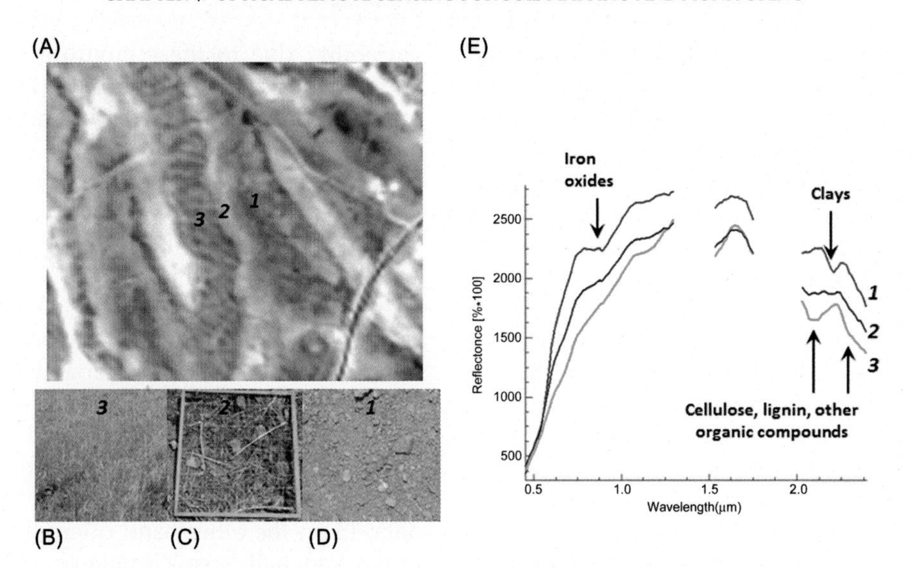

FIGURE 4.3 HyMap image in true color composition (A) showing a transect of an agricultural area with different proportions of dry vegetation on top of the soil surface, darkest areas in the image (3) represent areas fully covered by vegetation (B) yellowish (2) represents areas partially covered by dry vegetation (C) and reddish colors (1) represent eroded areas with bare soils rich in iron oxides (D). The spectral signatures of these covers are represented in (D). Red dark grey in print version: bare soil; blue (black in print versions): bare soil + dry vegetation; green (light grey in print versions): dry vegetation (E).

the spectral signature of a soil decreasing the overall reflectance (Ji et al., 2016); the spatial variability of this factor reduces the prediction accuracy of several soil properties (Stevens et al., 2008). Similarly, soil roughness decreases reflectance due to microshades. The mixing problem also deserves mention, because under natural conditions the presence of several covers at the soil surface (including vegetation, litter, pebbles, and physical or biological soil crusts) is frequent.

Several authors have pointed out that even a small proportion of vegetation cover within a pixel (from 40% to less than 10%) can hamper the accuracy of soil prediction models (Franceschini et al., 2015). This is because when vegetation is present on the soil surface it totally or partially masks the spectral traits of the soils (Fig. 4.3). Fig. 4.3 shows an example of a soil rich in iron oxides in an area located in Almería province (Southeast of Spain). The presence of iron oxides is spectrally distinguished because of an absorption located around 900 nm (Fig. 4.3), the presence of less than 30% dry vegetation on top of the soil surface minimizes this absorption and therefore modeling iron content in this area can lead to an underestimation of iron content if vegetation it is not correctly identified. On the SWIR wavelengths (1100–2500 nm), it is also obvious that vegetation diminishes the absorption feature attributable to clays around 2200 nm. The spectral signature of the pixel containing both dry vegetation and bare soil has spectral traits of both of its constituents.

For soil mapping purposes, pixels containing more than a given percentage of vegetation (typically 30%) should be removed from the analysis. Nevertheless interesting advances have been made in this field unraveling the contribution of soil properties to the total reflectance of a pixel when it is partially covered by vegetation. Ouerghemmi et al. (2016) have been able to accurately estimate clay in

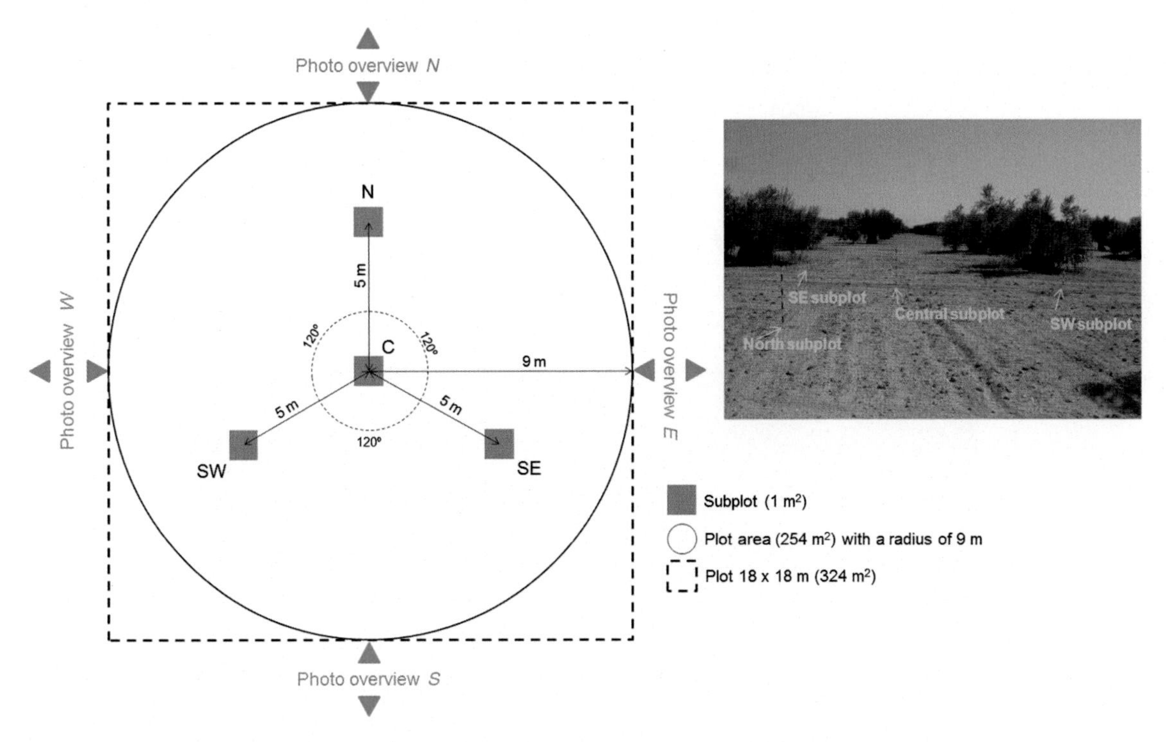

FIGURE 4.4 Field plot area with corresponding subplots. *Source: From Schmid, T., Rodríguez-Rastrero, M., Escribano, P., Palacios-Orueta, A., Ben-Dor, E., Plaza, A., et al., 2016. Characterization of soil erosion indicators using hyperspectral data from a Mediterranean rainfed cultivated región. IEEE J. Select. Top. Appl. Earth Observ. Remote Sens. 9 (2), 845–860.*

partially vegetated areas in France. Similarly, Franceschini et al. (2015) also predicted clay, sand, and CEC with IS for Oxisols (in Brazil) that were partly vegetated using an unmixing approach.

Another factor that must be kept in mind when analyzing IS data is the spatial scale. Upscaling from point to image it is not an easy task but must be performed correctly to assure that information collected on a pixel basis can be compared with that of field/laboratory based data. Geometrical correction of the image must be properly done; nevertheless, even when this is done adequately a pixel cannot be fully described by point measurements. It is common for 3 × 3 pixels to be aggregated and both the spectral and chemical information for validation should be averaged. Schmid et al. (2016) proposed a field validation schema for minimizing the errors that this upscaling might cause (Fig. 4.4). The plots were selected sites with an extension that was adapted to the spatial resolution of the hyperspectral data image and where detailed observations as well as soil samples were obtained (Fig. 4.4). The design had four subplots, each of $1\,m^2$, within an area of $324\,m^2$ that were arranged using a central point C and situating the N, SE, and SW subplots a distance of 5 m with an angle of 120° between the corresponding subplots. This configuration was easily set up using a survey pole, tape measure, and compass.

When comparing soil spectra from laboratory, field and image data of a study site in Camarena, Spain, following the validation schema proposed in Schmid et al. (2016)

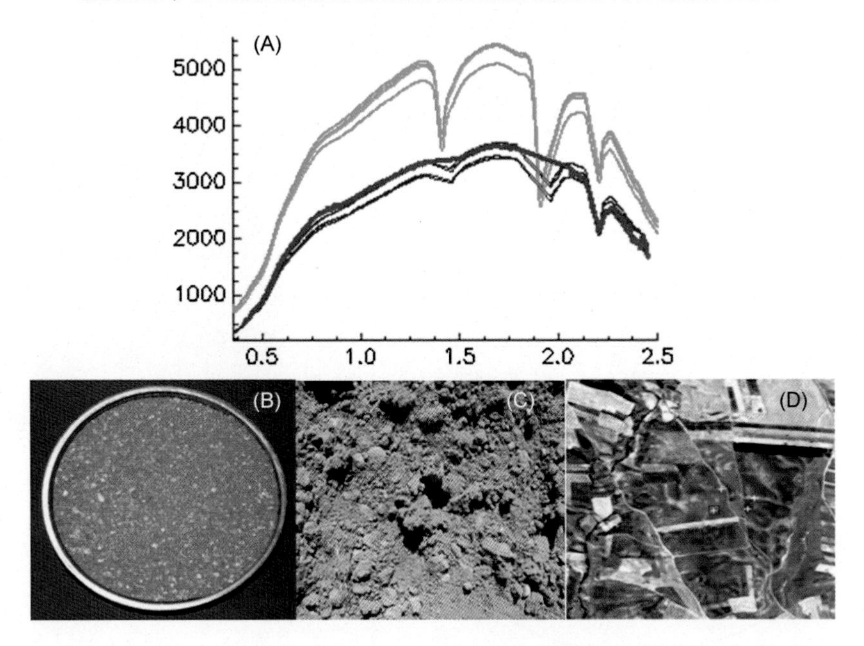

FIGURE 4.5 Spectral signatures of a soil sampled in Camarena, Toledo (Spain), at laboratory (green, light gray in print versions), field (blue, black in print versions) and image (red, dark gray in print versions) spatial scales (A). Pictures taken of the sample in the laboratory (B) and field (C). Hyperspectral image (AISA-Hawk) in true color composition (D); the red (dark gray in print versions) square represents the location of the sample. Laboratory and field spectra were retrieved with an ASD Field Spec pro espectro-radiometer.

(Fig. 4.4) it can be seen that the main spectral traits remain constant but differences in overall reflectance are obvious (Fig. 4.5). These data were collected in the field and imaged in August 2011, when soils in the area were completely dry. Therefore most of the variability in soil reflectance should be caused by differences in roughness between controlled laboratory conditions (after grinding and sieving) and natural conditions. It is known that roughness highly influences soil spectra decreasing reflectance due to microshades and therefore grinding has a profound effect on the spectral signature by increasing reflectance (Stenberg et al., 2010).

The use of IS is actually limited because of the high cost of aircraft flights, which sometimes makes it impossible to use IS data on a regular basis. Nevertheless, the development of new sensors that use less expensive and smaller new platforms (like drones) makes it possible to use these data on a regular basis for small areas. Upcoming spaceborne sensors are currently being built like EnMAP from Germany and HISUI (Japanese Hyperspectral Imager Suite) from Japan, both planned to be launched in 2019. SHALOM (Applicative Land and Ocean Mission), a joint initiative of Italy and Israel, HypXIM (Hyperspectral-X Imager) from France and HypsIRI (Hyperspectral Infra-Red Imager) from the United States are in the design phase (Staenz et al., 2013). The upcoming availability of these high signal-to-noise ratio spaceborne imaging spectroscopy data is expected to provide a major step toward the operational quantitative monitoring of soil surfaces at large scales. Indeed, these instruments could therefore provide global spectroscopic data for mapping quantitative soil properties at low costs and could allow accurate assessment and monitoring of issues such as soil erosion or carbon loss.

SOIL PROPERTIES MAPPING BY MEANS OF IMAGING SPECTROSCOPY

Building upon the approaches successfully developed for laboratory and FS in the VNIR and SWIR region, IS has shown the potential to map and quantify soil surface properties in many studies (Haubrock et al., 2008b; Ben-Dor et al., 2008; Stevens et al., 2010; Gerighausen et al., 2012; Bayer et al., 2012; Gomez et al., 2012). The most successful studies were the applications for soil properties that are directly related to soil chromophores such as iron oxides, clay, carbonates, water content, or SOC because they have a direct quantifiable influence on the optical reflectance signature. Most studies were successful at local and regional scales, when the soils were exposed at the surface, and vegetation cover and moisture content were low. IS allows direct identification of the main mineralogical composition based on reflectance signal analyses and allows quantitative determination of soil properties.

Methodologies for the Retrieval of Soil Properties Based on Imaging Spectroscopy Data

Thorough reviews of existing algorithms for the retrieval of soil properties from IS data have seldom been performed we summarize here the findings of Groom et al. (2010), and Minu et al. (2016). Due to the complexity and nonlinearity that often characterize the relationship between optical remote sensing measurements and the soil properties of interest, the retrieval of soil parameters from IS data is not a straightforward issue. Indeed, many effects have to be considered such as the mixed nature of the electromagnetic signal received by the sensor that integrates all elements in the field-of-view, or the influence of external disturbing factors such as illumination factors, sensor noise, and environmental conditions (heterogeneous soil moisture or vegetation cover). For this reason, there are many established techniques for the retrieval of soil parameters based on soil reflectance spectroscopy that cover a wide range of methodologies and expertise and that are used on a case-by-case basis. No simple best-practice rule could be observed that related methodology used with sensor resolution and key soil parameter of interest. The techniques used can be classified into two groups: physically based and empirically based algorithms.

Physically based algorithms include all approaches using direct signal analysis of the spectral reflectance to predict a soil property. For example, wavelength ratios, spectral indices, continuum analyses, extraction of absorption features, analyses of spectral features (depth, area, etc.), visible and NIR analyses are commonly used physically based methods. These methods have been used to map, e.g., SOC content (Bartholomeus et al., 2008), iron oxide (Richter et al., 2007, 2009), clay content (Lagacherie et al., 2008), soil moisture (Haubrock et al., 2008b), gravel coverage (Crouvi et al., 2006), EC (Ben-Dor et al., 2002), and many others relevant for land degradation applications (Chabrillat, 2006). The most popular methods are the most straightforward ones, associated with spectral indices or analysis of absorption depth such as the continuum-removed absorption depth (CRAD) in the Hyperspectral SOil MApper (HYSOMA) toolbox (see "Example of Soil Properties Mapping Applications at Airborne Scale" section), due to their robustness, transferability, and as they present an easy to use access for nonexpert users of IS. These methods can be used without or with ground data for calibration, for the determination of semiquantitative (relative abundance maps) or fully quantitative (absolute abundance maps) soil maps, respectively.

Empirically based algorithms are more complex in their use and mostly used by the soil spectroscopy community for quantitative prediction of soil properties. They do require ground data for model calibration and validation, and some fine-tuning. These approaches include machine learning methods such as support vector machine regression (SVMR), partial least-square regression (PLSR), artificial neural network (ANN), and multiple linear regression (MLR). PLSR is by far the most successfully used method for soil property determination in quantitative soil spectroscopy. PLSR has been used to predict and map SOC content (Stevens et al., 2010), texture (Selige et al., 2006; Castaldi et al., 2016), Al, Fe, silt (Hively et al., 2011), carbonates (Lagacherie et al., 2010), and more. SVMR has also been used to map SOC (Stevens et al., 2010) and erosion stages (Schmid et al., 2016), ANN to map EC (Farifteh et al., 2007), and MLR to map, e.g., EC, Ca, Mg, Na, and Cl (De Tar et al., 2008).

A few studies have focused on comparing the use of both types of techniques, physically and empirically based algorithms, e.g., for clay and carbonate content prediction (Gomez et al., 2008), clay, iron, and SOC content prediction in a degraded south African ecosystem (Bayer et al., 2012) and in agricultural bare fields (Chabrillat et al., 2014). Finally, both types of methodologies present different conditions of use and a priori expected performances which dictate their use depending on the science goals and data available. PLSR methods are very much in use and allow accurate modeling of soil properties but are based on the need for local ground data for model calibration and need manual fine-tuning. Physically based methods present the advantage that they are more generic and can be implemented in software interfaces on an automatic basis. Then repeatability and robustness of the soil prediction models can be better provided. Furthermore, physically based methods can be used for fully automatic derivation of semiquantitative soil maps that do not require the need for ground data. On the other hand, for the delivery of quantitative soil products, all types of techniques need ground data for model calibration.

Hyperspectral Soil Mapper Software Interface for Operational Soil Properties Mapping

In preparation for the scientific exploitation of current airborne and forthcoming high-quality spaceborne IS data, much effort has been devoted to the development of several open source software interfaces and toolboxes to provide current and upcoming soil science users of IS with adequate tools to derive thematic maps. In particular, higher performing soil algorithms were in development as demonstrators for end-to-end processing chains with harmonized quality measures. As a result, the HYSOMA software interface was developed as part of the EU-FP7 EUFAR (European Facility for Airborne Research)/JRA2-HYQUAPRO project (Reusen et al., 2010) and the EnMap satellite science program (Guanter et al., 2015). The HYSOMA is an experimental platform for soil mapping applications of IS that allows easy implementation in the IS and non-IS community and gives a choice of multiple algorithms for each soil parameter (Chabrillat et al., 2011, 2016b). The main motivation for HYSOMA development was to provide expert and nonexpert users with a suite of tools that can be used for soil applications. The algorithms focus on the fully automatic generation of semiquantitative soil maps for key soil parameters such as soil moisture, SOC, and soil minerals (iron oxides, clay minerals, carbonates) using physically based approaches (Table 4.1).

SWIR FI, Short-Wave Infrared Fine Particles Index; *CRAD*, Continuum-Removed Absorption Depth; *RI*, Redness Index; *NSMI*, Normalized Soil Moisture Index; *SMGM*, Soil Moisture Gaussian Modeling.

The main HYSOMA features for soil mapping include:

- the selection of dominant soil pixels by masking and excluding water pixels and vegetation pixels, in both vital and dry conditions;

TABLE 4.1 Overview of HYSOMA Automatic Soil Functions for Identification and Semiquantification

Soil Chromophores	Soil Algorithm	Spectral Region (nm)	Estimated Soil Parameters
Clay minerals Al-OH content	Clay index (SWIR FI)	2209, 2133, 2225	Clay mineral content (Levin et al., 2007)
	Clay CRAD	2120–2250	Clay mineral content
Iron oxides Fe_2O_3 content	Iron index (RI)	477, 556, 693	Hematite content (Madeira et al., 1997; Mathieu et al., 1998)
	Iron CRAD 1	450–630	Iron oxide content (Fe-VIS absorption band)
	Iron CRAD 2	750–1040	Iron oxide content (Fe-NIR absorption band)
Carbonates Mg-OH content	Carbonate CRAD	2300–2400	Carbonate content
Soil moisture content	Moisture index (NSMI)	1800, 2119	Soil moisture content (Haubrock et al., 2008a, 2008b)
	Gaussian modeling (SMGM)	~1500–2500	Soil moisture content (Whiting et al., 2004)
Soil organic carbon	Band analysis SOC 1	400–700	Organic matter content (Bartholomeus et al., 2008)
	Band analysis SOC 2	400, 600	Organic matter content (Bartholomeus et al., 2008)
	Band analysis SOC 3	2138–2209	Indirect organic matter content (Bartholomeus et al., 2008)

Source: After Chabrillat, S., Eisele, A., Guillaso, S., Rogaß, C., Ben-Dor, E., Kaufmann, H., 2011. HYSOMA: an easy-to-use software interface for soil mapping applications of hyperspectral imagery. In: Proceedings 7th EARSeL SIG Imaging Spectroscopy Workshop, Edinburgh, Scotland; Chabrillat, S., Guillaso, S., Eisele, A., Ben-Dor, E., 2016b. HYSOMA: a software interface for the derivation of soil maps and determination of common soil properties from point and imaging spectroscopy. Environ. Modell. Softw., submitted for publication.

- the use of established methods in the literature for soil mapping that can be implemented in an automatic context such as spectral indices and SMGM;
- the development of the HYSOMA CRAD tool for the automatic detection of spectral feature edges and calculation of absorption depth for hematite, iron oxide, clay, and carbonate content mapping;
- the implementation of additional soil analyses tools that allow in particular the input of in situ data for soil model calibration and derivation of fully quantitative maps of soil properties;
- the derivation of a soil quality layer that takes into account associated errors in the modeling due to bad quality pixels (preprocessing quality layer), and to the effect of disturbing factors (detection of small amount of vegetation and/or soil water content in the pixel).

The HYSOMA is available:

- as a stand-alone version distributed for free under the idl-virtual machine at www.gfz-potsdam.de/hysoma. Since the web site release in 2012, more than 100 users from all over the world have downloaded a plug-in of the HYSOMA software interface, demonstrating the interest of the airborne community for this software (Chabrillat et al., 2016b). The HYSOMA stand-alone version does not include visualization tools for the soil maps.

- as an open source version distributed for free within the EnMAP-Box software (Van der Linden et al., 2015) at www.enmap.org, under the umbrella of the EnSOMAP (EnMAP soil mapper) algorithms (Chabrillat et al., 2014). The EnMAP-Box software also includes tools for imaging spectroscopy data preprocessing and analyses, and visualization tools for the imagery and the soil maps.

Example of Soil Properties Mapping Applications at Airborne Scale

Fig. 4.6 shows the spatial mapping of common soil surface properties that are important indicators for soil fertility and soil degradation such as SOC, clay, iron, and carbonate content in bare and semibare fields in the Cabo de Gata-Níjar Natural Park (southern Spain) and Camarena area (Central Spain). The soil maps were obtained based on the automatic processing of the soil toolboxes without the need for ground data or manual intervention. Soil water content, which is another surface soil property map produced by the HYSOMA and EnSoMAP tools, is not shown here due to the dry conditions and lack of variability in soil moisture content at time of overflight. The Cabo de Gata soil mapping was based on airborne imagery from the HyMap sensor (Chabrillat et al., 2016a), and Camarena soil mapping was based on airborne imagery from the AISAeagle and AISAhawk sensors (Schmid et al., 2016). One can see, e.g., in Fig. 4.6B carbonate map there were some technical challenges. Here the carbonate content could not be satisfactorily extracted due to the lower signal quality of the AISAhawk sensor at longer wavelengths (>2.3 μm).

Quantitative soil spectroscopy analyses for the determination of soil maps in absolute abundance was further performed using available geochemical in situ datasets for soil model calibration. For example, in the Cabo de Gata area, 50 soil samples were collected at the 0–3 cm depth in the region of Cortijo del Fraíle which has highly variable soil iron oxide and clay content due to the transition from volcanic to carbonatic substrata. The soil samples were collected from areas with cultivated, abandoned crops, and autochthonous vegetation on the upper soil surface and covering all main soil types (Richter et al., 2009). Fig. 4.7 shows the prediction of clay and iron surface content based on HyMap airborne images compared with ground-truth data, using spectral features analyses as the property predictor.

The results show that airborne imaging spectroscopy HyMap data over the Cabo de Gata area were adequate to predict and quantitatively map soil surface properties such as clay and iron oxide as shown in other similar studies. The accuracy of the soil prediction models was very high to high (R^2 between 0.83 and 0.57). The comparison of the quantitative prediction accuracy depending on physically based versus machine learning-based approaches was performed in Chabrillat et al. (2014) and showed that in general PLSR and SVMR modeling performed very well to predict top-soil properties in Cabo de Gata HyMap imagery, with variable model performances depending on method used and soil properties of interest.

Global Soil Mapping and Monitoring with Imaging Spectroscopy Data: Potential and Challenges

The upcoming availability of high signal-to-noise ratio satellite imaging spectroscopy data in the next 5 years (EnMAP, SHALOM, PRISMA-Hyperspectral Precursor of the Application Mission-) is expected to provide a major step toward the operational quantitative monitoring of soil surfaces at large scales. These satellites will provide regional to global scale (spatial resolution from 10 to 30 m) soil spectral information at different times. Nonetheless

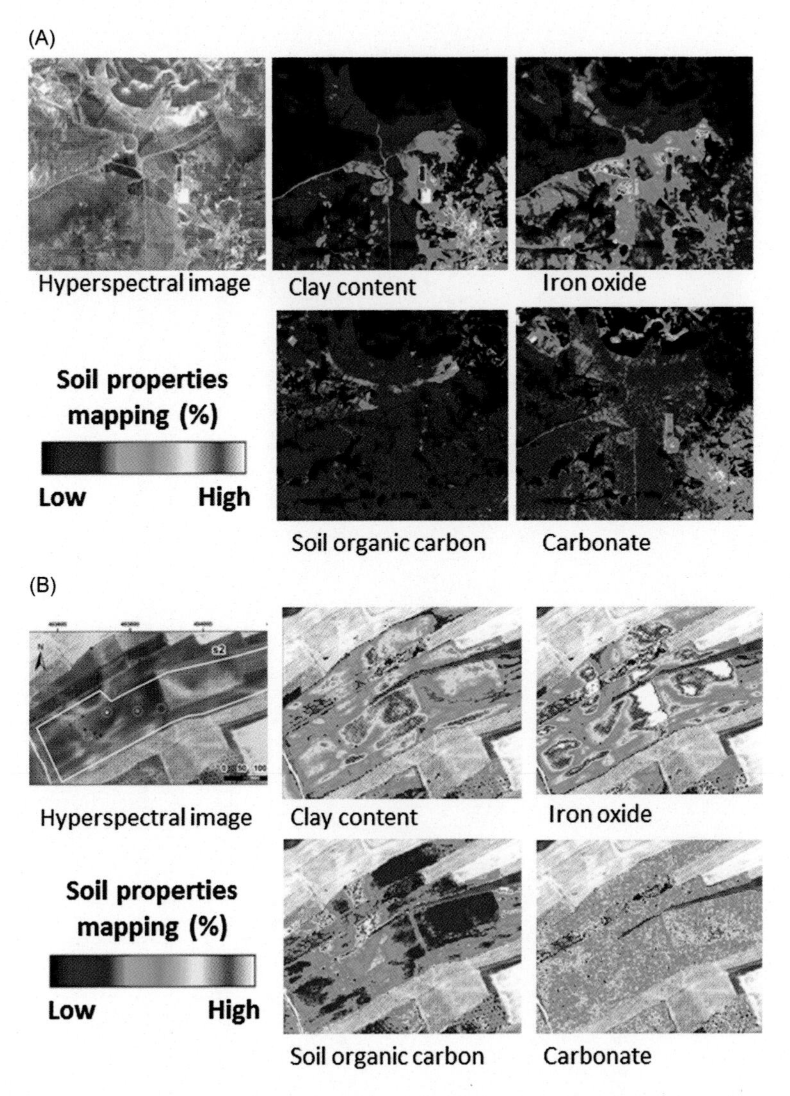

FIGURE 4.6 Mapping of common soil surface properties with airborne imaging spectroscopy data using HYSOMA software automatic outputs (A) Cabo de Gata-Níjar Natural Park, region of Cortijo del Fraile, (B) Camarena agricultural fields. *Source: Modified from Chabrillat, S., Guillaso, S., Eisele, A., Ben-Dor, E., 2016b. HYSOMA: a software interface for the derivation of soil maps and determination of common soil properties from point and imaging spectroscopy. Environ. Modell. Softw., submitted for publication.*

advances are still necessary to fully develop imaging spectroscopy soil products that can support, in a credible manner, global digital mapping and monitoring of soils. The real accuracy to be expected from the upcoming satellite sensors and the operationability of the prediction linked with harmonized methodologies for applications at regional to global scale is still to be demonstrated. For this, recent studies looked at the potential of upcoming EnMAP and other satellite missions for soil mapping using PLSR (Gómez et al., 2015; Steinberg et al., 2016) or other methods (Chabrillat et al., 2014). As an example, Fig. 4.8

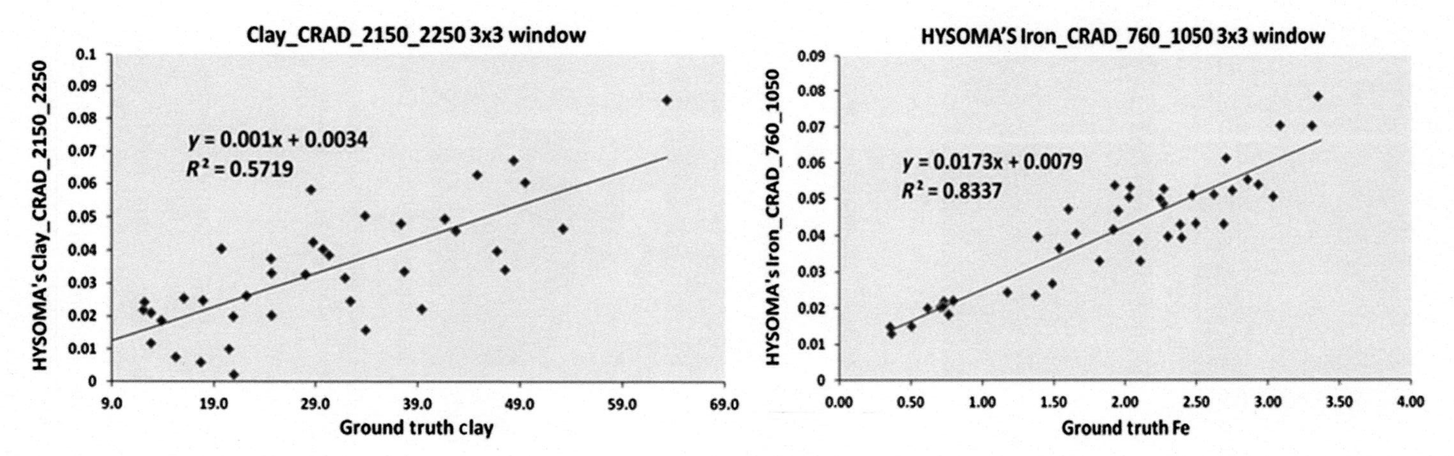

FIGURE 4.7 Prediction of clay and iron oxide surface content (%) based on airborne HyMap clay and Fe-NIR CRAD feature respectively at ~2.2 and 0.9 μm versus laboratory geochemical measurements. *Source: After Chabrillat et al., 2011, EUFAR Deliverable DJ2.4.4, unpubl.*

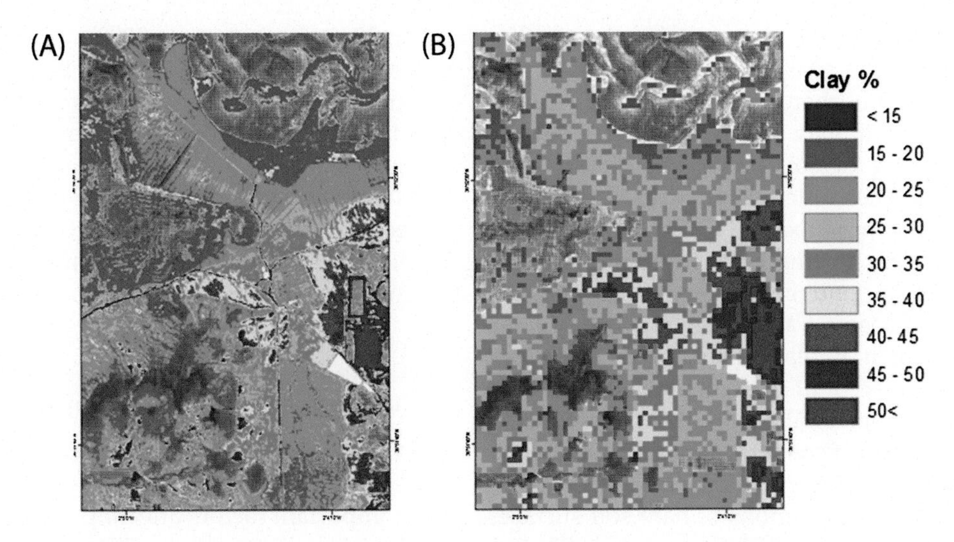

FIGURE 4.8 Potential of upcoming satellite missions for global soil mapping and monitoring: surface clay mapping in the Cabo de Gata-Nijar Natural Park, Spain, from (A) airborne HyMap imagery (pixel size 4.5 m) and (B) simulated EnMAP imagery (pixel size 30 m). *Source: Modified Steinberg, A., Chabrillat, S., Stevens, A., Segl, K., Foerster, S., 2016. Prediction of common surface soil properties based on Vis-NIR airborne and simulated EnMAP imaging spectroscopy data: prediction accuracy and influence of spatial resolution. Remote Sens., 8, 613; http://dx.doi.org/10.3390/rs8070613.*

shows a simulation of an EnMAP soil product compared with the airborne mapping that can be obtained over the Cortijo del Fraile area in Cabo de Gata Natural Park using spectral feature approach as the predictor.

In general, these studies demonstrated the high potential of upcoming spaceborne hyperspectral missions for soil science studies but have also shown the need for future adapted strategies to cope with the larger spatial resolution. Nevertheless, compared with airborne soil maps at much finer scale, simulated EnMAP images at 30 m scale with good spectral resolution and estimated signal-to-noise ratio similar to sensor tests were able to deliver regional soil maps that are consistent with previous analyses in the region.

SOIL EROSION MAPPING

Soil erosion by water is an extensive and increasing problem throughout Europe. The mean soil loss rate in the European Union's erosion-prone lands (agricultural, forests, and seminatural areas) was found to be 2.46 $t\,ha^{-1}\,year^{-1}$, resulting in a total soil loss of 970 Mt annually. The mean soil loss rate in the EU exceeds the average soil formation rate by a factor of 1.6 (Panagos et al., 2015).

The Mediterranean region is particularly affected by erosion. This is because it is subject to long dry periods followed by heavy bursts of erosive rainfall, falling on steep slopes with fragile soils, resulting in considerable amounts of erosion. In parts of the Mediterranean region, erosion has reached a stage of irreversibility and in some places erosion has practically ceased because there is no more soil left. With a very slow rate of soil formation, any soil loss of more than 1 $t\,ha^{-1}\,year^{-1}$ can be considered as irreversible within a time span of 50–100 years (Grimm et al., 2002). Losses of 20–40 $t\,ha^{-1}$ in individual storms, that may happen once every 2 or 3 years, are measured regularly in Europe with losses of more than 100 $t\,ha^{-1}$ in extreme events (Morgan, 2005).

IS can provide the necessary information for carrying out the assessment and monitoring of soil erosion (Shoshany et al., 2013). The effects of soil erosion may be described by several characteristics of soil surface conditions and properties. These include chemical (pH, SOC, texture, free iron oxides, $CaCO_3$) and physical (structure, texture, coarse fragments) soil properties, ground cover (with fine textured minerals to coarse fragments, and with organic elements, such as plant debris and vegetation) taking into account the dimension of space and time (Pinet et al., 2006; Boardman, 2007). These soil spectra properties are easily obtained within Mediterranean regions where it is common to have large regions of bare soil surfaces (Bartholomeus et al., 2007; Gomez et al., 2012).

Multispectral sensors record data in fewer bands, resulting in a coarser spectral resolution compared to hyperspectral sensors. Multispectral data such as LANDSAT and SPOT have been used to carry out land use and cover (Fung, 1990; Lanjeri et al., 2001; Pilon et al., 1988), land degradation and soil erosion studies (Dewitte et al., 2012; Hill and Schütt 2000; Lacaze et al., 1996; Lo Curzio and Magliulo, 2009; Schmid et al., 2001) and mapping of soil properties (Bartholomeus et al., 2007; Mulder et al., 2011; Schmid et al., 2008). Landsat is among the widest used satellites in soil erosion studies, because it has the longest time series of data of currently available satellites (De Jong, 1994; Dhakal et al., 2002; Vrieling, 2006). Further advanced multispectral sensors such as ASTER, Landsat8, and Sentinel2 have also been used to study soil properties and erosion (Hubbard et al., 2003; Hubbard and Crowley, 2005; Van der Werff and Van der Meer, 2015; Vrieling et al., 2008).

Studies combining field, airborne, and/or satellite-borne sensors with different spectral and spatial resolutions using spectroradiometer, hyperspectral, and multispectral data, respectively, can be implemented to determine a detailed spatial distribution of surface soil properties on a detailed local scale (Ben-Dor et al., 2006; Schmid et al., 2008). When carrying out soil erosion studies, it is not unusual to use a combination of sensors at different spatial and spectral resolutions to be able to extrapolate specific soil information obtained from field plots to a wider region using hyperspectral and/or multispectral sensors, respectively (Corbane et al., 2008; Hill and Schütt, 2000; Mathieu et al., 2007; Schmid et al., 2016; Vrieling et al., 2008).

Soil Erosion Stages Within a Rainfed Cultivated Area of Central Spain

In the Mediterranean region, plant cover, land uses, and topography are considered the most important factors affecting the intensity of soil erosion (García-Ruiz, 2010). Furthermore, widespread tillage activities contribute to the transformation of soil landscapes within these regions (De Alba, 2001). Therefore soil erosion in Mediterranean agricultural environments is often a combined effect of tillage practices and water erosion processes (Lindstrom et al., 1992).

The soil transformation may involve the whole or partial loss of fertile soil or a mixture or inversion of the differentiated layers that constitute the soil, known as soil horizons. Soils that have been eroded as a result of plowing will expose different surface (A) or subsurface (B and C) horizons with significant differences in their physical and chemical properties, such as color, pH, SOM, texture, structure, consistence, coarse fragments, free iron oxides, $CaCO_3$, and/or clay minerals (De Alba, 2001; Ortega, 1984). These types of soil properties can be considered as soil erosion indicators representative in Mediterranean regions (Schmid et al., 2016). The effect of tillage-induced soil erosion can be shown as an example of mapping soil erosion stages using high-resolution remote sensing data along with morphological and physicochemical field and laboratory data within central Spain (Schmid et al., 2016). In this case the properties of different soil horizons emerging at the surface are identified using FS and

TABLE 4.2 Soil Erosion and Accumulation Stages

Stage	Description		Surface Description
am1	*Accumulation stage*: deposits in footslope/toeslope positions		Sandy deposits
am2			Clayey and organic matter-rich deposits
es1	*Erosion stages*	*Stage 1*: slightly eroded soil.Presence of A horizon	A horizon with subsequent B and/or C horizons
es2a		*Stage 2*: moderately eroded soil. Loss of A horizon, presence of subsurface B horizon	B horizon weathered (*cambic*)
es2b			B horizon with clay accumulation (*argillic*, brown colors)
es2c			B horizon with clay and Fe_2O_3 accumulation (*argillic*, reddish colors)
es3a		*Stage 3*: strongly eroded soil. Loss of A and B horizons. Outcropping of C horizons (soil parent material)	C horizon (arkose)
es3b			Ck horizon (*calcic*, marls)
es3c			Ck horizon; (*calcic*), R (limestone)

Source: Modified from Schmid, T., Rodríguez-Rastrero, M., Escribano, P., Palacios-Orueta, A., Ben-Dor, E., Plaza, A., et al., 2016. Characterization of soil erosion indicators using hyperspectral data from a Mediterranean rainfed cultivated región. IEEE J. Select. Top. Appl. Earth Observ. Remote Sens. 9 (2), 845–860.

hyperspectral airborne data to obtain a spatial distribution of different soil erosion and accumulation stages.

In this case, soil erosion indicators according to the soil properties exposed on the surface can be defined taking into account that tillage-induced soil erosion. This brings about the progressive removal of soil horizons and the corresponding accumulation of soil materials at the base of the slope (Previtali, 2014). Hyperspectral data acquired with airborne sensor and field spectra obtained with spectroradiometers of the type ASD covering the spectral range of 350–2500nm can be used to map soil properties.

A qualitative procedure is based on the assumption that physico-chemical and morphological properties related to the different horizons constitute the main soils present at the terrain surface as a consequence of soil erosion induced by tillage. Erosion stages for slightly, moderately, and strongly eroded soils and two accumulation zones can be established (Table 4.2). These erosion stages are related with the properties of different soil surface characteristics and associated to the corresponding spectral features of the FS and hyperspectral data.

A spatial distribution of the accumulation zones and soil erosion stages for two selected areas within rainfed cultivated areas in Central Spain serve as examples using hyperspectral airborne data with a spatial resolution of 6 by 6m (Fig. 4.9). The soils of the first area are found on arkose sediments and have a higher iron oxide content, acid or neutral pH, coarse texture, higher abundance of coarse fraction and harder consistencies. In the second area the soils have significant amounts of $CaCO_3$ and a basic pH. The soils of both the areas have a low SOM content which is characteristic of Mediterranean soils. In these two cases, the erosion stages are present on elevated areas where minimal erosion is found on slightly sloping surfaces and maximum erosion on the slope shoulder. Accumulation areas form deposits of soil material transported to the bottom of the valleys.

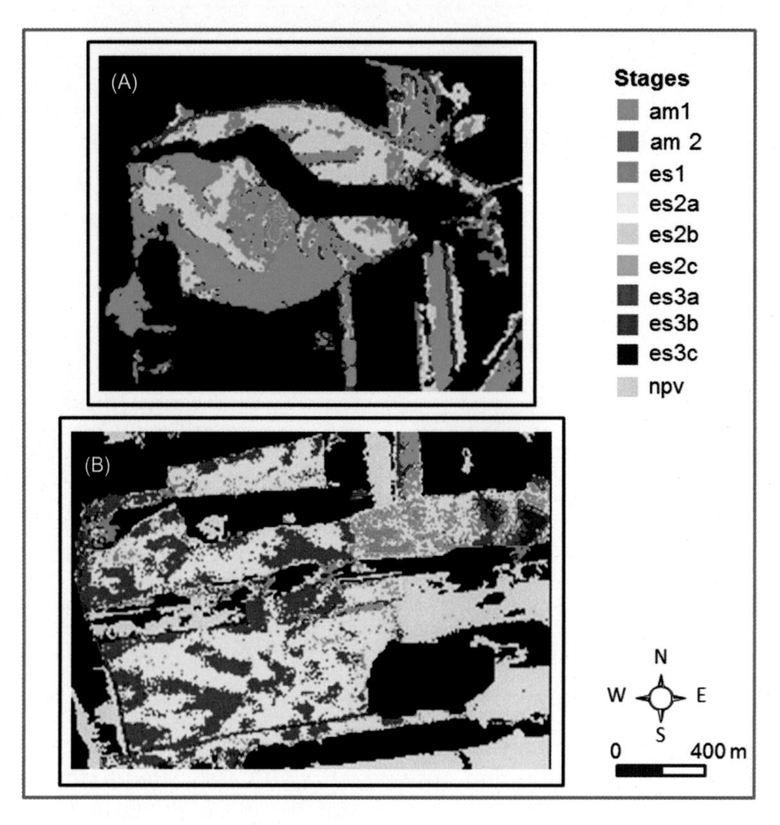

FIGURE 4.9 Distribution of soil erosion and accumulation stages for (A) soil on coarse and medium arkose sediments and (B) on carbonated fine arkoses and lutite sediments. *Source: Modified from Schmid, T., Rodríguez-Rastrero, M., Escribano, P., Palacios-Orueta, A., Ben-Dor, E., Plaza, A., et al., 2016. Characterization of soil erosion indicators using hyperspectral data from a Mediterranean rainfed cultivated regíón. IEEE J. Select. Top. Appl. Earth Observ. Remote Sens. 9 (2), 845–860.*

Improved identification and characterization of individual soil erosion stages and accumulation zones was achieved by the use of field spectral data together with soil analyses and field observation data. Such a proposed method is potentially applicable to other Mediterranean regions with rainfed cultivation.

MAPPING LAND DEGRADATION WITH ECOHYDROLOGICAL REMOTE SENSING INDICATORS

Land degradation produces a sustained reduction in functionality and complexity of terrestrial ecosystems and includes socio-economic, biological, and climatic aspects (UNCCD, 1996). Ecosystem functionality is determined by the flows of water, energy, and nutrients through the arrangement of biotic and abiotic components (Chapin et al., 2002). A theoretical continuum of functionality exists from ecosystems that efficiently use resources to those characterized by a severe degradation where most of the resources are lost (Ludwig and Tongway, 2000). The causes behind land degradation are related to the climate and human activities. Disturbance events such as fire, droughts, or land use changes are events taking place out of the range of natural variability of the ecosystems (Rykiel, 1985). If disturbance events press ecosystems beyond resilience thresholds to new irreversible states characterized by lower functionality land degradation has taken place (Puigdefabregas, 1995).

Several biophysical and biochemical processes are affected such that assessment of land degradation should include consideration of a combination of indicators/scales (Weinzierl et al., 2016). Remote sensing can monitor soil properties related to soil degradation such as salinity (Elhag, 2016), organic matter (Bartholomeus et al., 2008), and vegetation-soil patterns (Prince et al., 1998; Asner et al., 2003). However, to detect losses of functionality and resilience we need indicators that are also capable of tracking ecosystem flows and their responses to disturbance events (Maestre et al., 2016). A majority of large scale remote sensing studies of land degradation have focused on monitoring carbon fluxes, in particular aboveground net primary productivity (ANPP) with vegetation indices like the NDVI as proxies (Anyamba and Tucker, 2005; Eklundh and Olsson, 2003). This approach presents problems when vegetation is stressed and/or sparse (Garcia and Ustin, 2001), and this problem is aggravated when using large pixel sizes like the 8km GIMMS (Global Inventory Modeling and Mapping Studies) dataset (Goetz and Prince, 1999).

To separate climatic and human causes of land degradation, the rainfall use efficiency (RUE) was developed. The RUE reflects how much water is needed to produce the ANPP (Table 4.3). In the Sahel the lack of a temporal trend in RUE contributed to the dismissal of a prevalent hypothesis of a general human-induced-land degradation by showing the ecosystem's resilience after the longest drought recorded on Earth (Prince et al., 1998). The RUE has been improved to account for soil water redistribution (Huber et al., 2011) but there is still some controversy due to its sensitivity to annual rainfall variability (Wessels et al., 2008; Hein and De Ridder, 2006; Fensholt and Rasmussen, 2011; Prince et al., 2007). Although the need to use functional indicators related with ecohydrological processes to diagnose and understand land degradation in drylands has been pointed out (Maestre et al., 2006; 2016), indicators tracking ET fluxes are not yet as widely implemented as RUE or NDVI based indices due in part to the relative complexity of having accurate estimates of ET from remote sensing (Kalma et al, 2008). As ET links together the water, energy, and carbon cycles it is a key variable that can serve as an ecosystem indicator integrating the effects from multiple stresses (Fig. 4.10). ET indicators can overcome some of the RUE problems related with the accuracy of precipitation datasets (Mbow et al., 2015) and the use of NDVI as a proxy for ANPP. The water use efficiency (WUE), the ratio between net primary productivity and transpiration or the nonevaporative fraction (NEF), which explains how much of the surface energy is dissipated as heat rather than as water, have been successfully used to map sites at risk of becoming land degradation hotspots (Wang and Takahashi, 1998; Garcia et al., 2008, 2009) (Table 4.3). Current remote sensing efforts are focused on estimating ET partition into transpiration and evaporation (Henderson-Sellers et al., 2008) as this ratio is linked to carbon assimilation, groundwater recharge, infiltration, and runoff processes (Newman et al., 2010; D'Odorico et al., 2013).

Disentangling Natural Ecosystem Variability and Disturbance

One critical need is to characterize the natural variability associated with a healthy and fully functional ecosystem and distinguish it from the behavior of ecosystems under disturbed conditions. A disturbance event should generate a signal greater than that of the natural variability (Mildrexler et al., 2009). The level of natural variability in ecosystem fluxes can be estimated using (1) simulations from

TABLE 4.3 Ecohydrological Remote Sensing Indicators That Can Be Used to Track Ecosystem Function to Map Land Degradation Hotspots

Indicators	**Description (Units)**	**Remote Sensing Method**	**Land Degradation Remote Sensing Examples**	**Causes of Changes**
Aboveground Net Primary Production (ANPP)	Net fixed CO_2 from the atmosphere as aboveground biomass during a period of time ($g\,C\,m^{-2}\,year^{-1}$)	Vegetation indices (*NDVI, EVI*) or Process-based models (e.g., Monteith light use efficiency model)	Anyamba and Tucker (2005); Eklundh and Olsson (2003)	Climate and phenology
Rainfall Use Efficiency (RUE)	Unit of aboveground biomass produced per unit of precipitation P (annual and longer time scales)	$RUE = ANPP/P$ (annual and longer time scales)	Prince et al. (1998); Huber et al. (2011); Fensholt and Rasmussen (2011); del Barrio et al. (2010)	Land use and land cover changes Land degradation
Non-evaporative Fraction (NEF)	Proportion of energy (Rn-G) dissipated as sensible heat (*H*), rather than as water vapor (LE) (adimensional)	$NEF = 1\text{-}\lambda E/(Rn\text{-}G)$ Surface energy balance equation (Rn-G = H + LE)	Garcia et al. (2008, 2009); Wang and Takahashi (1998); Mildrexler et al. (2009)	Biodiversity (e.g., invasive species) Soil degradation (fertility, erosion)
Evaporative Fraction (EF)	Proportion of energy (*Rn-G*) dissipated as water vapor, rather than as sensible heat (adimensional)	$EF = \lambda E/(Rn\text{-}G)$ Surface energy balance equation (Rn-G = H + LE)		Hydrology (canches in phreatic level, inflows)
Water Use Efficiency (WUE)	Carbon assimilated per unit of water transpired *T* ($gC\,kg^{-1}$ water)	$WUE = ANPP/T$ Surface energy balance	Lu and Zhang (2010)	Vegetation stress (drought, pests)
Evaporation Ratio	Amount of water that evaporates per unit of precipitation (adimensional)	$E = ET/P$ (annual and longer time scales)	Boer et al. (2005)	

process-based models (Arribas et al., 2003) (2) time series analysis to detect anomalies (Mildrexler et al., 2009), trends (Prince et al., 1998), recovery rates after a disturbance event, changes in autocorrelation or abrupt shifts (De Keersmaecker et al., 2014; Horion et al., 2016). When the time series is not long enough a novel approach replaces time with space by assuming that within a climatic region disturbed sites and nondisturbed sites coexist (Boer and Puigdefábregas, 2005; del Barrio et al., 2010). This is straightforward for human-induced disturbances but more problematic when assessing large-scale climatic effects and requires careful consideration of the spatial window of analysis.

Remote Sensing of Soil Moisture and Evapotranspiration

To calculate ecohydrological indicators such as the NEFs, WUE, or E/P (Table 4.3), a

Land-atmosphere interactions: Surface energy balance

$$Rn - G = H + \lambda E$$

Nonevaporative fraction

$$\text{NEF} = \frac{H}{Rn - G} = 1 - \frac{\lambda E}{Rn - G}$$

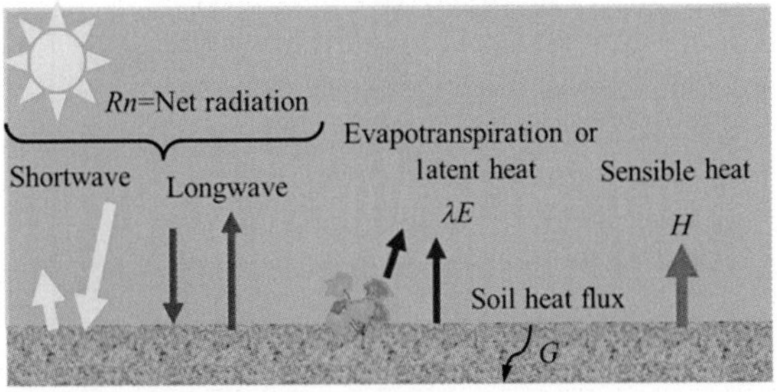

FIGURE 4.10 The surface energy balance equation states that all the incoming and outgoing energy fluxes from a surface should be balanced. *Rn* is net radiation, λE is latent heat flux or ET (expressed in mass units), *G* is soil heat flux and *H* is the sensible heat flux. The amount of the available energy of the surface (*Rn-G*) that is dissipated as sensible heat flux (*H*) or as latent heat flux depends on the water available in the soil.

combination of remote sensing data in both the optical and thermal ranges works best. In broad terms the energy emitted in the thermal range depends on the energy absorbed in the optical range and the water available for ET (Fig. 4.10). Variables such as radiometric temperature (remotely sensed), albedo, vegetation índices, or soil moisture proxies can be incorporated into different biophysical models that estimate ET (Garcia et al., 2013). For example, soil moisture has been estimated using thermal data in combination with vegetation indices such as in triangle methods (Sandholt et al., 2002), or in soil thermal inertia approaches (Sobrino et al., 1998; Verstraeten et al., 2006; Cai et al., 2007). As ET is part of both the energy cycle (named latent heat flux lE) and the water cycle (see Fig. 4.10), remotely sensed ET methods rely on using the surface energy balance equation rather than a water balance approach (Kalma et al., 2008) getting rid of the need to have estimates of rainfall, not always available at relevant spatial and temporal scales. ET modeling also requires meteorological datasets such as air temperature or solar irradiance (Anderson et al., 2011). Remote sensing ET can be estimated for mixed pixels (single-source models) (Chehbouni, 2001) or by separating the soil (evaporation) and vegetation (transpiration) fluxes (Kustas and Norman, 1999; Kustas et al., 2016; Morillas et al., 2013). Some methods focus on (1) solving the sensible heat flux, *H* (Brutsaert, 1982; Su, 2002) and others (2) directly solve ET using the Penman-Monteith (PM) equation (Leuning et al., 2008; Morillas et al., 2013; Zhang et al., 2010) or the simplified Priestley & Taylor equation (Fisher et al., 2008; Garcia et al., 2013). Errors in ET estimates from remote sensing data tend to be in the range of 15–30% (Kalma et al., 2008). Global ET datasets are in high demand for several applications and currently those from NASA MODIS (Mu et al., 2011) and other global products (Jiménez et al., 2011) can be freely downloaded, but because their accuracy differs greatly between sites they should be used with caution (Jiménez et al., 2011).

Example of Evapotranspiration Ratios Detecting Land Degradation Risk in Drylands

Application of ET ratios (Table 4.3) to monitor ecosystem function and degradation risk was done in a pilot study in a mountainous area in south Spain (Sierra de Gádor). The nonevaporative fraction, NEF (Table 4.3), was estimated from ASTER in the optical (30m pixel) and thermal (90m pixel) domains. NEF was rescaled or standardized by climate type (NEFs) following a spatial approach (Boer and Puigdefábregas, 2005) to compare dry and wetter regions that logically will differ in their maximum value of undisturbed ET and NEF on equal terms. The NEFs quantifies the decrease in water evapotranspired with respect to that of an undisturbed level of the ecosystem for each pixel. For example, an NEFs = 0.8 indicates that the ecosystem transpires 80% less water compared to an undisturbed site in the same climatic conditions. The undisturbed site will have NEFs = 0 and the severely disturbed site NEFs = 1. The ability of a spatial method to find extremely disturbed and undisturbed sites based on Fig. 4.11 was validated in the field with (1) severely disturbed human sites, including a burn scar, an active limestone quarry, an abandoned mine, and ploughed orchards and (2) undisturbed sites such as oak relicts and old pine forests (Fig. 4.12).

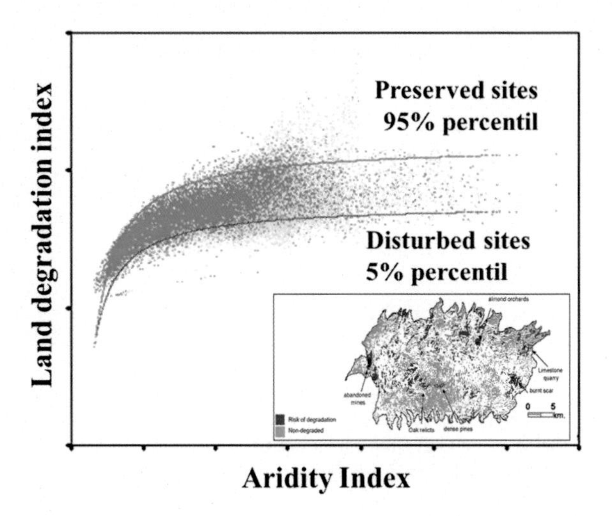

FIGURE 4.11 Spatial method to establish reference levels of maximum degradation/conservation. On a given date the scatterplot of a land degradation index versus a climatic index, like the aridity index, allows the extraction of extreme values for undisturbed and disturbed sites for each climate type. The observed land degradation index at each pixel is rescaled between the extreme values so that it varies between 0 and 1 (figure modified from del Barrio et al., 2010). The enclosure shows a classification in Sierra de Gador using the spatial method for extremely disturbed and undisturbed clases. *Source: After Garcia, M., Oyonarte, C., Villagarcia, L., Contreras, S., Domingo, F., Puigdefabregas, J., 2008. Monitoring land degradation risk using ASTER data: the non-evaporative fraction as an indicator of ecosystem function. Remote Sens. Environ. 112, 3720–3736.*

A second set of ground truth sites was selected based on soil erosion processes. The entisolization index was used in the field as an indicator of soil erosion associated with the disappearance of the Mollic diagnostic soil horizon (Dazzi and Monteleone, 2007). The rationale is that due to erosion, deeper, more developed soil typologies tend to be replaced by poorly developed ones (Entisols) (Grossman, 1983; Oyonarte et al., 2008), (Fig. 4.13). The presence of a Mollic epipedon requires stable conditions at the surface favoring accumulation of SOM of at least 18-cm of horizon depth, and the organic fraction should be bound to the mineral fraction generating stable aggregates (Soil Survey Staff, 1990).

The NEFs were successful to detect significant differences between field sites with and without the Mollic epipedon in limestone lithology (Fig. 4.13B). The field plots with higher NEFs also presented significantly lower total water availability (TWA) (Fig. 4.13A), thus showing the potential of ET ratios such as NEFs to track not only changes in evaporation due to land degradation but also changes in soil depth and/or soil hydraulic properties (Garcia et al., 2008).

BIOCRUST MAPPING

Open soils, and especially these from drylands regions, are covered by complex communities of cyanobacteria, algae, lichen, and mosses that are commonly known as biological soil crust or biocrust (Elbert et al., 2012). Although these cryptogamic communities

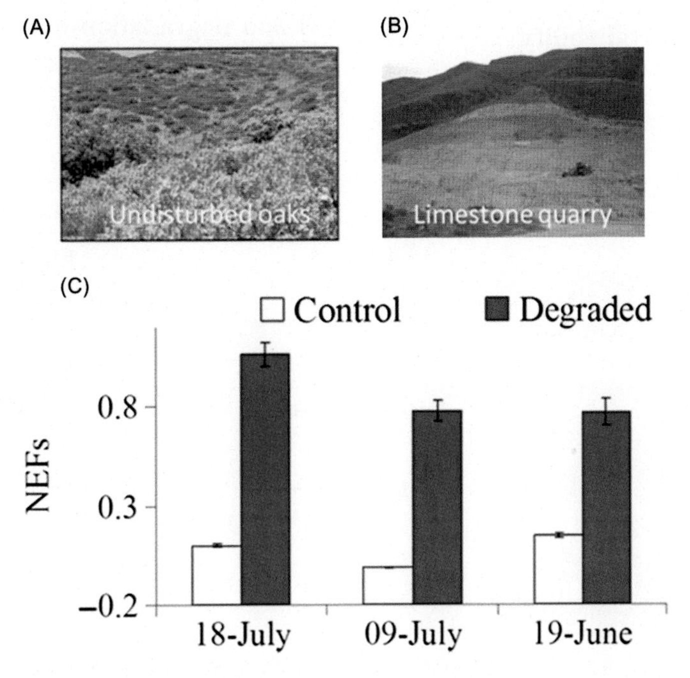

FIGURE 4.12 Validation of the spatial method to find reference conditions for the nonevaporative fraction standardized by climate (NEFs) in Sierra de Gador (Spain) on three dates using ASTER data. (A) Example of an undisturbed site (NEFs = 0); (B) example of an extremely disturbed site (limestone quarry); (C) NEFs at control sites (undisturbed) and disturbed areas. Extremely disturbed sites (limestone quarry, burnt areas, abandoned mines, tilled farming) showed values of NEFs between 0.8 and 1, meaning those areas evapotranspired 80–100% less than the undisturbed site.

represent an insignificant fraction of the soil profile and were traditionally unappreciated, they are one of the main phototrophic communities and C pools in hot, cool, and cold arid and semiarid regions around the world (Elbert et al., 2012). Moreover, they play a key role in ecosystem functioning (Maestre et al., 2012), modulating the N cycle (Elbert et al., 2012; Delgado-Baquerizo et al., 2013; Lenhart et al., 2015; Weber et al., 2015) and contributing to the maintenance of soil microbial biodiversity (Büdel et al., 2009). As result of their physiological activity, biocrusts produce extracellular polymeric substances that bind soil particles and modify some key soil surface properties like microtopography (Rodríguez-Caballero et al., 2012), soil texture and porosity (Miralles-Mellado et al., 2011; Felde et al., 2014), soil cohesion (Thomas and Dougill 2007) and stability (Wang et al., 2014), with strong implications for hydrological processes (Chamizo et al., 2016) and water and wind erosion (Belnap et al., 2014; Rodríguez-Caballero et al., 2015a).

As the soil epidermis, biocrusts also modify surface spectral response by the secretion of photosynthetic pigments and sunscreen products, like chlorophyll, carotenoids, or scytonemim, that absorb solar radiation at specific wave-lengths on the optical region of the spectra (Couradeau et al., 2016) and mask soil spectral properties (Weber et al., 2008). This complicates the soils characterization based on nondestructive spectral measurement and sometimes produces a spectral overlap between open areas and vascular vegetation (Escribano

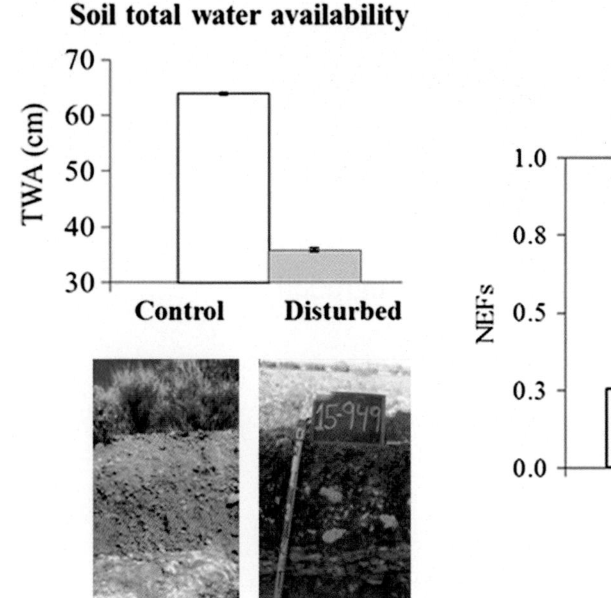

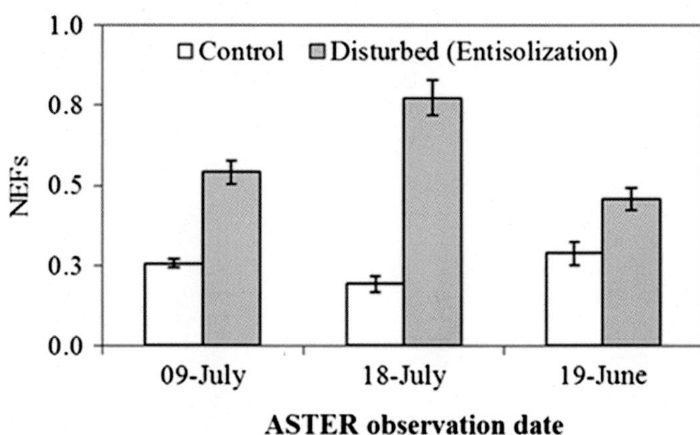

FIGURE 4.13 Analysis of soil erosion in Sierra de Gador (Spain). (A) Average field value of TWA based on soil water holding capacity and soil depth for undisturbed and disturbed soils (with and without a Mollic epipedon). (B) Mean differences of the ET ratio NEFs at undisturbed sites ($n = 80$ pixels) and disturbed sites ($n = 80$ pixels) in limestone geology estimated from ASTER data on three dates. Differences are significant at $P < 0.05$ (modified from García et al., 2008). The nonevaporative fraction standardized by climate, NEFs, quantifies the decrease in water evapotranspired with respect to that of undisturbed sites. For example, NEFs = 0.8 (July 18) indicates that the ecosystem transpires 80% *less* water compared to an undisturbed site (with a Mollic epipedon) in the same climatic conditions.

et al., 2010). For this reason their presence needs to be considered when soil spectra properties are analyzed using its spectral response.

Biocrust Spectral Traits

Since Wessels and van Vuuren (1986) measured the spectral reflectance of lichen dominated biocrust, considerable effort has been devoted to the identification of the main spectral characteristics of different biocrust communities around the world (Graetz and Gentle, 1982; Green, 1986; O'Neill, 1994; Pinker and Karnieli, 1995; Karnieli and Sarafis 1996; Karnieli et al., 1996; Chen et al., 2005; Weber et al., 2008; Ustin et al., 2009; Chamizo et al., 2012; Rodríguez-Caballero et al., 2014; Casanovas et al., 2015; Fang et al., 2015; Rozenstein and Karnieli, 2015, among others). Most of these studies were compiled in the bibliography revision recently published about biocrust spectral properties (Weber and Hill, 2016). According to Weber and Hill (2016), several spectral absorptions, related to the presence of photosynthetic pigments and sunscreen products exudated by biocrusts, produced a decrease in reflectivity on the visible region of the spectra (Karnieli and Tsoar 1995; Rodríguez-Caballero et al., 2015b), which combined with increased surface emissivity at longer wave lengths (Rozenstein and Karnieli, 2015), modeled biocrust spectra and modified soil surface reflectance. The first detailed spectral analysis of soils covered by biocrust was done by O'Neill (1994), who analyzed different types of biocrusts from Australia and identified a common spectral absorption feature at 675nm, related with the presence of chlorophyll-*a*. This absorption was later described by several authors in regions all over the world that were covered by different types of biological soil crusts (Chen et al., 2005; Weber et al., 2008; Chamizo et al., 2012; Casanovas et al., 2015; Fang et al., 2015). Weber et al. (2008) described another important absorption feature on cyanobacteria dominated biocrusts from Soebatsfontain (South Africa), about 500nm, that was shallower than the spectral absorption observed on bare soil (Weber et al., 2008). A similar pattern was observed in other areas where cyanobacteria dominated biocrusts represent one of the main surface components, like the Negev. In this study area, cyanobacteria spectra showed higher reflectance in the blue region of the spectra than bare soil, and this was assumed to be caused by the presence of phycobilin (Karnieli and Sarafis, 1996). Chamizo et al. (2012) observed a different pattern for cyanobacteria dominated biocrust from the Tabernas Badlands in Almeria, with deeper absorption in the blue region of the spectra than bare soil spectra. This was assumed to be a consequence of the spectral absorption of cyanobacteria carotenoids, which play a key role in photosynthesis and protection against photooxidative damage of biocrusts. Differences between the results obtained by Chamizo et al. (2012) and previous studies seem to be more affected by differences in soil properties, like soil Fe^{3+} content that produced deeper absorption at a similar wavelength to carotenoids (Weber et al., 2008), than by differences in biocrust spectral response. As shown in Fig. 4.14, similar biocrust communities from different regions show similar spectral responses that differ from the spectral responses of other biocrust types and mask soil spectral properties. Whereas white lichen dominated biocrusts show high reflectance values in the VISNIR regions of the spectra and an important absorption feature at 675nm (chlorophyll-*a* absorption), cyanobacteria reduce surface albedo and show another important absorption at 500nm (carotenoids absorption). However, biocrust communities rarely reach 100% coverage (Weber et al., 2015) and final spectral response of biocrust dominated areas is also affected by the spectral response of bare soil covering the open spaces between biocrust forming organisms, with increasing importance as biocrust coverage decreases. Bare soil from the Tabernas Badlands showed higher reflectance in the VISNIR regions of the spectra than soils from Las Amoladeras; the lichen and cyanobacteria biocrust from Tabernas showed higher albedo than lichen and cyanobacteria from Las Amoladeras (Fig. 4.14). Moreover, the spectral absorption of the bare soils from Las Amoladeras, where soils have an important Fe^{3+} content, interact with carotenoids absorption and produce deeper absorption about 500nm in cyanobacteria dominated biocrust than is observed for the same biocrust community at Tabernas Badlands, and a displacement of the spectral absorption to longer wave-length.

Moreover, spectral properties of biocrusts rapidly change after rainfall or dew events, when biocrusts are active (Zaady et al., 2007; Weber et al., 2008; Fang et al., 2015; Rodríguez-Caballero et al., 2015b). As observed in Fig. 4.15, surface reflectance of different biocrust communities decreased from dry to wet conditions, at the same time as spectral absorptions became deeper. Biocrust spectral properties response to water availability varies between different biocrust communities. Whereas an important increase in the spectral absorption about 500nm was observed for cyanobacteria dominated biocrusts when wet, this is almost negligible on lichen dominated biocrusts, which showed major changes at 670nm (Fig. 4.15B).

Thus differences in the spectral response between biocrusts type and water status

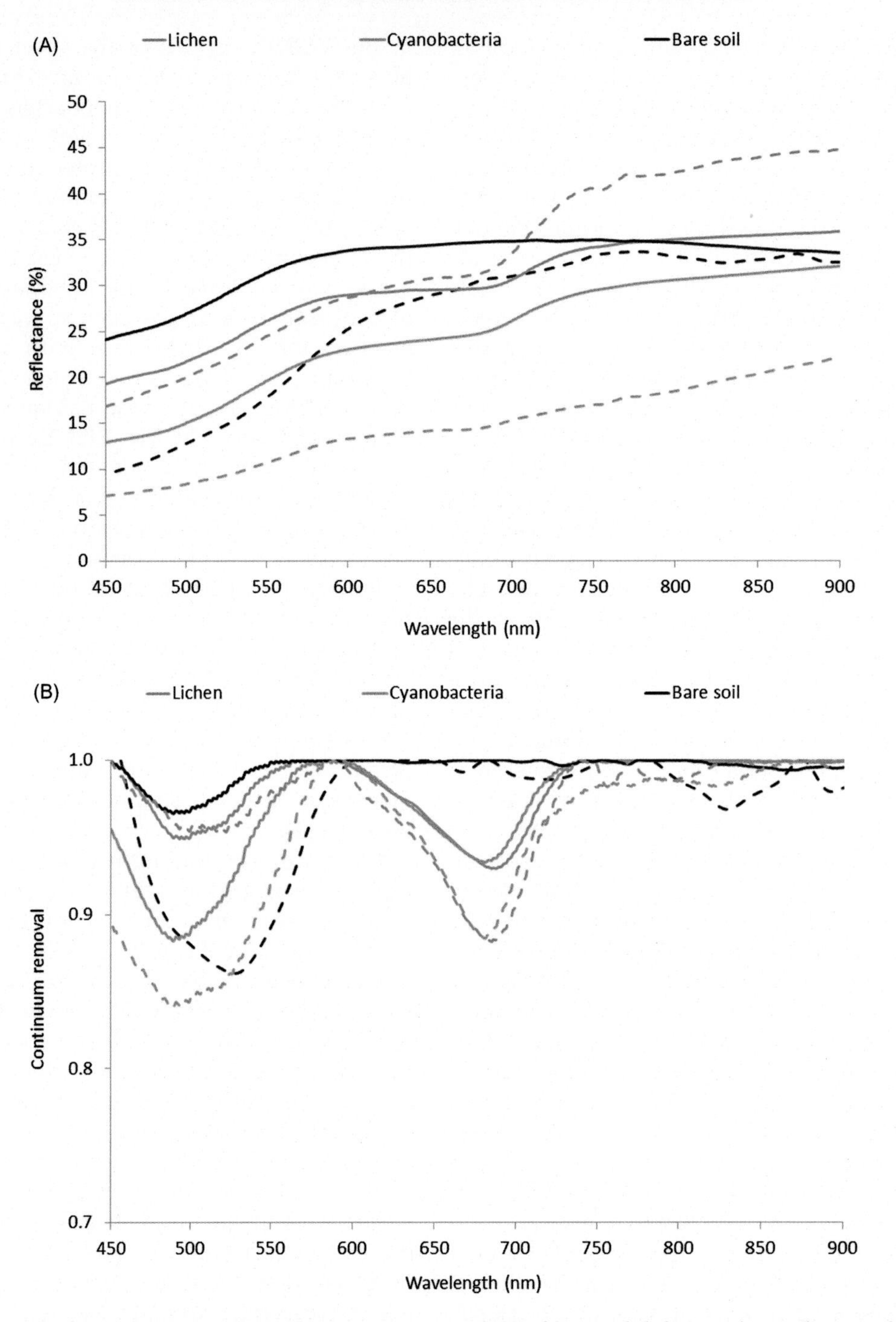

FIGURE 4.14 Spectral signature (A) and continuum removal (B) of bare soil, lichen- and cyanobacteria dominated biocrusts from two different semiarid areas located in southeast Spain: the Tabernas Badlands (continuous line), and Las Amoladeras (dashed lines).

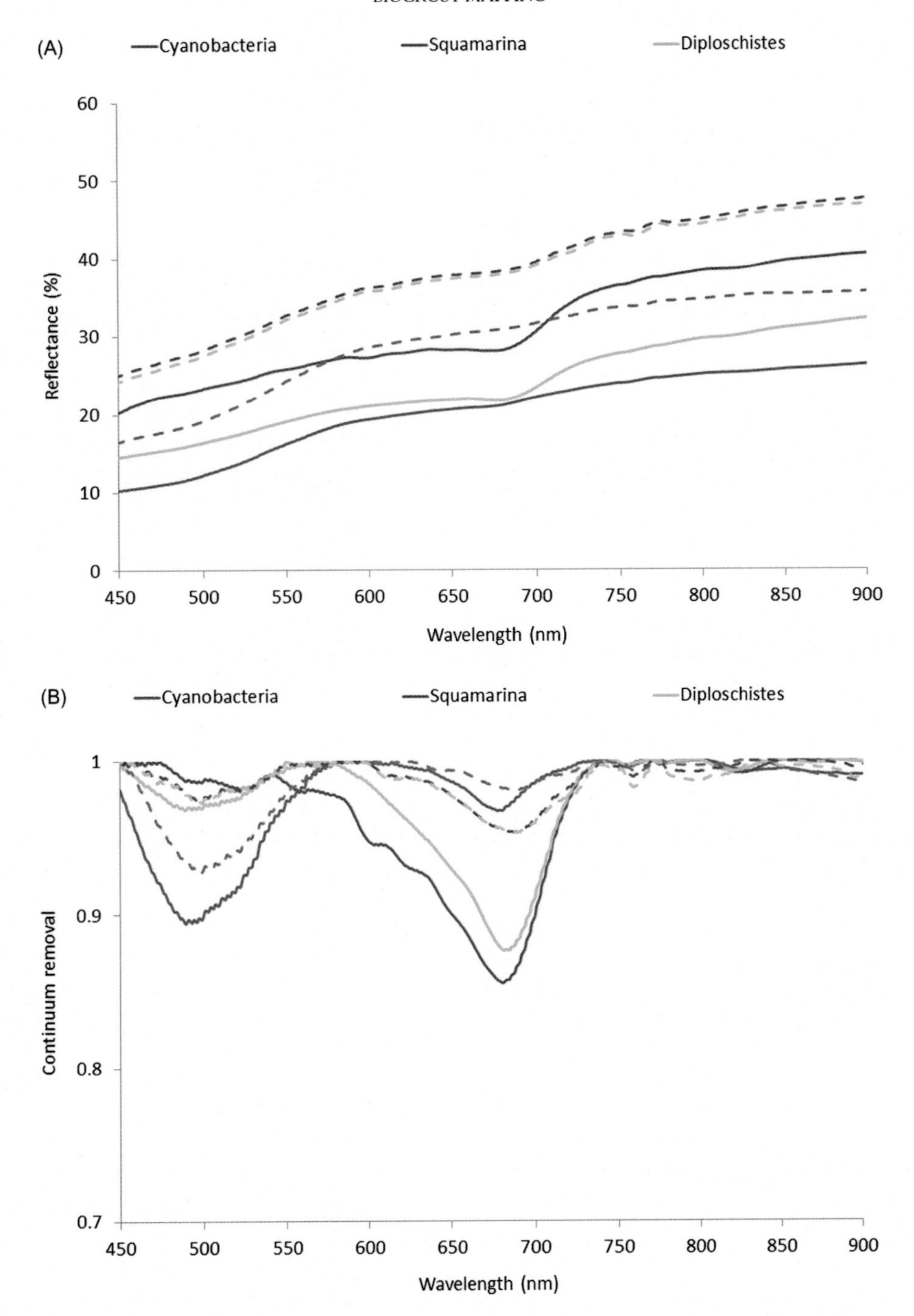

FIGURE 4.15 Spectral signature (A) and continuum removal (B) of cyanobacteria dominated biocrusts and two different lichen dominated biocrust communities (biocrusts mainly composed by the lichen species *Squamarina lentigera and Diploschistes diacapsis*), for wet (continuous line), and dry (dashed lines) conditions.

(Fig. 4.15) and their interaction with the spectral response of the underlying soil (Fig. 4.14), complicate the possibility to make general assumptions about biocrust effects on soil surface reflectance. However, it is clear that their presence needs to be considered when soil or vegetation properties are analyzed using remote sensing techniques and mapping methods are necessary to identify their presence and coverage at image scale.

Biocrust Mapping Methods

Given the importance of biocrusts on ecosystem processes controlling dryland functioning (Weber et al., 2016), and considering their effect on the general spectral response of these areas (Rodríguez-Caballero et al., 2015b), several biocrust mapping methods have been developed using optical reflectivity (Karnieli, 1997; Chen et al., 2005; Weber et al., 2008; Chamizo et al., 2012; Rodriguez-Caballero et al., 2014) and emissivity (Rozenstein and Karnieli, 2015). All of them were developed with the objective to identify and to map biocrusts at ecosystem scale, and they provide a wide range of opportunities to upscale biocrust effects on ecosystem functioning. However, they differ in complexity, information required, user experience needed and final product obtained. Based on these criteria we may consider three main types of biocrust mapping methods: (1) biocrust indices based on multispectral information (hereafter multispectral indices); (2) biocrust indices based on hyperspectral information (hereafter hyperspectral indices); and (3) biocrust surface cover quantification using linear mixture analysis (hereafter biocrust LMA).

Multispectral Indices to Map Biocrust

The first generation of biocrust mapping methods consisted of simple combinations of different spectral bands to identify pixels dominated by biocrust on binary maps (biocrust versus nonbiocrust). Following this approach, two different indices were developed: crust index (CI; Karnieli, 1997) and the biological soil crust index (BSCI; Chen et al., 2005). The CI was developed to map areas dominated by cyanobacteria in the Negev using multispectral optical information from LANDSAT ETM + images. It is based on previous studies, at the same study area, that described increased reflectance in the blue region of the spectra on sandy soils covered by cyanobacteria (Karnieli and Sarafis, 1996).

A similar approach was adopted by Chen et al. (2005), who developed the BSCI. As with the CI, BSCI also used multispectral LANDSAT ETM + images to identify soils covered by biocrust. However, this index was developed to identify areas dominated by lichen in the Gurbantonggut Desert (China) and they did not use the blue region of the spectra.

During the last year, Rozenstein and Karnieli (2015) developed a new multispectral index that, in contrast to previous indices that used VISNIR information, uses thermal emmisitivity to discriminate between sand areas and biocrust in the Negev region, the thermal crust index. This index improved the accuracy of previous methodologies especially when it was combined with CI or the NDVI (Rozenstein and Karnieli, 2015).

Hyperspectral Indices to Map Biocrusts

A different approach was followed by Weber et al. (2008) and Chamizo et al. (2012), who developed two different biocrust mapping indices based on hyperspectral information and the main biocrust spectral absorptions observed in Fig. 4.14: (1) the continuum removal crusts identification algorithm (CRCIA; Weber et al., 2008); and (2) the crust development index (CDI; Chamizo et al., 2012). As multispectral indices, these methodologies produce binary maps. But in contrast to multispectral indices, CRCIA and CDI use hyperspectral information and require more experience in remote sensing techniques. CRCIA was developed

TABLE 4.4 Kappa Coefficients Presented in the Different Studies and Study Areas Where Biocrust Mapping Indices Have Been Applied

	The Negev (Rozenstein and Karnieli, 2015)	Gurbangtun (Chen et al., 2005)	Soebatsfontain (Weber et al., 2008)	Las Amoladeras (Chamizo et al., 2012)	El Cautivo (Chamizo et al., 2012)	El Cautivo (Alonso et al., 2014)
CI	~0.2	–	0.411	–	–	0.49[a] 0.28
BSCI	–	0.82	0.43	–	–	0.48[a] 0.30
CRCIA	–	–	0.831	–	–	0.55[a] 0.55
CDI	–	–	–	1[a]	1[a]	0.81[a] 0.70

[a]*Classification based on field spectra, red letters represent studies in which different calibrations have been conducted in order to discriminate between different types of biocrust.*

to map cyanobacteria dominated biocrust in Soebatsfontain (South Africa), where it showed better capacity to map biocrust than multispectral indices. CDI was developed to be applied in two different semiarid areas in southeast Spain, and in contrast to all previous methodologies, it also discriminates between different biocrust types and developmental stages. Hyperspectral indices are able to identify the subtle spectral differences between sparse vegetation, bare soil, and biocrust (Table 4.4), and improve the accuracy of final biocrust classifications, as previously reported by several authors (Weber et al., 2008; Alonso et al., 2014). However, even hyperspectral indices result in important classification errors when they are applied in heterogeneous areas, in which each pixel is covered by a mixture of semiarid vegetation, bare soil, and biocrusts (Alonso et al., 2014), and they are not able to identify biocrust presence on pixels with more than 30% vegetation coverage (Weber et al., 2008). These occur as a consequence of the similarities in spectral reflectance between biocrust and pixels covered by a mixture of bare soil and sparse vegetation, with similar absorption to biocrust.

Biocrust Surface Cover Quantification

Biocrust-presence maps are useful for applications focused on the spatial distribution of biocrust; however, some applications need accurate estimations of biocrust coverage within a pixel. In order to solve this problem, Hill et al. (1999) performed a spectral mixture analysis in the Nizana region to quantify the fractional cover of biocrusts at subpixel image, and they found difficulties in discriminating biocrusts in the areas dominated by vegetation, which in the end resulted in an underestimation of biocrust coverage. A similar approach was adopted by Rodríguez-Caballero et al. (2014) at el Cautivo experimental area (Spain). In this study the authors used a hyperspectral image with high spatial resolution (1.5 m) to perform a multiple end member linear mixture analysis. However, in contrast to Hill et al. (1999), they divided the area into different land units, dominated by the same surface components, where they performed individual linear mixture analysis. The final accuracy for lichen and cyanobacteria dominated biocrusts was ~0.8, and they showed a clear spatial pattern (Fig. 4.16) that perfectly fit with field observations (Rodriguez-Caballero et al., 2014). These results were recently used to model the effect of biocrust on runoff generation and water erosion at the catchment scale (Rodríguez-Caballero et al., 2015a), showing the potential of biocrust surface cover quantification to advance the role of biocrusts in understanding key processes that control dryland functioning. However, they require

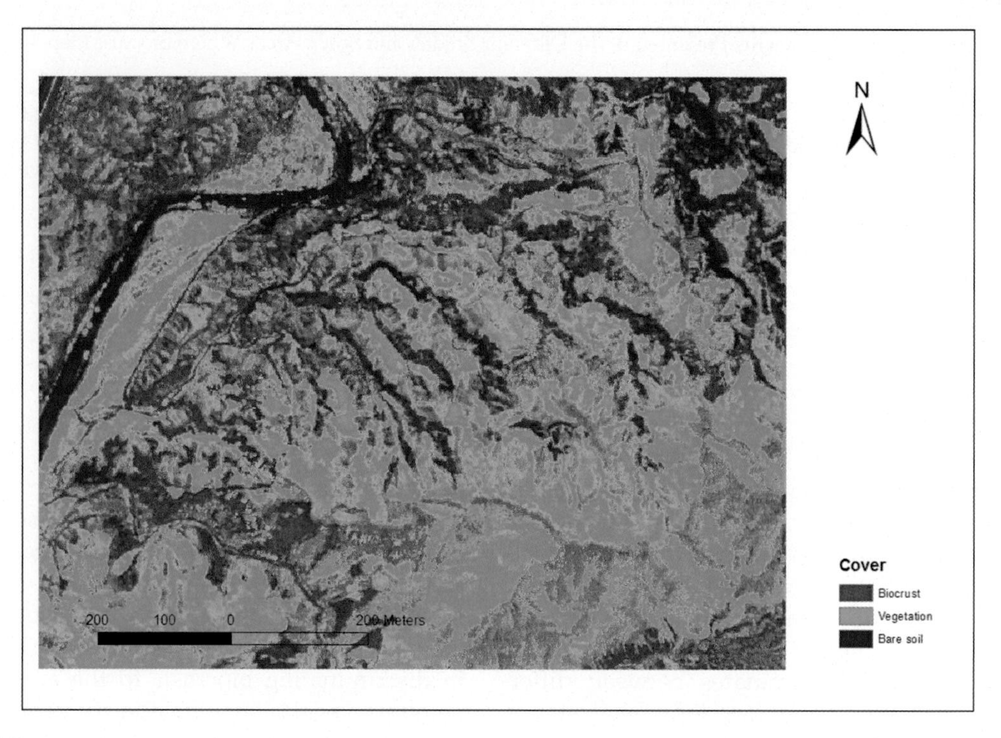

FIGURE 4.16 RGB composition of final biocrust, vegetation, and bare soil coverage at el Cautivo (Rodriguez-Caballero et al., 2014).

both extensive training in technical issues and detailed knowledge of the main coverages present at the field site and their spectral response.

Thus we may conclude that there is not a unique best methodology for biocrust mapping and it needs to be carefully selected for each objective, considering available information and technical knowledge. Moreover, all these methods have been developed for specific regions and difference in biocrust composition and soil spectral properties complicate their transferability. For this reason, they have to be tested in different regions considering the clear differences in the spectral response of the different biocrusts, vegetation, and underlying soils (Fig. 4.14). Moreover, the changes in biocrust spectral response when wet (Fig. 4.15) complicates this challenge and studies are necessary to identify the best conditions to discriminate between bare soil or vegetation and biocrust. Thus although there have been important advances in biocrust mapping during the last two decades, more effort is needed to develop a standard methodology to map different biocrust types all over the world.

CONCLUSIONS

Optical remote sensing in VISNIR and SWIR wavelengths is a good source of information for soil mapping and monitoring, ameliorating the cost and effort needed to have accurate and up-to-date soil maps at different spatial scales. At the laboratory scale the SSLs, containing soil spectra and laboratory data, open up the possibility to use calibration models for the prediction of soil properties exclusively by means of soil spectra. Hyperspectral and multispectral sensors on board aircraft, drones, or satellite platforms give the chance to obtain spatially explicit data avoiding interpolation methods typical of point-based measurements. They can also be used to target more intensive field campaigns for soil mapping and to design optimum sampling schemes using semivariograms or other spatial metrics.

Even though the general use of imaging spectroscopy is well recognized and gives accurate results for a variety of soil properties and processes, its use still is limited. The upcoming satellite missions as well as flexible UAVs (Unmanned Aerial Vehicle) carrying sensors with high spectral, spatial, and temporal resolutions and high signal-to-noise ratio will make this data available to use for soil studies on a regular basis. A synergic use of multispectral or hyperspectral data with other remote or proximal devices such thermal, Lidar, X-ray fluorescence meters, or ground penetrating radar, among others, will expand the information base on surface properties and counteract some of the current limitations of optical datasets. The integration of datasets from different sources is a current challenge for scientists, who should find new connections and design new joint methodologies that fully exploit the potential of such information.

Acknowledgments

The authors acknowledge funding of the data by the European Fleet for Airborne Research (EUFAR) and support from the Spanish Science R&D program with the project AGL2010-17505. Additional support from the EnMAP science program funded by the German Federal Ministry of Economics and Technology and internal support from GFZ in Potsdam are gratefully acknowledged. The authors gratefully acknowledge the Spanish Ministry of Economy and Competitiveness project CARBORAD (CGS2011-27493), the Max Planck Society and the Paul Crutzen Nobel Laureate Fellowship.

References

Adamchuk, V.I., Hummel, J.W., Morgan, M.T., Upadhyaya, S.K., 2004. On-the-go soil sensors for precision agriculture. Comput. Electron. Agric. 44 (1), 71–91.

Alonso, M., Rodríguez-Caballero, E., Chamizo, S., Escribano, P., Cantón, Y., 2014. Evaluación de los diferentes índices para cartografiar biocostras a partir de información espectral. Revista Española de Teledetección 42, 63–82.

Anderson, M.C., Kustas, W.P., Norman, J.M., Hain, C.R., Mecikalski, J.R., Schultz, L., et al., 2011. Mapping daily evapotranspiration at field to continental scales using geostationary and polar orbiting satellite imagery. Hydrol. Earth Syst. Sci. 15, 223–239.

Anyamba, A., Tucker, C.J., 2005. Analysis of Sahelian vegetation dynamics using NOAA-AVHRR NDVI data from 1981–2003. J. Arid Environ. 63, 596–614.

Arribas, A., Gallardo, C., Gaertner, A., Castro, M., 2003. Sensitivity of the Iberian Peninsula climate to a land degradation. Weather Forecast. 20, 477–489.

Asner, G.P., Borghi, C.E., Ojeda, R.A., 2003. Desertification in Central Argentina: changes in ecosystem carbon and nitrogen from imaging spectroscopy. Ecol. Appl. 13, 629–648.

Ballabio, C., Panagos, P., Monatanarella, L., 2016. Mapping topsoil physical properties at European scale using the LUCAS database. Geoderma 261, 110–123.

Bartholomeus, H., Epema, G., Schaepman, M., 2007. Determining iron content in Mediterranean soils in partly vegetated regions, using spectral reflectance and imaging spectroscopy. Int. J. Appl. Earth Observ. Geoinform. 9, 194–203.

Bartholomeus, H.M., Schaepman, M.E., Kooistra, L., Stevens, A., Hoogmoed, W.B., Spaargaren, O.S.P., 2008. Spectral reflectance based indices for soil organic carbon quantification. Geoderma 145, 28–36.

Bayer, A., Bachmann, M., Mueller, A., Kaufmann, H., 2012. A comparison of feature-based MLR and PLS regression techniques for the prediction of three soil constituents in a degraded South African ecosystem. Appl. Environ. Soil Sci. Article ID: 971252, 20 pages.

Belnap, J., Walker, B.J., Munson, S.M., Gill, R.A., 2014. Controls on sediment production in two U.S. deserts. Aeolian Res. 14, 15–24.

Ben-Dor, E., Patkin, K., Banin, A., Karnieli, A., 2002. Mapping of several soil properties using DAIS-7915 hyperspectral scanner data—a case study over clayey soils in Israel. Int. J. Rem. Sens. 23, 1043–1062.

Ben-Dor, E., Goldlshleger, N., Benyamini, Y., Agassi, M., Blumberg, D.G., 2003. The spectral reflectance properties of soil structural crusts in the 1.2 to 2.5 micrometers spectral region. Soil Sci. Soc. Am. J. 67 (1), 289–299.

Ben-Dor, E., Levin, T.N., Singer, A., Karnieli, A., Braun, O., Kidron, G.J., 2006. Quantitative mapping of the soil rubification process on sand dunes using an airborne hyperspectral sensor. Geoderma 131, 1–21.

Ben-Dor, E., Taylor, G.R., Hill, J., Demattê, J.A.M., Whiting, M.L., Chabrillat, S., et al., 2008. Imaging spectrometry for soil applications. Adv. Agron. J. 97, 321–392.

Ben-Dor, E., Chabrillat, S., Demattê, J.A.M., Taylor, G.R., Hill, J., Whiting, M.L., et al., 2009. Using imaging spectroscopy to study soil properties. Remote Sens. Environ. 113, S38–S55.

Ben-Dor, E., Ong, C., Lau, I.C., 2015. Reflectance measurements of soils in the laboratory: standards and protocols. Geoderma 245, 112–124.

Boardman, J., 2007. Soil erosion: the challenge of assessing variation through space and time. In: Geomorphological Variations. Goudie, A.S., Kalvoda, J. (Eds.), Nakladatelsti P3K, Prague, pp. 205–220.

Boer, M.M., Puigdefábregas, J., 2005. Assessment of dryland condition using spatial anomalies of vegetation index values. Int. J. Remote Sens. 26, 4045–4065.

Brevik, E.C., Fenton, T.E., Jaynes, D.B., 2003. Evaluation of the accuracy of a central iowa soil survey and implications for precision soil management. Precis. Agric. 4, 323–334.

Brevik, E.C., Calzolari, C., Miller, B.A., Pereira, P., Kabala, C., Baumgarten, A., Jordán, A., 2016. Soil mapping, classification, and modeling: history and future directions. Geoderma 264, 256–274. http://dx.doi.org/10.1016/j.geoderma.2015.05.017.

Brevik, E., Pereira, P., Munoz-Rojas, M., Miller, B., Cerda, A., Parras-Alcantara, L., et al., 2017. Historical perspectives on soil mapping and process modelling for sustainable land use management. In: Pereira, P., Brevik, E., Munoz-Rojas, M., Miller, B. (Eds.), Soil Mapping and Process Modelling for Sustainable Land Use Management. Elsevier, pp. 3–30.

Brutsaert, W., 1982. Evaporation into the Atmosphere—Theory, History, and Applications Reidel, Dordrecht, Holland. 299 p.

Büdel, B., Darienko, T., Deutschewitz, K., Dojani, S., Friedl, T., Mohr, K.I., et al., 2009. Southern african biological soil crusts are ubiquitous and highly diverse in drylands, being restricted by rajinfall frequency. Microb. Ecol. 57, 229–247.

Cai, G., Xue, Y., Hu, Y., Wang, Y., Guo, J., Luo, Y., et al., 2007. Soil moisture retrieval from MODIS data in Northern China Plain using thermal inertia model. Int. J. Remote Sens. 28, 3567–3581.

Casanovas, P., Black, M., Fretwell, P., Convey, P., 2015. Mapping lichen distribution on the Antarctic Peninsula using remote sensing, lichen spectra and photographic documentation by citizen scientists. Polar Res. 34, 25633. http://dx.doi.org/10.3402/polar.v34.2563.

Castaldi, F., Palombo, A., Santini, F., Pascucci, S., Pignatti, S., Casa, R., 2016. Evaluation of the potential of the current and forthcoming multispectral and hyperspectral imagers to estimate soil texture and organic carbon. Remote Sens. Environ. 179, 54–65.

Chabrillat, S., 2006. Land degradation indicators: spectral indices. Ann. Arid Zone 45 (3-4), 331–354. Special Issue Land Quality Assessment.

Chabrillat, S., Goetz, A.F.H., Krosley, L., Olsen, H.W., 2002. Use of hyperspectral images in the identification and mapping of expansive clay soils and the role of spatial resolution. Remote Sens. Environ. 82, 431–445.

Chabrillat, S., Eisele, A., Guillaso, S., Rogaß, C., Ben-Dor, E., Kaufmann, H., 2011. HYSOMA: an easy-to-use software interface for soil mapping applications of hyperspectral imagery. In: Proceedings 7th EARSeL SIG Imaging Spectroscopy Workshop, Edinburgh, Scotland.

Chabrillat, S., Foerster, S., Steinberg, A., Segl, K., 2014. Prediction of common surface soil properties using airborne and simulated EnMAP hyperspectral images: impact of soil algorithm and sensor characteristic. In: Proceedings 2014 IEEE International Geoscience and Remote Sensing Symposium, IGARSS 2014 & 35th Canadian Symposium on Remote Sensing, Québec City, QC, Canada, pp. 2914–2917.

Chabrillat, S., Naumann, N., Escribano, P., Bachmann, M., Spengler, D., Holzwarth, S., et al., 2016a. Cabo de Gata-Níjar Natural Park 2003-2005—A Multitemporal Hyperspectral Flight Campaign for EnMAP Science Preparatory Activities. EnMAP Flight Campaigns Technical Report, GFZ Data Services.

Chabrillat, S., Guillaso, S., Eisele, A., Ben-Dor, E., 2016b. HYSOMA: a software interface for the derivation of soil maps and determination of common soil properties from point and imaging spectroscopy. Environ. Modell. Softw., submitted for publication.

Chamizo, S., Stevens, A., Cantón, Y., Miralles, I., Domingo, F., Van Wesemael, B., 2012. Discriminating soil crust type, development stage and degree of disturbance in semiarid environments from their spectral characteristics. Eur. J. Soil Sci. 63, 42–53.

Chamizo, S., Cantón, Y., Rodríguez-Caballero, E., Domingo, F., 2016. Biocrusts positively affect the soil water balance in semiarid ecosystems. Ecohydrology. http://dx.doi.org/10.1002/eco.1719.

Chapin, F.S., Matson, P., Mooney, H.A., 2002. Principles of Terrestrial Ecosystem Ecology, Springer-Verlag, New York, USA.

Chehbouni, A., 2001. Estimation of surface sensible heat flux using dual angle observations of radiative surface temperature. Agric. Forest Meteorol. 108, 55–65.

Chen, J., Ming, Y.Z., Wang, L., Shimazaki, H., Tamura, M., 2005. A new index for mapping lichen-dominated biological soil crusts in desert areas. Remote Sens. Environ. 96, 165–175.

Corbane, C., Raclot, D., Jacob, F., Albergel, J., Andrieux, P., 2008. Remote sensing of soil surface characteristics from a multiscale classification approach. Catena 75, 308–318.

Couradeau, E., Karaoz, U., Lim, H., Nunes da Rocha, U., Northen, T., Brodie, E., et al., 2016. Bacteria increase arid-land soil surface temperature through the production of sunscreens. Nat. Commun., 7. http://dx.doi.org/10.1038/ncomms10373.

Crouvi, O., Ben-Dor, E., Beyth, M., Avigad, D., Amit, R., 2006. Quantitative mapping of arid alluvial fan surfaces using field spectrometer and hyperspectral remote sensing. Remote Sens. Environ. 104, 103–117.

D'Odorico, P., Bhattachan, A., Davis, K.F., Ravi, S., Runyan, C.W., 2013. Global desertification: drivers and feedbacks. Adv. Water Resour. 51, 326–344. ISSN: 0309-1708.

Dazzi, C., Monteleone, S., 2007. Anthropogenic processes in the evolution of a soil chronosequence on marly-limestone substrata in an Italian Mediterranean environment. Geoderma 141, 201–209.

De Alba, S., 2001. Modeling the effects of complex topography and patterns of tillage on soil translocation by tillage with mouldboard plough. J. Soil Water Conserv. 56, 335–345.

De Jong, S.M., 1994. Derivation of vegetative variables from a Landsat TM image for modelling soil erosion. Earth Surf. Process. Landform. 19 (2), 165–178.

De Keersmaecker, W., Lhermitte, S., Honnay, O., Farifteh, J., Somers, B., Coppin, P., 2014. How to measure ecosystem stability? An evaluation of the reliability of stability metrics based on remote sensing time series across the major global ecosystems. Global Change Biol. 20, 2149–2161.

De Tar, W.R., Chesson, J.H., Penner, J.V., Ojala, J.C., 2008. Detection of soil properties with airborne hyperspectral measurements of bare fields. Am. Soc. Agric. Biol. Eng. 51, 463–470.

del Barrio, G., Puigdefabregas, J., Sanjuan, M.E., Stellmes, M., Ruiz, A., 2010. Assessment and monitoring of land condition in the Iberian Peninsula, 1989–2000. Remote Sens. Environ. 114, 1817–1832.

Delgado-Baquerizo, M., Maestre, F.T., Gallardo, A., Bowker, M., Wallenstein, D., Quero, J.L., et al., 2013. Decoupling of soil nutrient cycles as a function of aridity in global drylands. Nature 502, 672–676.

Dewitte, O., Jones, A., Elbelrhiti, H., Horion, S., Montanarella, L., 2012. Satellite remote sensing for soil mapping in Africa: an overview. Prog. Phys. Geogr. 36 (4), 514–538.

Dhakal, A.S., Amada, T., Aniya, M., Sharma, R.R., 2002. Detection of areas associated with flood and erosion caused by a heavy rainfall using multitemporal Landsat TM data. Photogrammet. Eng. Remote Sens. 68 (3), 233–239.

Diouf, A., Lambin, E.F., 2001. Monitoring land-cover changes in semi-arid regions: remote sensing data and field observations in the Ferlo, Senegal. J. Arid Environ. 48 (2), 129–148.

Eklundh, L., Olsson, L., 2003. Vegetation index trends for the African Sahel 1982–1999. Geophys. Res. Lett. 30, 131–134.

Elbert, W., Weber, B., Burrows, S., Steinkamp, J., Büdel, B., Andreae, M.O., et al., 2012. Contribution of cryptogamic covers to the global cycles of carbon and nitrogen. Nat. Geosci. 5, 459–462.

Elhag, M., 2016. Evaluation of different soil salinity mapping using remotesensing techniques in arid ecosystems, Saudi Arabia. J. Sens. 2016, 1–8.

Escribano, P., Palacios-Orueta, A., Oyonarte, C., Chabrillat, S., 2010. Spectral properties and sources of variability of ecosystem components in a Mediterranean semiarid environment. J. Arid Environ. 74 (9), 1041–1051.

Fajardo, M., McBratney, A., Whelan, B., 2016. Fuzzy clustering of Vis–NIR spectra for the objective recognition of soil morphological horizons in soil profiles. Geoderma 263, 244–253.

Fang, S., Yu, W., Qi, Y., 2015. Spectra and vegetation index variations in moss soil crust in different seasons, and in wet and dry conditions. Int. J. Appl. Earth Observ. Geoinform. 38, 261–266.

Farifteh, J., Van der Meer, F.D., Atzberger, C.G., Carranza, E.J.M., 2007. Quantitative analysis of salt-affected soil reflectance spectra: a comparison of two adaptive methods (PLSR and ANN). Remote Sens. Environ. 110, 59–78.

Felde, V.J.M.N.L., Peth, S., Uteau-Puschmann, D., Drahorad, S., Felix-Henningsen, P., 2014. Soil microstructure as an under-explored feature of biological soil crust hydrological properties: Case study from the NW negev desert. Biodiver. Conserv. 23, 1687–1708.

Fensholt, R., Rasmussen, K., 2011. Analysis of trends in the Sahelian 'rain-use efficiency' using GIMMS NDVI, RFE and GPCP rainfall data. Remote Sens. Environ. 115, 438–451. 1-758.

Fisher, J.B., Tu, K.P., Baldocchi, D.D., 2008. Global estimates of the land-atmosphere water flux based on monthly AVHRR and ISLSCP-II data, validated at 16 FLUXNET sites. Remote Sens. Environ. 112, 901–919.

Franceschini, M.H.D., Demattê, J.A.M., da Silva Terra, F., Vicente, L.E., Bartholomeus, H., de Souza Filho, C.R., 2015. Prediction of soil properties using imaging spectroscopy: considering fractional vegetation cover to improve accuracy. Int. J. Appl. Earth Observ. Geoinform. 38, 358–370.

Fung, T., 1990. An assessment of TM imagery for land-cover change detection. IEEE Trans. Geosci. Remote Sens. 28 (4), 681–684.

Garcia, M., Ustin, S.L., 2001. Detection of interannual vegetation responses to climatic variability using AVIRIS data in a Coastal Savanna in California. IEEE Trans. Geosci. Remote Sens. 39, 1480–1490.

Garcia, M., Oyonarte, C., Villagarcia, L., Contreras, S., Domingo, F., Puigdefabregas, J., 2008. Monitoring land degradation risk using ASTER data: the non-evaporative fraction as an indicator of ecosystem function. Remote Sens. Environ. 112, 3720–3736.

Garcia, M., Contreras, S., Domingo, F., Puigdefábregas, J., 2009. Mapping land degradation risk: potential of the non-evaporative fraction using ASTER and MODIS data. In: Roeder, A., Hill, J. (Eds.), Recent Advances in Remote Sensing and Geoinformation Processing for Land Degradation Assessment. Taylor & Francis, London.

Garcia, M., Sandholt, I., Ceccato, P., Ridler, M., Mougin, E., Kergoat, L., et al., 2013. Actual evapotranspiration in drylands derived from in-situ and satellite data: Assessing biophysical constraints. Remote Sens. Environ. 131, 103–118.

García, M., Oyonarte, C., Villagarcía, L., Contreras, S., Domingo, F., Puigdefábregas, J., 2008. Monitoring land degradation risk using ASTER data: the non-evaporative fraction as an indicator of ecosystem function. Remote Sens. Environ. 112 (9), 3720–3736. http://books4.elsevierproofcentral.com/authorproofs/1481d89d787f0034d81c430e9a7105cf/MC.

García-Ruiz, J.M., 2010. The effects of land uses on soil erosion in Spain: a review. Catena 81, 1–11.

Gerighausen, H., Menz, G., Kaufmann, H., 2012. Spatially explicit estimation of clay and organic carbon content in agricultural soils using multi-annual imaging spectroscopy data. Appl. Environ. Soil Sci. 2012 Article ID: 868090, 23 pages.

Goetz, S.J., Prince, S.D., Goward, S.N., Thawley, M.M., Small, J., 1999. Satellite remote sensing of primary production: an improved production efficiency modeling approach. Ecol. Model. 122 (3), 239–255.

Gomez, C., Lagacherie, P., Coulouma, G., 2008. Continuum removal versus PLSR method for clay and calcium carbonate content estimation from laboratory and airborne hyperspectral measurements. Geoderma 148, 141–148.

Gomez, C., Lagacherie, P., Coulouma, G., 2012. Regional predictions of eight common soil properties and their spatial structures from hyperspectral Vis–NIR data. Geoderma 189, 176–185.

Gómez, C., Oltra-Carrió, R., Bacha, S., Lagacherie, P., Briottet, X., 2015. Evaluating the sensitivity of clay content prediction to atmospheric effects and degradation of image spatial resolution using hyperspectral VNIR/SWIR imagery. Remote Sens. Environ. 164, 1–15.

Graetz, R.D., Gentle, M.R., 1982. The relationships between reflectance in the Landsat wavebands and the composition of an Australian semi-arid shrub rangeland. Photogrammet. Eng. Remote Sens. 48 (11), 1721–1730.

Green, G.M., 1986. Use of SIR-A and Landsat MSS data in mapping shrub and intershrub vegetation at Koonamore, South Australia. Photogrammet. Eng. Remote Sens. 52 (5), 659–670.

Grimm, M., Jones, R., Montanarella, L., 2002. Soil Erosion Risk in Europe. European Soil Bureau, Institute for Environment & Sustainability, JRC Ispra.

Groom, S., Chabrillat, S., Eisele A., Knaeps E., Raymaekers D., Reusen I., et al., 2010. DJ2.4.1—Report on Higher Performing Water and Soil Algorithm Development: Version 1, EUFAR FP7 Deliverable.

Grossman, R.B., 1983. Entisols. Pedogenesis and soil taxonomy. Part II. The soil orders. In: Wilding, L.P., Smeck, N.E., Hall, G.F. (Eds.), Developments in Soil Science. Elsevier, Amstersdam, pp. 55–90.

Guanter, L., Kaufmann, H., Segl, K., Förster, S., Rogaß, C., Chabrillat, S., et al., 2015. The EnMAP spaceborne imaging spectroscopy mission for earth observation. Remote Sens. 7 (7), 8830–8857.

Hartemink, A.E., Minasny, B., 2014. Towards digital soil morphometrics. Geoderma 230, 305–317.

Haubrock, S.-N., Chabrillat, S., Kuhnert, M., Hostert, P., Kaufmann, H., 2008b. Surface soil moisture quantification and validation based on hyperspectral data and field measurements. J. Appl. Remote Sens. 2 023552.

Haubrock, S.-N., Chabrillat, S., Lemmnitz, C., Kaufmann, H., 2008a. Surface soil moisture quantification models from reflectance data under field conditions. Int. J. Remote Sens. 29 (1), 3–29.

Hein, L., De Ridder, N., 2006. Desertification in the Sahel: a reinterpretation. Global Change Biol. 12, 75.

Henderson-Sellers, A., Irannejad, P., Mcguffie, K., 2008. Future desertification and climate change: the need for land-surface system evaluation improvement. Global Planet. Change 64, 129–138.

Hill, J., Schütt, B., 2000. Mapping complex patterns of erosion and stability in dry Mediterranean ecosystems. Remote Sens. Environ 74, 557–569.

Hill, J., Udelhoven, T., Schütt, B., Yair, A., 1999. Differentiating biological soil crusts in a sandy arid ecosystem based on multi- and hyperspectral remote sensing data. In: Schaepmann, M., Schläpfer, D., Itten, K. (Eds.), 1st EARSEL Workshop on Imaging Spectroscopy. Proceedings of the EARSEL Workshop, Zürich, 6–8 October 1998. EARSEL Secretariat, Paris, pp. 427–436.

Hively, W.D., McCarty, G.W., Reeves, J.B., Lang, M.W., Oesterling, R.A., Delwiche, S.R., 2011. Use of airborne hyperspectral imagery to MAP soil properties in tilled agricultural fields. Appl. Environ. Soil Sci. Article ID: 358193, 20 pages.

Horion, S., Prishchepov, A.V., Verbesselt, J., de Beurs, K., Tagesson, T., Fensholt, R., 2016. Revealing turning points in ecosystem functioning over the Northern Eurasian agricultural frontier. Global Change Biol. 22 (8), 2801–2817.

Hubbard, B.E., Crowley, J.K., 2005. Mineral mapping on the Chilean-Bolivian Altiplano using co-orbital ALI, ASTER and hyperion imagery: data dimensionality issues and solutions. Remote Sens. Environ. 99 (1-2), 173–186.

Hubbard, B.E., Crowle, J.K., Zimbelman, D.R., 2003. Comparative alteration mineral mapping using visible to shortwave infrared (0.4–2.4 μm) Hyperion, ALI, and ASTER imagery. IEEE Trans. Geosci. Remote Sens. 41 (6 Paer I), 1401–1410.

Huber, S., Fensholt, R., Rasmussen, K., 2011. Water availability as the driver of vegetation dynamics in the African Sahel from 1982 to 2007. Global Planet. Change 76, 186–195.

Ji, W., Viscarra Rossel, R., Shi, Z., 2015. Accounting for the effects of water and the environment on proximally sensed vis–NIR spectra and their calibrations. Eur. J. Soil Sci. 66, 555–565.

Ji, W., Li, S., Chen, S., Shi, Z., Viscarra Rossel, R.A., Mouazen, A.M., 2016. Prediction of soil attributes using the Chinese soil spectral library and standardized spectra recorded at field conditions. Soil Till. Res. 155, 492–500.

Jiménez, C., Prigent, C., Mueller, B., Seneviratne, S.I., McCabe, M.F., Wood, E.F., et al., 2011. Global intercomparison of 12 land surface heat flux estimates. J. Geophys. Res. At. 116 (2), D02102.

Kalma, J.D., Mcvicar, T.R., Mccabe, M.F., 2008. Estimating land surface evaporation: a review of methods using remotely sensed surface temperature data. Surv. Geophy. 29, 421–469.

Karnieli, A., 1997. Development and implementation of spectral crust index over dune sands. Int. J. Remote Sens. 18, 1207–1220.

Karnieli, A., Dall'Olmo, G., 2003. Remote-sensing monitoring of desertification, phenology, and droughts. Manage. Environ. Q. Int. J. 14 (1), 22–38.

Karnieli, A., Sarafis, V., 1996. Reflectance spectrophotometry of cyanobacteria within soil crusts—a diagnostic tool. Int. J. Remote Sens. 17 (8), 1609–1615.

Karnieli, A., Tsoar, H., 1995. Spectral reflectance of biogenic crust developed on desert dune sand along the Israel–Egypt border. Int. J. Remote Sens. 16 (2), 369–374.

Karnieli, A., Shachak, M., Tsoar, H., Zaady, E., Kaufman, Y., Danin, A., et al., 1996. The effect of microphytes on the spectral reflectance of vegetation in semiarid regions. Remote Sens. Environ. 57, 88–96.

Kopačková, V., Ben-Dor, E., 2016. Normalizing reflectance from different spectrometers and protocols with an internal soil standard. Int. J. Remote Sens. 37 (6), 1276–1290.

Kustas, W.P., Norman, J.M., 1999. Evaluation of soil and vegetation heat flux predictions using a simple two-source model with radiometric temperatures for partial canopy cover. Agric. Forest Meteorol. 94, 13–29.

Kustas, W.P., Anderson, M.C., Alfieri, J.G., Prueger, J.H., Geli, H.M., Neale, C.M., 2016. Mapping evapotranspiration with high-resolution aircraft imagery over vineyards using one-and two-source modeling schemes. Hydrol. Earth Syst. Sci 20 (4), 1523.

Lacaze, B., Caselles, V., Coll, C., Hill, J., Hoff, C., De Jong, S., et al., 1996. Integrated Approaches to Desertification Mapping and Monitoring in the Mediterranean Basin. Final Report of the DeMon-1 Project. Joint Research Centre, European Commission., EUR 16448EN, 165 pp.

Lagacherie, P., Baret, F., Feret, J.-B., Madeira Netto, J.M., Robbez-Masson, J.M., 2008. Estimation of soil clay and calcium carbonate using laboratory, field and airborne hyperspectral measurements. Remote Sens. Environ. 112, 825–835.

Lagacherie, P., Gomez, C., Bailly, J.S., Baret, F., Coulouma, G., 2010. The use of hyperspectral imagery for digital soil mapping in Mediterranean areas. Digital Soil Mapping. Prog. Soil Sci. 2, 93–102.

Lanjeri, S., Melía, J., Segarra, D., 2001. A multi-temporal masking classification method for vineyard monitoring in central Spain. Int. J. Remote Sens. 22 (16), 3167–3186.

Lenhart, K., Weber, B., Elbert, W., Steinkamp, J., Clough, T., Crutzen, P., et al., 2015. Nitrous oxide and methane emissions from cryptogamic covers. Global Change Biol. 21 (10), 3889–3900.

Leuning, R., Zhang, Y.Q., Rajaud, A., Cleugh, H., Tu, K., 2008. A simple surface conductance model to estimate regional evaporation using MODIS leaf area index and the Penman-Monteith equation. Water Resour. Res. 44 http://dx.doi.org/10.1029/2007WR006562.

Levin, N., Kidron, G.J., Ben-Dor, E., 2007. Surface properties of stabilizing coastal dunes: combining spectral and field analyses. Sedimentology 54, 771–788.

Lindstrom, M.J., Nelson, W.W., Schumacher, T.E., 1992. Quantifying tillage erosion rates due to moldboard plowing. Soil Till. Res. 24, 243–255.

Lo Curzio, S., Magliulo, P., 2009. Soil erosion assessment using geomorphological remote sensing techniques: an example from southern Italy. Earth Surf. Process. Landform. 35 (3), 262–271.

Lu, X., Zhuang, Q., 2010. Evaluating evapotranspiration and water-use efficiency of terrestrial ecosystems in the conterminous United States using MODIS and AmeriFlux data. Remote Sens. Environ. 114 (9), 1924–1939.

Ludwig, J.A., Tongway, D.J., 2000. Viewing rangelands as landscape systems. In: Arnalds, O., Archer, S. (Eds.), Rangeland Desertification. Springer, Dordrecht, Netherlands.

Madeira, J., Bedidi, A., Cervelle, B., Pouget, M., Flay, N., 1997. Visible spectrometric indices of hematite (Hm) and goethite (Gt) content in lateritic soils: the application of a Thematic Mapper (TM) image for soil-mapping in Brasilia, Brazil. Int. J. Remote Sens. 18 (13), 2835–2852.

Maestre, F.T., Valladares, F., Reynolds, J.F., 2006. The stress-gradient hypothesis does not fit all relationships between plant–plant interactions and abiotic stress: further insights from arid environments. J. Ecol. 94, 17–22.

Maestre, F.T., Castillo-Monroy, A.P., Bowker, M.A., Ochoa-Hueso, R., 2012. Species richness effects on ecosystem multifunctionality depend on evenness, composition and spatial pattern. J. Ecol. 100 (2), 317–330.

Maestre, F.T., Eldridge, D.J., Soliveres, S., Kéfi, S., Delgado-Baquerizo, M., Bowker, M.A., et al., 2016. Structure and functioning of dryland ecosystems in a changing world. Ann. Rev. Ecol. Evol. Syst. 47, 215–237.

Manchanda, M.L., Kudrat, M., Tiwari, A.K., 2002. Soil survey and mapping using remote sensing. Trop. Ecol. 43 (1), 61–74.

Mathieu, R., Pouget, M., Cervelle, B., Escadafal, R., 1998. Relationships between satellitebased radiometric indices simulated using laboratory reflectancedata and typic soil color of an arid environment. Remote Sens. Environ. 66, 17–28.

Mathieu, R., Cervelle, B., Rémy, D., Pouget, M., 2007. Field-based and spectral indicators for soil erosion mapping in semi-arid mediterranean environments (Coastal Cordillera of central Chile). Earth Surf. Process. Landform. 32, 13–31.

Mbow, C., Brandt, M., Ouedraogo, I., de Leeuw, J., Marshall, M., 2015. What four decades of earth observation tell us about land degradation in the Sahel? Remote Sens. 7, 4048–4067.

Metternicht, G.I., Zinck, J.A., 2003. Remote sensing of soil salinity: potentials and constraints. Remote sens. Environ. 85 (1), 1–20.

Mildrexler, D.J., Zhao, M.S., Running, S.W., 2009. Testing a MODIS Global Disturbance Index across North America. Remote Sens. Environ. 113, 2103–2117.

Minu, S., Shetty, A., Gopal, B., 2016. Review of preprocessing techniques used in soil property prediction from hyperspectral data. Cogent Geosci. 2 (1), 1–7.

Miralles-Mellado, I., Cantón, Y., Solé-Benet, A., 2011. Two-dimensional porosity of crusted silty soils: indicators of soil quality in semiarid rangelands? Soil Sci. Soc. Am. J. 75, 1330–1342.

Morgan, R.P.C., 2005. Soil Erosion and Conservation. Blackwell Publishing Ltd.

Morillas, L., García, M., Nieto, H., Villagarcia, L., Sandholt, I., Gonzalez-Dugo, M.P., et al., 2013. Using radiometric surface temperature for surface energy flux estimation in Mediterranean drylands from a two-source perspective. Remote Sens. Environ. 136, 234–246.

Mouazen, A.M., Karoui, R., De Baerdemaeker, J., Ramon, H., 2006. Characterization of soil water content using measured visible and near infrared spectra. Soil Sci. Soc. Am. J. 70 (4), 1295–1302.

Mu, Q., Zhao, M., Running, S.W., 2011. Improvements to a MODIS global terrestrial evapotranspiration algorithm. Remote Sens. Environ. 115 (8), 1781–1800.

Mulder, V.L., De Bruin, S., Schaepman, M.E., Mayr, T.R., 2011. The use of remote sensing in soil and terrain mapping—a review. Geoderma 162 (1-2), 1–19.

Muñoz-Rojas, M., Jordán, A., Zavala, L.M., De la Rosa, D., Abd-Elmabod, S.K., Anaya-Romero, M., 2012. Organic carbon stocks in Mediterranean soil types under different land uses (Southern Spain). Solid Earth 3 (2), 375.

Munoz-Rojas, M., Pereira, P., Brevik, E., Cerdo, A., Jordan, A., 2017. Soil mapping and processes models for sustainable land management applied to modern challenges. In: Pereira, P., Brevik, E.C., Munoz-Rojas, M., Miller, B.A. (Eds.), Soil Mapping and Process Modeling for Sustainable Land Use Management. Elsevier, pp. 153–194.

Nachtergaele, F.O., 1999. From the soil map of the world to the digital global soil and terrain database: 1960–2002. In: Sumner, M.E. (Ed.), Handbook of Soil Science. CRC Press, Boca Raton, pp. H5–17.

Nanni, M.R., Demattê, J.A.M., 2006. Spectral reflectance methodology in comparison to traditional soil analysis. Soil Sci. Soc. Am. J. 70, 393–407.

Newman, B.D., Breshears, D.D., Gard, M.O., 2010. Evapotranspiration partitioning in a semiarid woodland: ecohydrologic heterogeneity and connecitvity of vegetation patches. Vadose Zone J. 9, 561–572.

Nocita, M., Stevens, A., van Wesemael, B., Aitkenhead, M., Bachmann, M., Barthes, B., et al., 2015. Soil spectroscopy: an alternative to wet chemistry for soil monitoring. Adv. Agron. 132, 139–159.

O'Neill, A.L., 1994. Reflectance spectra of microphytic soil crusts in semi-arid Australia. Int. J. Remote Sens. 15 (3), 675–681.

Ortega, C., 1984. Estudio Agrobiológico de la Provincia de Toledo. Consejo Superior de Investigaciones Científicas C.S.I.C.—Instituto Provincial de Investigaciones y Estudios Toledanos. Mapa de suelos E. 1:200.000.

Ouerghemmi, W., Gomez, C., Naceur, S., Lagacherie, P., 2016. Semi-blind source separation for the estimation of the clay content over semi-vegetated areas using VNIR/SWIR hyperspectral airborne data. Remote Sens. Environ. 181, 251–263.

Oyonarte, C., Aranda, V., Durante, P., 2008. Soil surface properties in Mediterranean mountain ecosystems: Effects of environmental factors and implications of management. Forest Ecol. Manage. 254, 156–165.

Panagos, P., Borrelli, P., Poesen, J., Ballabio, C., Lugato, E., Meusburger, K., et al., 2015. The new assessment of soil loss by water erosion in Europe. Environ. Sci. Policy 54, 438–447.

Pereira, P., Brevik, E., Munoz-Rojas, M., Miller, B., Smetanova, A., Depellegrin, D., et al., 2017. Soil mapping and process modelling for sustainable land management. In: Pereira, P., Brevik, E., Munoz-Rojas, M., Miller, B. (Eds.), Soil Mapping and Process Modelling for Sustainable Land Use Management. Elsevier, pp. 31–64.

Pilon, P.G., Howarth, P.J., Bullock, R.A., 1988. An enhanced classification approach to change detection in semi-arid environments. Photogramm. Eng. Remote Sens. 54 (12), 1709–1716.

Pinet, P.C., Kaufmann, C., Hill, J., 2006. Imaging spectroscopy of changing Earth's surface: a major step toward the quantitative monitoring of land degradation and desertification. Comptes Rendus Geosci. 338, 1042–1048.

Pinker, R.T., Karnieli, A., 1995. Characteristic spectral reflectance of a semi-arid environment. Int. J. Remote Sens. 16 (7), 1341–1363.

Previtali, F., 2014. Pedoenvironments of the Mediterranean countries: resources and threats. In: Kapur, S., Ersahin, S. (Eds.), Soil Security for Ecosystem Management. Springer, New York, NY, USA, pp. 61–82. Chapter 4.

Prince, S.D., Brown De Colstoun, E., Kravitz, L.L., 1998. Evidence from rain-use efficiencies does not indicate extensive Sahelian desertification. Global Change Biol. 4, 359–374.

Prince, S.D., Wessels, K.J., Tucker, C.J., Nicholson, S.E., 2007. Desertification in the Sahel: a reinterpretation of a reinterpretation. Global Change Biol. 13, 1308–1313.

Puigdefabregas, J., 1995. Desertification—stress beyond resilience, exploring a unifying process structure. Ambio 24, 311–313.

Reusen, I., Adarb, S., Bachmann, M., Beekhuizen, J., Ben-Dor, E., Biesemans, J., et al., 2010. Deployment of quality layers for airborne hyperspectral Imagery and end-to-end water and soil Products (HyQuaPro). In: Proceedings, ISPRS Commission VII Mid-Term Symposium (Vienna 2010).

Richter, N., Chabrillat, S., Kaufmann, H., 2007. Enhanced quantification of soil variables linked with soil degradation using imaging spectroscopy. In: Reusen, I., Cools, J., (Eds.), Proceedings 5th EARSeL Workshop, Bruges, Belgium, 23–25 April 2007, on CD-ROM.

Richter, N., Jarmer, T., Chabrillat, S., Oyonarte, C., Hostert, P., Kaufmann, H., 2009. Free iron oxide determination in Mediterranean soils using diffuse reflectance spectroscopy. Soil Sci. Soc. Am. J. 73 (1), 72–81.

Rodríguez-Caballero, E., Cantón, Y., Chamizo, S., Afana, A., Solé-Benet, A., 2012. Effects of biological soil crusts on surface roughness and implications for runoff and erosion. Geomorphology 145-146, 81–89.

Rodríguez-Caballero, E., Escribano, P., Cantón, Y., 2014. Advanced image processing methods as a tool to map and quantify different types of biological soil crust. ISPRS J. Photogramm. Remote Sens. 90, 59–67.

Rodríguez-Caballero, E., Cantón, Y., Jetten, V., 2015a. Biological soil crust effects must be included to accurately model infiltration and erosion in drylands: an example from Tabernas Badlands. Geomorphology 241, 331–342.

Rodríguez-Caballero, E., Knerr, T., Weber, B., 2015b. Importance of biocrusts in dryland monitoring using spectral indices. Remote Sens. Environ. 170, 32–39.

Rozenstein, O., Karnieli, A., 2015. Identification and characterization of biological soil crusts in a sand dune desert environment across Israel-Egypt border using LWIR emittance spectroscopy. J. Arid Environ. 112, 75–86.

Rykiel, E.J., 1985. Towards a definition of ecological disturbance. Aust. J. Ecol. 10 (3), 361–365.

Sandholt, I., Rasmussen, K., Andersen, J., 2002. A simple interpretation of the surface temperature/vegetation index space for assessment of surface moisture status. Remote Sens. Environ. 79, 213–224.

Schmid, T., Koch, M., Gumuzzio, J., 2001. Spectral and textural classification of multi-source imagery to identify soil degradation stages in semi-arid environments. In: Owe, M., D'Urso, G., Zilioli, E. (Eds.), Remote Sensing for Agriculture, Ecosystems, and Hydrology II, Proc. SPIE, vol. 4171, pp. 376–383.

Schmid, T., Koch, M., Gumuzzio, J., 2008. Application of hyperspectral imagery to map soil salinity. In: Metternicht, G., Zinck, A. (Eds.), Remote Sensing of Soil Salinization: Impact and Land Management. CRC Press, Taylor and Francis Publisher, pp. 113–139. Chapter 7.

Schmid, T., Rodríguez-Rastrero, M., Escribano, P., Palacios-Orueta, A., Ben-Dor, E., Plaza, A., et al., 2016. Characterization of soil erosion indicators using hyperspectral data from a Mediterranean rainfed cultivated region. IEEE J. Select. Top. Appl. Earth Observ. Remote Sens. 9 (2), 845–860.

Selige, T., Böhner, J., Schmidhalter, U., 2006. High resolution topsoil mapping using hyperspectral image and field data in multivariate regression modeling procedures. Geoderma 136, 235–244.

Shepherd, K.D., Walsh, M.G., 2002. Development of reflectance spectral libraries for characterization of soil properties. Soil Sci. Soc. Am. J. 66 (3), 988–998.

Shoshany, M., Goldshleger, N., Chudnovsky, A., 2013. Monitoring of agricultural soil degradation by remote-sensing methods: a review. Int. J. Remote Sens. 34 (17), 6152–6181.

Sobrino, J.A., El Kharraz, M.H., Cuenca, J., Raissouni, N., 1998. Thermal inertia mapping from NOAA-AVHRR data. Synergist. Use Multisens. Data Land Process. 22, 655–667.

Soil Survey Staff, 1990. Keys to soil taxonomy. SMSS Technical Monograph, N° 19, 422, Virginia Polytechnic and State University, Blacksburg, VA.

Soriano-Disla, J.M., Janik, L.J., Viscarra Rossel, R.A., Macdonald, L.M., McLaughlin, M.J., 2014. The performance of visible, near-, and mid-infrared reflectance spectroscopy for prediction of soil physical, chemical, and biological properties. Appl. Spectrosc. Rev. 49 (2), 139–186.

Staenz, K., Mueller, A., Heiden, U., 2013. Overview of terrestrial imaging spectroscopy missions. In: Proceedings 2013 IEEE International Geoscience and Remote Sensing Symposium, IGARSS 2013, Melbourne, Australia, pp. 3502–3505.

Steinberg, A., Chabrillat, S., Stevens, A., Segl, K., Foerster, S., 2016. Prediction of common surface soil properties based on Vis-NIR airborne and simulated EnMAP imaging spectroscopy data: prediction accuracy and influence of spatial resolution. Remote Sens. 8, 613. http://dx.doi.org/10.3390/rs8070613.

Stenberg, B., Viscarra Rossel, R.A., Mouazen, A.M., Wetterlind, J., 2010. Visible and near infrared spectroscopy in soil science. In: Donald, L.S. (Ed.), Advances in Agronomy. Academic Press, pp. 163–215.

Stevens, A., van Wesemael, B., Bartholomeus, H., Rosillon, D., Tychon, B., Ben-Dor, E., 2008. Laboratory, field and airborne spectroscopy for monitoring organic carbon content in agricultural soils. Geoderma 144 (1), 395–404.

Stevens, A., Udelhoven, T., Denis, A., Tychon, B., Lioy, R., Hoffmann, L., et al., 2010. Measuring soil organic carbon in croplands at regional scale using airborne imaging spectroscopy. Geoderma 158 (1), 32–45.

Stevens, A., Nocita, M., Tóth, G., Montanarella, L., Van Wesemael, B., 2013. Prediction of soil organic carbon at the European scale by visible and near infrared reflectance spectroscopy. PLoS ONE 8 (6), 1–13.

Stolte, J., Tesfai, M., Øygarden, L., Kværnø, S., Keizer, J., Verheijen, F., et al., 2016. Soil Threats in Europe; EUR 27607 EN; http://dx.doi.org/10.2788/488054 (print); http://dx.doi.org/10.2788/828742 (online).

Stoner, E.R., Baumgardner, M.F., 1981. Characteristic variations in reflectance of surface soils. Soil Sci. Soc. Am. J. 45, 1161–1165.

Su, Z., 2002. The Surface Energy Balance System (SEBS) for estimation of turbulent heat fluxes. Hydrol. Earth Syst. Sci. 6, 85–99.

Thomas, A.D., Dougill, A.J., 2007. Spatial and temporal distribution of cyanobacterial soil crusts in the Kalahari: implications for soil surface properties. Geomorphology 85, 17–29.

UNCCD, 1996. United Nations Convention to Combat Desertification in Those Countries Experiencing Serious Drought and/or Desertification, Particularly in Africa. U.N. Doc., A/AC.241/27, 71.

Ustin, S.L., Valko, P.G., Kefauver, S.C., Santos, M.J., Zimpfer, J.F., Smith, S.D., 2009. Remote sensing of biological soil crust under simulated climate change manipulations in the Mojave Desert. Remote Sens. Environ. 113, 317–328.

Van der Linden, S., Rabe, A., Held, M., Jakimow, B., Leitão, P.J., Okujeni, A., et al., 2015. The EnMAP-Box—a toolbox and application programming interface for EnMAP data processing. Remote Sens. 7 (9), 11249–11266.

Van der Werff, H., Van der Meer, F., 2015. Sentinel-2 for mapping iron absorption feature parameters. Remote Sens 7, 12635–12653.

Vaudour, E., Gilliot, J.M., Bel, L., Lefevre, J., Chehdi, K., 2016. Regional prediction of soil organic carbon content over temperate croplands using visible near-infrared airborne hyperspectral imagery and synchronous field spectra. Int. J. Appl. Earth Observ. Geoinform. 49, 24–38.

Verstraeten, W.W., Veroustraete, F., Van Der Sande, C.J., Grootaers, I., Feyen, J., 2006. Soil moisture retrieval using thermal inertia, determined with visible and thermal spaceborne data, validated for European forests. Remote Sens. Environ. 101, 299–314.

Viscarra Rossel, R.A., Adamchuk, V.I., Sudduth, K.A., McKenzie, N.J., Lobsey, C., 2011. Proximal soil sensing: an effective approach for soil measurements in space and time. Adv. Agron. 113, 243–291.

Viscarra Rossel, R.A., Cattle, S.R., Ortega, A., Fouad, Y., 2009. In situ measurements of soil colour, mineral composition and clay content by vis–NIR spectroscopy. Geoderma 150 (3), 253–266.

Viscarra Rossel, R.A., Behrens, T., Ben-Dor, E., Brown, D.J., Demattê, J.A.M., Shepherd, K.D., et al., 2016. A global spectral library to characterize the world's soil. Earth Sci. Rev. 155, 198–230.

Vrieling, A., 2006. Satellite remote sensing for water erosion assessment: a review. Catena 65, 2–18.

Vrieling, A., De Jong, S.M., Sterk, G., Rodrigues, S.C., 2008. Timing of erosion and satellite data: a multi-resolution approach to soil erosion risk mapping. Int. J. Appl. Earth Obs. Geoinf. 10 (3), 267–281.

Wang, B., Zhang, G.H., Zhang, X.C., Li, Z.W., Su, Z.L., Yi, T., et al., 2014. Effects of near soil surface characteristics on soil detachment by overland flow in a natural succession grassland. Soil Sci. Soc. Am. J. 78, 589–597.

Wang, Q., Takahashi, H., 1998. Regional hydrological effects of grassland degradation in the Loess Plateau of China. Hydrol. Process. 12, 2279–2288.

Weber, B., Hill, J., 2016. Remote sensing of biological soil crusts at different scales. In: Biological Soil Crusts: An Organizing Principle in Drylands. Springer International Publishing, pp. 215–234.

Weber, B., Olehowski, C., Knerr, T., Hill, J., Deutschewitz, K., Wessels, D.C.J., et al., 2008. A new approach for mapping of Biological Soil Crusts in semidesert areas with hyperspectral imagery. Remote Sens. Environ. 112, 2187–2201.

Weber, B., Wu, D., Tamm, A., Ruckteschler, N., Rodríguez-Caballero, E., Steinkamp, J., et al., 2015. Biological soil crusts accelerate the nitrogen cycle through large NO and HONO emissions in drylands. Proc. Natl. Acad. Sci. U.S.A. 112, 15384–15389.

Weber, B., Büdel, B., Belnap, J., 2016. Biological Soil Crusts: An Organizing Principle in Drylands. Springer, Switzerland.

Wessels, D.C.J., van Vuuren, D.R.J., 1986. Landsat imagery its possible use in mapping the distribution of major lichen communities in the Namib Desert, South West Africa. Madoqua 14 (4), 369–373.

Wessels, K.J., Prince, S.D., Reshef, I., 2008. Mapping land degradation by comparison of vegetation production to spatially derived estimates of potential production. J. Arid Environ. 72, 1940–1949.

Whiting, M.L., Li, L., Ustin, S.L., 2004. Predicting water content using Gaussian model on soil spectra. Remote Sens. Environ. 89, 535–552.

Xie, Y., Sha, Z., Yu, M., 2008. Remote sensing imagery in vegetation mapping: a review. J. Plant Ecol. 1 (1), 9–23.

Zaady, E., Karnieli, A., Shachak, M., 2007. Applying a field spectroscopy technique for assessing successional trends of biological soil crusts in a semi-arid environment. J. Arid Environ. 70 (3), 463–477.

Zhang, Y., Leuning, R., Hutley, L.B., Beringer, J., Mchugh, I., Walker, J.P., 2010. Using long-term water balances to parameterize surface conductances and calculate evaporation at 0.05° spatial resolution. Water Resour. Res. 46, W05512.

Further Reading

Cudahy, T., Caccetta, M., Lau, I., Rodger, A., Laukamp, C., Ong, C., et al., 2012. Satellite ASTER Geoscience Map of Australia, vol. 1. CSIRO. Data Collection.

Domingo, F., Serrano-Ortiz, P., Were, A., Villagarcía, L., García, M., Ramírez, D.A., et al., 2011. Carbon and water exchange in semiarid ecosystems in SE Spain. J. Arid Environ. 75, 1271–1281.

Ibáñez, J.J., Bello, A., García-Álvarez, A., 2005. La conservación de los suelos Europeos. Un análisis crítico de la actual estrategia de la Unión Europea. Protección del Suelo y el Desarrollo Sostenible, Serie Medio Ambiente, vol. 6, Publicaciones del IGME, MEC, Madrid, pp. 133–161.

Priestley, C.H.B., Taylor, R.J., 1972. On the assessment of surface heat flux and evaporation using large-scale parameters. Mon. Weather Rev. 100, 81–92.

Schlesinger, W.H., Reynolds, J.F., Cunningham, G.L., Huenneke, L.F., Jarrell, W.M., Virginia, R.A., et al., 1990. Biological feedbacks in global desertification. Science 247, 1043–1048.

Weinzierl, T., Wehberg, J., Böhner, J., Conrad, O., 2016. Spatial assessment of land degradation risk for the okavango river catchment, Southern Africa. Land Degrad. Dev. 27, 281–294.

CHAPTER

5

Geographic Information Systems and Spatial Statistics Applied for Soil Mapping: A Contribution to Land Use Management

Bradley A. Miller[1,2]

[1]Iowa State University, Ames, IA, United States [2]Leibniz Centre for Agricultural Landscape Research (ZALF), Müncheberg, Germany

INTRODUCTION

Soil mapping provides important information about the characteristics and condition of the land. There are generally two levels to a soil map for land use management. The first is an inventory of soil properties, which by themselves describe the condition of the soils when they were mapped. The second level consists of interpretations. This additional information, synthesized from the soil properties and site context, has a key role in guiding land owners on how to wisely manage their land. The interpretations provided in soil surveys can range from site limitations for installing septic systems to the suitability of different crops for a particular soil to appropriate conservation practices (Aandahl, 1957; FAO, 1967; Soil Survey Staff, 1993).

Although the first soil surveys were largely for the purpose of land valuation and taxation (Krupenikov, 1993), the explosion in soil surveying at the beginning of the 20th century was more focused on the suitability of land for certain crops and how to make better use of the land (Miller and Schaetzl, 2014). After the Dust Bowl environmental disaster in the United States, soil survey programs turned their attention to conservation practices and even retirement of land that was not suitable for the production of certain crops (Helms, 2008). The key theme running through these goals of soil survey is that soil properties and the environment that they reflect determine what land use management will be sustainable at a particular site. Fundamental to that information is, of course, location; place matters. Therefore we need the best maps possible for sustainable

Soil Mapping and Process Modeling for Sustainable Land Use Management.
DOI: http://dx.doi.org/10.1016/B978-0-12-805200-6.00005-0

land use management policies and local landowner decisions (Pereira, 2017; see Chapter 3, Goal Oriented Soil Mapping: Applying Modern Methods Supported by Local Knowledge).

The variability of the soil landscape is the reason that modern soil science has its roots in soil mapping (Miller and Schaetzl, 2015). We have learned a lot by observing the landscape and learning which soil conditions can best support different land uses. When the people are well connected to the land, many local residents are able to identify the appropriateness of land uses for different areas around their community (WinklerPrins, 1999). Even though they may not understand much about the soil itself, they recognized the importance of different soil conditions and properties across their landscape. Too often today, people are less connected to the land, but our demands for services provided by the soil have never been higher.

A unique element of the modern demands on soil maps is the variety of scales needed. To meet the demands of past centuries, it was sufficient to have generalized soil maps of nations or regions and more detailed maps as needed for local extents. This was driven by the logistics of cartography, but also met the needs of the time. Modern environmental issues require large extent climate and watershed modeling, with the resolution, precision, and accuracy of the input soil maps greatly affecting the quality of model predictions (Sanchez et al., 2009; Miller, 2012). In addition, to increase crop production efficiency, including the reduction of excess applications of pesticides and soil amendments, agronomic systems are managing land at finer and finer resolutions. These more precise management decisions can have both environmental and economic benefits. Technology available today allows farmers to vary management practices at the subfield scale based on a synthesis of fine resolution data from past crop yields, past management practices, and other auxiliary information (Bongiovanni and Lowenberg-DeBoer, 2004; McBratney et al., 2005). Soil maps are vital to those calculations, but like the environmental models, the quality of the model outcomes relies on the quality of the input data. It is in this context that legacy soil maps are no longer sufficient. However, together with more remote and proximal soil sensing data as well as spatial statistics, geographic information systems (GIS) are providing the tools we need to meet these rising demands on soil maps.

Spatial Prediction Required for Soil Mapping

An ever present reality for soil mapping is that it is only possible to directly observe a fraction of the soil landscape. Traditional soil mapping mapped large areas by using relatively few samples and applying the principle of spatial association (Hole and Campbell, 1985). However, as we experiment with more quantitative, digital methods, the results of transferring a model calibrated in one area to another are often unsatisfactory because it becomes more difficult to accurately account for all the interactions and histories (Miller et al., 2016). In some ways, this was compensated for in traditional soil mapping by individual soil scientists calibrating their mental model to local conditions. Yet these mental models generally fit within the framework of the widely accepted factors of soil formation (Jenny, 1941) and the soil-landscape paradigm (Hudson, 1992). In any case, traditional soil mapping extended relationships observed at limited locations to produce useful soil maps at the landscape scale.

Using the quantitative tools that we now have available from GIS and statistics, there are a variety of methods we can use to detect patterns and use those patterns to make predictions about unobserved sites. However, inherent within any of these methods are certain assumptions and biases that influence the characteristics of the resulting map. For example, spatial prediction methods can be classified as exact or inexact. Exact models produce values at the observation points equal to the value measured there, which is also sometimes referred to as honoring the data. Inexact models do not have to produce a prediction surface (map) that matches values at the observation points (Smith et al., 2015). Although using a model that is exact may sound better, it would actually depend on the circumstances, such as confidence in the accuracy of measurement at those observation points. There is no single answer for the best method to use GIS and spatial statistics to produce the soil maps needed for sustainable land management. However, understanding the main concepts behind the different approaches can guide the mapper to optimize their methods for serving the mapping goals. The objective of this chapter is to provide an overview of what distinguishes various digital soil mapping approaches from start to finish, thus providing a beginning framework for choosing the approach with the greatest likelihood for success.

SAMPLING DESIGN

Whenever creating a map, some direct observation of the attribute being mapped has to be done. However, because soil is a continuous body and is impractical to directly observe in every location, the direct observation has to be done by limited sampling (Webster and Oliver, 1990). Therefore an important question is how to best strategically sample the area being mapped. The answer is really dependent on the spatial model that will be used. However, there is some flexibility and the important thing is to not extrapolate much beyond what the data and model support. In spatial modeling, there are two primary principles that are used to make spatial predictions: spatial association and spatial autocorrelation. Which principle is used in the model has bearing on which sampling design is optimal and what constitutes extrapolation.

Spatial autocorrelation is the geographic principle that all things are related, but near things are more related than things that are farther apart (Tobler, 1970). The reliability of predictions using this principle decreases with distance from the points of observation. Examples of spatial models relying on this principle include spline, inverse distance weighting (IDW), and kriging. In contrast, spatial association is the geographic principle of things covarying together in space. Spatial models relying on this principle include any model that utilizes spatial covariates, such as decision trees and regression models. Rather than interpolating across locational space, spatial association approaches interpolate in the feature space. In other words, for spatial association approaches, the interpolation is between the minimum and maximum values of the observed relationship between the predictor and target variables. In this way, it is possible for a spatial prediction outside the extent of the samples to be reasonably predicted while a location within the sampling extent could actually be an extrapolation due to extreme values. While the problem of induction always allows for the possibility of unpredictable error (Hume, 1739/2001; Popper, 1959/2005), understanding the concepts behind the spatial prediction model to be used can guide sampling design to utilize the strengths of the model.

Optimal sampling designs for spatial autocorrelation models minimize the distance between sampling points while fully covering the extent needing to be mapped. Generally, a grid sampling design works well for this. However, if the area being mapped is irregular, then some additional analysis could help optimize sampling locations. One approach could be to use the centroids of *k*-means clustering to optimize the spacing and coverage of the area being mapped.

Optimizing sampling design for spatial association approaches means optimizing the coverage of the feature space (Gessler et al., 1995;

Hengl et al., 2003). A priori knowledge of which available covariates best represent that feature space is greatly beneficial. Purposive sampling could be done by an expert's knowledge of how the target variable is distributed in the mapping area and the range of related environmental conditions. However, using covariate data and GIS tools, more statistical approaches can be used to make the sampling systematic. Random stratified sampling is a probabilistic approach that splits the population into strata to focus sampling on categories important for the research. In the case of sampling in support of spatial regression, the goal is to have the samples be well distributed across the feature space of the covariates. With continuous variables as covariates, the covariates can be categorized by quantiles. The quantile categories form zones on the map in which random locations can be selected. Sampling only within certain quantile categories to capture low, medium, and high values is often used to increase sampling efficiency.

The challenge with a basic approach to stratified random sampling is that each additional covariate considered multiplies the number of sampling zones and complicates the selection of sampling points. Latin hypercube sampling selects points from a multidimensional distribution, where each covariate is considered to be a separate dimension. It is a more complex method of insuring that samples cover the feature space of the provided covariates, but is capable of handling a large number of covariates. Despite this advantage, it should be noted that Latin hypercube sampling considers the distribution of each covariate separately, but not the combinations of covariates. With the more manual approach to stratified random sampling the mapper can select different category combinations to sample (e.g., high covariate A with high covariate B and high covariate A with low covariate B). A demonstration of stratified random sampling is presented later in this chapter. For an example of a conditioned Latin hypercube sampling applied to digital soil mapping, see Minasny and McBratney (2006).

Another approach, available in the SPSANN package in R (Samuel-Rosa, 2016), is to use spatial simulated annealing (SPSANN). In metallurgy, annealing is the process by which a metal is heated to a specific temperature and then cooled slowly. The longer amount of time for cooling allows the atoms in the metal to arrange themselves in the crystal lattice more optimally by trial and error. The SPSANN algorithm mimics that process. This approach has advantages in that it can be set to optimize sample locations for different objectives. For example, the objective could be to optimize the sample locations for the best estimation of the semivariogram, a key component to kriging models that will be discussed later in the section on spatial autocorrelation approaches. In the case of spatial association-based models, however, the objective function could be set to optimally locate samples for reproducing the marginal distribution of the covariates.

There is some cross-over in strategies that help the respective sampling designs. If a sampling design perfectly covers the mapping extent and minimizes distance between points, but has no points in a small feature (e.g., a narrow band of sand), then the resulting model cannot reflect that feature's presence. Conversely, it is not extrapolation to make spatial predictions outside the spatial extent of sampling for a spatial association model. However, the farther in distance the predicted area is from the sampling area, the more likely that enough conditions would change that the relationships calibrated on the sampling points will not be as accurate.

Logistically one should consider the ease of locating planned sample points. From a GIS it is tempting to quantitatively design the location of each sample point. However, if the area being sampled has rough terrain or the potential for a weak global positioning system (GPS) signal, then a simpler system for identifying sample locations could have major benefits. Using a GPS to accurately record where a sample was actually taken—versus hunting for a prespecified waypoint—can be more time efficient, especially in difficult conditions for navigating. It is noteworthy that many GPS devices can record a more accurate location after being stationary at that spot for more time. Also, some GPS devices have the feature to average readings across a period of time, which can provide greater confidence in an accurate measure of location. One way to minimize reliance on the GPS to locate sampling points is to use a regularly spaced grid system. If not spaced too far apart, subsequent sample points in the grid can be quickly measured off of a previous sampling point and directions maintained from a regular compass or landmarks. If the sampling design is based on stratifying landscape variables, then identifying sampling zones can take some pressure off of actually locating the preidentified sample point. For example, with stratified random sampling, a random location is preselected within the zones needed to capture the feature space range. Some additional randomness stemming from only approximately locating the preplanned point will likely not affect the quality of the sampling. The important part is to record the location of the actual sample as accurately as possible.

SPATIAL ASSOCIATION

Spatial association is a broad category of spatial modeling approaches, which includes even qualitative relationships. The principle has been used since ancient times to predict the spatial distribution of various things that may be difficult to observe directly by observing the spatial distribution of something more easily seen. Related to land suitability for cultivation, two centuries before the common era, Diophanes of Bithynia is quoted in saying that, "you can judge whether land is fit for cultivation or not, either from the soil itself or from the vegetation growing on it" (Cato, 1934, p. 205). As understanding of the factors that produce variability in the soil landscape improved, Dokuchaev identified five qualitative covariates useful for creating better soil maps: climate, organisms, relief, parent material, and time (Dokuchaev, 1967). Later, with the production of more detailed soil maps, the covariates of parent material and relief were further reasoned out to formulate the soil-landscape paradigm (Hudson, 1992). Although terminology changes to highlight exciting developments, it is important to recognize that many of the spatial statistics techniques used for producing digital soil maps are rooted in concepts developed by early field observations of the soil landscape (Florinsky, 2012; Miller and Schaetzl, 2014).

The large, and growing, amount of potential covariates for predicting soil-landscape variation are increasing the opportunities for applying the spatial association principle in quantitative ways. Modern use of the spatial association principle is usually in the application of some form of regression. More specifically, because the covariates used are usually connected to the soil environment, the regression is called environmental correlation (McKenzie and Ryan, 1999; Brown, 2006). However, in most circumstances, a single, linear regression equation will not adequately account for the variability of any soil property. One reason that more complex models are needed is because many relationships in the soil environment are nonlinear. The other reason that more sophisticated models are needed is because there are many possible factors influencing soil variability, each interacting with negative or positive feedback loops.

With the large quantity of potential covariates and the many model structures to choose from, it is not a small task to sort through them all to derive the best prediction model. To assist with this problem, most spatial regression approaches employ some type of data mining. The process of predictive data mining for mapping has four primary steps: (1) collection of sample and covariate data, (2) pattern identification to build the model, (3) application of model covariates to produce a map, and (4) validation. Popular methods for data mining to build predictive models include classification and regression trees, generalized linear models, and neural network analyses.

Data Mining

When a data set includes more covariates than is practical to include in the model, some feature selection is usually required. To accomplish this, most predictive data mining procedures select a subset of predictors from the pool of candidate predictors that by statistical tests appear the most likely to be related to the target variable. These statistical tests do not assume that the relationships are linear or monotone.

By statistical data mining standards the data sets used for most soil mapping projects are relatively small. In general, soil surveying projects are typically managing only hundreds of samples, where most data mining algorithms behave better with thousands of data points. Although digital soil mapping projects for large extents often aggregate data sets to get the coverage and sample quantity needed (e.g., Adhikari et al., 2013; Hengl et al., 2014), that approach comes with its own challenges of incongruities between sampling methods and the factor of when samples were taken. In any case, when using data mining algorithms to identify patterns in relatively small data sets for later use in spatial prediction, the models tend to be unstable. Defining a minimum quantity of samples for data mining is difficult because the models gradually become more stable with an increasing quantity of training points. Yet the predictability of that stability gradient is affected by the wide variety of factors and interactions producing soil variation combined with the odds that a particular sampling set includes all of the relationships needed to predict soil variability in the validation. Because of this instability, one should be aware that results can differ slightly with the addition or subtraction of only a few samples. The degree of instability can be revealed using *k*-fold cross-validation. The less model performance is reduced by a cross-validation test, the more that we can say that the model is robust. A model that is found to not be robust can still be used for spatial prediction. However, one must keep in mind that the identified relationships have dependencies on the selection of samples, which increases the uncertainty in the predictions.

Some models address the issue of unstable results using techniques referred to as bagging and boosting. Bagging combines predictions from multiple models to derive a result that we can have greater statistical confidence in. Models that use this approach are also sometimes caused ensemble models. In the case of predicting classification, each model essentially votes how a target point should be classified. The classification that is voted for by the most models is used as the end result. For continuous target variables the end result is the average of model predictions. An example of a classification and regression tree approach that utilizes bagging is random forests (Breiman, 2001). By using random feature selection the random forest algorithm develops multiple decision trees using different subsets of the training data. The result is a model that is less likely to be over-fitted to the training data and tends to have improved prediction performance. Although there is some loss of interpretability of the model, this is somewhat compensated for by a ranking of variable importance.

Boosting adds a level of sophistication to bagging by deriving weights for the combination of predictions (Hastie et al., 2009). For example, weights could be added to points in the training data set that were difficult to predict by the model created by the previous run. Each iteration of this creates a model that is more focused on predicting the samples that were the most problematic for the previous run. That series of expert models for different parts of the data are then combined for the final result.

Deployment

After identifying the predictive relationships in the training data, the act of using those equations to make predictions is referred to as deployment. For spatial data, this is usually done with some form of map algebra, where the spatially exhaustive covariates are combined together as variables in the predictive model equation/s. This is a core function of GIS and is most easily done with all of the covariates in the raster format. Although many GIS software packages can make necessary map projection changes on the fly, it is always the best practice to have all spatial data sets in the same map projection and coordinate system before conducting any kind of spatial analysis. However, matching resolutions is not as critical. One approach for dealing with multiple resolutions among raster covariates would be to resample the coarser rasters to match the resolution of the finest resolution covariate. Another approach would be to use the setting in the GIS that creates the output raster in the same resolution as the finest resolution covariate used (Miller et al., 2015a). In both cases, some border shifting is possible when the coarser resolution size is not evenly divisible by the finer resolution size. Nonetheless the results are essentially the same.

Most models using spatial association are more complicated that a single, multiple linear regression (MLR) equation for the whole study area. If the model is rule based, then some classification of the specified covariates is needed to identify where in the study area the respective regression equation will be applied. This can be accomplished with a series of conditional statements in the map algebra or similar tool. It is often beneficial to produce a rule map first, then to use that as the variable that directs where the different regression equations will be applied. This makes the process easier by simplifying the steps. The rule map is also useful for mapping uncertainty. Each regression equation will have measures of fitting performance, which can be mapped as estimates of uncertainty by associating them with the rule zone where the respective regression equation was applied (Miller et al., 2015b).

When using bagging, the process for creating the final map is more tedious due to the multiple maps that need to be produced and subsequently combined. Doing this manually can be time-consuming and impractical for more than 10 map realizations. However, increasingly data mining algorithms are being combined with GIS to automate the process of deploying the map realizations and combining them to produce the final map. Particular progress in this regard has been in the R statistical software environment (www.r-project.org), allowing the combination of even thousands of map realizations. Nonetheless, when creating maps from data mining procedures it is useful to examine the covariates selected, the relationships in the regression equations, and the resulting patterns in the maps. If they match previously identified soil forming processes, then there can be some confidence that there is some substance behind the predicted patterns. Conversely, unexpected relationships can be productive for identifying processes that require additional investigation in the field. However, caution must be used when interpreting the relationships of individual variables in a MLR. The signs and coefficients are in the context of the other variables and do not directly reflect the correlation with the target variable.

Example of a Spatial Association Approach for Soil Mapping

To illustrate the use of these concepts for mapping with spatial association, the following demonstrates a spatially balanced, stratified random sampling design used in combination with Cubist (Quinlan, 1994; Kuhn et al., 2015) for data mining and model building. The goal is to produce a map of the percent soil organic carbon ($SOC_{\%}$) for the 0–30 cm layer. The example site is located in northeastern Germany in a field that is being intensively studied for carbon dynamics as part of the CarboZALF project (Sommer et al., 2016). Satellite imagery showing the landscape context, and the map extent boundary is shown in Fig. 5.1. Available covariates with exhaustive coverage of the study area were apparent electrical conductivity (ECa), a leaf area index (LAI) derived from the Quickbird satellite, and a digital elevation model (DEM). In order to reduce the effect of low LAI values between crop rows, that grid was smoothed using a moving-average low pass filter. The DEM was produced at a 1 m resolution from LiDAR and was subsequently analyzed to produce a large quantity of terrain derivatives. For the scale-dependent derivatives, such as slope gradient, curvature, and aspect, a wide range of analysis scales were utilized. In this case the resolution was always maintained at 1 m, but the neighborhood size was varied to obtain the range of analysis scales. Note, for the spatial analysis of rasters, analysis scale equals the resolution multiplied by the neighborhood size. For more on how considering multiple analysis scales for digital terrain derivatives as covariates can improve regression models for digital soil mapping, see Miller et al. (2015a).

As part of the stratified random sampling, three covariates were selected by the mappers as likely to represent the factors influencing the range of $SOC_{\%}$ in the field: LAI, ECa (vertical mode), and the topographic position index (TPI) (calculated at an 80 m analysis scale). For the extent of the study area the three covariates were partitioned into eight quantiles and the resulting zones were intersected in a GIS (Fig. 5.2). To increase sampling efficiency, zones resulting from the third and sixth quantiles were ignored. In each of the remaining intersected zones, sampling locations were randomly selected for a total of 80 points. More information about this sampling design is provided in Theobald et al. (2007). A separate set of 107 points were collected on a 20 m grid covering the same field, which for this example will be used as an independent validation set. It is unusual to have such an intensive data set for validation, but will allow this example to provide a detailed examination of the model's performance.

FIGURE 5.1 The example mapping area is highlighted in yellow (white in print versions). It is located in a hummocky, agricultural landscape. The shape of the mapping area was determined by the planned CarboZALF research site that was subsequently developed. Satellite imagery was taken in 2002 (Google Earth, 2016), which shows the site as it was when sampling was conducted, prior to the research site being heavily instrumented.

At both the training and validation sample points, cores that were 10 cm in diameter were extracted using a hydraulic probe. For the purpose of the present map a representative mixed mass section taken across the top 30 cm of each

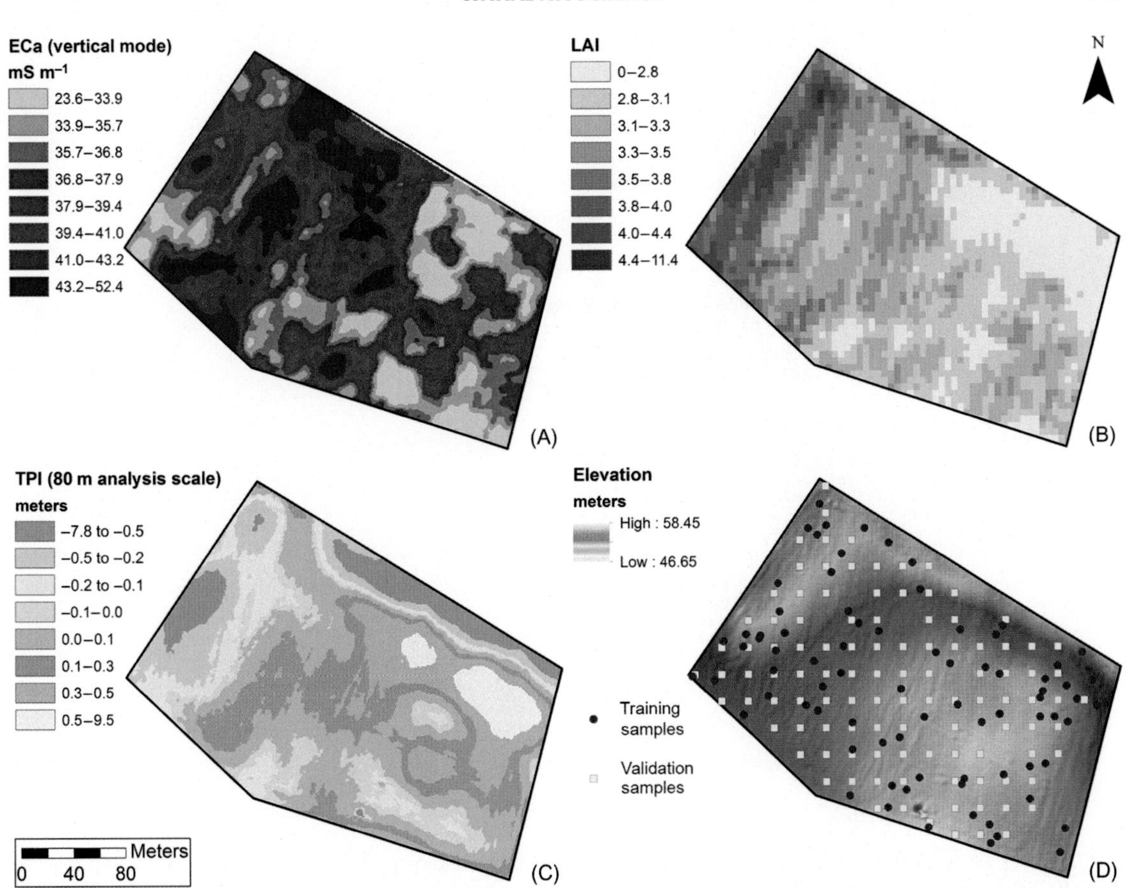

FIGURE 5.2 Covariates used to identify sample locations with stratified random sampling. The three covariates used to stratify the study area were (A) ECa measured in vertical mode, (B) LAI, and (C) TPI calculated on an 80 m analysis scale. Random sampling within the intersection of all but the third and sixth of eight quantiles for those covariates resulted in (D) the distribution of sampling points.

core was air dried and verified to contain a minimum of 500 g of soil. Laboratory analysis was conducted for percent total carbon and inorganic carbon, with $SOC_{\%}$ calculated as the difference between the two.

A shapefile of the training sample points with the respective laboratory data was loaded into a GIS and intersected with all of the covariate raster grids. This process is sometimes called "extract values to points." The resulting shapefile's attribute table was then exported and converted into a format that could be read by Cubist. From the pool of 297 potential covariates, Cubist selected four to be used in a model consisting of two rules or branches of the decision tree (Fig. 5.3). Summarizing the fitting performance of the model to the training data, Cubist reported an average error of 0.07, a relative error of 0.55, and a correlation coefficient of 0.87.

The average error is simply the mean of the absolute residuals, which are the absolute differences between the model and the observations

```
Model:

  Rule 1: [11 cases, mean 1.0232111, range 0.596632 to 1.46373, est err 0.0966560]

    if
        rel5200m <= -7.3
    then
        SOC_Ap = -1.4888796 - 0.2864 rel5200m - 0.009 eastness_3m
                 + 0.009 eastness_9m + 0.01 LAI5mAg

  Rule 2: [69 cases, mean 0.7394417, range 0.429662 to 1.007101, est err 0.0763053]

    if
        rel5200m > -7.3
    then
        SOC_Ap = 0.0921606 + 0.219 LAI5mAg - 0.065 eastness_3m
                 + 0.061 eastness_9m

Setting number of nearest neighbors to 4
Recommend using rules and instances


Evaluation on training data (80 cases):

    Average  |error|              0.0698252
    Relative |error|                   0.55
    Correlation coefficient            0.87

        Attribute usage:
          Conds  Model

          100%     14%    rel5200m
                  100%    eastness_3m
                  100%    eastness_9m
                  100%    LAI5mAg


Evaluation on test data (107 cases):

    Average  |error|              0.0694274
    Relative |error|                   0.61
    Correlation coefficient            0.82


Time: 0.1 secs
```

FIGURE 5.3 Summary of output from Cubist based on this chapter's example of a spatial association approach. Specifically, Cubist builds a classification and regression tree type model, where the MLR equations are only applied to areas that meet the criteria of the respective rules. Information about the cases in the training data that fall within those rules is provided, followed by a summary of the models overall fitting performance and use of covariates. Because this example included an independent validation test, those results are provided at the end.

at the training points. Note that if the absolute residuals were not used, the mean of the residuals would be expected to be zero for an unbiased model. Instead, Cubist's reporting of the average error is actually mean absolute error (MAE), which is a common model diagnostic. The MAE is beneficial for considering the magnitude of errors in the same units as the target variable and whether that magnitude is acceptable for the intended map use. Cubist's reporting of the relative error is different than the standard definition of the relative error. Relative error in this case is actually a comparison of the Cubist model's MAE with the MAE that would result from using the simplest statistical model, the mean model. The mean model is the use of the observed sample mean as a constant predictor. Because of that model's simplicity, it is commonly used as a benchmark for assessing how much a more complex model has improved predictions. Comparisons of an experimental model with the mean model are usually called model efficiency (e.g., Nash and Sutcliff, 1970). However, in those model efficiency measures, greater model performance is indicated by higher values. In Cubist the relative error is the ratio of its

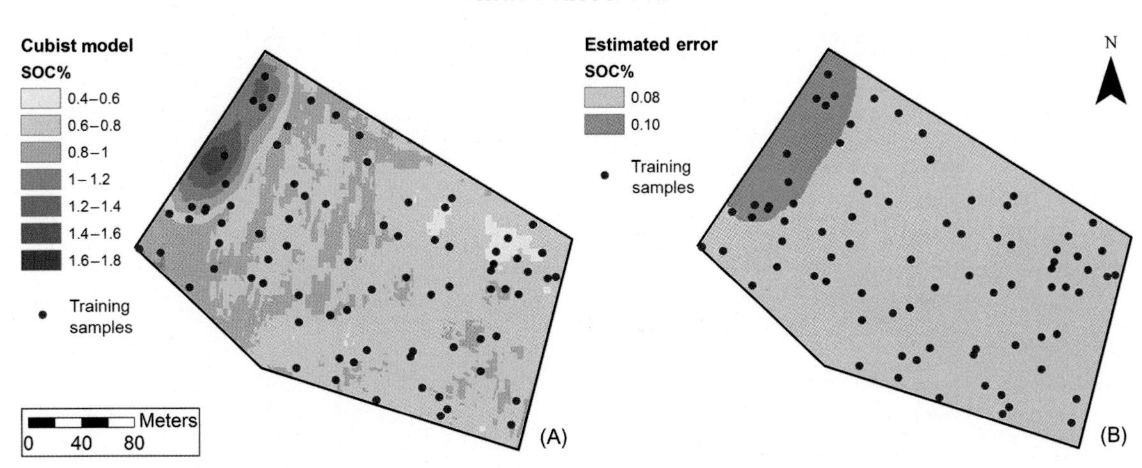

FIGURE 5.4 Applying the model generated by Cubist to the covariate raster data via map algebra resulted in (A) a continuous map of $SOC_{\%}$. By associating the estimated error of the MLR equations' fit under each rule with the areas classified by the Cubist-based model rules, a quantification of uncertainty can be mapped, as shown in (B) the estimated error map. Because only two rules were applied to the map extent, this approach produces a discrete map of uncertainty zones.

average error to the mean model's average error. Thus the lower the value, the better the Cubist model performed. Relative error values greater than one result when the mean model actually outperformed the more complex Cubist model. Finally, it should be noted that many research papers in digital soil mapping use the coefficient of determination (R^2) as a metric of performance, whereas Cubist reports the correlation coefficient (r), two related, but different statistical measures.

Using map algebra to apply the Cubist generated model to the raster data for the covariates resulted in the map shown in Fig. 5.4A. For each rule, Cubist calculated an empirically estimated error, which can also be spatially represented by assigning that estimated error to all cells that a respective rule was applied (Fig. 5.4B). Together, we have both a spatially exhaustive prediction of $SOC_{\%}$ for our example mapping area, plus a quantified indicator of the uncertainty we have in those predictions for different areas of the map.

The statistical measures for the model's ability to predict the independent validation points is shown at the bottom of the Cubist output (Fig. 5.3). The similarity of the values for fitting performance and for prediction importance indicate that the fitting performance values did provide a reasonable assessment of the model's ability to represent the spatial variation of $SOC_{\%}$ in that field. However, the relative error did increase and the correlation coefficient did decrease some. This decrease in performance is to be expected whenever making predictions and highlights the difference between fitting performance and an actual validation test.

To further evaluate the model generated by Cubist a cross-validation analysis can be conducted. The results of a 10-fold (10 iterations) cross-validation were averaged together to calculate an average error of 0.1, a relative error of 0.75, and a correlation coefficient of 0.75. All of these measures show a lower fitting performance for the model. However, this should be expected when fewer cases are used in the model building runs and tested against different validation points. The differences between the models built on subsets of the training data are noteworthy. Folds 8 and 10 shown in Fig. 5.5 resemble the model built on the full set

```
[ Fold 8 ]

Model:

  Rule 1: [62 cases, mean 0.7428688, range 0.429662 to 1.007101, est err 0.0736622]

    if
        rel5200m > -7.3
    then
        SOC_Ap = -0.1073864 - 0.0247 rel5600m + 0.239 LAI5mAg + 0.02 relc500m
                 + 0.058 slp39m + 0.0117 relc2800m - 2.99 cslp

  Rule 2: [10 cases, mean 1.0257493, range 0.596632 to 1.46373, est err 0.0998275]

    if
        rel5200m <= -7.3
    then
        SOC_Ap = -1.542318 - 0.297 rel5200m - 0.0004 relc2800m
                 + 0.0004 relc2000m + 0.004 LAI5mAg

Recommend using rules only

Evaluation on hold-out data (8 cases):

    Mean |error|   0.0792898


[ Fold 9 ]

Model:

  Rule 1: [14 cases, mean 0.9696238, range 0.596632 to 1.46373, est err 0.0776690]

    if
        rel4000m <= -6.89
    then
        SOC_Ap = -1.1678644 - 0.2589 rel4000m

  Rule 2: [58 cases, mean 0.7386996, range 0.429662 to 1.007101, est err 0.0880575]

    if
        rel4000m > -6.89
    then
        SOC_Ap = 0.7940818 + 4.59 plc135m - 0.0007 relc2800m

Recommend using rules and instances

Evaluation on hold-out data (8 cases):

    Mean |error|   0.0934358


[ Fold 10 ]

Model:

  Rule 1: [11 cases, mean 1.0232111, range 0.596632 to 1.46373, est err 0.0964939]

    if
        rel5200m <= -7.3
    then
        SOC_Ap = -1.55887 - 0.2975 rel5200m

  Rule 2: [61 cases, mean 0.7435138, range 0.473621 to 1.007101, est err 0.0837390]

    if
        rel5200m > -7.3
    then
        SOC_Ap = 0.8063363 - 0.0182 rel5200m + 0.0155 rel600m - 1.68 cslp

Recommend using rules and instances

Evaluation on hold-out data (8 cases):

    Mean |error|   0.1067151


Summary:

    Average  |error|              0.0974180
    Relative |error|                   0.75
    Correlation coefficient            0.75
```

FIGURE 5.5 Excerpt from running a 10-fold cross-validation for the example of using Cubist for digital soil mapping. As each fold uses a somewhat different subset of the original training data for model building and validation, different models are generated. The summary at the end of the output averages the hold-out (independent) evaluations from each fold, provided an indication of how robust this modeling process is to using different sampling points for training and validation.

of training data, but add catchment slope (cslp) as a predictive variable. Fold 9 is even more dissimilar by using different analysis scales of relative elevation (rel) and introducing the use of plan curvature (plc). The minimal difference between the original fitting performance metrics and the cross-validation results indicate that the model is fairly robust. In other words, if different sample points had been selected, we can have confidence that a comparably performing model would have been built. When sample data is limited, as it usually is for soil mapping, it is generally considered acceptable to only use a cross-validation to test the predictive performance of a model. Nonetheless, it should be noted that an independent validation and a cross-validation test really evaluate the prediction abilities of different models that share mostly the same training data.

SPATIAL AUTOCORRELATION

Spatial autocorrelation is utilized by any spatial model where proximity increases the influence of an observation point on the prediction. In contrast to the spatial regression approaches described above, no covariates are required; only the respective locations of observations made of the attribute being predicted is needed. Because the spatial predictions are done on the basis of location and proximity, the interpolation is truly in the spatial dimension. Deterministic approaches for spatial interpolation include IDW, radial basis functions, and polynomial/spline fitting. The other main group of spatial interpolation methods are referred to as geostatistics, which uses the interpolation method of kriging.

Deterministic Spatial Interpolation

The simplest deterministic approach is the nearest neighbor method, where unobserved locations are assigned the same value observed at the closest sample point. Although this utilizes the concept of spatial autocorrelation, it creates a stepped prediction surface that is not satisfactory for most mapping applications. Recognizing that there is a distance gradient to spatial autocorrelation, IDW is a weighted average of the points in the designated search neighborhood. The weighting is the inverse of the distance (d) to the respective observed point, modified by a power exponent (p) to shape how quickly the relative weight decays with distance ($1/d^p$). The search neighborhood can be defined by a distance or number of points. There is no theoretical basis for setting the power exponent or the search radius. Essentially the user should experiment with the settings until a satisfactory map is produced. The typical default for the power exponent is two. Regardless of how the search neighborhood is defined, generally including at least five neighborhood points is a reasonable starting point. Fig. 5.6 illustrates the influence of the power exponent by comparing interpolations produced from the same set of sample points and a search radius of 12 points, but different values for p.

Like many of the approaches described in this chapter, there are a variety of models that use the same central concept. These different models have adjustments in their design, attempting to overcome one or more issues with the original design. For example, the modified Shepard's method is Renka's (1988) improvement upon Shepard's (1968) algorithm for IDW. Some data sets may not have problems with the issues that the later versions of a model seek to address. Among the differences between Shephard's method and the modified Shepard's method is the improved ability to work with large data sets and reduced flatness near sample points (nodes). Although some guiding principles can help narrow down which model is right for a certain application, there is almost always value in some data exploration and experimentation to identify the optimal model and settings.

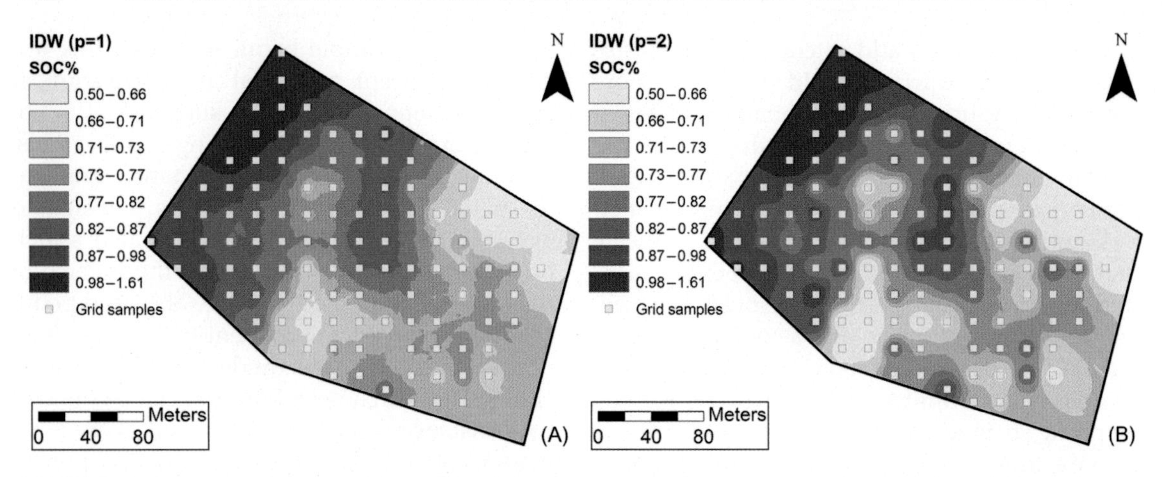

FIGURE 5.6 Side-by-side comparison of inverse distance weighted (IDW) interpolations based on the same data, but with different power (*p*) values set. A smoother prediction surface is produced with (A) p set to 1, while a prediction surface varying more with the sample points is produced with (B) p set to 2. Note the relationship of the spatial pattern in the interpolated maps with the sample points, which can be especially recognized in the southeast corner where there is some spatial extrapolation beyond the sampling points.

Another deterministic approach to spatial interpolation is to fit a polynomial through the data points. A common method in this regard is a piecewise polynomial called a spline. The term spline comes from draftsmen's use of a flexible ruler to draw lines connecting dots. Essentially a spline seeks to minimize bending while still passing through all of the data points (knots). By requiring predictions to pass through knots, spline interpolation is an exact interpolation model. One of the most common spline interpolations is a cubic spline, which uses a third order polynomial. This version of spline interpolation has a lower possibility of producing wild oscillations than higher degree polynomials. Like IDW, there are certain parameters that need to be set by the mapper, such as weight and number of neighborhood points, which do not have a theoretical basis for identifying their correct setting. The mapper must still experiment with the settings and use their expert knowledge to assess the best result. For digital soil mapping, it is more common for spline interpolation to be used to predict the distribution of soil properties vertically, in a soil profile (Odgers et al., 2012), than horizontally across a landscape.

Geostatistics

Kriging is in a class of its own because it uses a form of Bayesian inference that produces both a deterministic prediction and a standard error that can be used to quantify confidence intervals. Most notably, kriging quantifies spatial autocorrelation, removing the need for the mapper to guess the distance weighting that should be applied. However, the mapper still has control over the selection of the model that will determine those weighs. The quantification of the spatial autocorrelation's strength and reach in the observed data is calculated with a semivariogram. The empirical semivariogram is estimated by summarizing the differences (i.e., variances) between points at different lags, which are classes of distance (Fig. 5.7). In the earlier section on sampling designs, SPSANN was mentioned as having the ability to optimize for certain objectives. Although semivariograms can be constructed from most spatial

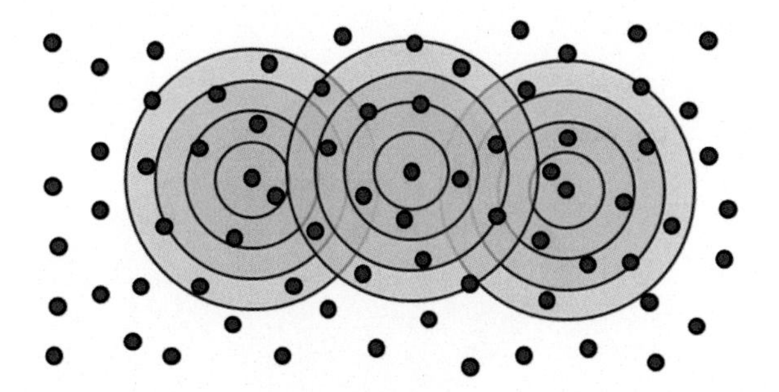

FIGURE 5.7 Illustration of lag configuration for three points. Lags are concentric rings away from central points of equal distances. The first lag being the closest, the second lag being the next ring out, and so on. Variances between a central point and points in its respective lags are calculated. This process is repeated using all of the points as a central point and summarized across all of the equivalent lags.

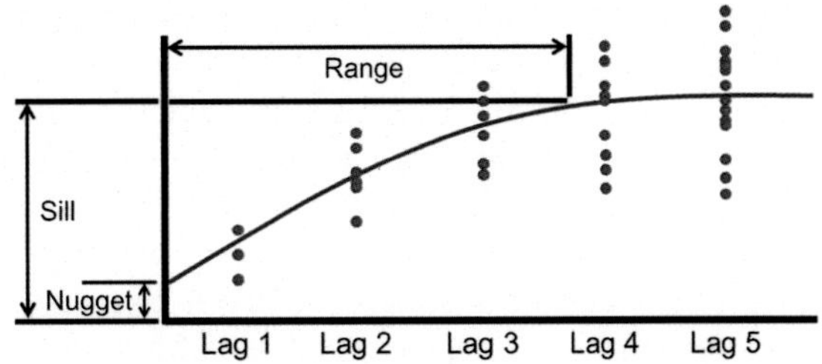

FIGURE 5.8 Schematic of a conceptual semivariogram. The red (gray in print versions) dots represent the variety of semivariances observed at each of the lag distances. The blue (black in print versions) curving line is the fitted model, summarizing the trend in semivariance with distance. Important components about the shape of the model include where it intersects the y-axis (nugget), where it approaches being level (sill), and the distance at which that leveling occurs (range). The ratio between the nugget and the sill indicates the strength of spatial autocorrelation and the range indicates the reach of spatial autocorrelation for the variable of interest.

point distributions, SPANN can optimize the sample point locations for estimating the semivariogram by using the objective of producing a uniform distribution of point-pairs per lag (Samuel-Rosa, 2016).

The semivariogram is a plot of points, which due to spatial autocorrelation, tend to increase in semivariance (y-axis) with increasing distance (Fig. 5.8). When we fit a function to those points, we have a model for the rate of decay of spatial autocorrelation's strength and an estimation for its reach or range. Fitting a model for a semivariogram and interpreting its shape is one of the most important ways that the mapper interacts with this spatial model (Oliver and Webster, 2014). Where the model intersects, the y-axis is called the nugget. This represents the variance that is expected from repeatedly measuring the same point. It can largely be thought of as representing measurement error, but can also be caused by microscale effects. The model graph should then follow the points with increasing semivariance with increasing distance. Because there is theoretically some limit to the variance a variable can have, the model graph will likely level off to form what is called the sill.

By examining the geometry of the fitted model, we can identify two key characteristics about the observed spatial autocorrelation. First the distance (x-axis) at which the model reaches the sill is considered the range for the influence of spatial autocorrelation. Second the difference between the sill and the nugget is called the partial sill, which represents the strength of the observed spatial autocorrelation effect. To better describe the relative strength of the spatial autocorrelation, the nugget to sill ratio is regularly used. Interpretation of that ratio can roughly be described as $<25\%$ showing strong spatial dependence, 25%–75% showing moderate spatial dependence, and $>75\%$ showing weak spatial dependence. However, there are some issues of scale in detecting the spatial autocorrelation. A flat semivariogram (high nugget to sill ratio) could be caused by samples being too far apart to detect spatial autocorrelation occurring more locally.

The two basic implementations of kriging are simple and ordinary kriging (OK). They differ in the assumption of stationarity, which accepts that the mean, variance, and autocorrelation structure remain the same across the map extent. Simple kriging fully relies on this assumption, while OK assumes a constant unknown mean only over a search neighborhood. Other commonly used forms of kriging begin to integrate information from covariates and are thus hybrid approaches utilizing the principles of both spatial association and spatial autocorrelation. Hybrid spatial models are discussed later in this chapter. For a more in depth explanation, including the mathematics that is kriging and more varieties of kriging not mentioned here, please see Clark (1979), Cressie (1993), or Goovaerts (1997).

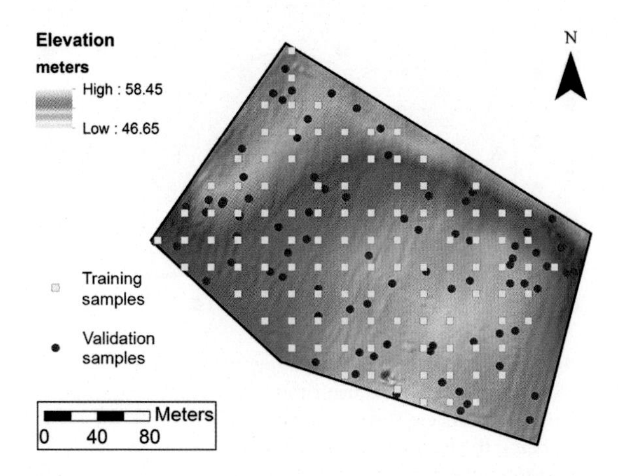

FIGURE 5.9 Spatial distribution of points used in the example of OK for mapping $SOC_{\%}$. Note that the training and validation points are reversed from the earlier example using a classification and regression tree approach. Using the grid sampling for OK is beneficial because it minimizes the distance between points and thus reduces areas of higher uncertainty. The grid sampling data set also has more than 100 points, a regularly emphasized threshold for adequately estimating a semivariogram.

Example of a Spatial Autocorrelation Approach for Soil Mapping

To illustrate the use of these concepts for mapping with spatial autocorrelation the following section demonstrates a regular grid sampling design used in combination with OK. The example site and target variable are the same used for the previous example of spatial association. In fact the training and independent validation data sets have been simply reversed. Thus in this example the 107-point, 20 m-spaced grid is now the training data and the 80 point, stratified random distribution will now be used to provide an independent validation test (Fig. 5.9). As is common for $SOC_{\%}$ data, the histogram of values was skewed. This distribution in the training data is a problem for most statistical methods because they often rely on the assumption that the data has a normal distribution. Skewed training data were not as much of an issue for the Cubist example because the data were divided into rule categories which had normal distributions. Kriging, on the other hand, relies on the normal distribution and calculation of the mean from the original data set. For these reasons, this example used a log transformation of the $SOC_{\%}$ data before kriging.

For this example the geostatistical analyst extension in ESRI's ArcMap 10.4 will be used to conduct the kriging. This software is often not favored among geostaticians, and there are open source software packages, such as R, with many benefits. Nonetheless the ubiquity and easy user interface of ArcMap make it a common choice for the casual spatial interpolation. Therefore it is a reasonable choice for this introductory chapter. There were some issues with the default settings, such as the geostatistical wizard selecting a lag size of 5 m, which on a 20 m-spaced grid was very noisy. Adjusting the lag size to 10 m produced a more suitable semivariogram (Fig. 5.10). Using an exponential model to fit the averaged points a small nugget of 0.005 was calculated. The partial sill, sill minus nugget, was calculated to be 0.027 at a range of 120 m.

Continuing the OK using the chosen semivariogram model resulted in the $SOC_{\%}$ map

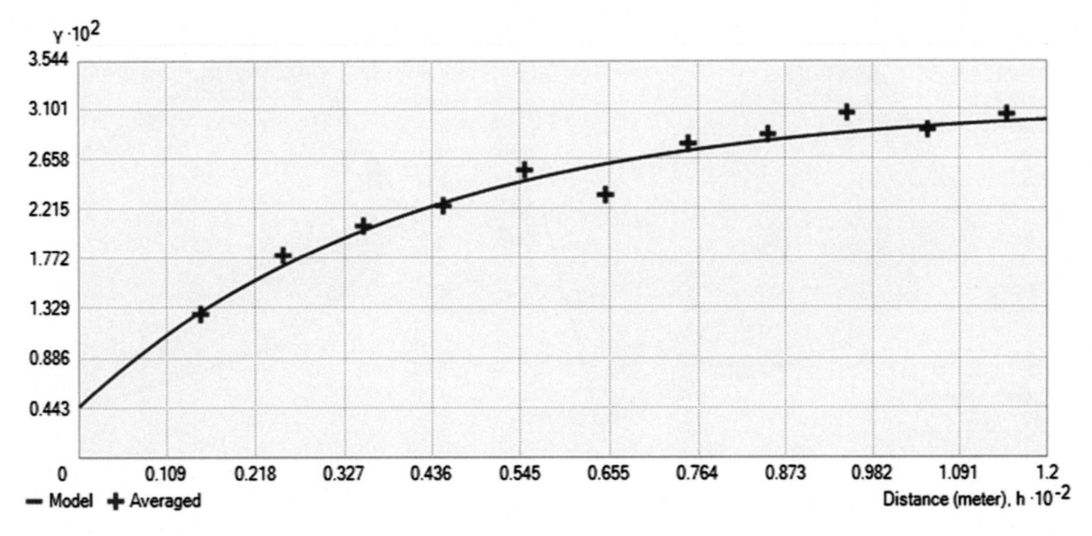

FIGURE 5.10 Semivariogram used for the example OK. The semivariances within each of 12 lags at widths of 10m were averaged to create the points shown on the graph. An exponential type model was selected to best fit those points.

in Fig. 5.11A. Of all of the example approaches demonstrated in this chapter the OK map is the smoothest. That patterning is likely driven by the long range modeled by the semivariogram. In comparison to the IDW examples, predictions are less influenced by the local points and more influenced by distant points. The map of the prediction standard error, shown in Fig. 5.11B, presents a continuous map of uncertainty. The pattern of the uncertainty in relation to the distribution of observation points highlights the role of spatial autocorrelation in this spatial modeling method. Uncertainty increases with distance from observed points.

In order to conduct an independent validation test on this OK map of $SOC_{\%}$ the data set not used as input points to the model (i.e., the 80 points distributed by stratified random sampling) will be used. The independent validation data set points were intersected with the OK map to extract the OK predictions at those points. The table was then exported to Microsoft Excel for some basic statistical analysis. For comparability with the Cubist results the same performance metrics will be used. The validation results were as follows: average error (MAE) = 0.07, relative error = 0.54, and correlation coefficient (r) = 0.86. Because the training and validation data sets were different between the two examples, one cannot make a direct comparison between the results of the Cubist-based and OK spatial models. However, with training sample sets that were designed to support the respective approaches strengths, the prediction performance of the resulting maps appears to be relatively similar. Despite the similarity in the performance metrics, there are clear differences in the resulting maps. Many of those differences are artefacts from the respective modeling approaches, such as the coarser resolution of the LAI covariate used in the Cubist model or the smoothing resulting from the OK interpolation. Performance statistics are useful indicators of map quality, but they cannot tell the full story about differences between maps.

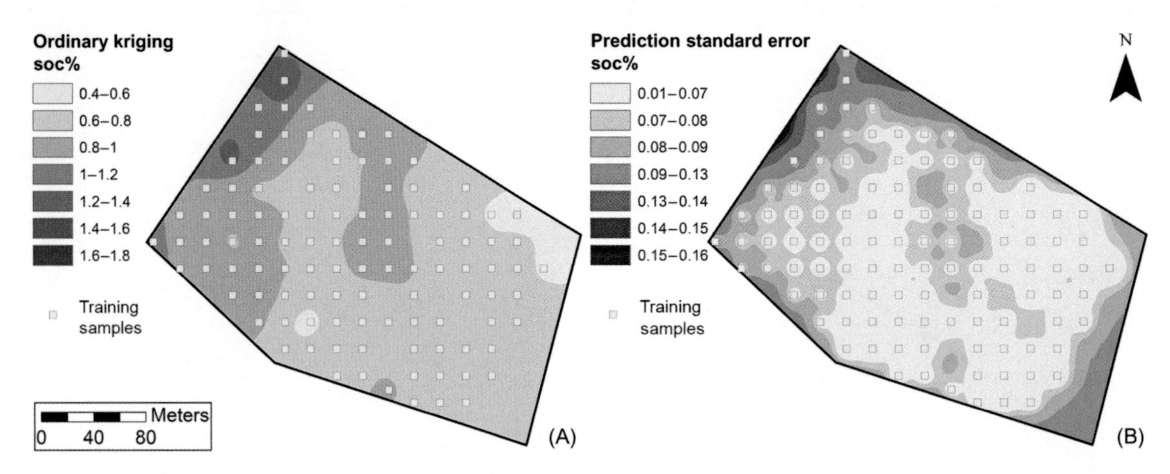

FIGURE 5.11 Using OK with the grid sampling points resulted in (A) a continuous map of $SOC_{\%}$. The statistics behind kriging also provide the ability to produce (B) a continuous map of estimated error. Note how the reliance on spatial autocorrelation manifests in the pattern of modeled uncertainty. The farther an unobserved point is away from an observed point, the greater the uncertainty.

HYBRID APPROACHES

Several approaches to creating soil maps with GIS and spatial statistics leverage both principles for spatial prediction. Cokriging, a variant of OK, utilizes the semivariogram for the variable being predicted, the semivariogram for a covariate more intensely sampled, and a covariogram between the two variables (Journel and Huijbregts, 1978; Matheron, 1978; Vauclin et al., 1983). The additional information provided by the covariate's semivariogram and its relationship to the target variable helps the kriging model to better estimate unknown spatial autocorrelation parameters. It is noteworthy that cokriging does not require the covariate to be spatially exhaustive, nor does it require the covariate to be measured at all of the same locations that the target variable was observed.

There are several other variations of kriging that utilize covariates, but probably the most popular are kriging methods that use covariates to model trends in the target variable then separately model the stochastic component. Universal kriging (Matheron, 1969), kriging with a trend model/external drift (Deutsch and Journel, 1992; Hudson and Wackernagel, 1994), and regression kriging (Ahmed and De Marsily, 1987; Odeh et al., 1995), which has also been called empirical best linear unbiased predictor (EBLUP; Minasny and McBratney, 2007), all use this strategy. The differences between these methods are subtle, but can cause differences in results. Unfortunately the naming of these procedures has not always been consistent. Therefore it is always advisable to consult the software's documentation to learn the details of how any of its functions actually perform the calculations.

There are a variety of studies in the literature comparing different forms of spatial prediction models, including the hybrid methods listed above, on different data sets (e.g., Knotters et al., 1995; Odeh et al., 1995; Zhu and Lin, 2010; Adhikari et al., 2013). For the purposes of this chapter it is most useful to summarize the differences in these hybrid approaches for spatial prediction. Universal kriging was originally intended to only use the x and y coordinates

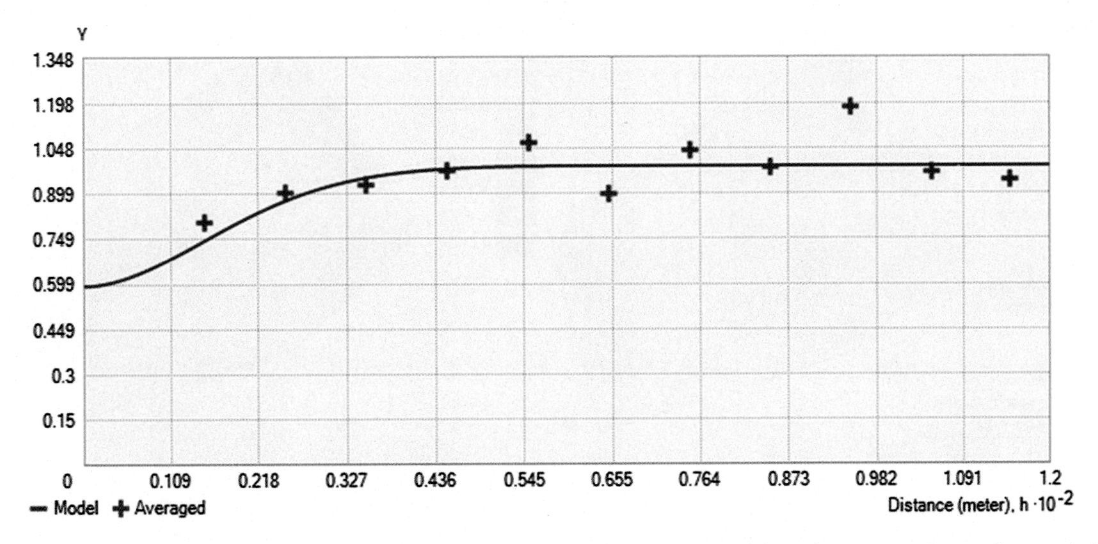

FIGURE 5.12 Semivariogram used for the OK of the error in the Cubist model predictions, as observed on an independent validation set of points. In this case the semivariogram shows a range for spatial autocorrelation that was almost too short to be detected by the 20m spacing of the points upon which it is based. The nugget is 60% of the sill, so the strength of spatial autocorrelation detected is only moderate.

as covariates to remove the apparent trend in the data. Kriging with a trend model/external drift extends the covariance matrix of residuals with the covariates. Finally regression kriging completely separates the modeling of the deterministic component from the stochastic. That separation makes regression kriging extremely flexible and more of a mapping strategy than a specific mapping method. Any spatial association model could be used to predict the deterministic component of the target variable. The kriging is used as a second step to spatially predict the prediction error. Technically the difference between the modeled value and the observed value is called the residual. However, as we are dealing with the differences between the modeled value and observations at independent validation points, the term prediction error seems more appropriate. By subtracting the prediction error map from the deterministic map of the target variable, the accuracy of the resulting map is improved by correcting for what could not be predicted by the available covariates.

Example of a Hybrid Approach for Soil Mapping

Starting with the spatial association example provided earlier, we have a deterministic prediction of $SOC_{\%}$ based on the relationship between $SOC_{\%}$ and four covariates observed at 80 points. We also have an independent validation set of 107 points that provide us with a spatial sampling of observed prediction error from the Cubist generated model. Therefore we can apply the regression kriging approach by interpolating a map of the observed error and adjusting the original map appropriately.

In this case, we will use OK to create the error map. A normal score transformation was used to convert the observed error values to a normal distribution. In fitting a model to the semivariogram, it seemed appropriate to again use 12 lags of 10m. Although there were not many points to base the model fitting on before it leveled out, a stable type model was selected. This model resulted in a nugget of 0.59, partial sill of 0.40, and a range of 40m (Fig. 5.12).

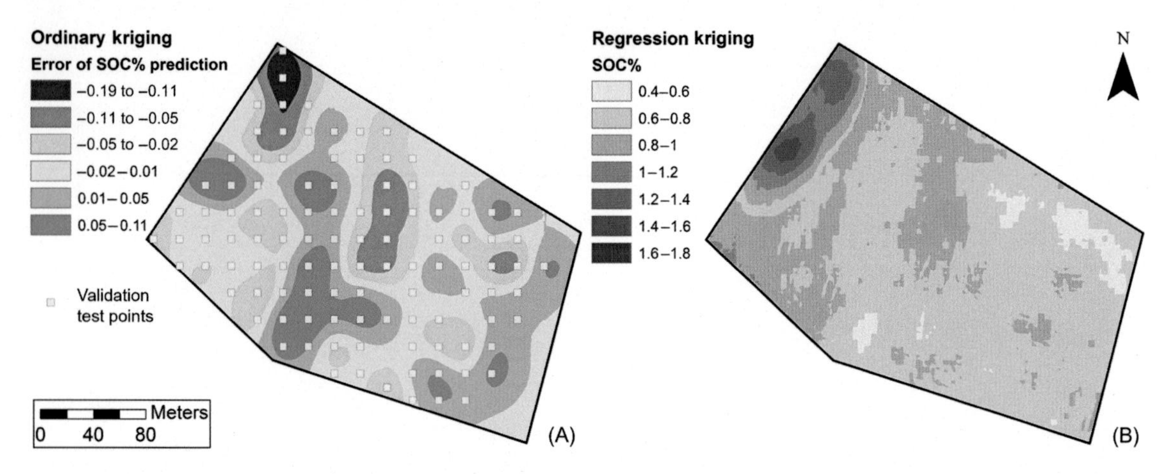

FIGURE 5.13 Process of regression kriging, starting from Cubist-based regression model used to produce the $SOC_{\%}$ map in Fig. 5.4A. The prediction error observed in the independent validation test was used as a basis for (A) OK a continuous map of prediction error. That information was then used to adjust the $SOC_{\%}$ values in the original Cubist-based map to increase the accuracy of the (B) final map.

The resulting continuous map of prediction error was then subtracted from the map produced by the Cubist model (Fig. 5.13). Thus areas seen to be underestimating $SOC_{\%}$ were raised and vice versa. This final adjustment likely improves the accuracy of the overall map by exhausting every tool available to extract information from the available data. The removal of some of the streaks of lower $SOC_{\%}$ from the original Cubist-based map is encouraging because they were artefacts of the drop in LAI between crop rows. Finally there may still be some more to be learned from these last analysis products. If the pattern identified in the last kriging step matches any kind of pattern recognizable from field experience, then finding a way to include that covariate in future work could be fruitful.

CONCLUSIONS

Farmers, engineers, hydrologists, climatologists, and many other users of soil information are demanding soil maps with finer resolution and greater accuracy than what is available in legacy soil maps. Of particular interest to the environmental modelers is for the maps to consistently cover large extents. Those maps also need to be in a georeferenced format that stores a large amount of data efficiently, which usually means a raster data model. Although field sampling will always be limited, the utility of those samples can be extend with the proper use of spatial statistics to create these digital soil map products. With better soil information, all the users of soil maps will have an increased ability to make land management decisions that increase land use efficiency and sustainability.

Geostatistics has made a tremendous impact on digital soil mapping, but it is clear that methods depending on spatial autocorrelation—by definition—are not suitable for spatial extrapolation. For small extents this is not a problem because it is usually practical to collect enough samples to cover the full area and with enough spatial density to keep prediction error low. However, as the mapping area increases, a limited number of samples means that the sampling points will become more spread out. Using kriging alone, areas in the map with greater distances from sampling points will have higher uncertainties. Quantitative approaches to spatial association, such as classification and regression trees, have the potential to make useful spatial predictions in locations farther away from individual sampling points. However, the relationships between covariates and the target soil property must be the same. In other words the soil environment must match conditions observed in the training data set. Nonetheless the map built on covariates predicting the soil variation can still have a remaining pattern of error. To make the most accurate map possible a tool-like kriging can then be brought in to model the spatial distribution of the prediction error and that error map can be used to adjust the soil map as needed.

Another way to sort through the different spatial statistics methods is to consider the type of data available. If only points observing the soil property of interest are available, then only methods like IDW, spline, simple kriging, or OK can be used. If some covariate information is available, but not spatially exhaustive for the mapping area, then cokriging is likely the best option. If covariates that cover the full mapping area are available, then their relationships with the soil property of interest can be used to make predictions. The family of methods that include universal kriging, kriging with drift/trend, and emperical best linear unbiased predictor (EBLUP) can utilize those relationships within the kriging framework. However, data mining methods such as Cubist and Random Forest are among the most sophisticated for finding those relationships and building models to make the spatial predictions. The data mining approaches have the additional benefit of having the ability to compare identified patterns in the data with processes of soil formation to either support the prediction or call for additional investigation.

The key to effectively using statistics to make the spatial predictions needed for digital soil mapping is understanding the principles and assumptions behind the methods. Remember to always consult the software package's documentation. Software tools or algorithms with the same name do not necessarily mean that their methods are the same. Conversely, tools or algorithms with different names may not be all that different in their approach. This situation is largely a result of spatial statistics methods continuing develop, especially in terms of understanding how the different approaches interact with different sampling designs, availability of new covariates, and the complex soil environment. It is for this reason that the literature is filled with studies comparing spatial prediction methods with seemingly contrasting results. However, it is difficult to compare those studies because they differ in sampling design and/or soil environment. The goal of this chapter was to provide a framework to assist the reader in sorting through the many options for how to use GIS and spatial statistics for digital soil mapping in support of sustainable land use management. Nonetheless there is always value in some experimentation with any new data set to identify exactly which method best meets the map users' needs.

References

Aandahl, A.R., 1957. Soil survey interpretation—theory and purpose. Soil Sci. Soc. Am. J. 22 (2), 152–154. http://dx.doi.org/10.2136/sssaj1958.03615995002200020016x.

Adhikari, K., Kheir, R.B., Greve, M.B., Greve, M.H., 2013. Comparing kriging and regression approaches for mapping soil clay content in a diverse Danish landscape. Soil. Sci. 178, 505–517. http://dx.doi.org/10.1097/SS.0000000000000013.

Ahmed, S., De Marsily, G., 1987. Comparison of geostatistical methods for estimating transmissivity using data on transmissivity and specific capacity. Water Resour. Res. 23 (9), 1717–1737. http://dx.doi.org/10.1029/WR023i009p01717.

Bongiovanni, R., Lowenberg-DeBoer, J., 2004. Precision agriculture ad sustainability. Precis. Agric. 5, 359–387. http://dx.doi.org/10.1023/B:PRAG.0000040806.39604.aa.

Breiman, L., 2001. Random forests. Mach. Learn. 45 (1), 5–32. http://dx.doi.org/10.1023/A:1010933404324.

Brown, D.J., 2006. A historical perspective on soil-landscape modeling. In: Grunwald, S. (Ed.), Environmental Soil-Landscape Modeling: Geographic Information Technologies and Pedometrics. Taylor & Francis Group, Boca Raton, pp. 61–98.

Cato, M.P., 1934. On Agriculture (Hooper, W.D., Trans.). Harvard University Press, Cambridge, MA.

Clark, I., 1979. Practical Geostatistics. Applied Sciences Publishers, London.

Cressie, N., 1993. Statistics for Spatial Data, Revised Edition. John Wiley & Sons, New York.

Deutsch, C., Journel, A., 1992. Geostatistical Software Library and User's Guide. Oxford University Press, New York.

Dokuchaev, V.V., 1967. The Russian Chernozem (Kaner, N., Trans.). Israel Program for Scientific Translations, Jerusalem (Original work published in 1883).

FAO, 1967. Soil Survey Interpretation and its Use. Food and Agriculture Organization of the United Nations. Soils Bulletin 8.

Florinsky, I.V., 2012. The Dokuchaev hypothesis as a basis for predictive digital soil mapping (on the 125th anniversary of its publication). Eurasian Soil Sci. 45 (4), 445–451. http://dx.doi.org/10.1134/S1064229312040047.

Gessler, P.E., Moore, I.D., McKenzie, N.J., Ryan, P.J., 1995. Soil-landscape modelling and spatial prediction of soil attributes. Int. J. Geogr. Inf. Syst. 9, 421–432. http://dx.doi.org/10.1080/02693799508902047.

Goovaerts, P., 1997. Geostatistics for Natural Resources Evaluation. Oxford University Press, Inc., New York, NY.

Hastie, T., Tibshirani, R., Friedman, J.H., 2009. The Elements of Statistical Learning: Data Mining, Inference, and Prediction, second ed. Springer, New York.745. http://dx.doi.org/10.1007/978-0-387-84858-7.

Helms, D., 2008. Hugh Hammond Bennett and the creation of the Soil Conservation Service, September 19, 1933—April 27, 1935. USDA-NRCS Historical Insights Number 9. Washington, DC.

Hengl, T., Rossiter, D., Stein, A., 2003. Soil sampling strategies for spatial prediction by correlation with auxiliary maps. Aust. J. Soil Res. 41, 1403–1422. http://dx.doi.org/10.1071/SR03005.

Hengl, T., Mendes de Jesus, J., MacMillan, R.A., Batjes, N.H., Heuvelink, G.B.M., Ribeiro, E., et al., 2014. SoilGrids1km—global soil information based on automated mapping. PLoS ONE. 9 (12), e114788. http://dx.doi.org/10.1371/journal.pone.0114788.

Hole, F.D., Campbell, J.B., 1985. Soil Landscape Analysis. Rowman & Allanheld, Totowa, NJ, 196 pp.

Hudson, B.D., 1992. The soil survey as a paradigm-based science. Soil Sci. Soc. Am. J. 56, 836–841.

Hudson, G., Wackernagel, H., 1994. Mapping temperature using kriging with external drift: theory and an example from Scotland. Int. J. Climatol. 14 (1), 77–91.

Hume, D., 1739/2001. A Treatise of Human Nature. Oxford University Press, New York.

Jenny, H., 1941. Factors of Soil Formation. Dover Publications, Inc, New York, NY.

Journel, A.G., Huijbregts, C.J., 1978. Mining Geostatistics. Academic Press, New York.

Krupenikov, I.A., 1993. History of Soil Science, (A.K. Dhote, Trans.). A.A. Balkema Publishers; Brookfield, VT.

Knotters, M., Brus, D.J., Oude Voshaar, J.H., 1995. A comparison of kriging, co-kriging and kriging combined with regression for spatial interpolation of horizon depth with censored observations. Geoderma 67, 3–4. http://dx.doi.org/10.1016/0016-7061(95)00011-C.

Kuhn, M., Weston, S., Keefer, C., Coulter, N., 2015. Package 'Cubist', CRAN.

Matheron, G., 1969. Le krigeage universel. Part 1 of Cahiers du Centre de Morphologie Mathématique. École Nationale Supérieure des Mines de Paris, Fontainebleau.

Matheron, G., 1978. Recherche de simplification dans un problème de cokrigeage. Centre de Geostatistique, Fontainebleau.

McBratney, A., Whelan, B., Ancev, T., Bouma, J., 2005. Future directions of precision agriculture. Precis. Agric. 6 (1), 7–23. http://dx.doi.org/10.1007/s11119-005-0681-8.

McKenzie, N.J., Ryan, P.J., 1999. Spatial prediction of soil properties using environmental correlation. Geoderma 89, 67–94. http://dx.doi.org/10.1016/S0016-7061(98)00137-2.

Miller, B.A., 2012. The need to continue improving soil survey maps. Soil Horizons 53, 11. http://dx.doi.org/10.2136/sh12-02-0005.

Miller, B.A., Schaetzl, R.J., 2014. The historical role of base maps in soil geography. Geoderma 230-231, 329–339. http://dx.doi.org/10.1016/j.geoderma.2014.04.020.

Miller, B.A., Schaetzl, R.J., 2015. History of soil geography in the context of scale. Geoderma. http://dx.doi.org/10.1016/j.geoderma.2015.08.041.

Miller, B.A., Koszinski, S., Wehrhan, M., Sommer, M., 2015a. Impact of multi-scale predictor selection for modeling soil properties. Geoderma 239-240, 97–106. http://dx.doi.org/10.1016/j.geoderma.2014.09.018.

Miller, B.A., Koszinski, S., Wehrhan, M., Sommer, M., 2015b. Comparison of spatial association approaches for landscape mapping of soil organic carbon stocks. Soil, 217–233. http://dx.doi.org/10.5194/soil-1-217-2015.

Miller, B.A., Koszinski, S., Hierold, W., Rogasik, H., Schröder, B., Van Oost, K., et al., 2016. Towards mapping soil carbon landscapes: Issues of sampling scale and transferability. Soil Till. Res 156, 194–208. http://dx.doi.org/10.1016/j.still.2015.07.004.

Minasny, B., McBratney, A.B., 2006. A conditioned Latin hypercube method for sampling in the presence of ancillary information. Comput. Geosci. 32 (9), 1378–1388. http://dx.doi.org/10.1016/j.cageo.2005.12.009.

Minasny, B., McBratney, A.B., 2007. Spatial prediction of soil properties using EBLUP with the Matérn covariance function. Geoderma 140 (4), 324–336. http://dx.doi.org/10.1016/j.geoderma.2007.04.028.

Nash, J.E., Sutcliffe, J.V., 1970. River flow forecasting through conceptual models: part 1—a discussion of principles. J. Hydrol. 10, 282–290.

Odeh, I.O.A., McBratney, A.B., Chittleborough, D.J., 1995. Further results on prediction of soil properties from terrain attributes: heterotopic cokriging and regression-kriging. Geoderma 67 (3-4), 215–226. http://dx.doi.org/10.1016/0016-7061(95)00007-B.

Odgers, N.P., Libohova, Z., Thompson, J.A., 2012. Equal-area spline functions applied to a legacy soil database to create weighted-means maps of soil organic carbon at a continental scale. Geoderma 189-190, 153–163. http://dx.doi.org/10.1016/j.geoderma.2012.05.026.

Oliver, M.A., Webster, R., 2014. A tutorial guide to geostatistics: Computing and modelling variograms and kriging. Catena 113, 56–69. http://dx.doi.org/10.1016/j.catena.2013.09.006.

Pereira, P., 2017. Goal oriented soil mapping: applying modern methods supported by traditional know how. In: Pereira, P., Brevik, E.C., Munoz-Rojas, M., Miller, B.A. (Eds.), Soil Mapping and Process Modeling for Sustainable Land Use Management. Elsevier 65–88.

Popper, K.R., 1959/2005. The Logic of Scientific Discovery. Routledge, New York.

Quinlan, J.R., 1994. C4.5: programs for machine learning. Mach. Learn 16, 235–240.

Renka, R.J., 1988. Multivariate interpolation of large data sets of scattered data. ACM Trans. Math. Softw. 14 (2), 139–148.

Samuel-Rosa, A., 2016. Spsann: optimization of sample configurat ions using spatial simulated annealing. Available from: https://cran.r-project.org/web/packages/spsann/vignettes/spsann.pdf.

Sanchez, P.A., Ahamed, S., Carré, F., Hartemink, A.E., Hempel, J.W., Huising, J., et al., 2009. Digital soil map of the world. Science 325, 680–681.

Shepard, D., 1968. A two-dimensional interpolation function for irregularly-spaced data. In: Proc. 23rd National Conference ACM, ACM, pp. 517–524.

Smith, M.J.D., Goodchild, M.F., Longley, P.A., 2015. Geospatial Analysis: A Comprehensive Guide to Principles, Techniques and Software Tools. Available from: www.spatialanalysisonline.com.

Soil Survey Staff, 1993. Soil Survey Manual. U.S. Department of Agriculture Handbook No. 18, Washington, DC.

Sommer, M., Augustin, J., Kleber, M., 2016. Feedbacks of soil erosion on SOC patterns and carbon dynamics in agricultural landscapes—the CarboZALF experiment. Soil Till. Res. 156, 182–184. http://dx.doi.org/10.1016/j.still.2015.09.015.

Theobald, D.M., Stevens, D.L., White, J.D., Urquhart, N.S., Olsen, A.R., Norman, J.B., 2007. Using GIS to generate spatially-balanced random survey designs for natural resource applications. Environ. Manage. 40, 134–146. http://dx.doi.org/10.1007/s00267-005-0199-x.

Tobler, W.R., 1970. A computer movie simulating urban growth in the Detroit region. Econ. Geogr. 46, 234–240.

Vauclin, M., Vieira, S.R., Vachaud, G., Nielsen, D.R., 1983. The use of cokriging with limited field observations. Soil Sci. Soc. Am. J. 47 (2), 175–184.

Webster, R., Oliver, M.A., 1990. Statistical Methods in Soil and Land Resource Survey: Spatial Information Systems. Oxford University Press, Oxford, UK.

Winklerprins, A.M.G.A., 1999. Insights and applications local soil knowledge: a tool for sustainable land management. Soc. Nat. Resour. 12 (2), 151–161. http://dx.doi.org/10.1080/089419299279812.

Zhu, Q., Lin, H.S., 2010. Comparing ordinary kriging and regression kriging for soil properties in contrasting landscapes. Pedosphere 20, 594–606. http://dx.doi.org/10.1016/S1002-0160(10)60049-5.

Further Reading

Boettinger, J.L., Howell, D.W., Moore, A.C., Hartemink, A.E., and Kienast-Brown, S., (Eds.), *Digital Soil Mapping: Bridging Research, Environmental Application, and Operation*, Springer; Netherlands, http://dx.doi.org/10.1007/978-90-481-8863-5.

Lagacherie, P., McBratney, A., Voltz, M. (Eds.), 2016. Digital Soil Mapping: An Introductory Perspective. Elsevier http://dx.doi.org/10.1016/S0166-2481(06)31061-6.

CHAPTER

6

Soil Mapping and Processes Models for Sustainable Land Management Applied to Modern Challenges

Miriam Muñoz-Rojas[1,2], Paulo Pereira[3], Eric C. Brevik[4], Artemi Cerdà[5] and Antonio Jordán[6]

[1]The University of Western Australia, Crawley, WA, Australia [2]Kings Park and Botanic Garden, Perth, WA, Australia [3]Mykolas Romeris University, Vilnius, Lithuania [4]Dickinson State University, Dickinson, ND, United States [5]University of Valencia, Valencia, Spain [6]University of Seville, Sevilla, Spain

INTRODUCTION

In the context of global change and increasing population the need for a sustainable use of available resources becomes critical. The rise in agricultural production and food availability set the conditions for the continuous increase in population that emerged in the middle of the 20th century. Today, world population is 7.1 billion and is expected to reach 9.7 and 11.3 billion in 2050 and 2100, respectively (United Nations Population Division, 2015). Although this increased agricultural production favored the global economy and population growth, land use changes towards more intensive land management practices have often had detrimental results for ecosystems worldwide. Predicted environmental risks include loss of biodiversity and soil quality due to increased rates of soil erosion and reduced levels of soil organic carbon, which may largely affect food security in the near future (Brevik, 2013a; Lal, 2007).

There is no question that achieving sustainable agricultural systems under current conditions represents a significant challenge going forward. Meeting this challenge will require a truly transdisciplinary approach, one that brings together natural scientists, social scientists, policy makers, land managers, and the general public to determine what approaches should be taken and that addresses questions not previously considered across the board by these various groups (Bouma, 2015; Brevik et al., 2016). Furthermore, to achieve a truly

Soil Mapping and Process Modeling for Sustainable Land Use Management.
DOI: http://dx.doi.org/10.1016/B978-0-12-805200-6.00006-2

sustainable agriculture it will be critically important that soil maps and other information generated by professional soil scientists be utilized by a wide range of stakeholders, going well beyond the traditional soil science community (Bouma et al., 2012; Bouma, 2014).

As agreed in Paris in December 2015, global average temperature is to be limited to "well below 2°C above pre-industrial levels" and efforts will be made to "limit the temperature increase to 1.5°C above pre-industrial levels" (UNFCCC, 2015). Reducing greenhouse gas emissions (GHG) in all sectors, therefore, becomes crucial and sustainable land management must play its part to meet climate goals. Thus mitigation and adaptation measures need to be taken and strategies such as increasing carbon sequestration in soils need to be properly addressed (Paustian et al., 2016; Smith, 2016). Carbon sequestered in the soil in the form of both organic and inorganic compounds contributes to a variety of key biological, physical, and chemical processes, with strong connections between each of them (Brevik, 2012; Keesstra et al., 2016; Lal, 2004a).

More than 25% of the total global land surface is degraded to some extent (Brevik et al., 2015). In these degraded areas, soil resources are being continuously depleted with a serious risk of biodiversity loss and food insecurity. Hence, to avoid further land degradation and maintain soil productivity and functionality, restoration of these damaged environments becomes critical at the local, regional, and global levels (Keesstra et al., 2016; Kildisheva et al., 2016; Muñoz-Rojas et al., 2016a). Monitoring land degradation status, predicting climate change impacts on land systems, or providing more accurate estimates of C pools in current and future scenarios are some of the issues that need to be tackled to ensure sustainability. In this context, mapping and modeling approaches provide indispensable tools and information to face these current challenges.

In this chapter we highlight the "state of the art" of sustainable soil management research, summarize current practices and potentials in soil and modeling to address key issues, identify gaps in data and knowledge, and suggest ways to close such gaps through new developments in soil mapping and process models.

CLIMATE CHANGE IMPACTS AND MITIGATION AND ADAPTATION STRATEGIES

Although several uncertainties remain regarding the causes, consequences, and extent of climate change, it has become evident in the last few decades that the global climate is changing largely due to human contributions (Smith et al., 2008, 2016). Global warming is mainly caused by increases in atmospheric GHG emissions, particularly CO_2 derived from fossil fuel use, agriculture, and land use changes (Wreford, 2010). The concentration of CO_2 in the atmosphere has increased to approximately 30%–40% above natural background levels and will continue to rise in the future. Between 1750 and 2011, anthropogenic CO_2 emissions to the atmosphere were 2040 ± 310 Gt CO_2 and 50% of these emissions have occurred only in the last years (Fig. 6.1). Approximately 40% of these emissions have remained in the atmosphere whereas the rest have been removed from the atmosphere and stored in terrestrial ecosystems (plants and soils) and the ocean (IPCC, 2014).

As a consequence, CO_2 accumulation in the atmosphere has affected the fluxes between the different C pools such as photosynthesis, plant respiration, litterfall, and soil respiration (Kutsch et al., 2009). These alterations are predicted to have serious consequences in agricultural and forest systems worldwide and the role that soil plays in climate regulation is critical to sustainable land use and management

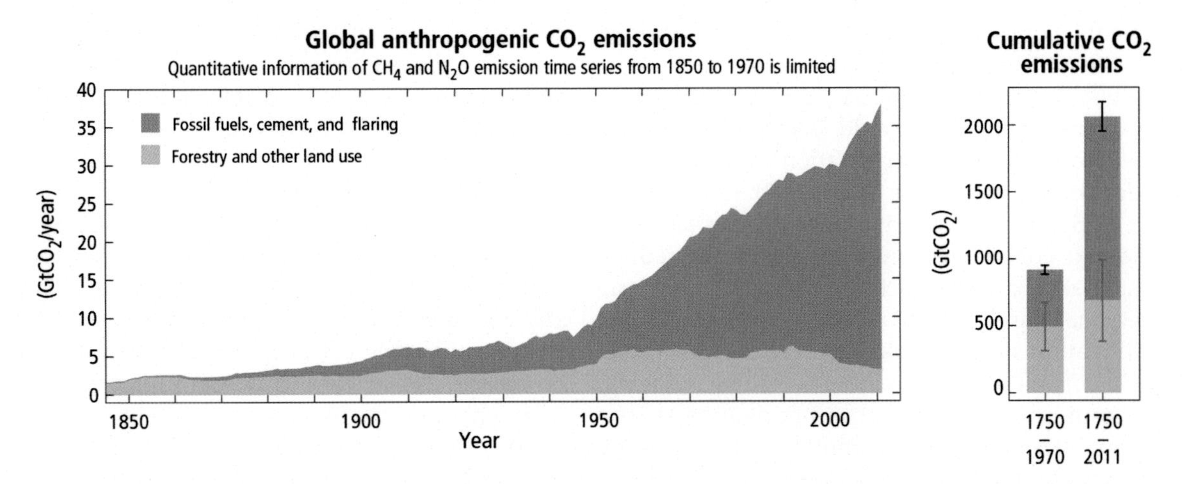

FIGURE 6.1 Global anthropogenic CO_2 emissions from forestry and other land uses as well as from burning of fossil fuel, cement production, and flaring. Cumulative emissions of CO_2 from these sources and their uncertainties are shown as bars and whiskers, respectively, on the right-hand side. *Source: IPCC, 2014. Climate Change 2014. Synthesis Report. Contribution of Working Groups I, II and III to the Fifth Assessment Report of the Intergovernmental Panel on Climate Change (Core Writing Team, Pachauri, R.K., Meyer, L.A.). IPCC, Geneva.*

(Fitter et al., 2010; IPCC, 2014; Smith, 2004). In this section the main impacts of climate change on soils and key adaptation and mitigation strategies to face these impacts are discussed.

Climate Change Impacts and Adaptation

Climate change impacts such as increases in temperature, changes in precipitation patterns, and changes in the occurrence of floods and droughts are predicted to have a large influence on forest and agricultural systems (Álvaro-Fuentes et al., 2012; Anaya-Romero et al., 2015; Conant et al., 2011; Lal, 2004b; Muñoz-Rojas et al., 2013). However, interrelations between climate changes and modifications in terrestrial ecosystems, e.g., changes in soil characteristics and processes, are complex and major research gaps remain for addressing the extent of these impacts (Brevik et al., 2015; Keesstra et al., 2016).

Previous studies suggested that continued rises in atmospheric CO_2 would entail increased photosynthetic rates and plant productivity, hence resulting in soil organic C (SOC) accumulation due to litter incorporation into the soil. This increased productivity would be accompanied by higher levels of microbial activity including mycrorrhizal colonization and symbiotic and root-zone N fixers exudates (Lovett et al., 2006; Nowak et al., 2004). Nevertheless recent studies have shown that this phenomenon, known as the CO_2 fertilization effect, might be actually less significant than initially expected due to increases in ozone levels. In fact, rising levels of ozone could offset the CO_2 fertilization effect leading to reduced plant productivity under elevated CO_2 (Camarero et al., 2015; Long et al., 2006). Through changes in climate and land use and management, many of the soils worldwide are expected to become more sensitive to soil erosion by water and wind due to increases in runoff (Anaya-Romero et al., 2015; Cerdá et al., 2005, 2010; Panagos et al., 2015). In the humid tropics, increased intensities of rainfall events could cause an increased occurrence of temporary flooding or water-saturation, whereas in other areas, such as arid or semi-arid environments, it could lead to further land degradation or desertification (IPCC, 2014; Martttínez-Casasnovas et al., 2002).

Mitigation and adaptation strategies are considered complementary approaches for reducing risks of climate change impacts at local, regional, and global scales (Wreford, 2010). Mitigation strategies focus on reducing the rate and magnitude of climate change by reducing its causes. Complementary to mitigation, adaptation strategies aim to minimize impacts and maximize the benefits of new opportunities. Adaptation measurements can be applied at different scales: temporal, e.g., short-term modifications or long-term adaptation, and spatial, e.g., plot level, regional, or national policy level. Short-term adjustments may include changes in land management such as modification of planting dates or incorporation of water conserving practices. Long-term adaptations involve major structural changes, e.g., changes in land use and land allocation (Aguilera et al., 2013; Muñoz-Rojas et al., 2011; Smith et al., 2008).

The adoption of adaptation practices will require developing system models to integrate and extrapolate anticipated climate changes. To predict climate change impacts on soil characteristics and processes, global climate models and regional climate models are essential to simulate global climate and generate projections of precipitation, temperature, and other climate variables (Lugato and Berti, 2008). However, changes in temperature and rainfall as a result of global warming are subject to large uncertainties due to the inconsistencies in current global circulation models (Wreford, 2010). These uncertainties in assessing impacts and responses of climate change are associated with physical and biological processes. For example, the role of soils and the feedbacks from soil processes that transform climate processes, e.g., the effects of soil water processes on the occurrence of extreme heat waves and droughts, have been underestimated in global models (Seneviratne et al., 2014; Trenberth et al., 2015).

Scenario analysis has been broadly used for assessing future changes and projecting the impacts of climate change on different issues such as agricultural production or water and forest resources (e.g., Pásztor et al., 2017, Chapter 9 in this volume). These scenarios can be comprised of various driving forces of climate change which consist of different future scenarios that might influence GHG sources and sinks, including the energy system and land use changes. Because of the large uncertainties concerning the evolution of such driving factors, there is a broad range of possible emissions paths of GHG (IPCC, 2014). Thus the representative concentration pathways (RCPs) make projections and describe four different 21st century pathways of GHG emissions, atmospheric concentrations, air pollutant emissions, and land use. The RCPs include a severe mitigation scenario (RCP2.6), two transitional scenarios (RCP4.5 and RCP6.0) and one scenario with very high GHG emissions (RCP8.5). Scenarios that do not bring additional efforts to restrict emissions ("baseline scenarios") lead to pathways that range between RCP6.0 and RCP8.5. The RCP2.6 scenario aims to keep global warming below 2°C above preindustrial temperatures (Fig. 6.2).

Integrating climate models driven by these IPCC scenarios in soil process models has allowed the investigation of potential changes and threats in soil characteristics and functions in future climate scenarios. These approaches can provide essential information to support decision-making and problem solving in land management as well as climate adaptation and mitigation strategies (Aguilera et al., 2013; Anaya-Romero et al., 2011; Smith et al., 2008, 2016).

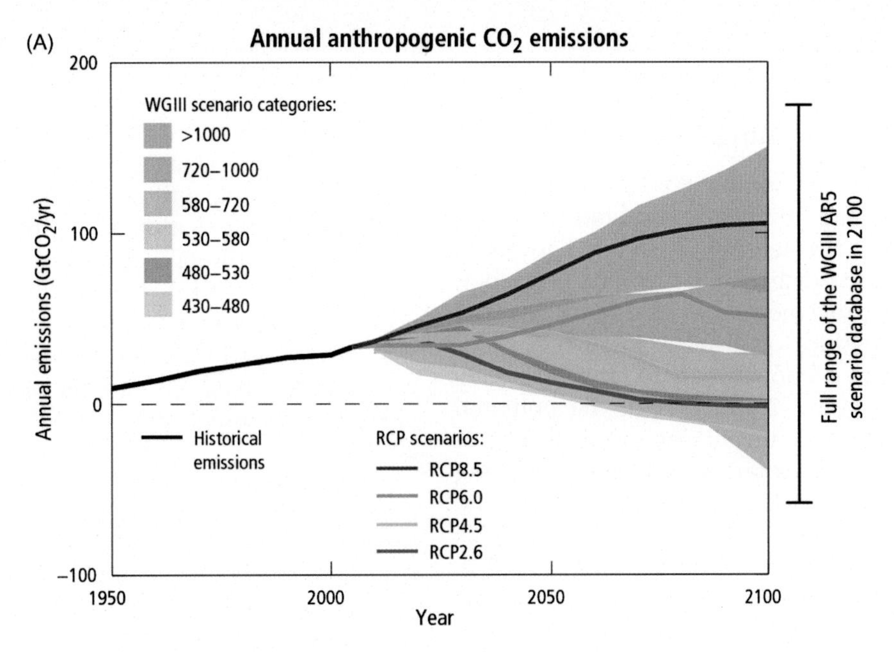

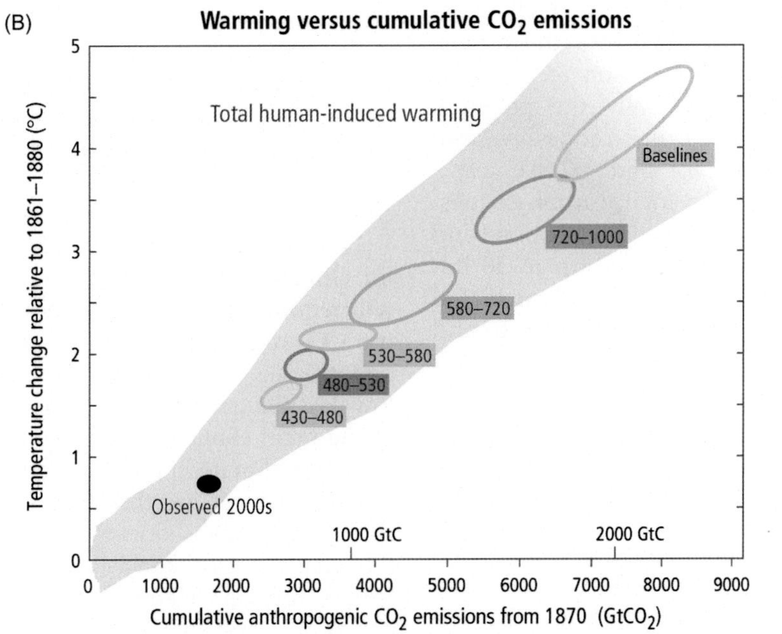

FIGURE 6.2 (A) Emissions of carbon dioxide (CO_2) alone in the RCPs (lines) and the associated scenario categories used in WGIII (colored areas show 5–95% range). (B) Global mean surface temperature increase at the time global CO_2 emissions reach a given net cumulative total, plotted as a function of that total, from various lines of evidence. Colored plume shows the spread of past and future projections from a hierarchy of climate-carbon cycle models driven by historical emissions and the four RCPs overall times out to 2100, and fades with the decreasing number of available models. Ellipses show total anthropogenic warming in 2100 versus cumulative CO_2 emissions from 1870 to 2100 from a simple climate model (median climate response) under the scenario categories used in WGIII. The width of the ellipses in terms of temperature is caused by the impact of different scenarios for non-CO_2 climate drivers. The filled black ellipse shows observed emissions to 2005 and observed temperatures in the decade 2000–09 with associated uncertainties. *Source: IPCC, 2014. Climate Change 2014. Synthesis Report. Contribution of Working Groups I, II and III to the Fifth Assessment Report of the Intergovernmental Panel on Climate Change (Core Writing Team, Pachauri, R.K., Meyer, L.A.). IPCC, Geneva.*

Climate Change Mitigation

The list of technically possible options for climate change mitigation in land use and management is extensive (Table 6.1). However, some of these practices, e.g., farming efficiency, genetic improvement, or alternative energy sources to reduce or displace fossil fuel emissions, entail additional costs to land managers (Alexander et al., 2015). One of the potential mechanisms to mitigate increased levels of atmospheric CO_2 is carbon sequestration using the soil–microorganism–plant system (Brevik, 2013a; Crow et al., 2016; Gaudinski et al., 2000; Janzen, 2004; Tian et al., 2016; Yu et al., 2016). Carbon sequestration in terrestrial ecosystems consists of the net removal of CO_2 from the atmosphere and the avoidance of CO_2 emissions from these ecosystems into the atmosphere (Lal, 2004b, 2007; Powlson et al., 2011; Stringer et al., 2012). Globally the potential capacity for soil carbon sequestration has been estimated at between 0.4 and 1.2 $Gt C year^{-1}$ (Paustian et al., 2016). In total, carbon sequestration coupled with biological carbon and bioenergy mitigation could save up to 38 billion tonnes of carbon and 3%–8% of calculated energy consumption by 2050 (Canadell and Schulze, 2014).

New strategies and policies within the international framework highlight the potential of improved agriculture and forestry management practices to increase carbon sequestration in soils, reducing net GHG emissions (Smith et al, 2008). Although the use and value of soils worldwide are frequently associated with agricultural production, they are also relevant for the provision of a large number of ecosystem services (Anaya-Romero et al., 2016; Fitter et al., 2005; Keesstra et al., 2016; Marañón et al., 2012a). Soil C contents and dynamics are key determinants of the quantity and quality of these services which includes enhancing cation exchange capacity, soil aggregation, water retention, and supporting soil biological activity. Furthermore, SOC promotes resistance to soil erosion and helps to regulate flooding by increasing infiltration, reducing runoff, and slowing water movement from upland to lowland areas (Brevik et al., 2015; Cerdá, 1997; Keesstra et al., 2016). Nonetheless, among all ecosystem services provided by soils, the role that SOC plays in climate regulation is critical to sustainable land use and management (Fitter et al., 2010; Marañon et al., 2012b; Stringer et al., 2012).

Acknowledging the need to monitor CO_2 emissions and carbon sequestration, the United Nations Framework Convention on Climate Change (UNFCCC) called for the annual monitoring of carbon by signatory nations. This monitoring mechanism, known as a National Greenhouse Gas Inventory (NGHGI), contains an accounting of carbon emissions and removals (i.e., sequestration). The development of new methods and the use of existing tools for soil carbon monitoring and accounting have therefore become critical in a global change context (Scharlemann et al., 2014).

Implementing effective soil-based strategies for climate change mitigation on a large scale will require the capacity to monitor and measure CO_2 reductions with satisfactory accuracy, quantifiable uncertainty, and at moderately low cost (Paustian et al., 2016). Currently, some of the main challenges with soil carbon assessment for climate change mitigation and adaptation programs are the large amount of work required, the consequently elevated costs, and the challenge of consistency between monitoring rounds. Hence, combining modeling approaches and spatial databases with monitoring programs can potentially reduce the work and costs associated with monitoring programs (Jandl et al., 2014).

To address the calculation of potential emissions of CO_2 from soils under global change scenarios (land use and climate change) several estimates of SOC contents have been published over the last 15–20 years based on the

TABLE 6.1 Land Management Practices for Soil Carbon Sequestration in Forest and Agricultural Systems

Measure	Effect
FORESTS	
Afforestation	Increases the C pool in the aboveground biomass and replenishes the soil C pool
Protection of existing forests	Preserves existing SOC stocks and prevents emissions due to biomass burning and land clearing
Reforestation	Increasing tree cover density in degraded forests increases C accumulation
Tree species selection	At identical biomass volumes, trees with high wood density (deciduous tree species) accumulate more C than trees with light wood (coniferous species)
Stand management	Harvest residues on the soil surface increase C stocks of the forest floor but disturb soil structure and lead to soil C loss
Site improvement	N fertilization stimulates biomass production, but leads to GHG emissions
GRASSLANDS AND CROPLANDS	
Zero or reduced tillage	Decreases the accelerated decomposition of organic C (and depletion of SOC) associated to intensive tillage. Prevents the breakage of soil aggregates that protect C
Mulching/residue management/ composting	Enhances soil moisture and prevents soil erosion. Crop residues prevent soil C loss. In flooded soils mulching can increase CH4 emissions
Introduction of earthworms	Improve aeration and organic matter decomposition
Application of inorganic fertilizers and manure	Adding manures and fertilizers stimulate biomass production. Increases plant productivity and thus SOC. However, chemical fertilizers are nonenvironmentally friendly and result in N_2O emissions
Water management	It can improve plant productivity and production of SOC. However, energy used for irrigation is associated to GHG emissions, and nutrient leaching can affect water quality. Carbon costs of producing fertilizer and pumping irrigation water should be considered
Improved rotations	Rotations with perennial pastures can increase biomass returned to the soil and therefore enhance SOC. Integration of several crops at the same time can increase organic material, soil biodiversity and soil health, as well as increasing food production
Site specific management	It may reduce the risk of crop failure and thus improve overall productivity, improving SOC stocks
Use of improved crop varieties	Increase productivity above and below ground and crop residues, thereby enhancing SOC

Source: Jandl, R., Lindner, M., Vesterdal, L., Bauwens, B., Baritz Hagedorn, F., Johnson, D.W., et al., 2007. How strongly can forest management influence soil C sequestration? Geoderma 137, 253–268. http://dx.doi.org/10.1016/j.geoderma.2006.09.003; Smith, P., 2004. Soils as carbon sinks: the global context. Soil Use Manag. 20, 212–218. http://dx.doi.org/10.1111/j.1475-2743.2004.tb00361.x.

representative soil profile information and soil maps (Arrouays et al., 2001; Bradley et al., 2005; Eswaran et al., 1995; Muñoz-Rojas et al., 2012). Soil C maps can help identify potential areas where management practices that promote C sequestration will be productive and guide the formulation of policies for climate change mitigation and adaptation strategies (Scharlemann et al., 2014). Different methods and approaches have been developed over the last four decades to tackle the calculation of SOC stocks at global (Bohn, 1976, 1982; Batjes, 1996; Buringh, 1984; FAO, 2009; Hiederer and Köchy, 2011), national (Arrouays et al., 2001; Batjes, 2005; Bradley et al., 2005; Rodriguez-Murillo, 2001; Viscarra Rossel et al., 2014), and regional levels (Batjes and Dijkshoorn, 1999; Muñoz-Rojas et al., 2012; Schwartz and Namri, 2002).

At the global scale, SOC contents of terrestrial ecosystems are difficult to determine because of the high spatial variability and the number of factors influencing soil C dynamics (FAO, 2009; Fig. 6.3). At this level, spatial distribution of SOC most closely reflects rainfall and temperature distribution, with most of the SOC stored in the northern hemisphere and greater accumulations of C in more humid and cold areas (Kapos et al., 2008). However, factors of both climate and topography are key determinants of SOC variability (Lozano-García et al., 2016; Miller et al., 2015; Phachomphon et al., 2010; Schulp et al., 2008). The pattern of macroclimate influencing soil properties is modified by patterns of parent material, microclimate, and topography at more local scales. Additionally, one of the main factors affecting SOC stocks is land use, which can largely alter the balance between C losses and C sequestration (Liebens, 2003; Meersmans et al., 2008; Smith, 2008). Thus estimates of SOC stocks may be particularly inaccurate in areas with diverse land use patterns, such as Mediterranean landscapes (Muñoz-Rojas et al., 2012, Willaarts et al., 2016).

Some studies have been undertaken at the continental level to provide more accurate estimates of the soil C pools (Batjes, 2005). In Europe, information on the spatial distribution

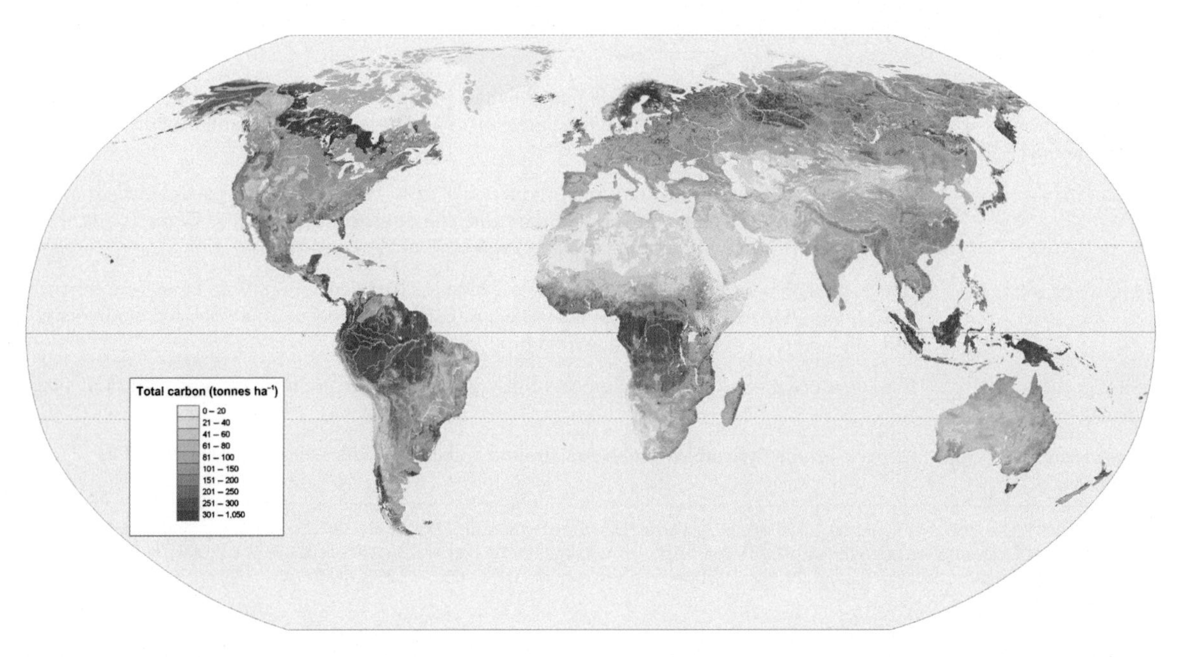

FIGURE 6.3 Global soil carbon map. *Source: FAO, 2009. International Institute for Applied Systems Analysis, International Soil Reference and Information Centre, Institute of Soil Science-Chinese Academy of Sciences, Joint Research Centre of the European Commission. Harmonized World Soil Database (Version 1.1). FAO, Rome, Italy; International Institute for Applied Systems Analysis, Laxenburg, Austria.*

of SOC content is currently offered by the Joint Research Center (Jones et al., 2004). The European Topic Centre for Spatial Information and Analysis, ETC/SIA, supports the European Environment Agency (EEA) and has developed a methodology for soil C accounting in Europe (Weber, 2011). Their method is based on CORINE land cover maps and other remote sensing techniques.

In Australia, a recent initiative has been developed by Viscarra Rossel et al. (2014) to create a baseline map of organic C in Australian soils in order to support national carbon accounting and monitoring under climate change. They assembled and harmonized data from different sources, such as Australia's National Soil Carbon Research Programme (SCaRP) (Baldock et al., 2013). Spectroscopic estimates of organic C and bulk density were made with the Australian visible near infrared database (Viscarra Rossel and Webster, 2012) and Australian Soil Resource Information System (ASRIS) (Johnston et al., 2003). By combining a bootstrap model a decision tree with piecewise regression on environmental variables, and geostatistical modeling of residuals they produced a fine spatial resolution baseline map of organic C at the continental scale for the upper 30 cm of the soil profile.

In the United States, monitoring forest carbon has been under continuous improvement, combining both field and remote sensing data (Woodall et al., 2015). Regional studies in the United States have placed great effort on increasing geospatial resolution of inventory-based carbon accounting of agricultural land by using land cover defined by NASA's moderate resolution imaging spectroradiometer (MODIS) and by the United States Department of Agriculture's National Agricultural Statistics Service (USDA-NASS) cropland data layer (West et al., 2010).

In Africa, to produce a soil map of the whole continent at 250 m spatial resolution, two soil profile datasets were compiled by the AfSIS project (Hengl et al., 2015): the Africa Soil Profiles (legacy) database and the AfSIS Sentinel Site database. This compilation was carried out over the period 2008–14 and consisted of 28,000 sampling locations across Africa that generated predictions of several soil properties that included soil carbon at two or six standard soil depths. These predictions were obtained using an automated mapping framework (3D regression-kriging based on random forests).

In spite of extensive efforts to compile soil information and map soil C, many uncertainties remain in the determination of soil C stocks, and the reliability of these estimates depends upon the quality and resolution of the spatial datasets used for its calculation. Thus better estimates of soil C pools and dynamics are needed to advance understanding of the C balance and the potential of soils for climate change mitigation (Falloon and Smith, 2003; Jandl et al., 2014). Current estimates could be enhanced by systematic collection of additional data from soil profiles, a higher density of soil sampling points, and sampling to greater depths. In fact, soil depth and how those depth intervals are determined has proven to be important in the calculation of SOC stocks (Liebens and VanMolle, 2003; Parras-Alcántara et al., 2015). Most calculations of SOC pools are restricted to the topsoil, and soil measurements are in most cases taken in the upper layers. However, vertical processes have a considerable effect on SOC variability and a significant amount of SOC can be stored in deeper layers (Brevik and Homburg, 2004; Jobbagy and Jackson, 2000; VandenBygaart, 2006). SOC can be unevenly distributed over varying soil depths and a standardized approach for SOC estimations along the soil profile is necessary (Muñoz-Rojas et al., 2013, 2015). Existing soil C maps are often based on data that inadequately reflects the C pool of deeper soil horizons, partly because the impact of land use changes on deep soil C contents has been poorly addressed.

Soil C models are capable of evaluating present SOC stocks and dynamics at different scales and predicting soil C sequestration trends under projected scenarios (Kutsch et al., 2009). Several SOC models have been developed with different features and for diverse purposes (Table 6.2), but despite the exhaustive research on C dynamics and the continuous development of soil C models, substantial limitations exist in their application. Mechanistic SOC simulation models are expected to play a key role in monitoring programs since they can support estimation of temporal trends in SOC pools. However, their use for prediction of existing soil properties in space and time may not be adequate (Paustian et al., 2016).

There are a broad variety of methods used for modeling purposes, ranging from classic and simple statistical methods to sophisticated and computer-intensive techniques. However, the more complex data mining methods are not necessarily superior and, in fact, simple models can achieve better performance for certain datasets (Muñoz-Rojas et al., 2013; Scharlemann et al., 2014). For example, in land evaluation, statistical systems can be powerful and effective empirical methods for land suitability prediction based on land characteristics.

TABLE 6.2 Most Commonly Used Soil Carbon Models

Model	Input Data	Output Data (C Pools)	Depth	Reference
CarboSOIL	Seasonal temperature and total precipitation; total N, pH, cation exchange; soil texture; land use	SOC pool	75 cm	Muñoz-Rojas et al. (2013)
CENTURY	Temperature and total precipitation; plant N, P, and S content; soil texture; atmospheric and soil nitrogen inputs; and initial soil C, N and sulfur levels	Litter; SOM pools	20 cm	Parton et al. (1987)
EPIC	Daily air T and precipitation, radiation, texture, bulk density, C content	Litter; SOM pools	From topsoil (5 cm) to 1 m	Williams (1990)
DAYSY	Air temperature and precipitation; N input; texture; global radiation	SOM pools	40 cm	Mueller et al. (1996)
DNDC	Plant growth data, soil clay, bulk density, pH, air temperature, precipitation, atmospheric N decomposition rate, crop rotation timing and type, inorganic fertilizer timing, amount and type, irrigation timing and amount, residue incorporation timing and amount, and tillage timing and type	Two layers with five pools: very labile litter, labile litter, resistant litter, humads and humus	50 cm	Li et al. (1994)
ICBM	Enviornmental input to soil, humification coefficient; fraction of initial decomposition	Young and old carbon	25 cm	Andrén and Kattere (1997)
RothC	Clay, monthly precipitation, monthly open pan evaporation, average monthly mean air T and an estimate of the organic input	Litter; SOM pools	1 m	Coleman and Jenkinson (1996)
ROMUL	Litter; soil C in organic layer and in mineral layer; soil moisture; soil texture	Litter; SOM pools	Organic soil; 1 m mineral soil	Chertov et al. (2001)

SOC, soil organic carbon; *SOM*, soil organic matter.

Correlation and multiple regression analyses have been used to investigate the contributions of selected land characteristics on land suitability and vulnerability (Abd-Elmabod et al., 2012; Anaya-Romero et al., 2015; De la Rosa et al., 2004; Wang et al., 2010). For processes related to soil C dynamics, empirical models based on regression/correlation techniques may not be capable of explaining complex mechanisms within the soil system, but they can be useful tools to identify different drivers of SOC dynamics and perform projections of SOC stocks (Viaud et al., 2010).

SUSTAINABLE AGRICULTURAL PRODUCTION

According to Koohafkan et al. (2012), there is no current universally accepted definition of "sustainable agriculture." However, there are several broadly agreed upon components to the sustainable agriculture concept, including (1) the production of an abundant, healthy and affordable food supply for the world's population, (2) doing so in a way that is environmentally friendly and socially equitable, (3) transforming agricultural production so that it contributes to conservation of global biodiversity, and (4) allowing the agroecosystem to continue providing essential ecosystem and socioeconomic services (Koohafkan et al., 2012). This will all need to be achieved in a world where the arable land base is shrinking, water and nitrogen supplies are increasingly limited, global climate is changing, and human population is growing (IAASTD, 2009), which means there is increasing stress on the global soil system (Brevik et al., 2015) (Fig. 6.4). This is why soil system health, productivity, and quality are increasingly important to sustainability (Keesstra et al., 2016).

The broadly agreed upon components of sustainable agriculture identified by Koohafkan et al. (2012) can be fit into four areas, of which two have received considerable attention from soil scientists in recent years: (1) food security and (2) soil quality/health, and two have been investigated, but to a lesser extent: (3) soils and human health and (4) soils, society, and economics. This section will take a closer look at each of these four areas and the ways they relate to sustainable agriculture in the context of soil mapping and process models.

Food Security

According to the United Nations Food and Agriculture Organization (FAO) "Food security [is] a situation that exists when all people, at all times, have physical, social and economic access to sufficient, safe and nutritious food that meets their dietary needs and food preferences for an active and healthy life" (FAO, 2002). An examination of the aspects of sustainable agriculture identified by Koohafkan et al. (2012) demonstrates the importance of sustainable agriculture to long-term food security.

The Koohafkan et al. (2012) sustainable agriculture aspect 1 corresponds to the food security need for access to sufficient, safe, and nutritious food. Most of the essential nutrients important to crop growth come from soil; in fact, of the 17 elements essential for plant growth, 14 come from soil (Havlin et al., 2005). However, declining soil fertility with corresponding declines in yield has been identified as a problem in many parts of the world (Acharya et al., 2007; Sanchez et al., 1997; Vanlauwe et al., 2008; Yap et al., 2016). It is difficult to produce sufficient and nutritious food if soil fertility and crop yields are in decline. In the developed world, declining natural soil fertility has been addressed through the use of anthropogenically produced fertilizers, but in some cases this has also introduced heavy metals into the fertilized soils (Chen et al., 2009; Senesi et al.,

FIGURE 6.4 Soils are required to provide many services in the modern world, including the production of food (top left), feed (top right), fiber and fuel (bottom left), and to serve as the foundation of our cities and other developments (bottom right). Given the growing world population, these demands put a serious strain on global soil resources. *Source: Photographs courtesy of Eric Brevik.*

1999), a condition that can lead to unsafe food (Brevik, 2009a,b).

The Koohafkan et al. (2012) sustainable agriculture aspect 2 corresponds to the food security need for socially equitable access to food. The concept of socially equitable includes having access to healthy foods that are culturally acceptable and at a price that consumers can afford (FAO, 2002). Truly sustainable agriculture will provide healthy foods at affordable prices (Koohafkan et al., 2012; Malhi et al., 2009; Pothukuchi and Kaufman, 1999) that is culturally acceptable (Hamm and Bellows, 2003). The Koohafkan et al. (2012) sustainable agriculture aspect also corresponds to the food security need for safe food, in that the sustainable agriculture concept calls for agriculture to be conducted in an environmentally friendly way. This, in turn, is expected to produce better quality food, something that will enhance human health and food security (Carvalho, 2006; Spiertz, 2010). Therefore there are distinct links between sustainable agriculture and food security. The food production issue is of special interest for developing countries that do not have ready access to fertilizers and soil production and fertility is a key issue for the subsistence of the population.

Soil information has the ability to improve food security models. For example, in a model of food security status in Ghana, Nata et al. (2014) found that adopting soil-improving practices decreased the probability that households would suffer from food insecurity. Furthermore, Nata et al. (2014) recommended that policies to enhance soil quality should be implemented. It would seem that such implementation, and models that indicate the need for such implementation, would benefit from the inclusion of good soils data. However, soil information is often poorly integrated into food security models, in part because soil scientists often do not take part in the creation of these models, but also because of the lack of adequate, quantitative global soil data (Bouma and McBratney, 2013). Enhanced soil mapping and modeling that incorporates evolution of the soil system, including alterations driven by anthropogenic activities, would benefit food security studies (Grunwald et al., 2012).

Soil Quality/Health

For many the terms soil quality and soil health are essentially same thing, with scientists often preferring the term quality and farmers the term health (Doran and Zeiss, 2000; Karlen et al., 1997; Romig et al., 1996). However, the term soil health has seen increasing acceptance among scientists in recent years, particularly because of the implied inclusion of biological soil properties within it (Kibblewhite et al., 2008). There are also many parallels in the goals of the soil quality/health community and the relatively new concept of soil security (Brevik et al., 2017; McBratney and Field, 2015). The soil quality/health concept uses a number of soil biological, chemical, and physical properties to evaluate the ability of a given soil to perform a number of functions related to biological productivity, ecosystem services, and human health (Karlen et al., 1997; Muñoz-Rojas et al., 2016a). When compared to the aspects of sustainable agriculture identified by Koohafkan et al. (2012), it can be seen that soil quality/health is an important component of sustainable agriculture, with the maintenance or enhancement of soil quality/health being an important component of sustainable agriculture (Fig. 6.5). In fact, Doran and Zeiss (2000) concluded that soil health was synonymous with sustainability. Soil quality is related to soil management as it is dependent on sustainable use of the soil resource (Tesfahunegn, 2016).

Many attempts have been made to model soil quality/health using a variety of techniques (Grunwald, 2009; Liu et al., 2016; McBratney and Odeh, 1997), including multiple statistical techniques, fuzzy methods, pedodiversity analyses, and spatial analyses. However, there is still a considerable need to gather additional soil information in support of soil quality/health studies. In particular, there is a need for data on use-dependent and dynamic soil properties as a companion to currently existing databases on soil mapping, classification, and interpretation (Norfleet et al., 2003). Soil models need to incorporate dynamic soil properties (Grunwald et al., 2012), and the additional use of covariates, along with better models to relate these covariates to select soil properties, will be needed so that soil properties important to estimating soil quality/health can be determined and mapped in places that are currently lacking in soil information (Brevik et al., 2016). However, it will also be important to continue funding and conducting field studies to collect physical data that can be used to ground-truth estimates made with models and

FIGURE 6.5 Healthy soils are capable of supporting lush, high-yielding plant growth that will supply plentiful and nutritious food and feed (left, maize), abundant fiber (right, flax), and fuel. *Source: Photographs courtesy of the UDSA-NRCS.*

covariates so that the models can be checked and improved (Brevik et al., 2016).

Human Health

Soils directly affect human health through (1) food quantity and quality, (2) exposure to various chemicals in the soil system, and (3) exposure to pathogens (Burras et al., 2013). There has been increasing interest in the links between soils and human health over the last approximately 50 years (Zornoza et al., 2015). Soils influence the supply of adequate food quantity and high food quality through the nutrients they offer to crops grown in them. A healthy rooting environment, good structure and appropriate temperatures, water contents and aeration also contribute. Soils can also have indirect positive influences on human health. One example is through the supply of medications (Mbila, 2013); in fact, about 40% of prescription drugs originate from soil sources, including about 60% of new drugs approved between 1989 and 1995 (Pepper et al., 2009). Exposure to soil microorganisms may also help prevent the development of allergies and other immunity-related disorders in children (Haahtela et al., 2008; Matricardi and Bonini, 2000; Rook, 2010).

Soils can also have negative effects on human health. Heavy metals in the soil environment offer considerable risk to human health (Morgan, 2013; Roy and McDonald, 2015). They can be taken up by crops or ingested directly, particularly by children, potentially leading to problems such as itai-itai disease (cadmium) or high lead levels in blood, which can cause reduced IQ, coordination, gastrointestinal damage, hypertension, organ damage, increased cancer rates, and other problems (Brevik, 2013b). Exposure to organic chemicals that ended up in the soil due to the application of agricultural treatments (insecticides, herbicides, fungicides, etc.), (Fig. 6.6) can cause a host of problems depending on the chemicals and chemical mixes encountered (Burgess, 2013). Some human

FIGURE 6.6 The application of agriculture chemicals leads to chemical coverage of the soil surface as well as the crop, as can be seen by the chemical mist behind this tractor. *Source: Photograph courtesy of Artemi Cerdà.*

pathogens also inhabit soil environments, and exposure has the possibility of causing a wide range of diseases (Brevik and Burgess, 2013; Loynachan, 2013). Both the positive and negative aspects of soils and human health tie into all four aspects of Koohafkan et al. (2012) characteristics of sustainable agriculture. Soils and human health also tie into the concepts of soil quality/health and each of the five dimensions of soil security (Brevik et al., 2017).

Soil mapping and models have not yet been extensively used in support of human health studies, but mapping and modeling does have great potential to improve our understanding of issues such as nutrient deficiencies in crops, uptake of contaminants by crops, and location of sites where contact with pathogens might be likely. A good example of this is the study conducted by Tabor et al. (2011), who studied incidences of valley fever in Tucson, AZ, United States. These researchers used soil taxonomic units from the soil survey of Tucson to classify their study area into three geomorphic units, based on the hypothesis that geomorphic position would be an important predictor of soil conditions that were conducive to the growth of *Coccidioides*, the fungus that causes valley fever. Tabor et al. (2011) concluded that their geomorphic stratification improved the efficiency of their sampling; therefore demonstrating that soil maps have the potential to provide useful input in human health studies. Additional work like this is critical to provide for a more complete incorporation of soils information into human health studies. Conversely, better global soil mapping and enhanced soil models could provide human health experts better data to work with (Brevik et al., 2016).

Soils, Society, and Economics

Soils are an important part of determining which socioeconomic activities can be profitably pursued at any given location (Brevik et al., 2015). Humans were locating their settlements on soils that were prime for agricultural activities by the end of the Neolithic (Miller and Schaetzl, 2014), and soil degradation has contributed to the collapse of numerous civilizations as well (Montgomery, 2007), highlighting the reason that sustainable agriculture is so important. As explained previously in this chapter, soils are critical to human health, and good overall health is essential for a healthy society. Soils are a repository of cultural information, providing important archeological insights (Holliday, 2004), and serve as an important aspect of art. Therefore soils are socially important.

Soils provide many economic benefits, with some of them being fairly easy to quantify. For example, it is not hard to determine what the average net economic return is from a given field for a given crop over a number of years, or the economic value of a piece of urban real estate as a place to construct a building. An agroforestry business can easily calculate their net return per hectare from a wooded plot. Brevik et al. (2017) estimated the 2014 value of soil-based prescription drugs as approximately US$280 billion. However, soils provide a number of services that are much more difficult to quantify. What is the value of water purification as rainwater or snowmelt infiltrates through soil? How much is the carbon sequestered in a hectare of land worth, or what is the value of waste recycling, contaminant sequestration, or pest and disease control? While putting economic value on such things is difficult, there have been some attempts to do so, with results that vary widely. For example, McBratney et al. (2017) estimated an average ecosystem services value for all soils of the world at US$867 ha^{-1} $year^{-1}$. However, when looking only at agricultural soils, these values can be much higher. Dominati et al. (2014) estimated the ecosystem services provided by agricultural soils in New Zealand at NZ$16,390 ha^{-1} $year^{-1}$ (approximately US$13,110 ha^{-1} $year^{-1}$), and Sandhu et al. (2008) estimated a range in the value of ecosystem services for a set of agricultural fields in New Zealand as US$1270 to 19,420 ha^{-1} $year^{-1}$.

Many factors come together to explain the wide range in soil ecosystem services estimated above (Smith et al., 2015; Zhang et al., 2013). They include the value placed on a particular service by the people making the estimates, different values of various land uses, management practices, and soil properties. Therefore while there is much more to these estimates than just soil properties and their distribution, those factors do play important roles. Better soil mapping and modeling would provide higher quality information that could then be used as a part of economic models seeking to quantify the value of soil ecosystem services.

LAND DEGRADATION AND RESTORATION

Soil functionality refers to the capability of soils to provide key functions, e.g., biological productivity, nutrient cycling, or physical stability and support for plant growth in agricultural or forest ecosystems (Fitter et al., 2010). Although natural disturbances such as fire can alter soil structure and functioning, often, ecosystems affected by fire will recover without intervention. However, there are several human-induced threats to soil functionality, i.e., erosion and compaction, that can have serious consequences on the environment. Use of appropriate land management and restoration practices will be crucial to improve soil functions and services and to contribute to climate resilience and land sustainability (Brevik et al., 2015). This section focuses on the most updated modeling and mapping techniques to monitor and evaluate the main causes of land degradation and to assess potential restoration measures.

Fire

Fire is a global phenomenon and a natural element across ecosystems worldwide that may cause both positive and negative impacts to the environment (Pereira et al., 2016a). It is recognized that fire is one of the main causes of land degradation (Perez-Cabello et al., 2010; Tessler et al., 2016a) due to removal of vegetation cover and the combustion of soil organic matter that increases soil vulnerability to erosion agents and nutrient depletion (Moody et al., 2013; Shakesby, 2011).

Soil burn severity can be classified as "low," "moderate," and "high" (Bodi et al., 2014; Keeley, 2009; Parsons et al., 2010). The impacts of low severity fires are usually minimal or absent, and to some extent, fires may be beneficial to ecosystems because of inputs of nutrients and organic matter into the soil profile (Pereira et al., 2014). In high severity fires, where temperatures reached are extremely high, there is an increased risk of land degradation due to the great impact on soil properties and the losses of soil and water (Mataix-Solera et al., 2011; Moody et al., 2013; Francos et al., 2016; Bárcenas-Moreno et al., 2016; López-Serrano et al., 2016; Muñoz-Rojas et al., 2016b). These types of fires have become more frequent as a consequence of changes in social behavior (rural exodus, land abandonment), land use policy (fire suppression, urban sprawl into wildland areas), and climate change (more intense and frequent drought spells, higher temperatures) (Batllori, et al., 2013; Brotons et al., 2013; Nunes et al., 2016; Pereira et al., 2015a; Shakesby, 2011).

Despite changes induced in the landscape, the impacts of fire are often short term (especially in fire resilient ecosystems, e.g., Mediterranean ecosystems), independent of the different rates of recovery, fire severity, ecosystem type, and topography of the affected area (Granged et al., 2011; Hanan et al., 2016; Meng et al., 2015; Petropoulos et al., 2014; Zavala et al., 2009). This short-term impact is due to the high availability of mineralized nutrients in the ash (Pereira et al., 2015b). There are exceptions, such as areas strongly affected by short-term recurrent fires, where recovery of soil and vegetation may be relatively slow (Flores et al., 2016; Jordán et al., 2014; Lippitt et al., 2012; Tessler et al., 2016a, b).

Mapping soil cover, burn severity, and changes in soil properties in the period immediately after a fire is fundamental to assess the degree of soil protection and to identify the areas that might require intervention. Maps can be generated at plot scale or consider the total fire affected area, through the use of field survey or remote sensing methods (Outeiro et al., 2008; Pereira et al., 2015b; Robichaud et al., 2007; Woods et al., 2007) (Fig. 6.7).

Assessment of fire severity can be done by analyzing ash color, thickness and cover on the soil surface, degree of vegetation cover, soil organic matter consumption, presence of fine roots, soil water repellency, and soil structure and infiltration in the field. Although most burned areas show a pattern of different fire severities, with variations at small scales (Alanís et al., 2016; Gordillo-Rivero et al., 2014; Jiménez-Pinilla et al., 2015; Jordán et al., 2015), data collected in the field can be complemented with remote sensing data. This methodology is used by the Burned Area Emergency Response (BAER) team from the USDA to identify areas with higher risk where rehabilitation measures should be applied (Robichaud et al., 2007). Once this information is collected, it can be integrated into a geographic information system (GIS) environment allowing the production of

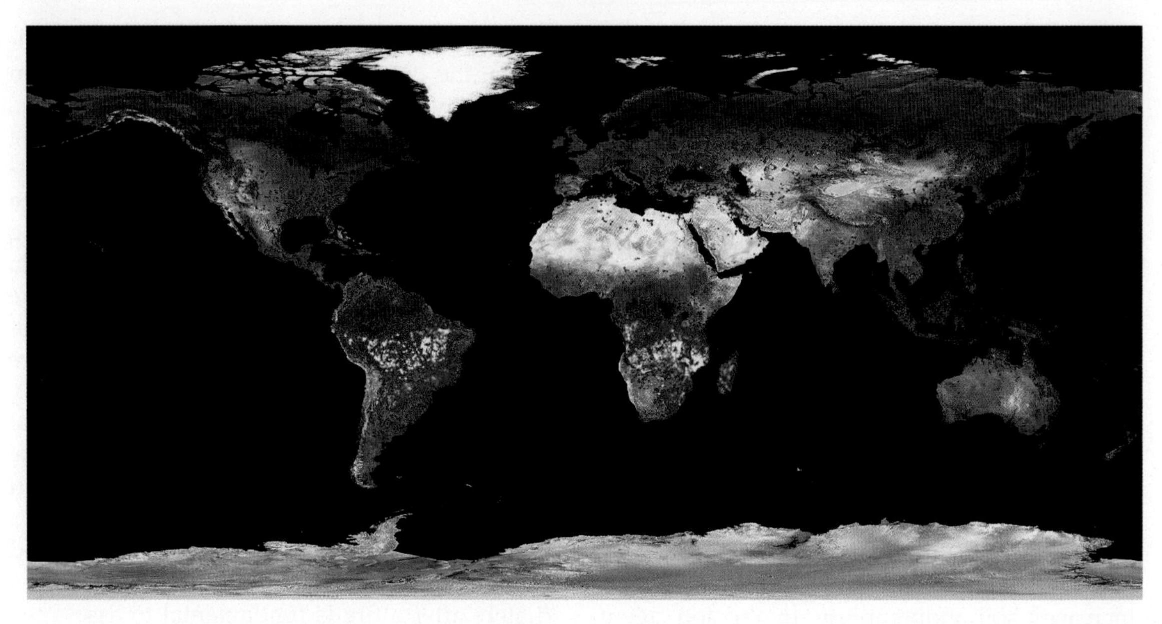

FIGURE 6.7 Moderate Resolution Imaging Spectroradiometer (MODIS) on NASA's Terra satellite showing fires around the world. Date: September 24, 2010. *Source: Credit: NASA.*

maps that represent the areas most vulnerable to erosional processes. These soil maps will be the base for intervention in the burned territory. Other methods that have commonly been used to assess fire severity include the Jain index (Jain et al., 2008) and the composite burned index (Kasischke et al., 2008). These methods are more comprehensive than those used by BAER.

Several tools have been used to provide data to models of postfire severity assessment including satellite sensors (e.g., MODIS, AVIRIS, and Landsat) (Barret and Kasischke, 2013; Kokaly et al., 2007; Veraverbeke et al., 2014; Fernandez-Manso et al., 2016). The most commonly used model is the normalized burn ratio (dNBR) that was designed to identify and classify burned areas according to fire severity. It uses a Landsat TM image from prior to the fire and other images immediately after the fire. Then, for a determined area it calculates changes in vegetation cover imposed by the fire from the reflectance, or radiance. Burn severity is assessed according to the difference between the prior and postfire images (Vlassova and Perez-Cabello, 2016; Vogeler et al., 2016). BAER uses Burned Area Reflectance Classification. This model evaluates burn severity using an image transformation by the NBR algorithm (Parsons et al., 2010). However, since these measurements have to be carried out in a short period after the fire, it is not always possible to obtain images with clear skies (Keeley, 2009). To have a better assessment of fire impacts on the landscape, it is essential to have field and remote sensing analysis. Several works have related field measurements using the composite burn index and remote sensing analysis and some of them found a good correlation (Parker et al., 2015; Wimberly et al., 2007), whereas others showed a poor relation (Hoy et al., 2008; Murphy et al., 2008).

The objectives of postfire restoration are to protect soil and reduce water losses, increasing the ecosystem recovery capacity, restoring the ecological functions and landscape aesthetics and managing the available fuel to prevent future wildfires. The implementation of these strategies is strongly dependent on the impact imposed by fire and the capacity of the ecosystems to recover (Alloza and Vallejo, 2006; Neris et al., 2016). Despite this, intervention in burned areas is a matter of discussion since in several fire affected areas natural recovery or partial cuts have proved to be more effective than any other interventions such as plantations, salvage logging, tree removal, or soil tilling (Beschetta et al., 2004; Beghin et al., 2010; Castro et al., 2011; Pereira et al., 2016b). These intervention practices involve strong soil disturbance, reducing the recovery of native species, favoring the spread of exotic species (Moreira et al., 2013), and increasing soil degradation (Moltttó et al., 2014). Some areas may need intervention; however, these should not involve activities that induce a high disturbance in the soil by using heavy machinery. On south-facing slopes with high inclination, fire severity is normally high and recuperation of the vegetation is low; therefore, specific restoration measures may be needed to reduce soil and water losses (Pausas et al., 2004; Pereira et al., 2016b). This is particularly problematic as carbon and nitrogen stocks tend to be lower in south-facing soils to begin with (Lozano-García et al., 2016).

Several methods can be applied to restore postfire affected areas, including erosion barriers (Robichaud, 2009), mulching (Bautista et al., 2009; Jordán et al., 2011), seeding woody plants, plant species selection, site preparation, and development of quality nursery stock (Vallejo et al., 2009). Among the techniques currently applied, the most effective are the ones that provide soil surface cover such as straw or hydromulching (Prosdocimi et al., 2016). However, the use of these techniques requires a great investment (MacDonald and Larsen, 2009), which is not always possible due to budget limitations. The effectiveness of these measurements also depends on the postfire meteorological conditions and normally their implementation depends on the socio-political and cultural issues that govern the affected communities. However, fire is an ecological element and its impacts are not necessarily coercive to the environment, so as previously discussed, intervention should be punctual and clearly identified (Wohlgemuth et al., 2009).

Soil Erosion

Soil erosion is one of the major causes, evidences of, and key variables used to assess and understand land degradation (Bone et al., 2014; Guessesse et al., 2015; Houyou et al., 2016). Soil erosion is a consequence of unsustainable land use (Cerdà et al., 2010) and other disturbances, such as fire, mining, or intensive agricultural uses (Cerdà and Doerr, 2005). The loss of soil may have serious impacts on the quantity and quality of soil ecosystem services, with serious economic, social, and political implications (De Vente et al., 2013; Panagos et al., 2016).

Mapping areas affected by soil erosion is essential to develop a proper assessment of areas that need to be restored and to promote awareness among farmers and decision makers (Prasuhn et al., 2013). The advance of GIS and remote sensing techniques have allowed the production of more accurate maps, often combining different data sources (Fernandez et al., 2016). The majority of studies that dealt with mapping water soil erosion have focused on risk (Kheir et al., 2006; Lu et al., 2004), susceptibility (Krishna Bahadur, 2009), and potential (Millward and Mersey, 1999). Other studies have focused on the contribution of the most vulnerable areas such as cultivated areas (Le Bissonnais et al., 2001; Martttínez-Casasnovas et al., 2002). Mapping wind erosion can be more complex than water erosion due to the high spatial variability of wind patterns, frequency, and intensity (Sterk, 1997). Despite the difficulties in studying wind erosion in the field, more efforts are needed to measure and map sediment deposition (Sterk, 1997) and transport (Visser et al., 2004). As in the case of water erosion, the majority of studies have focused on vulnerability (Mezosi et al., 2015), susceptibility (Borrelli et al., 2015, 2016), and risk (Asensio et al., 2016; Reiche et al., 2011) of wind erosion.

Nowadays, the great availability of field and remotely sensed data has allowed the creation of erosion vulnerability maps at a monthly temporal scale (Panagos et al., 2012a), which is extremely important for land managers to assess temporary soil erosion vulnerability and to identify areas that need to be restored. Nevertheless one of the main challenges is the scale of analysis, which is normally 1 km^2 (Pereria et al., 2017b, Chapter 2 in this volume), which limits to a large extent the identification of soil erosion vulnerability. Recently, efforts have been made at the European level to produce maps of soil erosion at a finer resolution (100 m for water erosion and 500 m for wind erosion), which are more suitable for land management and soil restoration assessment (Panagos et al., 2015) (Fig. 6.8).

A large number of studies have been carried out using geostatistical methods to estimate the spatial vulnerability of soil erosion by water using the universal soil loss equation (USLE) and the revised soil loss equation (RUSLE) (Buttafuoco et al., 2012; Baskan and Dengiz, 2008; Galdino et al., 2016; Ugur Ozcan et al., 2008) and trying to extrapolate these erosion indexes. However, these studies assumed that geostatistical techniques are the most suitable to interpolate soil erosion. As in previous works in other fields (Pereira et al., 2014, 2016a), kriging may not be the most suitable method for data interpolation, and a comparison with other methods is needed in order to find the best spatial predictor. An overview of the range of spatial prediction methods available is provided by Miller (2017). This is fundamental for a good assessment of the risks of and vulnerability to soil erosion, because potential errors of prediction can be high. More studies are needed to compare different interpolation techniques and produce better soil erosion risk and prediction maps. This is crucial from the point of view of land restoration since the maps produced could identify the most vulnerable areas and those where the efforts and resources should be allocated. An erroneous identification of these areas would have implications for society, the economy, and environment.

There are several models currently used to estimate soil erosion by water. The most common are Water Erosion Prediction Project (WEPP), USLE, RUSLE, Modified Universal Loss Equation (MUSLE), Soil Loss Estimation Model for South Africa (SLEMSA), Erosion Productivity Impact Calculator (EPIC), Pelletier's model, Pacific Southwest Inter-Agency Committee (PSIAC), Regional Modelling of Soil Erosion (MESALES), Risk European Soil Erosion Model (EUROSEM), Factorial Scoring Model (FSM), SPAtially Distributed Scoring Model (SPADS), SSY Index Model, spatially distributed model WATEM–SEDEM, Annualized AGricultural Non-Point Source Pollution Model (AnnAGNPS), Limburg Soil Erosion Model (LISEM), Pan-European Soil Erosion Risk Assessment (PESERA), and the Soil and Water Assessment Tool (SWAT) (De Vente et al., 2013; Laflen and Flanagan, 2013; Morgan et al., 1998). A detailed description of these models can be found in De Vente et al. (2013).

Recently, the Joint Research Centre developed a new model, the G2 model,[1] which maps soil loss estimates from local to regional level. This model is a step up from its predecesor since it includes the vegetation factor and a management parameter based on the combination of Gravilovic models and empirical tables for USLE. In addition, with this improved version it is possible to identify soil erosion

[1]http://eusoils.jrc.ec.europa.eu/library/themes/erosion/G2/data.html

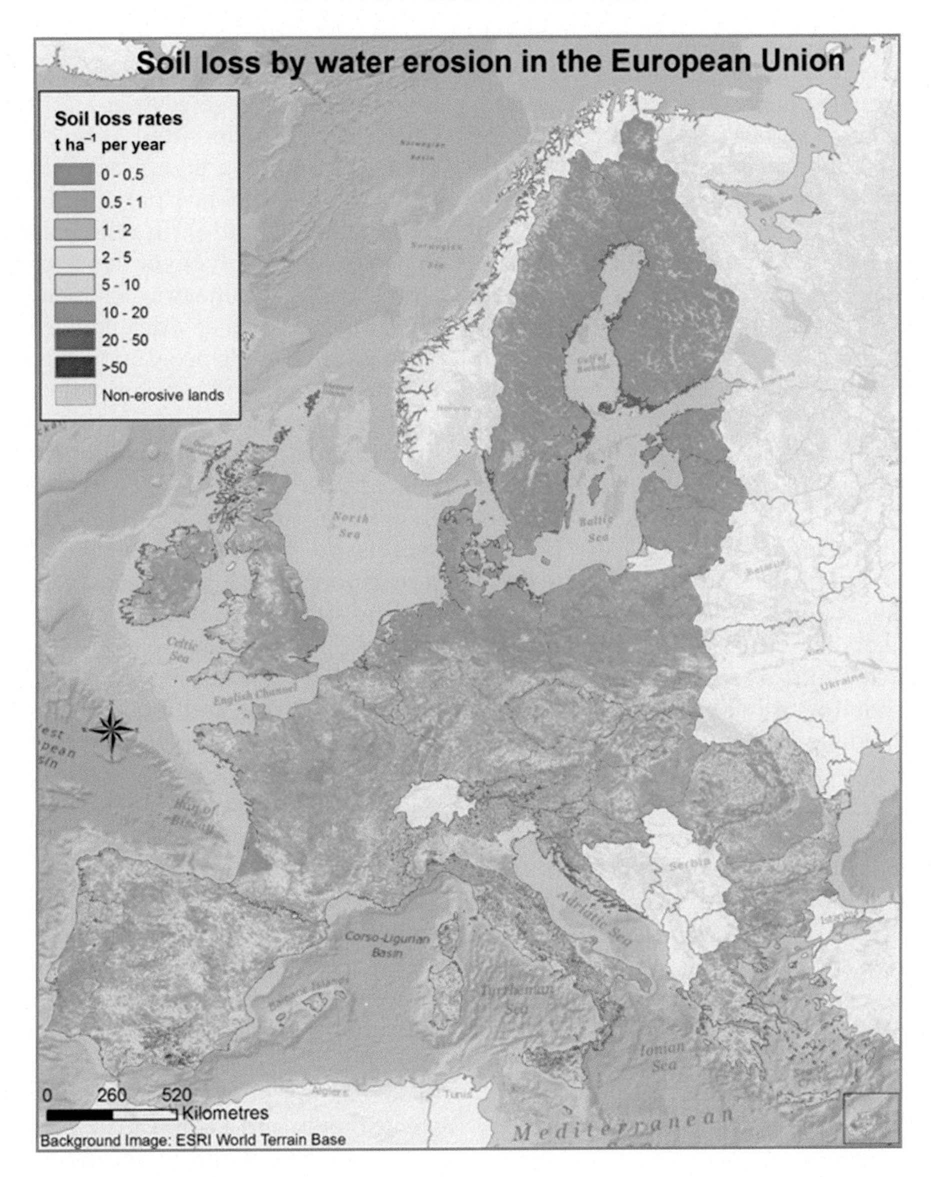

FIGURE 6.8 Soil erosion in European Union (reference year: 2010) based on RUSLE2015. *Source: Panagos, P., Borrelli, P., Poesen, J., Ballabio, C., Lugato, E., Meusburger, K., et al., 2015. The new assessment of soil loss by water and erosion. Environ. Sci. Policy 54, 438–447. http://dx.doi.org/10.1016/j.envsci.2015.08.012.*

according to land use type. Other advantages of this model include the use of USLE family models, the large data availability from European databases, its applicability to different scales of analysis (depending on the pixel size), and the possibility of considering different alternatives (Panagos et al., 2014a). This is crucial for land restoration, since it allows accurate identification of land uses and periods when vulnerability is high, making possible the implementation of techniques to restore land. A review of all the models used in soil erosion studies was carried out by Karydas et al. (2014).

As is the case for water soil erosion, there are a large number of models from field scale to global scales for wind erosion assessment (Hagen, 2005). The most commonly used are the Wind Erosion Equation (WEQ), Revised Wind Erosion Equation (RWEQ), Wind Erosion Prediction Systems (WEPS), Wind Erosion Stochastic Simulator (WEES), Texas Tech Erosion Analysis Model (TEAM), Wind Erosion Assessment Model (WEAM), Wind Erosion and European Light Soils (WEELS), Wind Erosion and Loss of Soil Nutrients in Semi-Arid Spain (WELSONS), and Dust Production Model (DPM) (Blanco-Canqui and Lal, 2010; Borrelli et al., 2016). Detailed information about these models can be found in Gomes et al. (2003) and Blanco-Canqui and Lal (2010). Recently the Joint Research Centre developed a new index of Land Susceptibility to Soil Erosion (ILSWE) and the Wind Erodible Fraction of Soil (EF),[2] joining data from several EU projects (Borrelli et al., 2016).

There are several issues that determine the optimal model to apply in a specific location depending on the modeling objectives, scale of analysis, type of soil erosion, sediment yield, and driving environmental variables. Thus performance of existing models is highly variable and depends on the area where they are applied and the situation they are applied to (De Vente et al., 2013).

Despite the large number of models used for soil erosion assessment, there are several limitations to their applicability. For example, validation of these models is needed and the results can be highly inconsistent (De Vente et al., 2013). At larger scales their application can be challenging and financially and technically not feasible (Bosco et al., 2015). Some studies presented a good validation of the respective models (Haregeweyn et al., 2013; Wu and Chen, 2012), others were very poor (De Roo et al., 1996), but a significant number of previous studies did not provide any validation (Panagos et al., 2016). In other cases, accuracy depends on the type of land use analyzed. Additionally, the majority of models require a large input of data at a detailed level that is not often available (Blanco-Canqui and Lal, 2010). For this reason, soil erosion projects developed by the Joint Research Centre are extremely important for the evaluation of land at risk of degradation that needs to be restored.

Erosion prevention is one of the main aspects in the reduction of site damage and restoration of degraded ecosystems and ecosystem services provision, as several studies have highlighted (Burylo et al., 2014; Trabucchi et al., 2014; Zhao et al., 2013). Thus accurate soil erosion maps and models are fundamental to assess the needs for and effectiveness of land restoration (Pereria et al., 2017b, Chapter 2 in this volume).

Soil Compaction

Soil compaction is a consequence of urbanization and other human activities such as forest harvesting, pipeline installation, construction, amenity land use, wildlife trampling, intensive use of heavy machinery, grazing, short crop rotations, and other types of deficient management (Brevik and Fenton, 2012). It can occur in different climates and its incidence is higher in soils grazed or otherwise disturbed with high moisture levels and low organic matter contents and with the use of tillage. Compaction has negative impacts on soil water infiltration, biota, energy transfer, gas diffusion, and organic carbon accumulation. It reduces root and plant development, increases the vulnerability of crops to diseases, and enhances soil erosion risks (Batey, 2009; Brevik et al., 2002; Hamza and

[2] http://esdac.jrc.ec.europa.eu/content/index-land-susceptibility-wind-erosion-ilswe-and-wind-erodible-fraction-soil-ef-europe

Anderson, 2005; Scalenghe and Marsan, 2009; Troldborg et al., 2013).

Effective restoration of soil compaction is only possible with correct and detailed spatial analysis of the problem. Mapping soil compaction has become extremely important, especially in precision agriculture. The spatial variability of soil compaction can have serious implications on plant growth and soil erosion (Hemmat and Adamchuck, 2008). Thus it is essential to identify where it is located and the causes responsible for it. Several types of sensors have been used to measure soil mechanical resistance, including water content sensors, soil strength sensors, and fluid permeability sensors developed for mapping. A review of sensors used in soil compaction studies can be found in Hemmat and Adamchuck (2008).

Several works have been carried out using on-site proximal sensing methods to directly or indirectly study soil compaction, such as observing the impacts of wheel traffic in corn fields (Duttmann et al., 2014; Naderi-Boldaji et al., 2014), potato crops (Shamal et al., 2016), road trails (Brevik and Fenton, 2004; Campbell et al., 2013), cereal crops (Naderi-Boldaji et al., 2013), wheat fields (Sirjacobs et al., 2002), vineyards (Rossi et al., 2013), rice paddies (Islam et al., 2014), and compaction due to different tillage methods (Fountas et al., 2013). Proximal sensors incorporate a GPS device, which immediately records the geographical coordinates and facilitates the task of mapping soil compaction. In some of the available studies, data were not interpolated, and the studied area was shown with point data (Campbell et al., 2013; Troldborg et al., 2013) or lines (Duttmann et al., 2014), making it difficult to completely understand the spatial distribution or patterns of soil compaction. Data interpolation are essential to estimate soil compaction in areas where measurements are not available.

Geostatistical models have been extensively applied to predict soil compaction, using bulk density measurements (Shamal et al., 2016; Yang et al., 2016), penetration resistance (Barik et al., 2014; Naderi-Boldaji et al., 2013), tillage resistance (van Bergeijk et al., 2001), cone penetration resistance (Veronesi et al., 2012), and electrical resistivity (Naderi-Boldaji et al., 2014; Rossi et al., 2013). More advanced spatial models have incorporated several covariates such as soil water content to increase the accuracy of soil compaction predictions (Yang et al., 2016). In Wallonia (Belgium), D'Or and Destain (2016) mapped the risk of precompression stress using several auxiliary variables, e.g., soil bulk density, organic matter content, air capacity, available water content, nonavailable water content, hydraulic conductivity, cohesion, and internal friction angle. Vrindts et al. (2005) mapped soil bulk density using other auxiliary variables (soil moisture, grain yield, soil cover and position of red edge inflection point) to identify the impacts of soil compaction on wheat production.

Similarly to soil erosion, the Joint Research Center has developed a map of natural susceptibility to soil compaction with a pixel size of 1 km^3 (Fig. 6.9) and bulk density derived from texture datasets (Ballabio et al., 2016) (Fig. 6.10), which can be an useful tool for researchers; however, finer scale maps are needed for land managers.

In the past 20–30 years the increase of data availability and the development of advanced statistical methods have improved the accuracy of soil compaction predictions. Soil compaction models aim to calculate the stress induced by a mechanical loading on the soil profile leading to failure at a determined soil status (e.g., moisture). This can be very useful for farmers and planners to identify areas that are more vulnerable and where traffic should be avoided. There are two main types of models used in soil compaction studies: analytical and finite models (Keller and Lemande, 2010). Analytical

[3] http://esdac.jrc.ec.europa.eu/content/natural-susceptibility-soil-compaction-europe

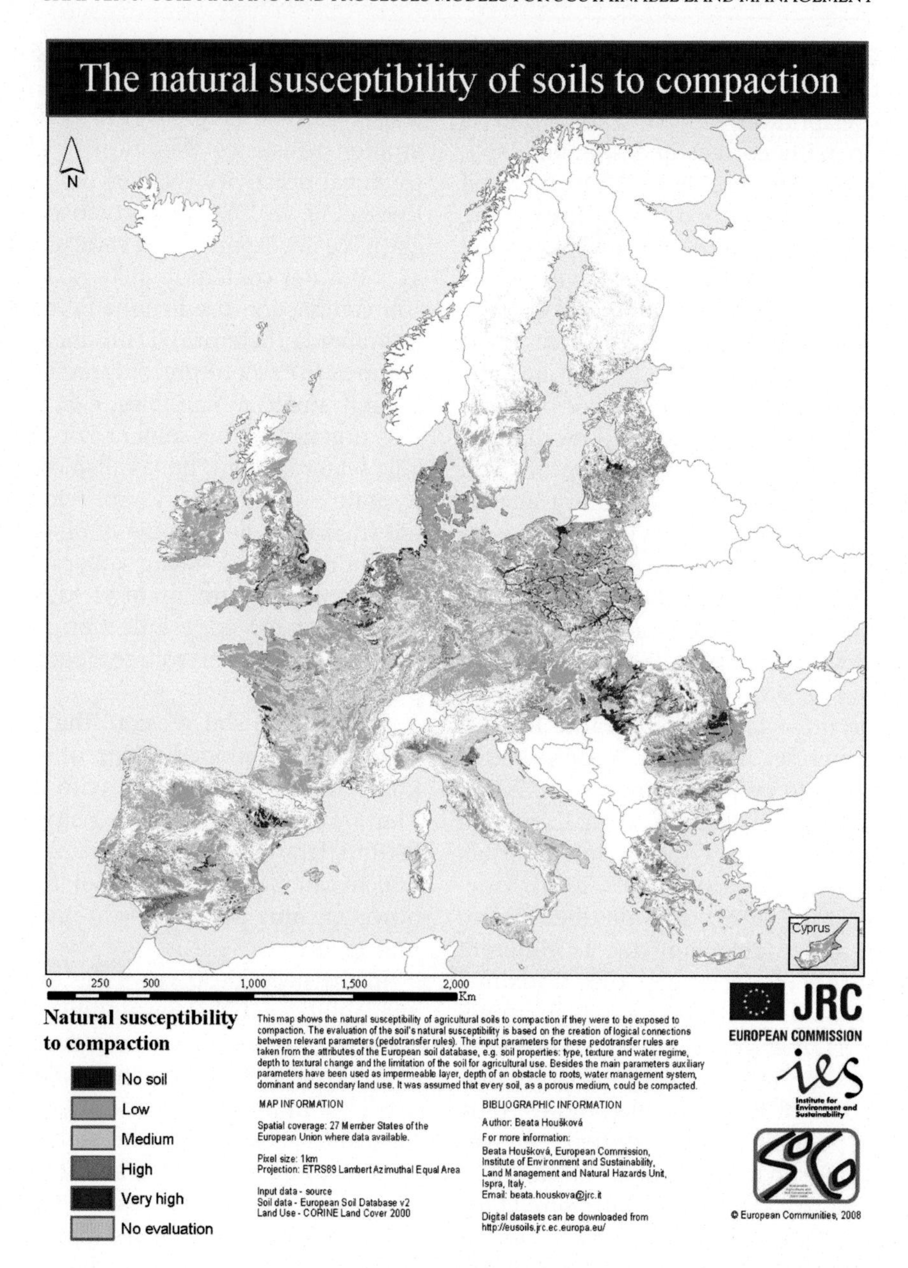

FIGURE 6.9 The natural susceptibility of soils to compaction. *Source: Panagos P., Van Liedekerke M., Jikones A., Montanarella L., 2012b. European Soil Data Centre: Response to European policy support and public data requirements. Land Use Policy 29, 329–338. http://dx.doi.org/10.1016/j.landusepol.2011.07.003.*

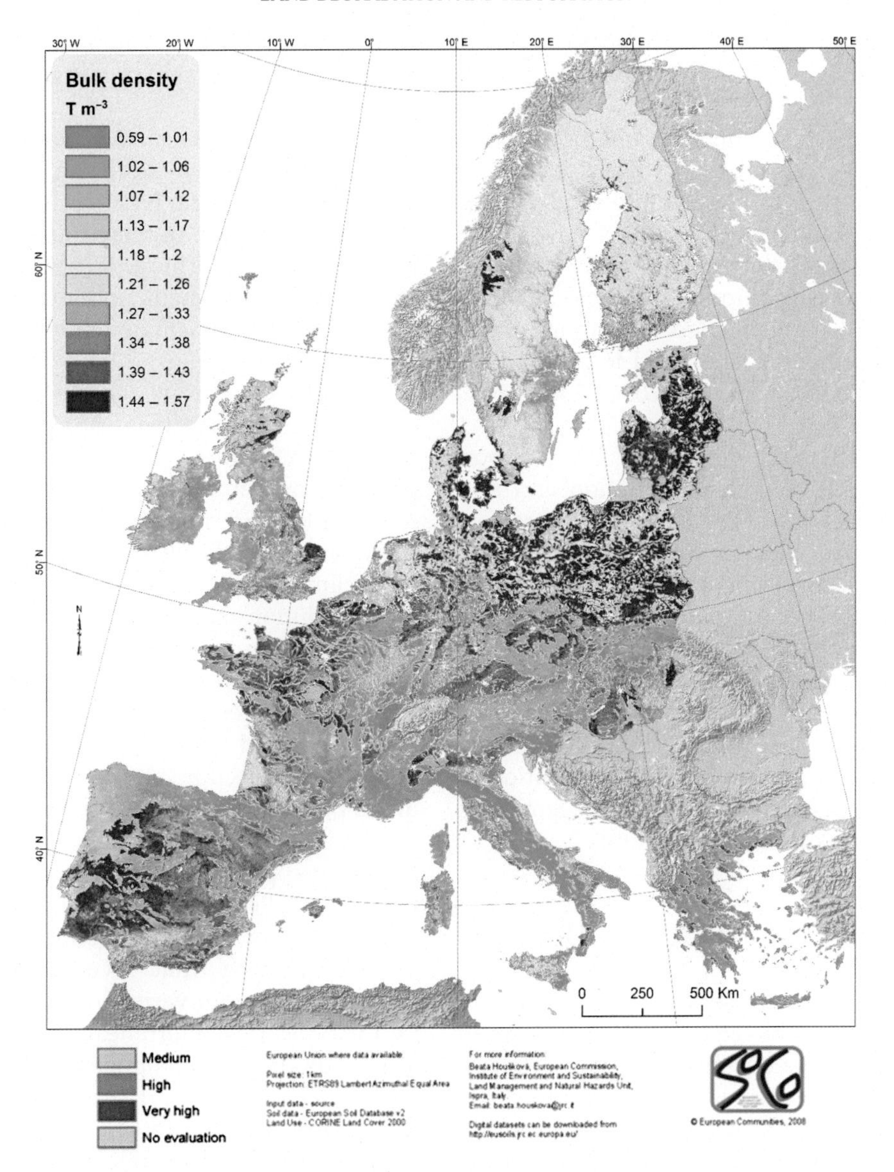

FIGURE 6.10 Bulk density derived from soil texture datasets. *Source: Ballabio, C., Panagos, P., Monatanarella, L., 2016. Mapping topsoil physical properties at European scale using the LUCAS database. Geoderma 261, 110–123. http://dx.doi.org/10.1016/j.geoderma.2015.07.006.*

models are based on the development of equations that estimate the propagation of the stress through elastic, homogeneous and semi-infinite media. Finite models are normally applied to continuum mechanics and are usually more appropriate to describe soil behavior. These models need a larger amount of data and parameters than analytical models, which can be difficult to obtain. Despite the limitations of analytical methods in relation to finite models, these are easier to apply and can be used by managers and agricultural advisors (Keller et al., 2007).

A number of analytical methods, e.g., Soilflex (Keller et al., 2007) and SOCOMO (van den Akker, 2004), and finite models, e.g., Prager/Cap model (Xia, 2011), have been developed in the last decade. With the development of numerical computers, 2D and 3D finite models have been generated to estimate traffic impacts on soil compaction and to identify the soil–tire interaction (Fervers, 2004; Tekin et al., 2008).

In summary, soil compaction is an important component of land degradation and several techniques have been applied to restore soil compaction such as complete cultivation, mega lift, industrial ripper (Sinnett et al., 2006), loosening operations (Spoor, 2006), subsoiling and ploughing (Chamen et al., 2015), and use of cover crops (Brevik, 2009a,b). In this context, soil mapping and modeling are key tools to assess the main causes of soil compactation and identify priority areas that need application of restoration measures or special management to avoid this form of soil degradation.

CONCLUDING REMARKS

The need to increase food production to meet the nutritional requirements of a growing population in a sustainable way, i.e., maintaining soil quality levels and preventing soil degradation, is a major challenge that we currently face. Soil degradation can include loss of soil organic matter, decline in soil fertility and structure, increased erosion, salinity, or compaction and decreased biodiversity. In turn, these declines in soil quality may lead to high levels of poverty, serious environmental, or human health problems, and can even contribute to civilization collapse. Increased agricultural intensification may result in further land degradation if sustainable practices are not widely adopted and, additionally, soil degradation may be accelerated or mitigated by climatic factors.

On-going climate change will have critical (and unpredictable) effects on land resources and the environment. New strategies of adaptation and mitigation of these impacts need to be urgently adopted. This will require appropriate monitoring, assessment, and analyses of potential scenarios at adequate temporal and spatial scales.

Appropriate use of geographical information systems, geospatial mapping, and remote sensing techniques are critical for adequate planning and implementation of effective and sustainable land management practices (Pino-Mejías et al., 2010). These tools allow us to integrate climate, soil, and land use data to generate high-resolution maps that will help to improve capabilities for modeling processes at different spatio-temporal scales. Process-based and empirical models offer differing but complementary approaches and both will be needed to support field data collection and generate predictions under potential scenarios of global change.

In the following chapters, several applications of mapping and process models to address modern challenges are shown. Abd-Elmabod et al. (Chapter 7 in this volume) presents an interesting study that investigates the influence of soil factors variability on land suitability for typical Mediterranean crops using a set of models that are components of an agro ecological decision support system. In their chapter, Geitner et al. (Chapter 8 in this volume) use innovative GIS and modeling approaches to soil mapping to favor data integration in land use planning and management applications. Pásztor et al. (2017, Chapter 9 in this volume) apply advance soil mapping techniques, e.g., digital soil mapping, in Hungary to assist in regional land planning activities. In their research, Pereira et al. (2017c, Chapter 10 in this volume) apply several modeling techniques and principal component analyses to map ash distribution after a fire. Richter and Burras (Chapter 11 in this volume) discuss the role of catenas in soil mapping and pedology in the 21st century; and finally Alaoui (Chapter 12 in this volume) introduces a new method for mapping soil vulnerability to floods.

References

Abd-Elmabod, S.K., Ali, R.R., Anaya-Romero, M., Jordán, A., Muñoz-Rojas, M., Abdelmagged, T.A., et al., 2012. Evaluating soil degradation under different scenarios of agricultural land management in mediterranean region. Nat. Sci. 10, 103–116.

Acharya, G.P., McDonald, M.A., Tripathi, B.P., Gardner, R.M., Mawdesley, K.J., 2007. Nutrient losses from rain-fed bench terraced cultivation systems in high rainfall areas of the mid-hills of Nepal. Land Degrad. Dev. 18, 486–499. http://dx.doi.org/10.1002/ldr.792.

Alanís, N., Hernández-Madrigal, V.M., Cerdà, A., Muñoz-Rojas, M., Zavala, L.M., Jordán, A., 2016. Spatial gradients of intensity and persistence of soil water repellency under different forest types in central Mexico. Land Degrad. Dev. http://dx.doi.org/10.1002/ldr.2544.

Aguilera, E., Lassaletta, L., Gattinger, A., Gimeno, B.S., 2013. Managing soil carbon for climate change mitigation and adaptation in Mediterranean cropping systems: a meta-analysis. Agric. Ecosyst. Environ. 168, 25–36. http://dx.doi.org/10.1016/j.agee.2013.02.003.

Alexander, P., Paustian, K., Smith, P., Moran, D., 2015. The economics of soil C sequestration and agricultural emissions abatement. Soil 1, 331–339. http://dx.doi.org/10.5194/soil-1-331-2015.

Alloza, J.A., Vallejo, R., 2006. Restoration of burned areas in forest management plans. In: Kepner, W.G., Rubio, J.L., Mouat, D.A., Pedrazzini, F. (Eds.), Desertification in the Mediterranean Region: A Security Issue. Springer, The Netherlands, pp. 475–488. http://dx.doi.org/10.1007/1-4020-3760-0-22.

Álvaro-Fuentes, J., Easter, M., Paustian, K., 2012. Climate change effects on organic carbon storage in agricultural soils of northeastern Spain. Agric. Ecosyst. Environ. 155, 87–94. http://dx.doi.org/10.1016/j.agee.2012.04.001.

Anaya-Romero, M., Abd-Elmabod, S.K., Muñoz-Rojas, M., Castellano, G., Ceacero, C.J., Alvarez, S., et al., 2015. Evaluating soil threats under climate change scenarios in the Andalusia region. Southern Spain. Land Degrad. Dev. 26, 441–449. http://dx.doi.org/10.1002/ldr.2363.

Anaya-Romero, M., Pino, R., Moreira, J.M., Muñoz-Rojas, De la Rosa, M., 2011. Analysis of soil capability versus land use change by using CORINE land cover and MicroLEIS. Int. Agrophys. 25, 395–398.

Anaya-Romero, M., Muñoz-Rojas, M., Ibáñez, B., Marañón, T., 2016. Evaluation of forest ecosystem services in Mediterranean areas. A regional case study in South Spain. Ecosyst. Serv, 82–90. http://dx.doi.org/10.1016/j.ecoser.2016.07.002.

Andrén, O., Kattere, T., 1997. ICBM: the introductory carbon balance model for exploration of soil carbon balances. Ecol. Appl. 7, 1226–1236. http://dx.doi.org/10.1890/1051-0761(1997)007[1226:ITICBM]2.0.CO;2.

Arrouays, D., Deslais, W., Badeau, V., 2001. The carbon content of topsoil and its geographical distribution in France. Soil Use Manag. 17, 7–11. http://dx.doi.org/10.1111/j.1475-2743.2001.tb00002.x.

Asensio, C., Lozano, F.J., Gallardo, P., Giménez, A., 2016. Soil wind erosion in ecological olive trees in the Tabernas desert (southeastern Spain): a wind tunnel experiment. Solid Earth 7, 1233–1242. http://dx.doi.org/10.5194/se-7-1233-2016.

Baldock, J.A., Sanderman, J., Macdonald, D., et al., 2013. Australian Soil Carbon Research Program, 1. CSIRO Data Collection. http://dx.doi.org/10.4225/08/5101F31440A36.

Ballabio, C., Panagos, P., Monatanarella, L., 2016. Mapping topsoil physical properties at European scale using the LUCAS database. Geoderma 261, 110–123. http://dx.doi.org/10.1016/j.geoderma.2015.07.006.

Bárcenas-Moreno, G., Garcia-Orenes, F., Mataix-Solera, J., Mataix-Beneyeto, J., 2016. Plant community influence on soil microbial response after a wildfire in Sierra Nevada Park (Spain Science of the Total Environment). Sci. Total. Environ. http://dx.doi.org/10.1016/j.scitotenv.2016.05.013.

Barik, K., Akskal, E.L., Islam, R.K., Angin, I., 2014. Spatial variability in soil compaction properties associated with traffic field oppertations. Catena 120, 122–133. http://dx.doi.org/10.1016/j.catena.2014.04.013.

Barret, K., Kasischke, E.S., 2013. Control on variations in MODIS fire radiative power in Alaskian boreal forests: Implications for fire severity conditions. Remote Sens. Environ. 130, 171–181. http://dx.doi.org/10.1016/j.rse.2012.11.017.

Baskan, O., Dengiz, O., 2008. Comparision of traditional and geostatistical methods to estimate soil erodibility factor. Arid Land Res. Manag. 22, 29–45. http://dx.doi.org/10.1080/15324980701784241.

Batjes, N.H., 1996. Documentation to ISRIC-WISE Global Data Set of Derived Soil Properties on a ½° by ½° Grid (Version 1.0). Working Paper and Reprint 96/05. International Soil Reference and Information Centre, Wageningen, The Netherlands.

Batjes, N.H., 2005. Organic carbon stocks in the soils of Brazil. Soil Use Manag. 21, 22–24. http://dx.doi.org/10.1111/j.1475-2743.2005.tb00102.x.

Batjes, N.H., Dijkshoorn, J.A., 1999. Carbon and nitrogen stocks in the soils of Amazon Region. Geoderma 89, 273–286. http://dx.doi.org/10.1016/S0016-7061(98)00086-X.

Batey, T., 2009. Soil compaction and soil management—a review. Soil Use Manag. 25, 335–345. http://dx.doi.org/10.1111/j.1475-2743.2009.00236.x.

Batllori, E., Parisien, M.A., Krawchuk, M.A., Moritz, M.A., 2013. Climate change-induced shifts in fire for Mediterranean ecosystems. Glob. Ecol. Biogeogr. 22, 1118–1129. http://dx.doi.org/10.1111/geb.12065.

Bautista, S., Robichaud, P., Balde, C., 2009. Post-fire mulching. In: Cerdà, A., Robichaud, P. (Eds.), Fire Effects on Soils and Restoration Measures. Science Publishers, Enfield, pp. 353–372.

Beghin, R., Lingua, E., Garbarino, M., Lonati, M., Bovio, G., Motta, R., et al., 2010. *Pynus sylvestris* forest regeneration under different post-fire restoration practices in the northwesthern Alps. Ecol. Eng. 36, 1365–1372. http://dx.doi.org/10.1016/j.ecoleng.2010.06.014.

Beschetta, R.L., Rhodes, J.J., Kauffmann, J.B., Gresswell, R.E., Minshall, G.W., Karr, J.R., et al., 2004. Post fire management on forested public lands in the western United States. Conserv. Biol. 18, 957–967. http://dx.doi.org/10.1111/j.1523-1739.2004.00495.x.

Blanco-Canqui, H., Lal, R., 2010. Principles of Soil Conservation and Management. Springer, New York.

Bodi, M., Martin, D.A., Santin, C., Balfour, V., Doerr, S.H., Pereira, P., et al., 2014. Wildland fire ash: production, composition and eco-hydro-geomorphic effects. Earth-Sci. Rev. 130, 103–127. http://dx.doi.org/10.1016/j.earscirev.2013.12.007.

Bohn, H.L., 1976. Estimate of organic carbon in world soils. Soil Sci. Soc. Am. J. 40, 468–470. http://dx.doi.org/10.2136/sssaj1976.03615995004000030045x.

Bohn, H.L., 1982. Estimate of organic carbon in world soils: II. Soil Sci. Soc. Am. J. 46, 1118–1119. http://dx.doi.org/10.2136/sssaj1982.03615995004600050050x.

Bone, J., Barraclough, D., Eggletton, P., Head, M., Jones, D.T., Voulvoulis, N., 2014. Prioritising soil quality assessment through the screening of sites. The use of publicly collected data. Land Degrad. Dev. 25, 251–266. http://dx.doi.org/10.1002/ldr.2138.

Borrelli, P., Panagos, P., Ballabio, C., Lugato, E., Weynants, M., Montanarella, L., 2016. Towards a Pan-European assessment of land susceptibility to wind erosion. Land Degrad. Dev. 27, 1093–1105. http://dx.doi.org/10.1002/ldr.2318.

Borrelli, P., Panagos, P., Montanarella, L., 2015. New insights into the geography and modelling of wind erosion in the European agricultural land. Application of a spatially explicit indicator of land susceptibility to wind erosion. Sustainability 7, 8823–8836. http://dx.doi.org/10.3390/su7078823.

Bosco, C., de Rigo, D., Dewitte, O., Poesen, J., Panagos, P., 2015. Modelling soil erosion at European scale: towards harmonization and reproducibility. Nat. Hazard Earth Syst. Sci. 15, 225–245. http://dx.doi.org/10.5194/nhess-15-225-2015.

Bouma, J., 2014. Soil science contributions towards Sustainable Development Goals and their implementation: linking soil functions with ecosystem services. J. Plant Nutr. Soil Sci. 177, 111–120. http://dx.doi.org/10.1002/jpln.201300646.

Bouma, J., 2015. Engaging soil science in transdisciplinary research facing "wicked" problems in the information society. Soil Sci. Soc. Am. J. 79, 454–458. http://dx.doi.org/10.2136/sssaj2014.11.0470.

Bouma, J., Broll, G., Crane, T.A., Dewitte, O., Gardi, C., Schulte, R.P.O., et al., 2012. Soil information in support of policy making and awareness raising. Curr. Opin. Environ. Sustain. 4, 552–558. http://dx.doi.org/10.1016/j.cosust.2012.07.001.

Bouma, J., McBratney, A., 2013. Framing soils as an actor when dealing with wicked environmental problems. Geoderma 200-201, 130–139. http://dx.doi.org/10.1016/j.geoderma.2013.02.011.

Bradley, R.I., Milne, R., Bell, J., Lilly, A., Jordan, C., Higgins, A., 2005. A soil carbon and land use database for the United Kingdom. Soil Use Manag. 21, 363–369. http://dx.doi.org/10.1079/SUM2005351.

Brevik, E.C., 2009a. Soil, food security, and human health. In: Verheye, W. (Ed.), Soils, Plant Growth and Crop Production. Encyclopedia of Life Support Systems (EOLSS), Developed under the Auspices of the UNESCO. EOLSS Publishers, Oxford. <http://www.eolss.net>.

Brevik, E.C., 2009b. Soil Health and Productivity. In: Verheye, W. (Ed.), Soils, Plant Growth and Crop Production. Encyclopedia of Life Support Systems (EOLSS), Developed under the Auspices of the UNESCO. EOLSS Publishers, Oxford, UK. <http://www. eolss. net>.

Brevik, E.C., 2012. Soils and climate change: gas fluxes and soil processes. Soil Horiz. 53 (4), 12–23. http://dx.doi.org/10.2136/sh12-04-0012.

Brevik, E.C., 2013a. The potential impact of climate change on soil properties and processes and corresponding influence on food security. Agriculture 3, 398–417. http://dx.doi.org/10.3390/agriculture3030398.

Brevik, E.C., 2013b. Soils and human health: an overview. In: Brevik, E.C., Burgess, L.C. (Eds.), Soils and Human Health. CRC Press, Boca Raton, pp. 29–56.

Brevik, E.C., Fenton, T.E., 2004. The effect of changes in bulk density on soil electrical conductivity as measured with the Geonics® EM-38. Soil Surv. Horiz. 45, 96–102.

Brevik, E.C., Fenton, T.E., 2012. Long-term effects of compaction on soil properties along the Mormon Trail, South-Central Iowa, USA. Soil Horiz. 53, 37–42. http://dx.doi.org/10.2136/sh12-03-0011.

Brevik, E.C., Homburg, J., 2004. A 5000 year record of carbon sequestration from a coastal lagoon and wetland complex, Southern California, USA. Catena 57, 221–232. http://dx.doi.org/10.1016/j.catena.2003.12.001.

Brevik, E.C., Burgess, L.C., 2013. The 2012 fungal meningitis outbreak in the United States: connections between soils and human health. Soil Horiz. 54, 1–4. http://dx.doi.org/10.2136/sh12-11-0030.

Brevik, E.C., Fenton, T.E., Moran, L.P., 2002. Effect of soil compaction on organic carbon amounts and distribution, South-Central Iowa. Environ. Pollut. 116, S137–S141. http://dx.doi.org/10.1016/S0269-7491(01)00266-4.

Brevik, E.C., Cerdà, A., Mataix-Solera, J., Pereg, L., Quinton, J.N., Six, J., et al., 2015. The interdisciplinary nature of SOIL. Soil 1, 117–129. http://dx.doi.org/10.5194/soil-1-117-2015.

Brevik, E.C., Calzolari, C., Miller, B.A., Pereira, P., Kabala, C., Baumgarten, A., et al., 2016. Soil mapping, classification, and modeling: history and future directions. Geoderma 264, 256–274. http://dx.doi.org/10.1016/j.geoderma.2015.05.017.

Brevik, E.C., Steffan, J.J., Burgess, L.C., Cerdà, A., Links between soil security and the influence of soil on human health. In: Field, D., Morgan, C.L., McBratney, A.B., (Eds.), Global Soil Security. Progress in Soil Science Series. Springer, 2017, Springer, Basel, 261–274.

Brotons, L., Aquilue, N., Caceres, M., Fortin, M.J., Fall, A., 2013. How fire history, fire suppression practices and climate change affect wildfire regimens in Mediterranean landscapes. PLoS One 8, e62392. http://dx.doi.org/10.1371/journal.pone.0062392.

Burgess, L.C., 2013. Organic pollutants in soil. In: Brevik, E.C., Burgess, L.C. (Eds.), Soils and Human Health. CRC Press, Boca Raton, pp. 83–106.

Buringh, P., 1984. Organic carbon in soils of the world. In: Woodwell, G.M. (Ed.), The Role of Terrestrial Vegetation in the Global Carbon Cycle: Measurement by Remote Sensing. John Wiley and Sons, Chichester, UK, pp. 91–109.

Burras, C.L., Nyasimi, M., Butler, L., 2013. Soils, human health, and wealth: a complicated relationship. In: Brevik, E.C., Burgess, L.C. (Eds.), Soils and Human Health. CRC Press, Boca Raton, pp. 215–226.

Burylo, M., Dutoit, T., Rey, F., 2014. Species traits as practical tools for ecological restoration of marly eroded lands. Restor. Ecol. 22, 633–640. http://dx.doi.org/10.1111/rec.12113.

Buttafuoco, G., Conforti, M., Aucelli, P.P.C., Robustelli, G., Scarciglia, F., 2012. Assessing spatial uncertainty in mapping soil erodibility factor using geostatistical stochastic simulation. Environ. Earth Sci. 66, 1111–1125. http://dx.doi.org/10.1007/s12665-011-1317-0.

Camarero, J.J., Gazol, A., Tardif, J.C., Conciatori, F., 2015. Attributing forest responses to global-change drivers: limited evidence of a CO_2-fertilization effect in Iberian pine growth. J. Biogeogr 42, 2220–2233. http://dx.doi.org/10.1111/jbi.12590.

Campbell, D.M.H., White, B., Arp, P.A., 2013. Modelling and mapping soil resistence to penetration and rutting using LiDAR-derived digital elevation data. J. Soil Water Conserv. 68, 460–473. http://dx.doi.org/10.2489/jswc.68.6.460.

Canadell, J., Schulze, E.D., 2014. Global potential of biospheric carbon management for climate mitigation. Nat. Commun. 5, 5282. http://dx.doi.org/10.1038/ncomms6282.

Carvalho, F.P., 2006. Agriculture, pesticides, food security and food safety. Environ. Sci. Policy 9, 685–692. http://dx.doi.org/10.1016/j.envsci.2006.08.002.

Castro, J., Allen, C.D., Molina-Morales, M., Marañón-Jiménez, S., Sánchez-Miranda, A., Zamora, R., 2011. Salvage logging versus the use of burnt wood as a nurse object to promote post-fire tree seedling establishment. Restor. Ecol. 19, 537–544. http://dx.doi.org/10.1111/j.1526-100X.2009.00619.x.

Cerdá, A., 1997. The effect of patchy distribution of *Stipa tenacissima* L. on runoff and erosion. J. Arid. Environ. 36, 37–51. http://dx.doi.org/10.1006/jare.1995.0198.

Cerdà, A., Doerr, S.H., 2005. Influence of vegetation recovery on soil hydrology and erodibility following fire: an 11-year investigation. Int. J. Wildland Fire 14, 423–437. http://dx.doi.org/10.1071/WF05044.

Cerdà, A., Lavee, H., Romero-Díaz, A., Hooke, J., Montanarella, J., 2010. Preface: soil erosion and degradation in Mediterranean type ecosystems. Land Degrad. Dev. 21, 71–74. http://dx.doi.org/10.1002/ldr.968.

Chamen, W.C.T., Moxey, A.P., Towers, W., Balana, B., Hallett, P.D., 2015. Mitigating arable soil compaction: A review and analysis of available cost and benefit data. Soil Tillage Res. 146, 10–25. http://dx.doi.org/10.1016/j.still.2014.09.011.

Chen, W., Wu, L., Chang, A.C., Hou, Z., 2009. Assessing the effect of long-term crop cultivation on distribution of Cd in the root zone. Ecol. Model. 220, 1836–1843. http://dx.doi.org/10.1016/j.ecolmodel.2009.04.036.

Chertov, O.G., Komarov, A.S., Nadporozhskaya, M., Bykhovets, S.S., Zudin, S.L., 2001. ROMUL—a model of forest soil organic matter dynamics as a substantial tool for forest ecosystem modeling. Ecol.

Model. 138, 289–308. http://dx.doi.org/10.1016/S0304-3800(00)00409-9.

Coleman, K., Jenkinson, D.S., 1996. ROTHC-26.3. A model for the turnover of carbon in soil. In: Powlson, D.S., Smith, P., Smith, J.O. (Eds.), Evaluation of Soil Organic Matter Models Using Existing Long-Term Datasets. Springer-Verlag, Belin, pp. 237–246. http://dx.doi.org/10.1007/978-3-642-61094-3_17.

Conant, R.T., Ryan, M.G., Ågren, G.I., Birge, H.E., Davidson, E.A., Eliasson, P.E., et al., 2011. Temperature and soil organic matter decomposition rates-synthesis of current knowledge and a way forward. Glob. Chang. Biol. 1, 3392–3404. http://dx.doi.org/10.1111/j.1365-2486.2011.02496.x.

Crow, S.E., Reeves, M., Turn, S., Taniguchi, S., Schubert, O.S., Koch, N., 2016. Carbon balance implications of land use change from pasture to managed eucalyptus forest in Hawaii. Carbon Manag. http://dx.doi.org/10.1080/17583004.2016.1213140.

Dominati, E., Mackay, A., Green, S., Patterson, M., 2014. A soil change-based methodology for the quantification and valuation of ecosystem services from agro-ecosystems: a case study of pastoral agriculture in New Zealand. Ecol. Econ., 119–129.

D'Or, D., Destain, M.F., 2016. Risk assessment of soil compaction in the Wallon Region in Belgium. Math. Geosci. 48, 89–103. http://dx.doi.org/10.1007/s11004-015-9617-7.

De la Rosa, D., Mayol, F., Diaz-Pereira, E., Fernández, M., De la Rosa Jr., D., 2004. A land evaluation decision support system (MicroLEIS DSS) for agricultural soil protection: with special reference to the Mediterranean region. Environ. Model. Softw. 19, 929–942. 10.1016/j.envsoft.2003.10.006.

De Roo, A.P.J., Offermans, R.J.E., Cremers, N.H.D.T., 1996. LISEM: A single event, physically based hydrological and soil erosion model for drainage basins. II. Sensitivity, validation and application. Hydrol. Process. 10, 1119–1129. http://dx.doi.org/10.1002/(SICI)1099-1085(199608)10:8<1119::AID-HYP416>3.0.CO;2-V.

De Vente, J., Poesen, J., Verstraeten, G., Govers, G., Vanmaercke, M., Van Rompaey, A., et al., 2013. Predicting soil erosion and sediment yield at regional scales: where do we stand? Earth-Sci. Rev. 127, 16–29. http://dx.doi.org/10.1016/j.earscirev.2013.08.014.

Doran, J.W., Zeiss, M.R., 2000. Soil health and sustainability: managing the biotic component of soil quality. Appl. Soil Ecol. 15, 3–11. http://dx.doi.org/10.1016/S0929-1393(00)00067-6.

Duttmann, R., Schwanebeck, M., Nolde, M., Horn, R., 2014. Predicting soil compaction risks related to field traffic during silage maize harvest. Soil Sci. Soc. Am. J. 78, 408–421. http://dx.doi.org/10.2136/sssaj2013.05.0198.

Eswaran, H., Van den Berg, E., Reich, P., Kimble, J., 1995. Global soil carbon resources. In: Lal, R., Kimble, J., Levine, E., Stewart, B.A. (Eds.), Soils and Global. CRC Press, Boca Raton, FL, pp. 27–43.

Falloon, P., Smith, P., 2003. Accounting for changes in soil carbon under the Kyoto Protocol: need for improved long-term data sets to reduce uncertainty in model projections. Soil Use Manag. 19, 265–269. http://dx.doi.org/10.1111/j.1475-2743.2003.tb00313.x.

FAO, 2002. The State of Food Insecurity in the World 2001. Food and Agriculture Organization of the United Nations, Rome.

FAO, 2009. International Institute for Applied Systems Analysis, International Soil Reference and Information Centre, Institute of Soil Science-Chinese Academy of Sciences, Joint Research Centre of the European Commission. Harmonized World Soil Database (Version 1.1). FAO, Rome, Italy; International Institute for Applied Systems Analysis, Laxenburg, Austria.

Fernandez, H.M., Martins, F.M.G., Isidoro, J.M.G.P., Zavala, L., Jordán, A., 2016. Soil erosion, Serra de Grândola (Portugal). J. Maps 12, 1138–1142. http://dx.doi.org/10.1080/17445647.2015.1135829.

Fernandez-Manso, A., Quintano, C., Roberts, D.A., 2016. Burn severity influence on post-fire vegetation cover resilience from Landsat MESMA fraction images time series in Mediterranean forest ecosystems. Remote Sens. Environ. 184, 112–123. http://dx.doi.org/10.1016/j.rse.2016.06.015.

Fervers, C.W., 2004. Improved FEM simulation model for tire-soil interaction. J. Terramech. 41, 87–100. http://dx.doi.org/10.1016/j.jterra.2004.02.012.

Fitter, A.H., Gilligan, C.A., Hollingworth, K., Kleczkowski, A., Twyman, R.M., Pitchford, J.W., 2005. Biodiversity and ecosystem function in soil. Funct. Ecol. 19, 369–377. http://dx.doi.org/10.1111/j.0269-8463.2005.00969.x.

Fitter, A., Elmqvist, T., Haines-Young, R., Potschin, M., Rinaldo, A., Setala, H., et al., 2010. An assessment of ecosystem services and biodiversity in Europe. Issue Environ. Sci. Technol. 30, 1–28. http://dx.doi.org/10.1039/9781849731058-00001.

Flores, B.M., Fagoaga-Sanchez, R., Nelson, B.W., Holmgren, M., 2016. Repeated fires trap Amazonian blackwater floodplains in open vegetation state. J. Appl. Ecol. 53, 1597–1603. http://dx.doi.org/10.1111/1365-2664.12687.

Fountas, S., Paraforos, D., Cavalaris, C., Karamoutis, C., Gemtos, T.A., Abu-Khalaf, N., et al., 2013. A five-point penetrometer with GPS for measure soil compaction variability. Comput. Electron. Agric. 96, 109–116. http://dx.doi.org/10.1016/j.compag.2013.04.018.

Francos, M., Pereira, P., Alcañiz, M., Mataix-Solera, J., Úbeda, X., 2016. Impact of an intense rainfall event on soil properties following a wildfire in a Mediterranean environment (North-East Spain). Sci. Total. Environ. http://dx.doi.org/10.1016/j.scitotenv.2016.01.145.

Galdino, S., Sano, E.E., Andrade, R.G., Grego, C.R., Nogueira, S.F., Bragantini, C., et al., 2016. Large scale modelling of soil erosion with RUSLE for conservationist planning of degraded cultivated Brazilian pastures. Land Degrad. Dev. 27, 773–787. http://dx.doi.org/10.1002/ldr.2414.

Gaudinski, J.B., Trumbore, S.E., Davidson, E.A., Zheng, S., 2000. Soil carbon cycling in a temperate forest: radiocarbon-based estimates of residence times, sequestration rates and partitioning of fluxes. Biogeochemistry 51, 33–69. http://dx.doi.org/10.1023/A:1006301010014.

Gomes, L., Arrúe, J.L., López, M.V., Sterk, G., Richard, D., Gracia, R., et al., 2003. Wind erosion in a semiarid agricultural area of Spain: the WELSONS project. Catena 52, 235–256. http://dx.doi.org/10.1016/S0341-8162(03)00016-X.

Gordillo-Rivero, Á.J., García-Moreno, J., Jordán, A., Zavala, L.M., Granja-Martins, F.M., 2014. Fire severity and surface rock fragments cause patchy distribution of soil water repellency and infiltration rates after burning. Hydrol. Process. 28, 5832–5843. http://dx.doi.org/10.1002/hyp.10072.

Granged, A.J.P., Jordán, A., Zavala, L.M., Muñoz-Rojas, M., Mataix-Solera, J., 2011. Short-term effects of experimental fire for a soil under eucalyptus forest (SE Australia). Geoderma 167-168, 125–134. http://dx.doi.org/10.1016/j.geoderma.2011.09.011.

Grunwald, S., 2009. Multi-criteria characterization of recent digital soil mapping and modeling approaches. Geoderma 152, 195–207. http://dx.doi.org/10.1016/j.geoderma.2009.06.003.

Grunwald, S., Thompson, J.A., Minasny, B., Boettinger, J.L., 2012. Digital soil mapping in a changing world. In: Minasny, B., Malone, B.P., McBratney, A.B. (Eds.), Digital Soil Assessments and Beyond: Proceedings of the 5th Global Workshop on Digital Soil Mapping, Sydney, Australia. CRC Press, Boca Raton, pp. 301–305. http://dx.doi.org/10.1201/b12728-60.

Guessesse, B., Bewket, W., Brauning, A., 2015. Model-based characterization and monitoring of runoff and soil erosion in response to land use/land cover changes in the Modjo watershed, Ethiopia. Land Degrad. Dev. 26, 711–724. http://dx.doi.org/10.1002/ldr.2276.

Haahtela, T., von Hertzen, L., Mäkelä, M., Hannuksela, M., Erhola, M., Kaila, M., et al., 2008. Finnish allergy programme 2008–2018-time to act and change the course. Allergy 63, 634–645. http://dx.doi.org/10.1111/j.1398-9995.2008.01712.x.

Hagen, L.J., 2005. Erosion by wind: modeling. In: Lal, R. (Ed.), Encyclopedia of Soil Science, second ed. New York, pp. 1–4. http://dx.doi.org/10.1081/E-ESS-120044016.

Hamm, M.W., Bellows, A.C., 2003. Community food security and nutrition educators. J. Nutr. Educ. Behav. 35, 37–43. http://dx.doi.org/10.1016/S1499-4046(06)60325-4.

Hamza, M.A., Anderson, W.K., 2005. Soil compaction in cropping systems: a review of the nature, causes and possible solutions. Soil Tillage Res. 82, 121–145. http://dx.doi.org/10.1016/j.still.2004.08.009.

Hanan, E.J., D'Antonio, C.M., Roberts, D.A., Schimel, J.P., 2016. Factors regulating nitrogen retention during the early stages of recovery from fire in coastal chaparral ecosystems. Ecosystems 19, 910–926. http://dx.doi.org/10.1007/s10021-016-9975-0.

Haregeweyn, N., Poesen, J., Verstraeten, G., Govers, G., De Vente, J., Nyssen, J., et al., 2013. Assessing the performance of spatially distributed soil erosion and sediment delivery model (WATEM/SEDEM) in Norther Ethiopia. Land Degrad. Dev., 188–204. http://dx.doi.org/10.1002/ldr.1121.

Havlin, J.L., Beaton, J.D., Tisdale, S.L., Nelson, W.L., 2005. Soil Fertility and Fertilizers: An Introduction to Nutrient Management, seventh ed. Pearson Prentice Hall, Upper Saddle River.

Hemmat, A., Adamchuck, V.I., 2008. Sensor systems for measuring soil compaction. Review and analysis. Sensor systems for measuring soil compaction. Rev. Anal. 63, 89–103. http://dx.doi.org/10.1016/j.compag.2008.03.001.

Hengl, T., Heuvelink, G.B.M., Kempen, B., Leenaars, J.G.B., Walsh, M.G., Shepherd, K.D., et al., 2015. Mapping soil properties of Africa at 250 m resolution: random forests significantly improve current predictions. PLoS One 10 (6), e0125814. http://dx.doi.org/10.1371/journal.pone.0125814.

Hiederer, R., Köchy, M., 2011. Global soil organic carbon estimates and the harmonized world soil database. EUR 25225 EN. Publications Office of the EU, Luxembourg. http://dx.doi.org/10.2788/13267.

Holliday, V.T., 2004. Soils in Archaeological Research. Oxford University Press, New York.

Houyou, Z., Bielders, C.L., Benhorama, H.A., Dellal, A., Boutemdjet, A., 2016. Evidence of strong land degradation by erosion as a result of rainfed cropping in Algerian steppe: a case study at Laghouat. Land Degrad. Dev. http://dx.doi.org/10.1002/ldr.2295.

Hoy, E.E., French, N.H.F., Turetsky, M.R., Trigg, S.N., Kasischke, E.S., 2008. Evaluating the potential of Landsat TM/EIM+ for assessing fire severity in Alaskian black spruce forests. Int. J. Wildland Fire 17, 500–514. http://dx.doi.org/10.1071/WF08107.

IAASTD, 2009. Agriculture at a crossroads. The global report. International assessment of agricultural knowledge, Science and Technology for Development Global Report. Island Press, Washington, DC.

IPCC, 2014. Climate Change 2014. Synthesis Report. Contribution of Working Groups I, II and III to the Fifth Assessment Report of the Intergovernmental Panel on Climate Change (Core Writing Team, Pachauri, R.K., Meyer, L.A.). IPCC, Geneva.

Islam, M.M., Meerschman, E., Saey, T., De Smedt, P., Van De Vijver, E., Delefortrie, S., et al., 2014. Characterizing compaction variability with an electromagnetic induction sensor in a puddled paddy rice field. Soil Sci. Soc. Am. J. 78, 579–588. http://dx.doi.org/10.2136/sssaj2013.07.0289.

Jain, T.B., Gould, W.A., Graham, R.T., Pillod, D.S., Lentile, L.B., Gonzalez, G., 2008. A soil burn severity index for understanding soil-fire relations in Tropical forests. Ambio: A J. Hum. Environ. 37, 563–568. http://dx.doi.org/10.1579/0044-7447-37.7.563.

Jandl, R., Lindner, M., Vesterdal, L., Bauwens, B., Baritz Hagedorn, F., Johnson, D.W., et al., 2007. How strongly can forest management influence soil C sequestration? Geoderma 137, 253–268. http://dx.doi.org/10.1016/j.geoderma.2006.09.003.

Jandl, R., Rodeghiero, M., Martínez, C., Cotrufo, M.F., Bampa, F., van Wesemael, B., Harrison, R.B., Guerrini, I.A., Richter Jr., D. deB., Rustad, L., Lorenz, K., Chabbi, A., Miglietta, F., 2014. Current status, uncertainty and future needs in soil organic carbon monitoring. Sci. Total Environ. 468-469, 376–383. http://dx.doi.org/10.1016/j.scitotenv.2013.08.026.

Janzen, H.H., 2004. Carbon cycling in earth systems—a soil science perspective. Agric. Ecosyst. Environ. 104, 399–417. http://dx.doi.org/10.1016/j.agee.2004.01.040.

Jiménez-Pinilla, P., Lozano, E., Mataix-Solera, J., Arcenegui, V., Jordán, A., Zavala, L.M., 2015. Temporal changes in soil water repellency after a forest fire in a Mediterranean calcareous soil: influence of ash and different vegetation type. Sci. Total Environ. http://dx.doi.org/10.1016/j.scitotenv.2015.09.121.

Jobbágy, E.G., Jackson, R.B., 2000. The vertical distribution of soil carbon and its relation to climate and vegetation. Ecol. Appl. 10, 423–436. http://dx.doi.org/10.1890/1051-0761(2000)010[0423:TVDOSO]2.0.CO;2.

Johnston, R.M., Barry, S.J., Bleys, E., Bui, E.N., Moran, C.J., Simon, D.A.P., et al., 2003. ASRIS: the database. Soil Res. 41, 1021–1036. http://dx.doi.org/10.1071/SR02033.

Jones, R.J.A., Hiederer, R., Rusco, E., Loveland, P.J., Montanarella, L., 2004. The map of organic carbon in topsoils in Europe, Version 1.2, September 2003: Explanation of Special Publication Ispra 2004 No.72 (S.P.I.04.72). European Soil Bureau Research Report No.17, EUR 21209 EN, 26pp. and 1 map in ISO B1 format. Office for Official Publications of the European Communities, Luxembourg.

Jordán, A., Zavala, L.M., Muñoz-Rojas, M., 2011. Mulching, effects on soil physical properties. In: Glinski, J., Horabik, J., Lipiec, J. (Eds.), Encyclopedia of Agrophysics. Springer, Berlin, pp. 492–496. http://dx.doi.org/10.1007/978-90-481-3585-1.

Jordán, A., Gordillo-Rivero, T.J., García-Moreno, J., Zavala, L.M., Granged, A.J.P., Gil, J., et al., 2014. Post-fire evolution of water repellency and aggregate stability in Mediterranean calcareous soils: a 6-year study. Catena 118, 115–123. http://dx.doi.org/10.1016/j.catena.2014.02.001.

Jordán, A., Zavala, L.M., Granged, A.J.P., Gordillo-Rivero, A.J., García-Moreno, J., Pereira, P., et al., 2015. Wettability of ash conditions splash erosion and runoff rates in the post-fire. Sci. Total Environ. http://dx.doi.org/10.1016/j.scitotenv.2015.09.140.

Kapos, V., Ravilious, C., Campbell, A., Dickson, B., Gibbs, H.K., Hansen, M.C., et al., 2008. Carbon and Biodiversity: A Demonstration Atlas. United Nations Environment Programme World Conservation Monitoring Centre, UNEP-WCMC, Cambridge, UK.

Karlen, D.L., Mausbach, M.J., Doran, J.W., Cline, R.G., Harris, R.F., Schuman, G.E., 1997. Soil quality: a concept, definition, and framework for evaluation. Soil Sci. Soc. Am. J. 61, 4–10. http://dx.doi.org/10.2136/sssaj1997.03615995006100010001x.

Karydas, C.G., Panagos, P., Gitas, I.Z., 2014. A classification of water erosion models according to their Geospatial characteristics. Int. J. Digital Earth 7, 229–250. http://dx.doi.org/10.1080/17538947.2012.671380.

Kasischke, E.S., Turetsky, M.R., Ottmar, R.D., French, N.H.F., Hoy, E.E., Kane, E.S., 2008. Evaluation of the cpmposite burn index for assessing fire severity in Alaskian black spruce forests. Int. J. Wildland Fire 17, 515–526. http://dx.doi.org/10.1071/WF08002.

Keeley, J.E., 2009. Fire intensity, fire severity and burn severity: a brief review and suggested use. Int. J. Wildland Fire 18, 116–126. http://dx.doi.org/10.1071/WF07049.

Keesstra, S.D., Bouma, J., Wallinga, J., Tittonell, P., Smith, P., Cerdà, A., et al., 2016. The significance of soils and soil science towards realization of the United Nations Sustainable Development Goals. Soil 2, 111–128. http://dx.doi.org/10.5194/soil-2015-88.

Keller, T., Defossez, P., Weisskopf, P., Arvidsson, J., Richard, G., 2007. SoilFlex: a model for prediction of soil stresses and soil compaction due agricultural field and traffic including a synthesis of analytical approaches. Soil Tillage Res. 93, 391–411. http://dx.doi.org/10.1016/j.still.2006.05.012.

Keller, T., Lemande, M., 2010. Challenges in the development of analytical compaction models. Soil Tillage Res. 111, 54–64. http://dx.doi.org/10.1016/j.still.2010.08.004.

Kheir, R.B., Cerdan, O., Abdallah, C., 2006. Regional erosion risk mapping in Lebanon. Geomorphology 3-4, 347–359. http://dx.doi.org/10.1016/j.geomorph.2006.05.012.

Kibblewhite, M.G., Ritz, K., Swift, M.J., 2008. Soil health in agricultural systems. Philis. Trans. R. Soc. B 363, 685–701. http://dx.doi.org/10.1098/rstb.2007.2178.

Kildisheva, O.A., Erickson, T.E., Merritt, D.J., Dixon, K.W., 2016. Setting the scene for dryland restoration: an overview and key findings from a workshop targeting seed enablement technologies. Restor. Ecol. 52, S36–S42. http://dx.doi.org/10.1111/rec.12392.

Kokaly, R.F., Rockwell, B.W., Haire, S.L., King, T.V.V., 2007. Characterization of post-fire surface cover, soils and burn severity at Cerro Grande Fire, New Mexico, using hyoperspectral and multispectral remote sensing. Remote Sens. Environ. 106, 305–325. http://dx.doi.org/10.1016/j.rse.2006.08.006.

Koohafkan, P., Altieri, M.A., Gimenez, E.H., 2012. Green Agriculture: foundations for biodiverse, resilient and productive agricultural systems. Int. J. Agric. Sustain., 1–13. http://dx.doi.org/10.1080/14735903.2011.610206.

Krishna Bahadur, K.C., 2009. Mapping soil erosion susceptibility using remote sensing and GIS: a case of the Upper Nam Wa Watersheed, Nan Province, Thailand. Environ. Geol. 57, 695–705. http://dx.doi.org/10.1007/s00254-008-1348-3.

Kutsch, W.L., Bahn, M., Heinemeyer, A., 2009. Soil carbon relations: an overview. In: Kutsch, W.L., Bahn, M., Heinemeyer, A. (Eds.), Soil Carbon Dynamics: An Integrated Methodology. Cambridge University Press, Cambridge, pp. 1–15.

Laflen, J.M., Flanagan, D.C., 2013. The development of U.S. soil erosion prediction and modelling. Int. Soil Water Conserv. Res. 1, 1–11. http://dx.doi.org/10.1016/S2095-6339(15)30034-4>.

Lal, R., 2004a. Soil carbon sequestration impacts on global climate change and food security. Science 304, 1623–1627. http://dx.doi.org/10.1126/science.1097396.

Lal, R., 2004b. Soil carbon sequestration to mitigate climate change. Geoderma 123, 1–22. http://dx.doi.org/10.1016/j.geoderma.2004.01.032.

Lal, R., Follett, F., Stewart, B.A., Kimble, J.M., 2007. Soil carbon sequestration to mitigate climate change and advance food security. Soil Sci. 172, 943–956. http://dx.doi.org/10.1097/ss.0b013e31815cc498.

Le Bissonnais, Y., Montier, C., Jamagne, M., Daroussin, J., King, D., 2001. Mapping erosion risk for cultivated soil in France. Catena 46, 207–220. http://dx.doi.org/10.1016/S0341-8162(01)00167-9.

Li, C.S., Frolking, S., Harriss, R., 1994. Modeling carbon biogeochemistry in agricultural soils. Glob. Biogeochem. Cycle 8, 237–254. http://dx.doi.org/10.1029/94GB00767.

Liebens, J., VanMolle, M., 2003. Influence of estimation procedure on soil organic carbon stock assessment in Flanders, Belgium. Soil Use Manag. 19, 364–371. http://dx.doi.org/10.1111/j.1475-2743.2003.tb00327.x.

Lippitt, C.L., Stow, D.A., O'Leary, J.F., Franklin, J., 2012. Influence of short-interval fire on post fire recovery of fire-phrone shrublands in California, USA. Int. J. Wildland Fire 22, 184–193. 10.1071/WF10099.

Liu, Y., Wang, H., Zhang, H., Liber, K., 2016. A comprehensive support vector machine-based classification model for soil quality assessment. Soil Tillage Res. 155, 19–26. http://dx.doi.org/10.1016/j.still.2015.07.006.

Long, S.P., Ainsworth, E.A., Leakey, A.D.B., Nosberger, J., Ort, D.R., 2006. Food for thought: lower-than-expected crop yield simulation with rising CO_2 concentrations. Science 312, 1918–1921. http://dx.doi.org/10.1126/science.1114722.

López-Serrano, F.R., Rubio, E., Dadi, T., Moya, D., Amdrés-Abellán, M., García-Morote, F.A., et al., 2016. Influences of recovery from wildfire and thinning on soil respiration of a Mediterranean mixed forest. Sci. Total Environ. http://dx.doi.org/10.1016/j.scitotenv.2016.03.242.

Lovett, G.M., Cole, J.J., Pace, M.M., 2006. Is net ecosystem production equal to ecosystem carbon accumulation? Ecosystems 9, 1–4. http://dx.doi.org/10.1007/s10021-005-0036-3.

Loynachan, T.E., 2013. Human disease from introduced and resident soilborne pathogens. In: Brevik, E.C., Burgess, L.C. (Eds.), Soils and Huan Health. CRC Press, Boca Raton, pp. 107–136. http://dx.doi.org/10.1201/b13683-8.

Lozano-García, B., Parras-Alcántara, L., Brevik, E.C., 2016. Impact of topographic aspect and vegetation (native and reforested areas) on soil organic carbon and nitrogen budgets in Mediterranean natural areas. Sci. Total Environ. 544, 963–970. http://dx.doi.org/10.1016/j.scitotenv.2015.12.022.

Lu, D., Li, G., Valladares, G.S., Batistella, M., 2004. Mapping soil erosion risk in Rondônia, Brazilian Amazonia: using RUSLE, remote sensing and GIS. Land Degrad. Dev. 15, 499–512. http://dx.doi.org/10.1002/ldr.634.

Lugato, E., Berti, A., 2008. Potential carbon sequestration in a cultivated soil under different climate change scenarios: a modelling approach for evaluating promising management practices in north-east Italy. Agric. Ecosyst. Environ. 128, 97–103. http://dx.doi.org/10.1016/j.agee.2008.05.005.

MacDonald, L.H., Larsen, I.J., 2009. Effects of forest fires and post-fire rehabilitation: a Colorado, USA case study. In: Cerdà, A., Robichaud, P. (Eds.), Fire Effects on Soils and Restoration Measures. Science Publishers, Enfield, pp. 423t–445t.

Malhi, L., Karanfil, Ö., Merth, T., Acheson, M., Palmer, A., Finegood, D.T., 2009. Places to intervene to make complex food systems more healthy, green, fair, and affordable. J. Hunger. Environ. Nutr. 4, 466–476. http://dx.doi.org/10.1080/19320240903346448.

Marañón, T., Ibáñez, B., Anaya-Romero, M., Muñoz-Rojas, M., 2012a. Estado y tendencia de los servicios de los ecosistemas forestales de Andalucía. In: Evaluación de Ecosistemas del Milenio en Andalucía. Consejería de Medio Ambiente, Junta de Andalucía. Sevilla.

Marañón, T., Ibáñez, B., Anaya-Romero, M., Muñoz-Rojas, M., Pérez- Ramos, I.M., 2012b. Oak trees and woodlands providing ecosystem services in Southern Spain. In: Rotherham, I.D., Handley, C., Agnoletti, M., Samojlik, T. (Eds.), Trees Beyond the Wood. An Exploration of Concepts of Woods, Forests Trees. Wildtrack Publishing, Sheffield, UK, pp. 369–378.

Martttínez-Casasnovas, J.A., Ramos, M.C., Ribes-Dasi, M., 2002. Soil erosion caused by extreme rainfall events: mapping and quantification in agricultural plots from very detailed digital elevation models. Geoderma 105, 125–140. http://dx.doi.org/10.1016/S0016-7061(01)00096-9.

Mataix-Solera, J., Cerdà, A., Arcenegui, V., Jordán, A., Zavala, L.M., 2011. Fire effects on soil aggregation: a review. Earth-Sci. Rev. 109, 44–60. http://dx.doi.org/10.1016/j.earscirev.2011.08.002.

Matricardi, P.M., Bonini, S., 2000. Mimicking microbial "education" of the immune system: a strategy to revert the epidemic trend of atopy and allergic asthma? Respir. Res. 1, 129–132. http://dx.doi.org/10.1186/rr22.

Mbila, M., 2013. Soil minerals, organisms, and human health: medicinal uses of soils and soil materials. In: Brevik, E.C., Burgess, L.C. (Eds.), Soils and Human Health. CRC Press, Boca Raton, pp. 199–213.

McBratney, A., Field, D., 2015. Securing our soil. Soil Sci. Plant Nutr. 61, 587–591. http://dx.doi.org/10.1080/00380768.2015.1071060.

McBratney, A.B., Morgan, C.L.S., Jarrett, L.E., The value of soil's contributions to ecosystem services. In: Field, D.J., McBratney, A.B., Morgan, C.L.S. (Eds.), Global Soil Security. Progress in Soil Science Series. Springer, 2017, Springer, Basel, 227–235.

McBratney, A.B., Odeh, I.O.A., 1997. Application of fuzzy sets in soil science: fuzzy logic, fuzzy measurements and fuzzy decisions. Geoderma 77, 85–113. http://dx.doi.org/10.1016/S0016-7061(97)00017-7.

Meersmans, J., De Ridder, F., Canters, F., De Baets, S., Van Molle, M., 2008. A multiple regression approach to assess the spatial distribution of soil organic carbon (SOC) at the regional scale (Flanders, Belgium). Geoderma 143, 1–13. http://dx.doi.org/10.1016/j.geoderma.2007.08.025.

Meng, R., Dennison, P.E., Huang, C., Moritz, M.A., D'Antonio, C., 2015. Effects of fire severity and post-fire climate on short-term vegetation recovery of mixed-conifer and red fir forests in the Sierra Nevada Mountains of California. Remote Sens. Environ. 171, 311–325. http://dx.doi.org/10.1016/j.rse.2015.10.024.

Mezosi, G., Blanka, V., Bata, T., Kovacs, F., Meyer, B., 2015. Estimation of regional differences in wind erosion sensivity in Hungary. Nat. Hazard Earth Syst. Sci. 15, 97–107. http://dx.doi.org/10.5194/nhess-15-97-2015.

Miller, B.A., 2017. GIS and spatial statistics applied for soil mapping. In: Pereira, P., Brevik, E.C., Muñoz-Rojas, M., Miller, B.A. (Eds.), Soil Mapping and Process Modeling for Sustainable Land Use Management. Elsevier, pp. 129–152.

Miller, B.A., Schaetzl, R.J., 2014. The historical role of base maps in soil geography. Geoderma 230-231, 329–339. http://dx.doi.org/10.1016/j.geoderma.2014.04.020.

Miller, B.A., Koszinski, S., Wehrhan, M., Sommer, M., 2015. Comparison of spatial association approaches for landscape mapping of soil organic carbon stocks. Soil 1, 217–233. http://dx.doi.org/10.5194/soil-1-217-2015.

Millward, A.A., Mersey, J.E., 1999. Adapting the RUSLE to model soil erosion potential in a mountainous tropical watershed. Catena 38, 109–129. http://dx.doi.org/10.1016/S0341-8162(99)00067-3.

Moody, J.A., Shakesby, R.A., Robichaud, P.R., Cannon, S.H., Martin, D.A., 2013. Current research issues related to post-wildfire runoff and erosion processes. Earth-Sci. Rev. 122, 10–37. http://dx.doi.org/10.1016/j.earscirev.2013.03.004.

Moltttó, J., Mataix-Solera, J., Arcenegui, V., Morugán, A., Girona, A., García-Orenes, F., 2014. Short-time effect of salvage-logging harvesting on microbial soil properties in a Mediterranean area affected by a wildfire: preliminary results. Geophysical Research Abstracts 16, EGU2014-3727.

Montgomery, D.R., 2007. Dirt: The Erosion of Civilizations. University of California Press, Berkeley.

Moody, J.A., Shakesby, R.A., Robichaud, P.R., Cannon, S.H., Martin, D.A., 2013. Current research issues related to post-fire runoff and erosion process. Earth-Sci. Rev. 122, 10–37. http://dx.doi.org/10.1016/j.earscirev.2013.03.004.

Moreira, F., Ferreira, A., Abrantes, N., Catry, F., Fernandes, P., Roxo, L., et al., 2013. Occurrence of native and exotic invasive trees in burned pine and eucalypt plantations: implications for post-fire forest conversion. Ecol. Eng. 58, 296–302. http://dx.doi.org/10.1016/j.ecoleng.2013.07.014.

Morgan, R., 2013. Soil, heavy metals, and human health. In: Brevik, E.C., Burgess, L.C. (Eds.), Soils and Human Health. CRC Press, Boca Raton, pp. 59–82.

Morgan, R.P.C., Quinton, J.N., Smith, R.E., Govers, G., Poesen, J.W.A., Austerwald, K., et al., 1998. The European soil erosion model (EUROSEM): a dynamic approach for predicting sediment transport from fields and small catchments. Earth Surf. Process. Landf. 23, 527–544. http://dx.doi.org/10.1002/(SICI)1096-9837(199806)23:6<527::AID-ESP868>3.0.CO;2-5.

Mueller, T., Jensen, L.S., Hansen, S., Nielsen, N.E., 1996. Simulating soil carbon and nitrogen dynamics with the soil-plant-atmosphere system model Daisy In: Powlson, D.S. Smith, P. Smith, J.U. (Eds.), Evaluation of Soil Organic Matter Models Using Existing Long-Term Datasets. NATO ASI Series I, vol. 38. Springer-Verlag, Heidelberg, pp. 275–282.

Muñoz-Rojas, M., De la Rosa, D., Zavala, L.M., Jordán, A., Anaya-Romero, M., 2011. Changes in land cover and vegetation carbon stocks in Andalusia, Southern Spain (1956–2007). Sci. Total Environ. 409, 2796–2806. http://dx.doi.org/10.1016/j.scitotenv.2011.04.009.

Muñoz-Rojas, M., Jordán, A., Zavala, L.M., De la Rosa, D., Abd-Elmabod, S.K., Anaya-Romero, M., 2012. Organic carbon stocks in Mediterranean soil types under different land uses. Solid Earth 3, 375–386. http://dx.doi.org/10.5194/se-3-375-2012.

Muñoz-Rojas, M., Jordán, A., Zavala, L.M., González-Peñaloza, F.A., De la Rosa, D., Pino-Mejias, R., et al., 2013. Modelling soil organic carbon stocks in global change scenarios: a CarboSOIL application. Biogeosciences 10, 8253–8268. http://dx.doi.org/10.5194/bg-10-8253-2013.

Muñoz-Rojas, M., Doro, L., Ledda, L., Francaviglia, R., 2015. Application of CarboSOIL model to predict the effects of climate change on soil organic carbon stocks in agro-silvo-pastoral Mediterranean management. Agric. Ecosyst. Environ. 202, 8–16. http://dx.doi.org/10.1016/j.agee.2014.12.014.

Muñoz-Rojas, M., Erickson, T.E., Dixon, K.W., Merritt, D.J., 2016a. Soil quality indicators to assess functionality of restored soils in degraded semi-arid ecosystems. Restor. Ecol. 24, S43–S52. http://dx.doi.org/10.1111/rec.12368.

Muñoz-Rojas, M., Erickson, T.E., Martini, D., Dixon, K.W., Merritt, D.J., 2016b. Soil physicochemical and microbiological indicators of short, medium and long term post-fire recovery in semi-arid ecosystems. Ecol. Indic. 63, 14–22. http://dx.doi.org/10.1016/j.ecolind.2015.11.038.

Murphy, K.A., Reynolds, J.H., Koltun, J.M., 2008. Evaluating the ability to the differenced Normalized Burn Ratio (dNBR) to predict ecologically significant burn severity in Alaskan boreal forests. Int. J. Wildland Fire 17, 490–499. http://dx.doi.org/10.1071/WF08050.

Naderi -Boldaji, M., Sarifi, A., Hemmat, A., Alimardani, R., Keller, T., 2014. Feasibility study on the potential of electrical conductivity sensor Veris® 3100 for field mapping of topsoil strength. Biosyst. Eng. 126, 1–11. http://dx.doi.org/10.1016/j.biosystemseng.2014.07.006.

Naderi-Boldaji, M., Sharifi, A., Alimardani, R., Hemmat, A., Keyhani, A., Loonstra, E.H., et al., 2013. Use of a triple-sensor fusion system for on-the-go measurement of soil compaction. Soil & Tillage Res. 128, 44–53. http://dx.doi.org/10.1016/j.still.2012.10.002.

Nata, J.T., Mjelde, J.W., Boadu, F.O., 2014. Household adoption of soil-improving practices and food insecurity in Ghana. Agric. Food Security 3, 17. http://dx.doi.org/10.1186/2048-7010-3-17.

Neris, J., Santamarta, J.C., Doerr, S.H., Prieto, F., Agulló-Pérez, J., García-Villegas, P., 2016. Post-fire soil hydrology, water erosion and restoration strategies in Andosols: a review of evidence from Canary Islands (Spain). iForest. http://dx.doi.org/10.3832/ifor1605-008.

Norfleet, M.L., Ditzler, C.A., Puckett, W.E., Grossman, R.B., Shaw, J.N., 2003. Soil quality and its relationship to pedology. Soil Sci. 168, 149–155. http://dx.doi.org/10.1097/01.ss.0000058887.60072.07.

Nowak, R.S., Ellsworth, D.S., Smith, S.D., 2004. Functional responses of plants to elevated atmospheric CO_2—do photosynthetic and productivity data from FACE experiments support early predictions? New Phytol. 162, 253–280. http://dx.doi.org/10.1111/j.1469-8137.2004.01033.x.

Nunes, A.N., Lourenco, L., Castro Meira, A.C., 2016. Exploring spatial patterns and drivers of forest fires in Portugal (1980–2014). Sci.Total. Environment.,. http://dx.doi.org/10.1016/j.scitotenv.2016.03.121.

Outeiro, L., Asperó, F., Úbeda, X., 2008. Geostatistical methods to study spatial variability of soil cations after a prescribed fire and rainfall. Catena 74, 310–320. http://dx.doi.org/10.1016/j.catena.2008.03.019.

Panagos, P., Karydas, C.G., Gitas, I.Z., Montanarella, L., 2012a. Monthly soil erosion monitoring based on remotly sensed biophysical parameters: a case study in Strymonas river basin towards a functional pan-Europe service. Int. J. Digital Earth 5, 461–487. http://dx.doi.org/10.1080/17538947.2011.587897.

Panagos, P., Van Liedekerke, M., Jikones, A., Montanarella, L., 2012b. European Soil Data Centre: Response to European policy support and public data requirements. Land Use Policy 29, 329–338. http://dx.doi.org/10.1016/j.landusepol.2011.07.003.

Panagos, P., Christos, K., Cristiano, B., Ioannis, G., 2014a. Seasonal monitoring of soil erosion at regional scale: an application of G2 model in Crete focusing on agricultural land uses. Int. J. Appl. Earth Observ. Geoinf. 27, Part B, 147–155. http://dx.doi.org/10.1016/j.jag.2013.09.012.

Panagos, P., Borrelli, P., Poesen, J., Ballabio, C., Lugato, E., Meusburger, K., et al., 2015. The new assessment of soil loss by water and erosion. Environ. Sci. Policy 54, 438–447. http://dx.doi.org/10.1016/j.envsci.2015.08.012.

Panagos, P., Imeson, A., Meusburger, K., Borrelli, P., Poesen, J., Alewell, C., 2016. Soil conservation in Europe: wish or relality. Land Degrad. Dev. http://dx.doi.org/10.1002/ldr.2538.

Parker, B.M., Lewis, T., Srivastava, S.K., 2015. Estimation and evaluation of multi-decadal fire severity patterns using Landsat sensors. Remote Sens. Environ. 170, 340–349. http://dx.doi.org/10.1016/j.rse.2015.09.014.

Parras-Alcántara, L., Lozano-García, B., Brevik, E.C., Cerdá, A., 2015. Soil organic carbon stocks assessment in Mediterranean natural areas: a comparison of entire soil profiles and soil control sections. J. Environ. Manag. 15, 155–215. http://dx.doi.org/10.1016/j.jenvman.2015.03.039.

Parsons, A., Robichaud, P.R., Lewis, S.A., Napper, C., Clark, J.T., 2010. Field guide for mapping post-fire burn severity. Gen. Tech. Rep. RMRS-GTR-243. Fort Collins, CO: U.S. Department of Agriculture, Forest Service, Rocky Mountain Research Station. Fort Collins, CO.

Parton, W.J., Schimel, D.S., Cole, C.V., Ojima, D.S., 1987. Analysis of factors controlling soil organic matter levels in Great Plains grasslands. Sci. Soc. Am. J. 51, 1173–1179.

Pásztor, L., Laborczi, A., Takács, K., Szatmári, G., Fodor, Y., Illés, G., et al., 2017a. Compilation of functional soil maps for the support of spatial planning and land management in Hungary. In: Pereira, P., Brevik, E.C., Munoz-Rojas, M., Miller, B.A. (Eds.), Soil Mapping and Process Modeling for Sustainable Land Use Management. Elsevier this volume.

Pausas, J.G., Ribeiro, E., Vallejo, R., 2004. Post-fire regeneration variability of *Pinus halepensis* in the eastern Iberian Peninsula. For. Ecol. Manag. 203, 251–259. http://dx.doi.org/10.1016/j.foreco.2004.07.061.

Paustian, K., Lehmann, J., Ogle, S., Reay, D., Robertson, G.P., Smith, P., 2016. Climate-smart soils. Nature 532, 49–57. http://dx.doi.org/10.1038/nature17174.

Pepper, I.L., Gerba, C.P., Newby, D.T., Rice, C.W., 2009. Soil: a public health threat or savior? Crit. Rev. Environ. Sci. Technol. 39, 416–432. http://dx.doi.org/10.1080/10643380701664748.

Pereira, P., Brevik, E.C., Cerdà, A., Úbeda, X., Novara, A., Francos, M., et al., 2017. Mapping ash $CaCO_3$, pH and extractable elements using principal component analysis. In: Pereira, P., Brevik, E.C., Muñoz-Rojas, M., Miller, B.A. (Eds.), Soil Mapping and Process Modeling for Sustainable Land Use Management. Elsevier, pp. 323–340.

Pereira, P., Úbeda, X., Mataix-Solera, J., Oliva, M., Novara, A., 2014. Short-term changes in soil Munsell colour value, organic matter content and soil water repellency after a spring grassland fire in Lithuania. Solid Earth 5, 209–225. http://dx.doi.org/10.5194/se-5-209-2014.

Pereira, P., Cerdà, A., Misiune, I., 2015a. Wildfires in Lithuania. In: Bento, A., Vieira, A. (Eds.), Wildland Fires-A Worldwide Reality. Nova Science Publishers Inc., Hauppauge, NY, pp. 185–198.

Pereira, P., Jordán, A., Cerdà, A., Martin, D., 2015b. The role of ash in fire affected areas. Catena 135, 337–339. http://dx.doi.org/10.1016/j.catena.2014.11.016.

Pereira, P., Rein, G., Martin, D.A., 2016a. Past and present post-fire environments. Sci. Total Environ. http://dx.doi.org/10.1016/j.scitotenv.2016.05.040.

Pereira, P., Oliva, M., Misiune, I., 2016b. Spatial interpolation of precipitation indexes in Sierra Nevada (Spain): comparing the performance of some interpolation methods. Theor. Appl. Climatol. 126, 683–698. http://dx.doi.org/10.1007/s00704-015-1606-8.

Pérez-Cabello, F., Ibarra, P., Echeverría, M.T., de la Riva, J., 2010. Post-fire land degradation of Pinus sylvestris L. woodlands after 14 years. Land Degrad. Dev. 21, 145–160. http://dx.doi.org/10.1002/ldr.925.

Petropoulos, G.P., Griffiths, H.M., Kalivas, D.P., 2014. Quantifying spatial and temporal vegetation recovery dynamics following a wildfire event in a Mediterranean landscape using EO and GIS. Appl. Geogr. 50, 120–131. http://dx.doi.org/10.1016/j.apgeog.2014.02.006.

Phachomphon, K., Dlamini, P., Chaplot, V., 2010. Estimating carbon stocks at regional level using soil information and easily accessible auxiliary variables. Geoderma 155, 372–380. http://dx.doi.org/10.1016/j.geoderma.2009.12.020.

Pino-Mejías, R., Cubiles-de-la-Vega, M.D., Anaya-Romero, M., Pascula-Acosta, A., Jordán-López, A., Bellinfante-Crocci, N., 2010. Predicting the potential habitat of oaks with data mining models and the R system. Environ. Model. Softw. 25, 826–836. http://dx.doi.org/10.1016/j.envsoft.2010.01.004.

Pothukuchi, K., Kaufman, J.L., 1999. Placing the food system on the urban agenda: the role of municipal institutions in food systems planning. Agric. Hum. Value 16, 213–224. http://dx.doi.org/10.1023/A:1007558805953.

Powlson, D.S., Whitmore, A.P., Goulding, K.W.T., 2011. Soil carbon sequestration to mitigate climate change: a critical re-examination to identify the true and the false. Eur. J. Soil Sci. 62, 42–55. http://dx.doi.org/10.1111/j.1365-389.2010.01342.x.

Prasuhn, V., Liniger, H., Gisler, S., Herweg, K., Candinas, A., Clement, J.P., 2013. A high resolution soil erosion risk map of Switzerland as strategic support system. Land Use Policy 32, 281–291. http://dx.doi.org/10.1016/j.landusepol.2012.11.006.

Prosdocimi, M., Jordán, A., Tarolli, P., Keesstra, S., Novara, A., Cerdà, A., 2016. The immediate effectiveness of barley straw mulch in reducing soil erodibility and surface runoff generation in Mediterranean vineyards. Sci. Total Environ. 547, 323–330. http://dx.doi.org/10.1016/j.scitotenv.2015.12.076.

Reiche, M., Funk, R., Zhang, Z., Hoffmann, C., Reiche, J., Wehrhan, M., et al., 2011. Application of satellite remote sensing for mapping wind erosion risk and dust emission deposition in Inner-deposition Mongolia grassland, China. Grassl. Sci. 58, 8–19. http://dx.doi.org/10.1111/j.1744-697X.2011.00235.x.

Robichaud, P., 2009. Using erosion barriers for post-fire stabilization. In: Cerdà, A., Robichaud, P. (Eds.), Fire Effects on Soils and Restoration Measures. Science Publishers, Enfield, pp. 337–352.

Robichaud, P.R., Lewis, S.A., Laes, D.Y.M., Hudak, A.T., Kokaly, R.F., Zamudio, J.A., 2007. Postfire soil burn severity mapping with hyperspectral image unmixing. Remote Sens. Environ. 108, 467–480. http://dx.doi.org/10.1016/j.rse.2006.11.027.

Rodríguez-Murillo, J.C., 2001. Organic carbon content under different types of land use and soil in peninsular Spain. Biol. Fertil. Soils 33, 53–61. http://dx.doi.org/10.1007/s003740000289.

Romig, D.E., Garlynd, M.J., Harris, R.F., 1996. Farmer-based assessment of soil quality: a soil health scorecard. In: Doran, J.W., Jones, A.J. (Eds.), Methods for Assessing Soil Quality. SSSA Special Publication 49. Soil Science Society of America, Madison, WI, pp. 39–60.

Rook, G.A.W., 2010. 99th Dahlem conference on infection, inflammation and chronic inflammatory disorders: Darwinian medicine and the "hygiene" or "old friends" hypothesis. Clin. Exp. Immunol. 160, 70–79. http://dx.doi.org/10.1111/j.1365-2249.2010.04133.x.

Rossi, R., Pollice, A., Diago, M.P., Oliveira, M., Millan, B., Bitella, G., et al., 2013. Using an automatic resistivity profiler soil sensor on-the-go in precision viticulture. Sensors 13, 1121–1136. http://dx.doi.org/10.3390/s130101121.

Roy, M., McDonald, L.M., 2015. Metal uptake in plants and health risk assessments in metal-contaminated smelter soils. Land Degrad. Dev. 26, 785–792. http://dx.doi.org/10.1002/ldr.2237.

Sanchez, P.A., Shepherd, K.D., Soule, M.J., Place, F.M., Buresh, R.J., Izac, A.M.N., 1997. Soil fertility replenishment in Africa: an investment in natural resource capital. In: Buresh, R.J., Sanchez, P.A., Calhounp, F. (Eds.), Replenishing Soil Fertility in Africa. SSSA Special Publication No. 51. Soil Science Society of America, Madison, WI, pp. 1–46.

Sandhu, H.S., Wratten, S.D., Cullen, R., Case, B., 2008. The future of farming: the value of ecosystem services in conventional and organic arable land. An experimental approach. Ecol. Econ., 835–848.

Scalenghe, R., Marsan, F.A., 2009. The anthropogenic sealing of soils in urban areas. Landsc. Urban Plan. 90, 1–10. http://dx.doi.org/10.1016/j.landurbplan.2008.10.011.

Scharlemann, J.P.W., Tanner, E.V.J., Hiederer, R., Kapos, V., 2014. Global soil carbon: understanding and managing the largest terrestrial carbon pool. Carbon Manag. 5, 81–91. http://dx.doi.org/10.4155/cmt.13.77.

Schulp, C.J.E., Nabuurs, G.-J., Verburg, P.H., 2008. Future carbon sequestration in Europe—effects of land use change. Agric. Ecosyst. Environ. 127, 251–264. http://dx.doi.org/10.1016/j.agee.2008.04.010.

Schwartz, D., Namri, M., 2002. Mapping the total organic carbon in the soils of the Congo. Glob. Planet. Chang. 33, 77–93. http://dx.doi.org/10.1016/S0921-8181(02)00063-2.

Senesi, G.S., Baldassarre, G., Senesi, N., Radina, B., 1999. Trace element inputs into soils by anthropogenic activities and implications for human health. Chemosphere 39, 343–377.

Seneviratne, S.I., Donat, M.G., Mueller, B., Alexander, L.V., 2014. No pause in the increase of hot temperature extremes. Nat. Clim. Chang. 4, 161–163. http://dx.doi.org/10.1038/nclimate2145.

Shakesby, R.A., 2011. Post-wildfire soil erosion in the Mediterranean: Review and future research directions. Earth-Sci. Rev. 105, 71–100. http://dx.doi.org/10.1016/j.earscirev.2011.01.001.

Shamal, S.A.M., Alhwaimel, S.A., Mouazen, A.M., 2016. Application of an on-line sensor to map soil packing density for specific cultivation. Soil Tillage Res. 162, 48–89. http://dx.doi.org/10.1016/j.still.2016.04.016.

Sinnett, D., Poole, J., Hutchins, R.T.R., 2006. The efficiency of three techniques to alleviate soil compaction at a restored sand and gravel quary. Soil Use Manag. 22, 362–371. http://dx.doi.org/10.1111/j.1475-2743.2006.00053.x.

Sirjacobs, D., Hanquet, B., Lebeau, F., Destain, M.F., 2002. On-line soil mechanical resistance mapping and correlation with soil physical properties for precision agriculture. Soil Tillage Res. 64, 231–242. http://dx.doi.org/10.1016/S0167-1987(01)00266-5.

Smith, P., 2004. Soils as carbon sinks: the global context. Soil Use Manag. 20, 212–218. http://dx.doi.org/10.1111/j.1475-2743.2004.tb00361.x.

Smith, P., Cotrufo, M.F., Rumpel, C., Paustian, K., Kuikman, P.J., Elliott, J.A., et al., 2015. Biogeochemical cycles and biodiversity as key drivers of ecosystem services provided by soils. Soil 1, 665–685. http://dx.doi.org/10.5194/soil-1-665-2015.

Smith, P., House, J.I., Bustamante, M., Sobocká, J., Harper, R., Pan, G., et al., 2016. Global change pressures on soils from land use and management. Glob. Chang. Biol. 22, 1008–1028. http://dx.doi.org/10.1111/gcb.13068.

Smith, P., Martino, D., Cai, Z., Gwary, D., Janzen, H., Kumar, P., et al., 2008. Greenhouse gas mitigation in agriculture. Philos. Trans. R. Soc. B 363, 789–813. http://dx.doi.org/10.1098/rstb.2007.2184.

Spiertz, J.H.J., 2010. Nitrogen, sustainable agriculture and food security. A review. Agron. Sustain. Dev. 30, 43–55. http://dx.doi.org/10.1051/agro:2008064.

Spoor, G., 2006. Alleviation of soil compaction: requirements, equipment and techniques. Soil Use Manag. 22, 113–122. http://dx.doi.org/10.1111/j.1475-2743.2006.00015.x.

Sterk, G., Stein, A., 1997. Mapping wind-blow mass transport by modelling variability in space and time. Soil Sci. Soc. Am. J. 61, 232–239. http://dx.doi.org/10.2136/sssaj1997.03615995006100010032x.

Stringer, L.C., Dougill, A.J., Thomas, A.D., Spracklen, D.V., Chesterman, S., Ifejika Speranza, C., et al., 2012. Challenges and opportunities in linking carbon sequestration, livelihoods and ecosystem service provision in drylands. Environ. Sci. Policy 19-20, 121–135. http://dx.doi.org/10.1016/j.envsci.2012.02.004.

Tabor, J.A., O'Rourke, M.K., Lebowitz, M.D., Harris, R.B., 2011. Landscape-epidemiological study design to investigate an environmentally based disease. J. Exposure Sci. Environ. Epidemiol. 21, 197–211. http://dx.doi.org/10.1038/jes.2009.67.

Tekin, Y., Kul, B., Okursoy, R., 2008. Sensing and 3D mapping of soil compaction. Sensors 8, 3447–3459. http://dx.doi.org/10.3390/s8053447.

Tesfahunegn, G.B., 2016. Soil quality indicators response to land use and soil management systems in northern Ethiopia's catchment. Land Degrad. Dev. 27, 438–448. http://dx.doi.org/10.1002/ldr.2245.

Tessler, N., Sapir, Y., Wittenberg, L., Greenbaum, N., 2016a. Recovery of Mediterranean vegetation after recurrent fires: insight from 2010 forest fire on Mount Carmel, Israel. Land Degrad. Dev. http://dx.doi.org/10.1002/ldr.2419.

Tessler, N., Wittenberg, L., Greenbaum, N., 2016b. Vegetation cover and species richness after recurrent fires in the Eastern Mediterranean ecosystem of Mount Carmel, Israel. Sci. Total Environ. http://dx.doi.org/10.1016/j.scitotenv.2016.02.113.

Tian, Z., Wu, X., Dai, E., Zhao, D., 2016. SOC storage and potential of grasslands from 2000 to 2012 in central and eastern Inner Mongolia, China. J. Arid. Land 8, 364–374. http://dx.doi.org/10.1007/s40333-016-0041-8.

Trabucchi, M., O'Farrell, P., Notivol, E., Comin, F.A., 2014. Mapping ecological processes and ecosystem services for prioritizing restoration efforts in a semi-arid Mediterranean river basin. Environ. Manag. 53, 1132–1145. http://dx.doi.org/10.1007/s00267-014-0264-4.

Trenberth, K.E., Fasullo, J.T., Shepherd, T.G., 2015. Attribution of climate extreme events. Nat. Clim. Chang. 5, 725–730. http://dx.doi.org/10.1038/nclimate2657.

Troldborg, M., Aalders, I., Towers, W., Hallett, P.D., McKenzie, B.M., Bengough, A.G., et al., 2013. Application of Bayesian belief networks to qualify and map areas at risk to soil threats: using soil compaction as an example. Soil Tillage Res. 132, 56–68. http://dx.doi.org/10.1016/j.still.2013.05.005.

Ugur Ozcan, A., Erpul, G., Basaran, M., Erdogan, H.M., 2008. Use of USLE/GIS technology integrated with geostatistics to assess erosion risk in different land uses of Indagui Mountain Pass-Çankiri, Turkey. Environ. Geol. 53, 1731–1741. http://dx.doi.org/10.1007/s00254-007-0779-6.

United Nations Population Division, 2015. World Population Prospects 2015. Report.

UNFCCC, 2015. Paris Agreement. UNFCCC, Report.

Vallejo, V.R., Serrasoles, I., Alloza, A., Baeza, M.J., Blade, C., Chirino, E., et al., 2009. Long-term restoration strategies and techniques. In: Cerdà, A., Robichaud, P. (Eds.), Fire Effects on Soils and Restoration Measures. Science Publishers, Enfield, pp. 374–398.

Van Bergeijk, J., Goense, D., Speelman, L., 2001. Soil tillage resistence as a tool to map soil type differences. J. Agric. Eng. Res. 79, 371–387. http://dx.doi.org/10.1006/jaer.2001.0709.

Van den Akker, J.J.H., 2004. SOCOMO: a soil compaction model to calculate soil stresses and subsoil carrying capacity. Soil Tillage Res. 79, 113–127. http://dx.doi.org/10.1016/j.still.2004.03.021.

VandenBygaart, A.J., 2006. Monitoring soil organic carbon stock changes in agricultural landscapes: issues and a proposed approach. Can. J. Soil Sci. 86, 451–463. http://dx.doi.org/10.4141/S05-105.

Vanlauwe, B., Kanampiu, F., Odhiambo, G.D., De Groote, H., Wadhams, L.J., Khan, Z.R., 2008. Integrated management of *Striga hermonthica*, stemborers, and declining soil fertility in western Kenya. Field Crop. Res. 107, 102–115. http://dx.doi.org/10.1016/j.fcr.2008.01.002.

Veraverbeke, S., Stravos, E.N., Hook, S.J., 2014. Assessing fire severity using imaging spectroscopy data from the Airborne Visible/Infrared imaging spectrometer (AVIRIS) and comparison with multispectral capabilities. Remote Sens. Environ. 154 http://dx.doi.org/10.1016/j.rse.2014.08.019.

Veronesi, F., Constranje, R., Mayr, T., 2012. Mapping soil compaction in 3D with depth functions. Soil Tillage Res. 124, 111–118. http://dx.doi.org/10.1016/j.still.2012.05.009.

Viaud, V., Angers, D.A., Walter, C., 2010. Towards landscape-scale modeling of soil organic matter dynamics in agroecosystems. Soil Sci. Soc. Am. J. 74, 1847–1860. http://dx.doi.org/10.2136/sssaj2009.0412.

Viscarra Rossel, R., Webster, R., Bui, E., Baldock, J., 2014. Baseline map of organic carbon in Australian soil to support national carbon accounting and monitoring under climate change. Glob. Chang. Biol. 20, 2953–2970. http://dx.doi.org/10.1111/gcb.12569.

Viscarra Rossel, R.A., Webster, R., 2012. Predicting soil properties from the Australian soil visible-near infrared spectroscopic database. Eur. J. Soil Sci. 63, 848–860. http://dx.doi.org/10.1111/j.1365-2389.2012.01495.x.

Visser, S., Sterk, G., Snepvangers, J.J.C., 2004. Spatial variation in wind-blow sediment transport in geomorphic units in Burkina Faso using a geostatistical map. Geoderma 120, 95–107. http://dx.doi.org/10.1016/j.geoderma.2003.09.003.

Vlassova, L., Perez-Cabello, F., 2016. Effects of post-fire wood management strategies on vegetation recovery and land surface temperature (LST), estimated from Landsat images. Int. J. Appl. Earth Observ. Geoinf. 44, 171–183. http://dx.doi.org/10.1016/j.jag.2015.08.011.

Vogeler, J.C., Yang, Z., Cohen, W.B., 2016. Mapping post-fire habitat characteristics through the fusion of remote sensing tools. Remote Sens. Environ. 173, 294–303.

Vrindts, E., Mouazen, A.M., Reyniers, M., Maertens, K., Maleki, M.R., Ramon, H., et al., 2005. Management zones based on correlation between soil compaction, yield and crop data. Biosyst. Eng., 419–428.

Wang, J., Chen, J., Ju, W., Li, M., 2010. IA-SDSS: a GIS-based land use decision support system with consideration of carbon sequestration. Environ. Model. Softw. 25, 539–553. http://dx.doi.org/10.1016/j.rse.2015.08.011.

Weber, J.L., 2011. An experimental framework for ecosystem capital accounting in Europe. European Environment Agency, Technical Report 13.

West, T.O., Brandt, C.C., Baskaran, L.M., Hellwinckel, C.M., Mueller, R., Bernacchi, C.J., et al., 2010. Cropland carbon fluxes in the United States: increasing geospatial resolution of inventory-based carbon accounting. Ecol. Appl. 20, 1074–1086. http://dx.doi.org/10.1890/08-2352.1.

Willaarts, B.A., Oyonarte, C., Muñoz-Rojas, M., Ibáñez, J.J., Aguilera, P.A., 2016. Environmental factors controlling soil organic carbon stocks in two contrasting Mediterranean climatic areas of Southern Spain. Land Degrad. Dev. 27, 603–611. http://dx.doi.org/10.1002/ldr.2417.

Williams, J.R., 1990. The erosion-productivity impact calculator (EPIC) model: a case history. Philos. Trans. R. Soc. London, Ser. B 329, 421–428.

Wimberly, M.C., Reilly, M.J., 2007. Assessment of fire severity and species diversity in the southern Appalachians using Landsat TM and ETM+ imagery. Remote Sens. Environ. 108, 189–197. http://dx.doi.org/10.1016/j.rse.2006.03.019.

Wohlgemuth, P.M., Beyers, J.L., Hubbert, K.R., 2009. Rehabilitation strategies after fire: The California, USA, experience. In: Cerdà, A., Robichaud, P. (Eds.), Fire Effects on Soils and Restoration Measures. Science Publishers, Enfield, pp. 511–535.

Woodall, C.W., Walters, B.F., Coulston, J.W., D'Amato, A.W., Domke, G.M., Russell, M.B., et al., 2015. Monitoring network confirms land use change is a substantial component of the forest carbon sink in the eastern United States. Sci. Rep. 5, 17028. http://dx.doi.org/10.1038/srep17028.

Woods, S.W., Birkas, A., Ahl, R., 2007. Spatial variability of soil hydrophobicity after wildfires in Montana and Colorado. Geomorphology 86, 465–479. http://dx.doi.org/10.1016/j.geomorph.2006.09.015.

Wreford, A., Moran, D., Adger, N., 2010. Climate Change and Agriculture: Impacts, Adaptation and Mitigation. OECD Publishing, Paris. http://dx.doi.org/10.1787/9789264086876-en.

Wu, Y., Chen, J., 2012. Modeling of soil erosion and sediment transport in the East River Basin in Southern China. Sci. Total Environ. 441, 159–168. http://dx.doi.org/10.1016/j.scitotenv.2012.09.057.

Xia, K., 2011. Finite element modelling of tire/terrain interaction: Application to predicting soil compaction and tire mobility. J. Terramech. 48, 113–123. http://dx.doi.org/10.1016/j.jterra.2010.05.001.

Yang, Q., Luo, W., Jiang, Z., Li, W., Yuan, D., 2016. Improve the prediction of bulk density by cokriging with predicted soil water content as auxiliary variable. J. Soils Sediment. 16, 77–84. http://dx.doi.org/10.1007/s11368-015-1193-4.

Yap, V.Y., de Neergaard, A., Bruun, T.B., 2016. 'To adopt or not to adopt?' Legume adoption in maize-based systems of Northern Thailand: constraints and potentials. Land Degrad. Dev. http://dx.doi.org/10.1002/ldr.2546.

Yu, Z., Li, Y., Jin, J., Liu, X., Wang, G., 2016. Carbon flow in the plant-soil-microbe continuum at different growth stages of maize grown in a Mollisol. Arch. Agron. Soil Sci. http://dx.doi.org/10.1080/03650340.2016.1211788.

Zavala, L.M., Jordán, A., Gil, J., Bellinfante, N., Pain, C., 2009. Intact ash and charred litter reduces susceptibility to rain splash erosion post-wildfire. Earth Surface Processes and Lnadforms 34, 1522–1532. http://dx.doi.org/10.1002/esp.1837.

Zhang, J.J., Fu, M.C., Zeng, H., Geng, Y.H., Hassani, F.P., 2013. Variations in ecosystem service values and local economy in response to land use: a case study of Wu'an, China. Land Degrad. Dev. 24, 236–249. http://dx.doi.org/10.1002/ldr.1120.

Zhao, G., Mu, X., Wen, Z., Wang, F., Gao, P., 2013. Soil erosion, conservation, and eco-environment changes in the Loess plateau of China. Land Degrad. Dev. 24, 499–510. http://dx.doi.org/10.1002/ldr.2246.

Zornoza, R., Acosta, J.A., Bastida, F., Domínguez, S.G., Toledo, D.M., Faz, A., 2015. Identification of sensitive indicators to assess the interrelationship between soil quality, management practices and human health. Soil 1, 173–185. http://dx.doi.org/10.5194/soil-1-173-2015.

Further Reading

Defossez, P., Richard, G., 2002. Models to compaction due to traffic and their evaluation. Soil Tillage Res. 67, 41–64. http://dx.doi.org/10.1016/S0167-1987(02)00030-2

Hättenschwiler, S., Handa, I.T., Egli, L., Asshoff, R., Ammann, W., Körner, C., 2002. Atmospheric CO_2 enrichment of alpine treeline conifers. New Phytol. 156, 363–375. http://dx.doi.org/10.1046/j.1469-8137.2002.00537.x

Jian, Y., Zhu, Z., Xiao, M., Yuan, H., Wang, J., Zou, D., et al., 2016. Microbial assimilation of atmospheric CO_2 into soil organic matter revealed by the incubation of paddy soils under ^{14}C-CO_2 atmosphere. Arch. Agron. Soil Sci. 62, 1678–1685. http://dx.doi.org/10.1080/03650340.2016.1171850.

Panagos, P., Meusburger, K., Van Liedekerke, M., Alewell, C., Hiederer, R., Montanarella, L., 2014b. Assessing soil erosion in Europe based on data collected through a European network. Soil Sci.Plant Nutr. 60, 15–29. http://dx.doi.org/10.1080/00380768.2013.835701.

Parton, W.J., McKeown, R., Kirchner, V., Ojima, D., 1992. CENTURY Users' Manual. Fort Collins: Natural Resource Ecology Laboratory, Colorado State University, Fort Collins, CO.

Pereira, P., Cerdà, A., Jordán, A., Zavala, L.M., Mataix-Solera, J., Arcenegui, V., et al., 2016c. Short-term vegetation recovery after grassland fire in Lithuania. The effects of fire severity, slope position and aspect. Land Degrad. Dev. 27, 1523–1534. http://dx.doi.org/10.1002/ldr.2498.

Vallejo, V.R., Arianoutsou, M., Moreira, F., 2012. Fire ecology and post-fire restoration. Approaches in southern Europe forest types. In: Moreira, F., Arianoutsou, F., Corona, P., De las Heras, J. (Eds.) Post-Fire Management and Restoration of Southern European Forests, Springer, London, pp. 93–119.

PART III

CASE STUDIES AND GUIDELINES

CHAPTER

7

Modeling Agricultural Suitability Along Soil Transects Under Current Conditions and Improved Scenario of Soil Factors

Sameh K. Abd-Elmabod[1,2], *Antonio Jordán*[2], *Luuk Fleskens*[3], *Jonathan D. Phillips*[4], *Miriam Muñoz-Rojas*[5,6], *Martine van der Ploeg*[3], *María Anaya-Romero*[7], *Soad El-Ashry*[1] *and Diego de la Rosa*[8]

[1]National Research Centre, Cairo, Egypt [2]University of Seville, Seville, Spain [3]Wageningen University and Research Centre, Wageningen, The Netherlands [4]University of Kentucky, Lexington, KY, United States [5]The University of Western Australia, Crawley, WA, Australia [6]Kings Park and Botanic Garden, Perth, WA, Australia [7]Evenor-Tech, Seville, Spain [8]Earth Sciences Section, Royal Academy of Sciences, Seville, Spain

INTRODUCTION

Enhancing food production and supporting civil/engineering structures for a growing and urbanizing global population have been the principal focus points of soil science research during the 20th century (Lal, 2008; Bekchanov et al., 2015; Recha et al., 2015; Wildemeersch et al., 2015; Hanh et al., 2016). To overcome the stress on soil from these increased demands, three general approaches are recommended: (1) to protect land from degradation processes such as soil erosion and contamination; (2) to improve the land productivity and support the intensification of agricultural production; and (3) the exploration of new productive areas to increase agricultural production (Sterk, 2003; Sonneveld, 2003; Branca et al., 2013; Lal, 1997, 2008, 2013). In many areas of the world the last option is not realistic, because expansion of agricultural land is restricted by environmental factors and, where possibilities for expansion

Soil Mapping and Process Modeling for Sustainable Land Use Management.
DOI: http://dx.doi.org/10.1016/B978-0-12-805200-6.00007-4

do exist, the impact of land use changes on the environment should be taken into consideration (Albaladejo et al., 2013; Anaya-Romero et al., 2011). A deeper knowledge of land use systems should lead to a better use of land resources. With such a background, it becomes clear that the assessment of land vulnerability and suitability for agricultural production is critical (De la Rosa et al., 2009; Beyene, 2015; Gessesse et al., 2015; Kalema et al., 2015; Wolfgramm et al., 2015).

OBJECTIVES

The general aim of this work is to evaluate the soil suitability for different Mediterranean crops along soil transects under current conditions and improved scenario of soil factors, by using the Micro Land Evaluation Information System (MicroLEIS DSS). To achieve the general goal the following specific objectives were established:

1. Compilation and harmonization of soil factors data from a detailed soil profile database (SDBm-Seville) that contains information about morphological description, chemical and physical analyses of the representative soil profiles of the studied transects.
2. Estimation of agricultural soil suitability for 12 Mediterranean crops, using the Almagra model, particularly along topographic transects.
3. Propose a recommended scenario in order to increase the soil suitably through the improvement of inhibiting/limiting soil factors as drainage, calcium carbonate, salinity, and exchangeable sodium percent.

LAND AND SOIL EVALUATION

Land and soil are important sources of wealth and the foundation on which many civilizations are constructed. Society must make sure that soil is not degraded and that it is used according to its capacity to satisfy human requirements for present and future generations, while also maintaining the sustainability of the Earth's ecosystems (Rossiter, 1996; Anaya-Romero et al., 2015; Beyene, 2015; Wolfgramm et al., 2015; Wilson et al., 2016). Land evaluation is a part of the solution to the land use problems, as it supports rational land use planning and the suitable and sustainable use of both natural and human resources. Land evaluation is defined as the assessment process of land performance for the specific purposes the land is being used for (Elsheikh et al., 2013; Anaya-Romero et al., 2015; Giger, et al., 2015; Yeshaneh, et al., 2015) or as all methods to explain or predict the use potential of land (Van Diepen et al., 1991). Rossiter (1996) indicated that land evaluation supplies the technical coefficients necessary for optimal land allocation. Once this potential has been determined, land use planning can proceed on a rational basis, at least with respect to what the land resources can offer (FAO, 1993). Therefore land evaluation is a tool for strategic land use planning. It predicts land performance, both in terms of the expected benefits from, and constraints to, productive land use, as well as the expected environmental degradation due to these uses. Sys (1980) designed methods for rating, (1) the land and soil characteristics ((a) related to the land characteristics as climate, topography, wetness, flooding, drainage; (b) related to soil characteristics as texture, depth, carbonate content, gypsum content, base saturation, cation exchange capacity, organic matter content, salinity) and (2) land qualities as ((a) internal qualities as water availability, oxygen availability, nutrient availability, absence of salinity and alkalinity, and availability of foothold for root; (b) external qualities as temperature regime, erosion, workability and absence of flood), in which the land characteristics are measurable properties of the physical environment related to land use.

Computerized land evaluation systems are an accepted way to predict land productivity and agriculture suitability, and to evaluate the consequences of land use change (FAO, 2007). Land evaluation includes qualitative and quantitative systems. Current IT knowledge still be difficult due to the evaluation provides limited value, unless the evaluator has the resources to collect a large amount of information (Olson, 1981; Giger et al., 2015; Kalema et al., 2015; Yami and Snyder, 2015). Thus most land evaluation systems are indirect. They suppose that certain soil and site properties influence the success of a particular land use in a reasonably predictable manner and that the quality of land can be assessed by observing those properties (Vink, 1983). According to De la Rosa (2005), land evaluation, which is used to predict land behavior for each specific use, is not the same as soil quality assessment, because biological parameters of the soil are not considered in land evaluation. Recently expand to indicate current research, e.g., link to work exploring the impact of management, ecosystem services to sustainable food production, where challenges are producing enough food and rational use of natural resources (Schulte et al., 2014).

MICROLEIS

MicroLEIS DSS is an agro-ecological decision support system (technology developed by CSIC-IRNAS and transferred to Evenor-Tech, www.evenor-tech.com) (De la Rosa et al., 2004). It is considered a particularly suitable tool to store soil and climatic attributes for a better identification of vulnerable areas and formulation of action programs (Anaya-Romero et al., 2011). One obvious reason for MicroLEIS' common use is the straightforward approach of the procedure, which uses simple models. MicroLEIS is able to predict the optimum land use and the best management practices, individualized for each soil type, to predict the optimum biomass productivity, minimum environmental vulnerability and maximum CO_2 sequestration (Muñoz-Rojas et al., 2013, 2015b, 2016).

MicroLEIS intends to help decision makers solve specific agro-ecological problems. It has been designed as a knowledge-based approach, which incorporates a set of information tools, which are linked to each other. Thus custom applications can be performed on a wide variety of problems related to land productivity and land degradation (Anaya-Romero et al., 2011; Muñoz-Rojas et al., 2013). MicroLEIS DSS has been used worldwide including Iran (Shahbazi et al., 2010), Egypt (Abd-Elmabod et al., 2012; Darwish et al., 2006), Italy (Farroni et al., 2002), Libya (Nwer, 2006), and in other countries as Brazil (Garcia et al., 2003), Germany (Kelgenbaeva and Buchroithner, 2007), Venezuela (Lugo-Morin, 2006; Lugo-Morin and Rey, 2009), Mexico (López García et al., 2006), Turkey (Erdogan et al., 2003), and Australia (Triantafilis et al., 2001) for many different purposes over the last 30 years.

In the initial development of MicroLEIS DSS, qualitative methods were widely used to predict the general land productivity of the most important crops and the specific suitability for individual crops or for a selection of forest species (i.e., Cervatana, Almagra, and Sierra models, respectively; De la Rosa et al., 1992). In the Almagra model, simple look-up tables are used to define qualitatively soil suitability classes for 12 traditional crops (wheat, corn, melon, potato, soybean, cotton, sunflower, sugar beet, alfalfa, peach, citrus, and olive) according to the principle of maximum limiting factor. The Marisma model also uses a qualitative methodology to estimate the limitations of a given soil according to selected soil indicators of natural fertility.

MicroLEIS DSS includes three databases: (1) soil database (SDBm), with physical, chemical, and morphological descriptions, the representative soil profiles; (2) climate database (CDBm), that contain mean monthly temperature,

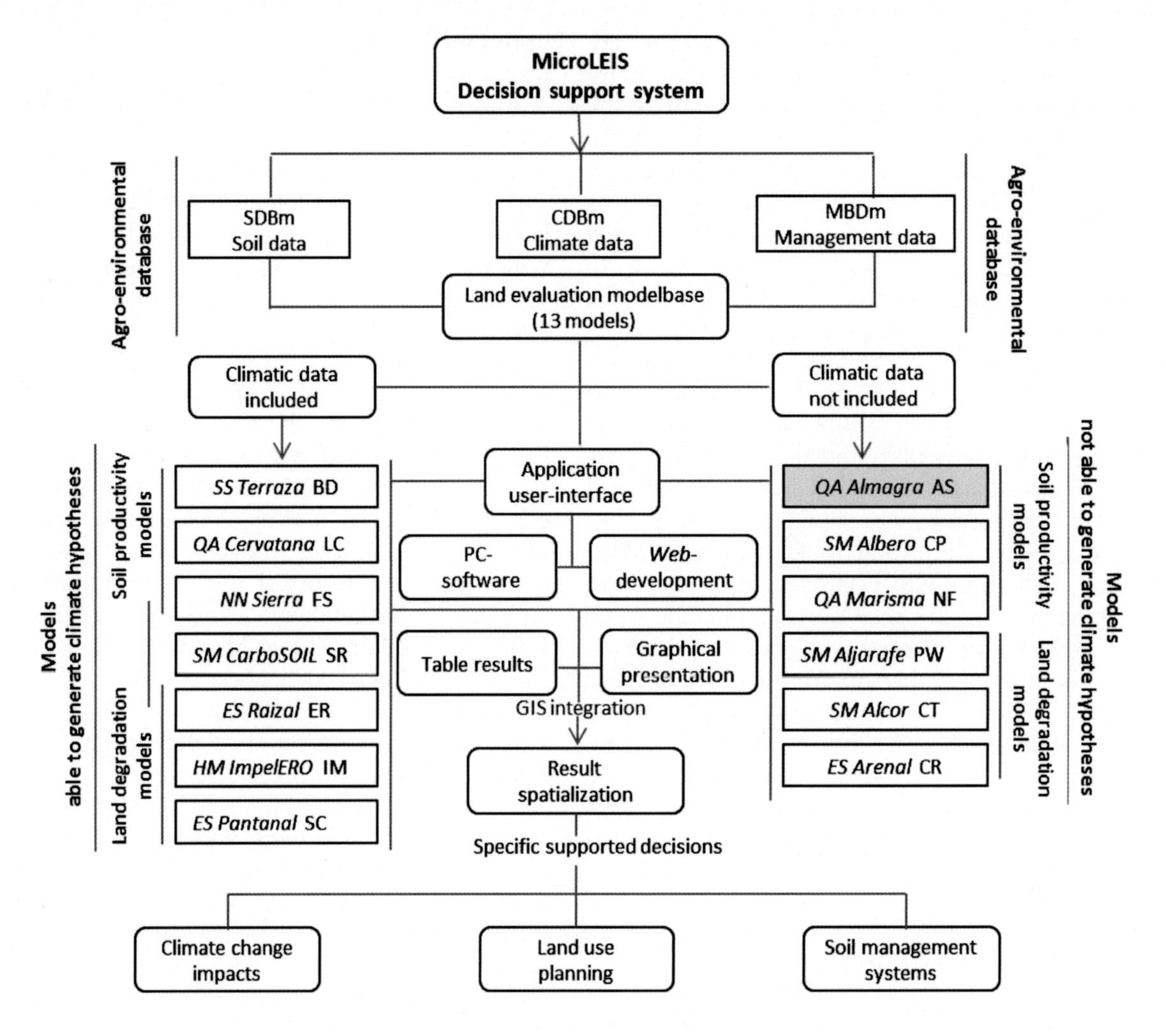

FIGURE 7.1 Conceptual design and components integration of the MicroLEIS DSS land evaluation decision support system. Environmental modeling: *SS*, simulation system; *QA*, qualitative approach; *NN*, neural network; *SM*, statistical model; *HM*, hybrid model; *ES*, expert system. Biophysical processes/outputs: *BD*, bioclimatic deficiency; *LC*, land capability; *FS*, forestry suitability (61 species); *SR*, soil carbon prediction; *ER*, erosion risk; *IM*, erosion impact/mitigation; *SC*, specific contamination; *AS*, agricultural suitability (12 crops); *CP*, crop productivity (three crops); *NF*, natural fertility; *PW*, plasticity/workability; *CT*, compaction/trafficability; *CR*, contamination risk.

maximum and minimum monthly rainfall; and (3) agriculture management database (MDBm), containing information about agricultural use and management of major crops (Fig. 7.1).

In addition, MicroLEIS DSS include 13 models; seven models of the system can apply in different hypothetical scenarios of climate and agriculture management:

- Terraza, Cervatana, and Sierra are models to evaluate soil productivity as bioclimatic deficiency, general land capability, and forestry land suitability, respectively. Raizal,

ImpelERO, and Pantanal are the models related to land degradation assessment as soil erosion risk, prediction of soil loss, and specific soil contamination risk, respectively; CarboSOIL is used to evaluate the carbon sequestration under different climate conditions.

- The other six models (Almagra, Albero, Marisma, Aljrafe, Alcor, and Arenal) are used to evaluate soil productivity and land degradation depending on physical, chemical, and pedological soil characteristics. All the results of evaluation models can be spatialized by Geographic Information Systems (GIS) integration (Fig. 7.1).

Applications of evaluation models are useful to assess global change impacts, planning of land use and intended for suitable soil.

As land evaluation increasingly focuses on global change, the methodology proposed by MicroLEIS DSS can be used to investigate the impact of new scenarios, such as climate change, on potentialities and vulnerabilities of the land (Anaya-Romero et al., 2015; Shahbazi et al., 2010). MicroLEIS DSS is described in detail by, for example, De la Rosa (2004), De la Rosa et al. (1981, 1992, 1993, 1999), Farroni et al. (2002), Horn et al. (2002), and Sánchez et al. (1982).

MEDITERRANEAN REGION

Mediterranean regions are areas of the world with a typical Mediterranean climate that distinguished by wet winters, and hot, dry summers. These regions are concentrated around the Mediterranean Sea, hence its name, and in limited parts of other continents, for example, California in North America, Chile in South America, Southwestern Australia, and at the southern fringes of South Africa (John et al., 2008). Climate, topography, parent material, biota, time, and humans influenced and still influence soil formation, so the soils of Mediterranean regions show considerable variation (Muñoz-Rojas et al., 2012; Oyonarte et al., 2008; Willaarts et al., 2016), though Mediterranean soils are, by definition, formed under Mediterranean climatic conditions.

MATERIALS AND METHODS

Location of the Study Area

Seville province is located in the southwest of Spain, in the autonomous region of Andalusia, of which the main city (Seville) is the capital. Seville is situated between latitudes 36°40′ and 38°05′N and longitudes 6°50′ and 4°60′ W and has nine natural regions with a total area of 14,056 km^2. The large natural regions are Campiña and Sierra Norte with an area 4555 and 3748 km^2, respectively, and the smallest are Alcores and Aljarafe with an area 217 and 592 km^2, respectively. Its mainland is bordered to the south by Cadiz and Malaga; to the west by Huelva; to the north by Extremadura and to the east by Cordoba (Fig. 7.2).

Most of the natural vegetation in Seville is forest dominated by trees such as oaks, pines, and firs, with dense riparian forests and Mediterranean shrubland (De la Rosa et al., 2009). Agriculture in Seville has traditionally been based on cereals, olive trees, and vineyards but in modern decades, traditional crops have been replaced with various intensive and extensive crops, e.g., sunflower, rice, cotton, and sugar beet (Muñoz-Rojas et al., 2012).

Elevation Data

The digital elevation models (DEMs) of the Seville province were obtained from the Shuttle Radar Topographic Mission (SRTM). The SRTM uses precisely positioned radar to map the Earth surface at intervals of 1-arc second

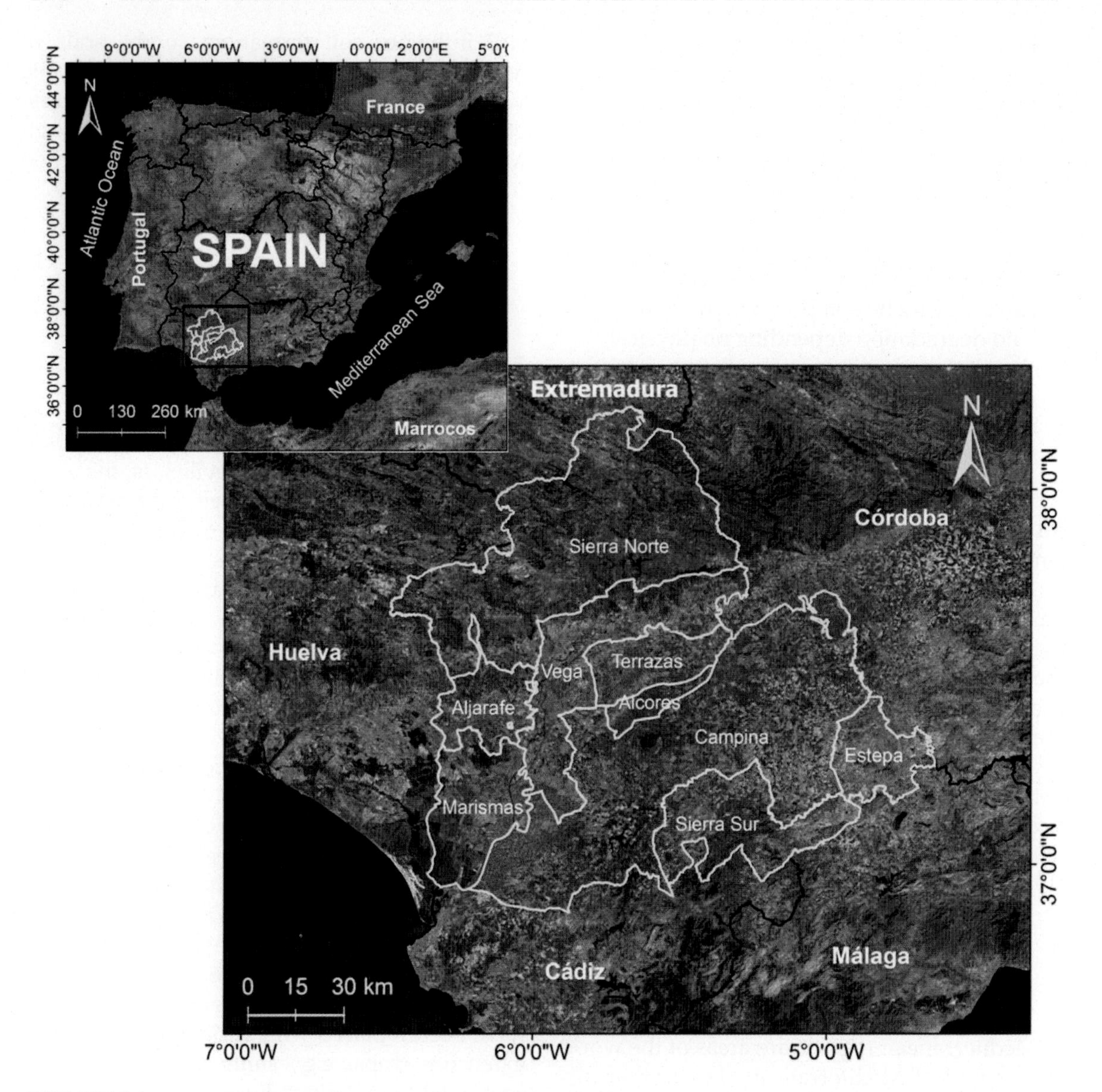

FIGURE 7.2 Location of the study area (Seville Province) within Spain.

(~30 m). According to the DEM the elevation values vary from zero in Marismas natural region (coastline and Guadalquivir River basin) to 1122 masl in Sierra Norte (Fig. 7.3). The DEM extracted from SRTM data can be used in conjunction with controlled imagery sources to provide better visualization of the terrain. DEMs have a key role in improving accuracy in soil and agricultural characterization (Matinfar et al., 2011).

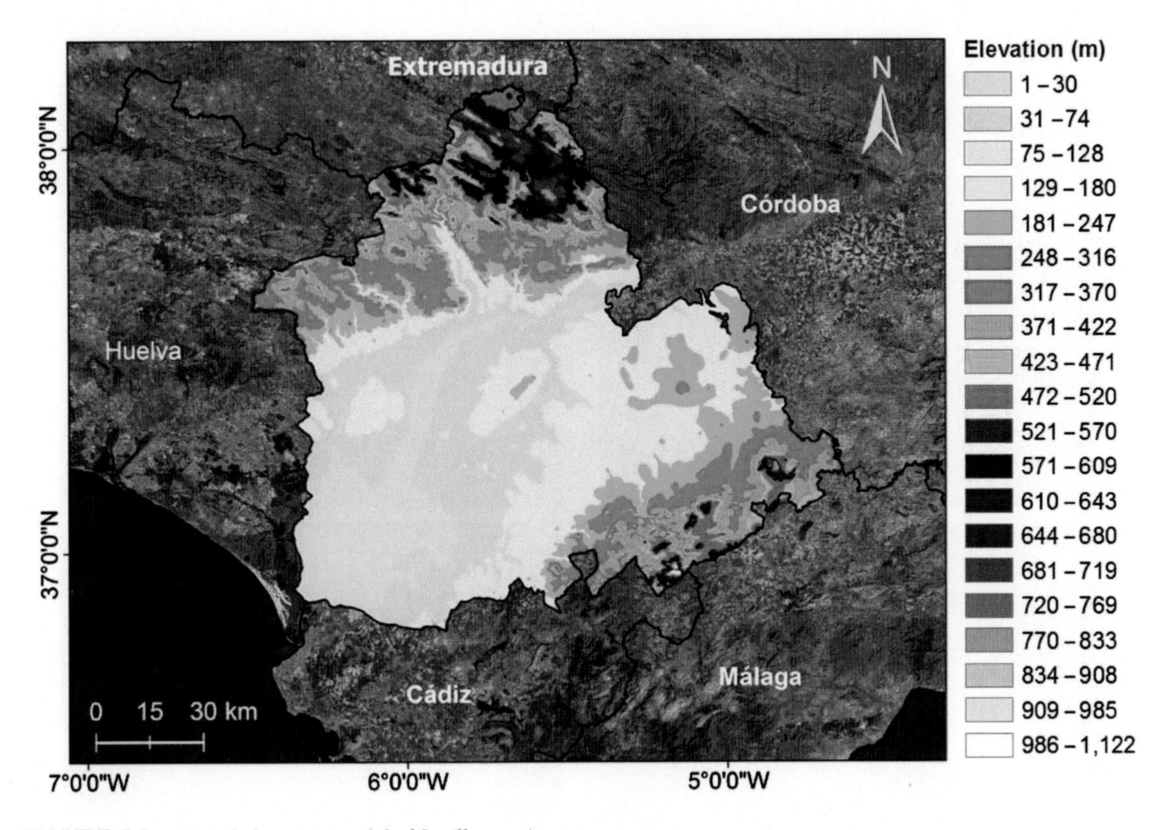

FIGURE 7.3 Digital elevation model of Seville province.

Analysis of Soil Transects

Data from selected soil profiles of Seville province, including morphological, chemical, and physical properties, were attached to the attribute table of the soil profiles database. To represent the variability in elevation, lithology, and soil type in this region, two transects (S-N and W-E) were considered (Fig. 7.4). These transects were subsequently represented by 41 soil profiles from the SDBm-Seville. In the first stage the exact location of 576 soil profiles from the database was represented (Fig. 7.4A). Then, a representative selected topographic transects were selected (Fig. 7.4B) and representative points at regular 4 km intervals were established (Fig. 7.4C). Data from the nearest soil profile were considered as representative of each point in addition to extracting the elevation data for each transect points and creating a buffer with a 6-km distance (Fig. 7.4D).

Soil transect TA included 31 points; 25 soil profiles have been chosen to represent TA due to the high variation in soil types. Soil transect TB has 32 points and 16 soil profiles to represents TB transect variation (Table 7.1). Fig. 7.5 shows the position of transects points and their elevation. Spatial data and interpolation

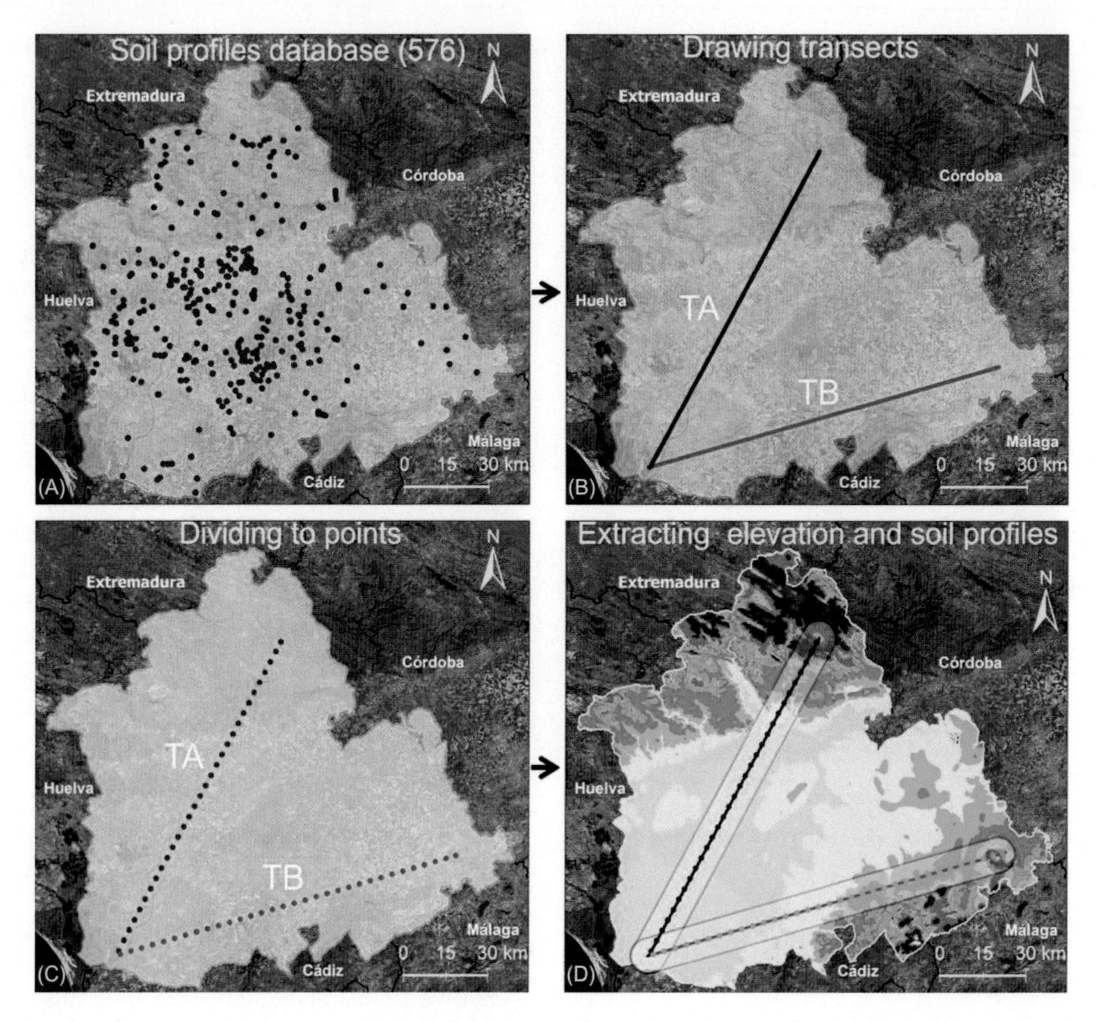

FIGURE 7.4 Steps for extracting the studied soil profile information using Arc.GIS 10.2. (A) Soil profiles database (576); (B) drawing transects; (C) dividing to points; (D) extracting elevation and soil profiles.

(inverse distance weighting) processing were carried out using ArcGIS 10.2 (ESRI, 2013).

Studying soils along transects helps to understand the relations between elevation and soil proprieties as physical, chemical, and mineralogical in order to assess their suitability for crop development, taking into account the climate parameters and the projected climatic changes (Abd-Elmabod, 2014; Brevik, 2013; Levine et al., 1994).

Climate Information

Current climate data (average data from 1960 to 2000) were calculated according to the global climate model CNRM-CM3 (Salas-Mélia et al., 2005) by extracting spatial monthly climate data. To represent the high spatial variation in the climatic parameters along with the 63 studied points, three points (starting, mid, and final) for each transect were selected. High

TABLE 7.1 Transects (T), Representative Soil Profiles (SDBm code), Horizons, Depth (cm), Effective Depth (cm), Transect Points (TP), Soil Type (USDA, 2010), and Present Land Use of TA and TB Transects Points. Both Transects Start in One Common Point so TA01 and TB01 are Identical

T	Code	Horizons	Depth	TP	Eff. Depth	Soil Type	Land Use
TA	SE0103	Ap-B2g-C1g-C2g	0–10–37–56	TA01	100	Typic Halaquept	Grassland
	SE0117	Ap-IIC1-IIC2-IVC3	0–20–45–70–111	TA02–TA05	>150	Aquic Xerofluvent	Wheat and cotton
	SE0602	Ap1-Ap2-C1-C2	0–20–32–52–80	TA06–TA07	>150	Typic Halaquept	Rice
	SE1021	Ap1-Ap2-C1-C2-C3	0–20–30–52–64–	TA08	>150	Typic Fluvaquent	Rice
	SE1066	Ap-A3g-Bg-gD	0–10–20–65–170	TA09	>150	Typic Haploxeralf	Olive
	SE1062	Ap11-Ap12-A3-B-CD-IIg2D	0–5–10–30–70–80–	TA10–TA11	>150	Typic Hapludalfs	Olive
	SE1065	Ap-Ap2g-g11-g12-g21-g22-Cgca	0–10–25–42–58–85–160	TA12	>150	Typic Xerofluvent	Olive
	SE0710	Ap-B-C	0–20–55–110	TA13	>150	Typic Haploxeralf	Olive
	SE1002	Ap-B1-C1ca-C2ca	0–20–45–65–	TA14	>150	Typic Calcixerolls	Olive
	SE1013	Ap-Ap2-IIB1g-IIB2g-IIICca	0–25–45–90–130–	TA15	>150	Aquic Haploxeralf	Settlements
	SE0412	Ap-Ap2-B2-B3ca-Cca	0–20–30–45–90–	TA16	>150	Typic Dystrudepts	Agriculture
	SE0993	Ap-AB-B1-BC-BCca	0–30–50–100–150–	TA17	>150	Chromic Dystruderts	Agriculture
	SE1010	Ap1-Ap2-B1-IIB2g-IIB22g -IICca-IIIC	0–20–38–60–95–115–165–	TA18	50–100	Aquic Haploxeralf	Vine
	SE1009	A1-A2-B1-IIB21g-IIB22g-IIIBg-IVC	0–12–32–42–100–135–165–	TA19	>150	Aquic Haploxeralf	Olive
	SE1000	Ap-B11-B12-Cca	0–25–70–85–	TA20	>150	Vertic Calcixeroll	Irrigation crops
	SE0016	Ap-B1-C1ca-C2ca-IICca	0–20–45–70–100–	TA21	100–150	Chromic Dystruderts	Olive
	SE0313	Ap-AC-C	0–30–100–	TA22	>150	EnticDystruderts	Wheat
	SE0309	A-R	0–20–	TA23	0–25	Lithic Xerorthents	Grassland
	SE0635	Ap-B11-B12-C	0–20–60–110	TA24–TA25	>150	Typic Haploxeralfs	Cotton
	SE0404	Ap-B2t-BC1-BC2-R	0–10–40–70–100–	TA26	>150	Typic Haploxeralfs	Forest
	SE0052	A1-B2t-BC1-BC2-R	0–10–40–70–100–	TA27	100	Typic Haploxeralfs	Forest

(Continued)

TABLE 7.1 Transects (T), Representative Soil Profiles (SDBm code), Horizons, Depth (cm), Effective Depth (cm), Transect Points (TP), Soil Type (USDA, 2010), and Present Land Use of TA and TB Transects Points. Both Transects Start in One Common Point so TA01 and TB01 are Identical (Continued)

T	Code	Horizons	Depth	TP	Eff. Depth	Soil Type	Land Use
	SE0930	AO-A1-B-C	(−5)–1–40–60–	TA28	100	Typic Xerorthents	Forest
	SE0072	A1-AB-B2t-BC-C	0–10–30–120–150	TA29	>150	Typic Rhodoxeralf	Olive
	SE0401	A1-A2-AB-B1-B2t-B3-C	0–8–15–30–55–220–250–	TA30	>150	Typic Palexerult	Cork oak
	SE0082	A1-A2-AB-B1-B2t1-B2t2	0–8–15–30–55–110–	TA31	>150	Typic Palexerult	Cork oak
TB	SE0103	Ap-B2g-C1g-C2g	0–10–37–56–	TB01–TB02	100	Typic Halaquept	Grassland
	SE1082	Ap-Bg1-Bg2-Cg	0–10–20–70–80	TB03	> 150	Chromic Haploxererts	Sugar beet
	SE0107	Ap-C	0–20–	TB04–TB05	50	Typic Xerorthents	Forest cropping
	SE0109	A1-Csa	0–25–	TB06	50	Halic Haploxererts	Grassland
	SE0104	Ap1-Ap2-B1-B2-Bca-C	0–15–40–65–100–110–	TB07–TB09	> 150	Typic Haploxererts	Cotton, maize
	SE0818	Ap-A/C-Ca/C	0–20–35–	TB10	100	Chromic Haploxererts	Olive
	SE0854	Ap1-Ap2-Ap3/C	0–10–50–	TB11–TB12	> 150	Typic Xerorthent	Olive
	SE0815	Ap-Ca/C-C	0–20–60–	TB13	100	Typic Xerorthent	Olive
	SE0821	Ap-Ca/C-C	0–20–80–	TB14, TB15	50	Typic Xerorthents	Olive
	SE0836	Ap-B1-B2-Ca/C	0–10–30–60–	TB16–TB18	> 150	Typic Xerorthents	Olive
	SE0856	Ap1-Ap2/C	0–30–	TB19	> 150	Typic Haploxerepts	Olive
	SE0855	Ap-Ca/C	0–60–	TB20	> 150	Calcic Haploxerepts	Olive
	SE0886	Ap-Ap (B1)-B2-B2ca-Cca	0–12–42–75–90–	TB21–TB22	> 150	Typic Xerorthent	Olive
	SE0229	Ap-C1-C2	0–20–80–	TB23–TB25	> 150	Vertic Xerofluvent	Wheat
	SE0142	Ap-B-Cca	0–20–50–	TB26–TB28	> 150	Calcic Haploxerepts	Agriculture crops
	SE0140	Ap-B2-B3 ca-Cca	0–20–40–70–	TB29–TB30	> 150	Calcic Haploxerepts	Olive
	SE0101	Ap-AC-C	0–25–35–	TB31–TB32	25	Entic Haploxeroll	Olive

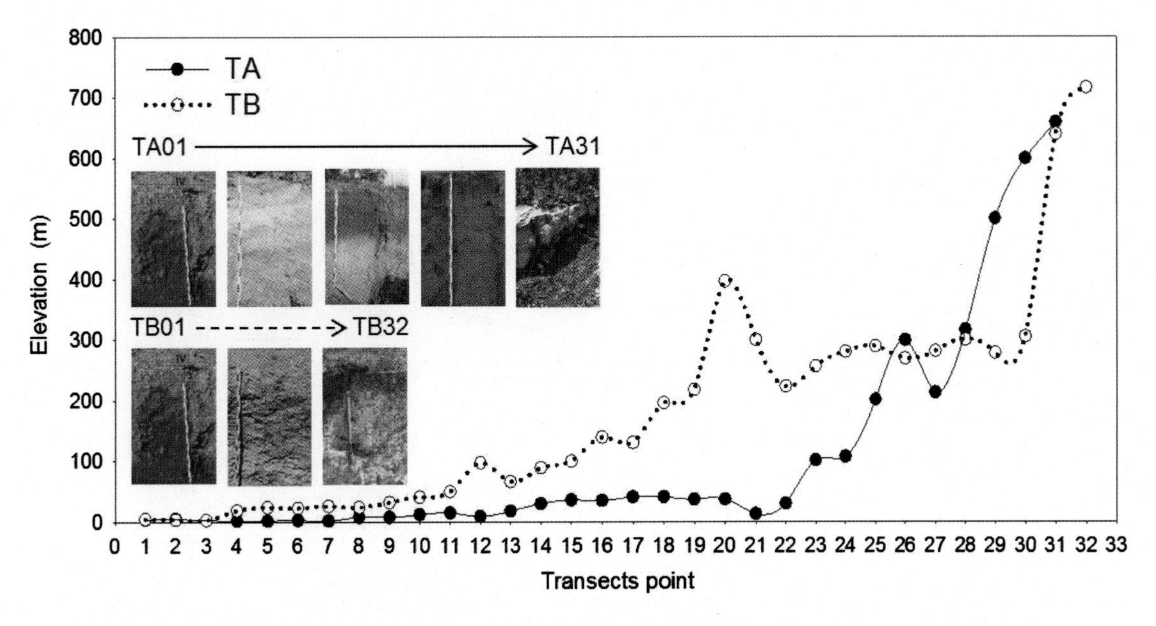

FIGURE 7.5 Elevation (m) and images of soil profiles along the studied topographic transects TA and TB.

variation in temperature and precipitation was observed, with the highest annual precipitation value (983 mm) located in TA31 and the lowest precipitation in TA01 with an annual precipitation of 378 mm. The lowest monthly mean temperature was 7.5°C in January in TB32, and the warmest month July at 27.3°C in TA15.

Fig. 7.6 shows the graphical output of climate conditions for six representative climatic points: TA01, TA15, and TA31 for transect A and TB01, TB16, and TB32 for transect B. Fig. 7.6 also represents the average variation of precipitation (P), temperature (T_m), potential evapotranspiration (ET_0), and aridity index (AR_i) showing which months are arid (actual precipitation lower than potential evapotranspiration).

Crop Suitability: Almagra Model

The Almagra model represents a biophysical evaluation that uses as diagnostic criteria those soil factors or conditions favorable for crop development (De la Rosa et al., 1977). The reference zone chosen for this work was the left bank of the lower Guadalquivir valley, northwest of the city of Seville. The area includes 690 km^2, and its characteristics are typical of a Mediterranean region (De la Rosa, 1974). Fig. 7.7 shows the level of generalization of Almagra model considering the soil factors according to Antoine et al. (1995): profile depth (p), texture (t), drainage (d), carbonate and pH (c), salinity (s), sodium saturation (a) and profile development (g), and the evaluation of these factors depending on the level of generalization and the different crops.

Calcium carbonate is a major component of many semiarid soils, as it accumulates in soils over time. Systematic morphological changes occur in these calcic soils, which affects the way water infiltrates in the soil profile. The soil pH is a measure of the acidity or basicity in soils; the pH of the soil solution is very important

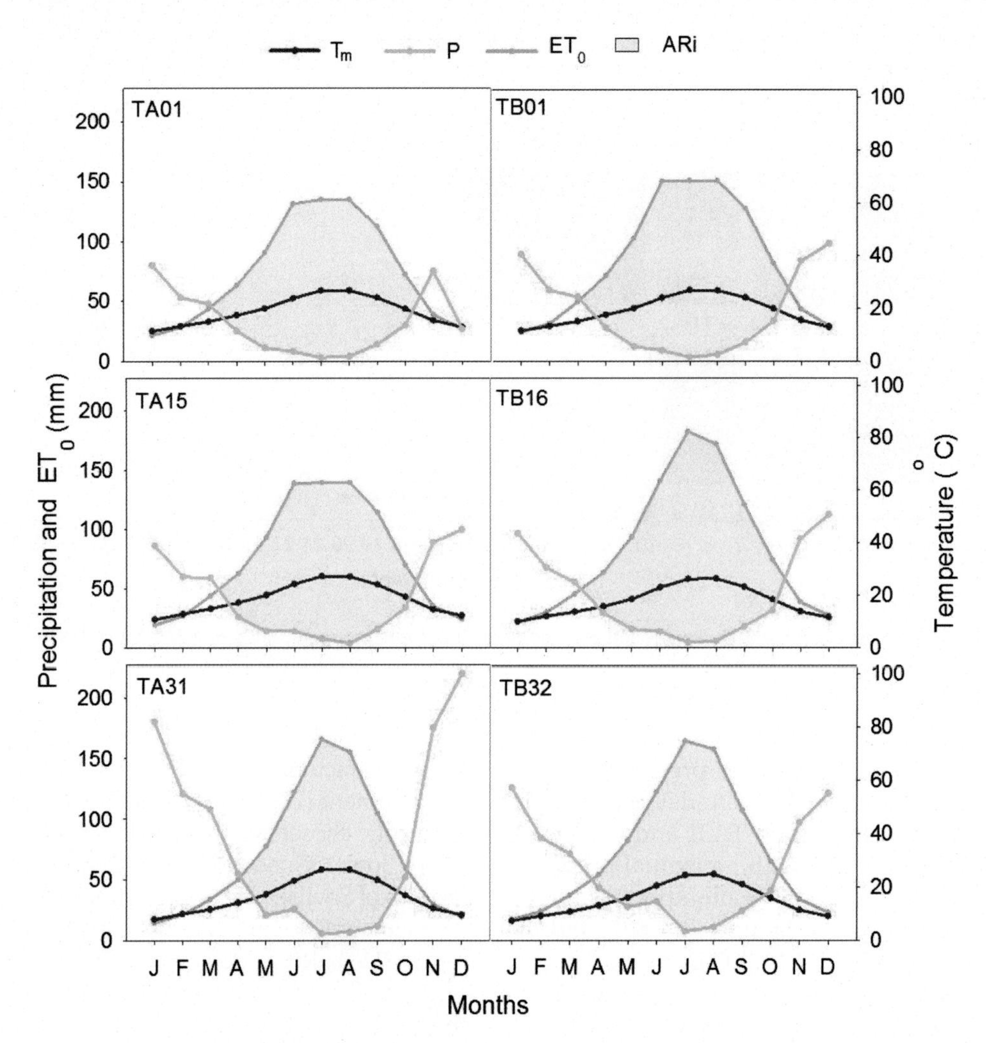

FIGURE 7.6 Graphical representation of climate conditions of starting points (TA01 and TB01), mid-points (TA15 and TB16), and final points (TA31 and TB32) for the two topographic transects. T_m: temperature mean in °C; P: precipitation in mm; ET_0: reference evapotranspiration in mm; ARi: aridity index expressed as the number of months per year in which the reference evapotranspiration exceeds the precipitation.

since it affects the solubility of nutrients such as N, K, and P and thereby affects the availability of these nutrients to plants. Plants need nutrients in specific amounts to grow, thrive, and fight off diseases (McCauley et al., 2009). According to Rengasamy (2006), soil salinity is the salt content of the soil. Natural processes such as a shallow saline water table can cause salinization and the presence of mineral sediments (may be of marine origin, volcanic sediments, or due to aridity). Soil sodicity is the term given to the amount of sodium

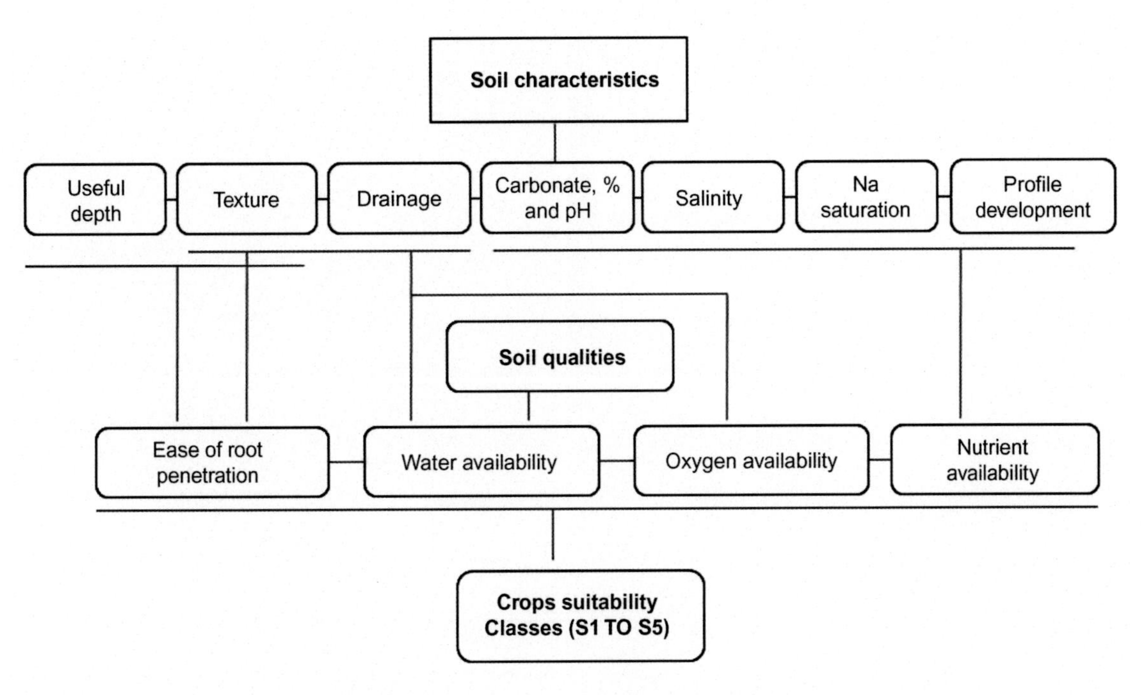

FIGURE 7.7 General scheme of Almagra model (Pro& Eco package of MicroLEIS).

held in the soil, this can be expressed in different ways: exchangeable sodium percentage (ESP), or sodium absorption ratio ($SAR = [Na^+]/\sqrt{(½([Ca^{2+}] + [Mg^{2+}]))}$, with concentrations in mequiv./L) (Warrence, 2002). Soil texture refers to the proportions of sand, silt, and clay in a soil. Soil drainage is the natural or artificial removal of water from the surface and subsurface of an area. Soil depth is another important soil property; it refers to the thickness of the soil materials that provide structural support, nutrients, and water for plants. Organic matter can act as a reservoir that will release nutrients into the soil solution upon decomposition. It behaves somewhat like a sponge, with the ability to absorb and hold up to 90% of its weight in water; also, it causes soil to clump and form soil aggregates, which improves soil structure (Overstreet, 2009). The relatively higher content can only be maintained under continuous cultivation when there is sufficient input of organic manure or other organic material such as crop residue.

The agricultural land suitability evaluation considered the following traditional crops: wheat (T), maize (M), melon (Me), potato (P), soybean (S), cotton (A), sunflower (G), and sugar beet (R) as annuals; alfalfa (Af) as semi-annual; and peach (Me), citrus fruits (C), and olive (O) as perennials. The control section of soil factors measured as texture, carbonates, salinity, and sodium characteristics were established by adapting the criteria developed for the differentiation of families and series in the soil taxonomy (USDA, 2014). Development, inputs, and validity of this model described in De la Rosa et al. (1992). The Almagra model has been applied on the studied soil transects in Seville. Fig. 7.8 shows screen captures of the Almagra model (Pro & Eco package), at the

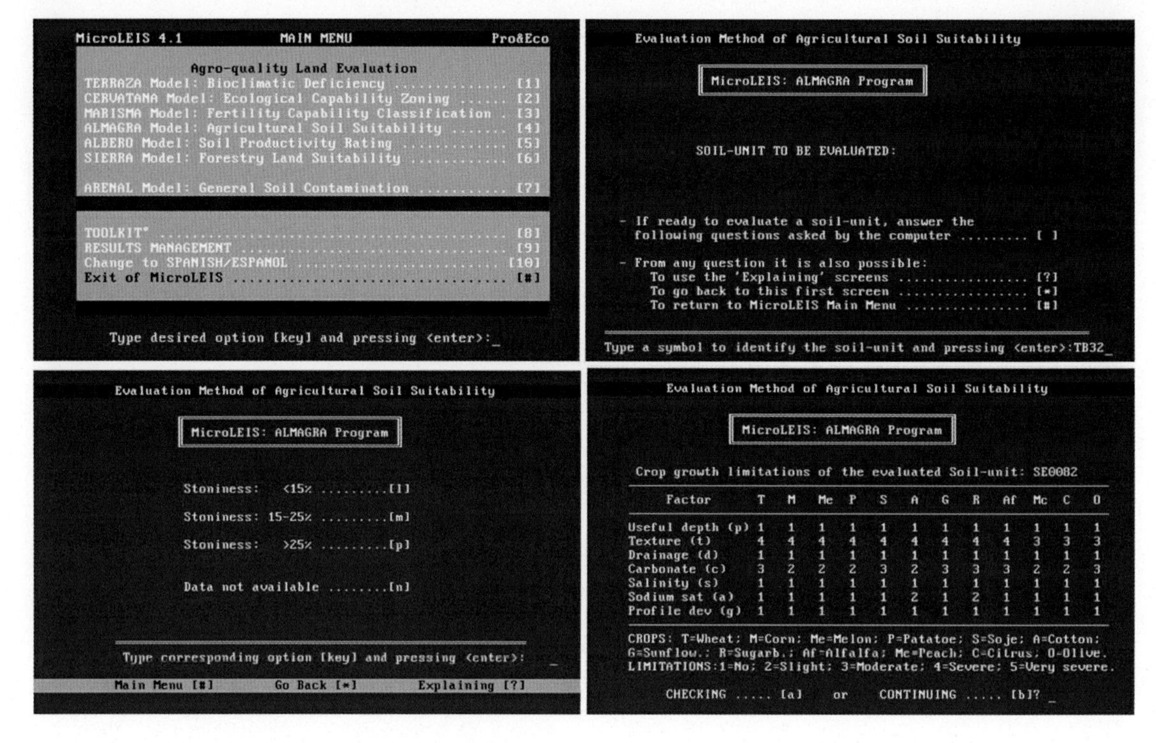

FIGURE 7.8 Screen capture of Almagra model.

stage of entering soil parameter information and evaluation results.

RESULTS AND DISCUSSIONS

Major Soil Factors Considered in Soil and Land Evaluation

Soil factors may be related directly to crop suitability, or indirectly via impacts on soil quality (Fig. 7.7). The selected major soil factors from the Seville-SDBm were pH, carbonate content, salinity and ESP, texture, drainage, depth of the soil profile, and soil organic matter content. Table 7.2 shows the ranges, dominant (in the case of soil qualitative characteristics), and the mean values (in case of soil quantitative characteristics) of the 42 benchmark soils that represent 63 transect (TA and TB) points for Seville.

Soil pH

Soil pH is considered one of the most important factors influencing the plant uptake of trace elements, with generally higher adsorption (and therefore lower availability) at higher soil pH (Kabata-Pendias, 2001; Safari et al., 2015; Diyabalanage et al., 2016; Hilt et al., 2016; Kim et al., 2016; Reijonen et al., 2016). In transect TA the lowest pH (5.41) is observed in the final parts (TA30 and TA31) where the soil type Typic Palexerult. The highest values are found in TA06 and TA07 that has a pH value of 8.42 with a soil type Typic Halaquept. A pH value above 8.0 was also observed in

TABLE 7.2 Ranges and Dominant Values of Land Characteristics of the 42 Benchmark Soils Along Seville Soil Transects

Land Characteristics		(Range) Dominant and Mean	
		TA	TB
Site related	Landform	(Plan-mountain) plan	(Plan-mountain) hill
	Slope gradient (%)	(0.0–19) 1	(0.0–24.0) 3
	Elevation (masl)	(1–902)	(1–845)
Soil-related characteristics	Useful depth (cm)	(25–160) 140	(25–160) 150
	Drainage	(v. poor-excessive) moderately well	(poor-excessive) well
	Particle size distribution	(Sandy–clayey) clayey	(Sandy loam-clay) clayey
	Clay[a]	(10.0–57.8) 37.1	(12.8–70.0) 35.5
	Silt[a]	(4.4–55.7) 27.9	(7.2–40.8) 27.1
	Sand[a]	(4.0–84.0) 34.9	(3.5–72.6) 37.3
	Superficial stoniness	(Nill–abundant) nill	(Nill–abundant) nill
	Organic matter[a] (%)	(0.20–1.61) 0.62	(0.30–1.71) 1.0
	pH[a]	(5.41–8.42) 7.10	(7.6–8.23) 7.85
	Calcium carbonate[a] (%)	(0–30.1) 5.3	(4–71.1) 33.1
	Sodium saturation[a] (%)	(0.4–12) 2.7	(0.4–12) 3.3

[a]*Soil parameters measured within the soil section 0–50 cm.*

TA08-Typic Fluvaquent (8.3) and in TA15-Aquic Haploxeralf (8.04). On the other hand the pH values along TB ranged from 7.6 (TB10-Chromic Haploxererts) and 8.23 in the end parts of the transect (TB31, TB32) that has a soil type of Entic Haploxeroll.

Carbonate Content

Carbonate affects the behavior of nutrients and it totally related with the soil buffering capacity (Villen-Guzman et al., 2015). The $CaCO_3$ particles are found in the coarse (primary) and fine fractions (secondary formation). Inactive $CaCO_3$ is more closely associated with the coarse and medium mineral fractions (sand and silt fractions). In transect A the carbonate content was undetectable in the final parts (TA30 and TA31) where the Typic Palexerults occur, and the highest value was observed in TA13-Typic Haploxeralf soil (30.1%). On the other hand, in TB the highest carbonate content was observed in soils of calcareous parent material. The concentration of carbonates reached 71.1% in TB29 and TB30 (Calcic Haploxerepts). The lowest values occurred in TB16, TB17, and TB18 (Typic Xerorthents).

Salinity and Exchangeable Sodium Percentage

In some cases, soils like Vertisols are naturally high in ESP throughout the soil profile (Syers, 2001). Salinization can also be caused by artificial processes such as irrigation, because poor irrigation water quality and low irrigation efficiency that contribute to increased soil salinity and sodicity. Spatial information on soil salinity in more detailed scale is needed, particularly for better soil management and land use planning (Rongjiang and Jingsong, 2010).

If excessive amounts of salt are taken up by the plant, the salts will accumulate and reach toxic levels, mainly in the older, transpiring, leaves, causing premature senescence, and reduced yields. The growth rate of plants under salt stress differs strongly between plant species, and sometimes also between cultivars (Qadir and Schubert, 2002; Jacobsen et al., 2012). In TA and TB, salinity problems have been observed in the natural land use areas (i.e., TA01 and TB01), which show a high concentration of salt (electric conductivity up to 7.3 $dS m^{-1}$) and are classified as Typic Halaquept. Apart from these points, there is no further salinity problem along both transects. Crescimanno et al. (1995) showed that an ESP above 15% is considered to negatively affect soil structure and hydraulic characteristics. Sodicity affects soils through two related phenomena: swelling and dispersion. In both transects the majority of soil types had low ESP values, except in TA06, TA07-Typic Halaquept soil type that had 12% and in TA08-Typic Fluvaquent that had a 10 ESP, and for TB the highest point in ESP was TB06-Halic Haploxererts where it had a value of 12%.

Soil Texture

Soil texture is an important soil property that drives crop production and field management (Greve et al., 2012). Soil particles may be either mineral or organic but in most soils, the largest proportions of particles are mineral and these soils are therefore referred to as "mineral soils." The texture based on the relative proportion of the particles fewer than 2 mm (fine earth). The basic elements of the soil texture are

- Sand defined as mineral soil particles that have diameters ranging from 2 to 0.02 mm. In TA the highest content of the sand was observed in TA19, with a proportion 84.0% and TA15 (79.5%) (both soils classified as Aquic Haploxeralfs). Lowest sand content was found in TA08 (typic fluvaquent: 4.0%). In TB the sand content ranged from 3.5% in TB06 (Halic Haploxererts) to 72.6% in the Typic Xerorthents that dominated the TB16, TB17, and TB18 sites.
- Silt can be defined as the mineral soil particles that range in diameter from 0.02 to 0.002 mm. In TA TA15 (Aquic Haploxeralf) showed the lowest silt content with 4.4%: at the other extreme, Aquic Xerofluvents had a silt content of 55.7% that represented the TA02, TA03, TA04, and TA05 points. In TB the lowest value was observed in Typic Xerorthents, especially in the points TB16, TB17, and TB18 where it was 7.2%, and the highest value (40.8%) was observed in TB14 and TB15 which were Typic Xerorthents.
- Clay is the soil texture component with soil particles that have diameters less than 0.002 mm (Greve et al., 2012). In TA the soil type TA19-Aquic Haploxeralf showed low clay content of 10%, on the other hand, clay content in TA10, TA11-Typic Hapludalfs was 57.8%. In TB the highest content of clay (70%) was observed in TB06-Halic Haploxererts soil type and the lowest (12.8%) in Typic Xerorthents situated in TB14 and TB15.

Drainage

Soil drainage may determine which types of plants grow best in an area. Many agricultural soils need good drainage to improve or sustain production or to manage water supplies (Haroun, 2004). Poor drainage (causing water-logged areas) can often be identified by examining the soil color. In zones dominated by longer periods of saturation, and thereby reducing conditions, there can be mottles that occupy small areas and that differ in color from the soil matrix.

In TA sample points drainage could be divided into different classes, those that had a good drainage (TA10, TA11-Typic Hapludalfs, TA12-Typic Xerofluvent, TA13-Typic Haploxeralf, TA14-Typic Calcixerolls), and other points with a very poor drainage; TA01-Typic Halaquept, and in Aquic Xerofluvents at points TA02, TA03, TA04, and TA05. In TB, very poor drainage was observed in TB06 (Halic Haploxerert) and in Typic Haploxererts at TB07, TB08, and TB09. A well-drained soil was found in TB21 and TB22 (Typic Xerorthents).

Soil Depth

Soil depth is very critical for plant growth. Any discontinuities in the soil profile, from layers of sand or gravel to even bedrock, can physically limit root penetration. It can also create problems when using irrigation. Soil macro- and mesobiota need enough soil to grow and increase physical fertility (Louis, 2011). In TA, shallow soil depths were observed in the animal husbandry land use within the soil type Lithic Xerorthents which was located in TA23 where depth was 25 cm. Also some areas along TB, depth did not reach 50 cm (TB31, TB32-Entic Haploxeroll, with a depth of 25 cm).

Organic Matter

It is noteworthy that soil organic matter influences the availability of micronutrients to plants; particularly those that are largely present as insoluble forms. Organic matter can increase their solubility through the effect on the soil redox potential (Lindsay, 1991; Zeng et al., 2011; Lehmann and Kleber, 2015; Moura et al., 2016; Reijonen et al., 2016). In addition, organic matter can be one of the main nutrient sources; therefore a strong relationship between nutrient status and organic matter content in the soil exists. In TA the soil type TA28-Typic Xerorthents showed the highest organic matter content of 1.61%, while the lowest value was found in TA13-Typic Haploxeralf with only 0.20% organic matter. In TB, soil Typic Xerorthents that represented the points TB04 and TB05 showed the highest organic matter content with 1.71% and the lowest content of OM were found in TB13-Typic Xerorthent (0.3%).

Crop Suitability

In the suitability classes of the Almagra model the main limiting factors in the soil category are the subclasses depth (p), texture (t), drainage (d), calcium carbonate content (c), salinity (s), sodium saturation (a), and profile's degree of development (g). The model application results are grouped into five soil suitability classes: S1 (optimum), S2 (high), S3 (moderate), S4 (marginal), and S5 (not suitable). Table 7.3 shows how the Almagra model classifies three points of 63 Seville transects points, where some limiting factors it is difficult to improve and manage in the recommended improved scenario as soil texture and the useful depth in SE0082 and SE0101, respectively, but other limiting factors as salinity in SE0103 it recommends to reach in the improved scenario.

Soil Suitability Under Current Situation

Almagra model was applied under current situation of soil limiting factors in both transects TA and TB. Fig. 7.9 shows the results of the soil suitability evaluation for the 12 Mediterranean crops under the current situation of soil factors. The results of soil suitability evaluation ranged from S1 to S5p and S5s; the final part of transect TA was classified as S4t subclass due to the high content of gravel and

TABLE 7.3 How the Almagra Model Calculates the Final Classifications: Each Soil Profile is Classified on a Scale From 1 (Best) to 5 (Worst) for Each Subclass, Depending on the Specific Requirements of Each Crop

		Crops											
SDBm Code	**Soil Factors/Classification**	**T**	**M**	**Me**	**P**	**S**	**A**	**G**	**R**	**Af**	**Mc**	**C**	**O**
SE0103	Useful depth (p)	1	1	1	1	1	1	1	1	1	1	1	1
	Texture (t)	2	2	2	2	2	2	2	2	2	4	4	4
	Drainage (d)	3	2	2	2	3	2	2	3	3	4	4	4
	Carbonate (c)	1	2	2	2	1	2	1	1	1	2	2	1
	Salinity (s)	4	4	4	4	4	3	4	3	3	5	5	3
	Sodium sat (a)	1	1	1	1	1	2	1	2	1	1	1	1
	Profile dev (g)	1	1	1	1	1	1	1	1	1	2	2	1
	Classification	S4s	S4s	S4s	S4s	S4s	S3s	S4s	S3ds	S3ds	S5s	S5s	S4td
SE0082	Useful depth (p)	1	1	1	1	1	1	1	1	1	1	1	1
	Texture (t)	4	4	4	4	4	4	4	4	4	3	3	3
	Drainage (d)	1	1	1	1	1	1	1	1	1	1	1	1
	Carbonate (c)	3	2	2	2	3	2	3	3	3	2	2	3
	Salinity (s)	1	1	1	1	1	1	1	1	1	1	1	1
	Sodium sat (a)	1	1	1	1	1	2	1	2	1	1	1	1
	Profile dev (g)	2	2	2	2	2	3	2	3	2	3	3	2
	Classification	S4t	S4t	S4t	S4t	S4t	S4t	S4t	S4t	S4t	S3tg	S3tg	S3tc
SE0101	Useful depth (p)	5	5	5	5	5	5	5	5	5	5	5	5
	Texture (t)	1	1	2	2	1	2	1	1	1	2	2	3
	Drainage (d)	1	1	1	1	1	1	1	1	1	1	1	1
	Carbonate (c)	2	3	3	3	2	3	2	2	2	3	3	2
	Salinity (s)	1	1	1	1	1	1	1	1	1	1	1	1
	Sodium sat (a)	1	1	1	1	1	2	1	2	1	1	1	1
	Profile dev (g)	1	1	1	1	1	1	1	1	1	1	1	1
	Classification	S5p	S5p	S5p	S5p	S5p	S5p	S5p	S5p	S5p	S5p	S5p	S5p

The final classification is determined by the worst subclasses (in red), which is indicated by their letter. The crops are: wheat (T), corn (M), melon (Me), potato (P), soybean (S), cotton (A), sunflower (G), sugar beet (R), alfalfa (Af), peach (Mc), citrus fruits (C), and olive (O).

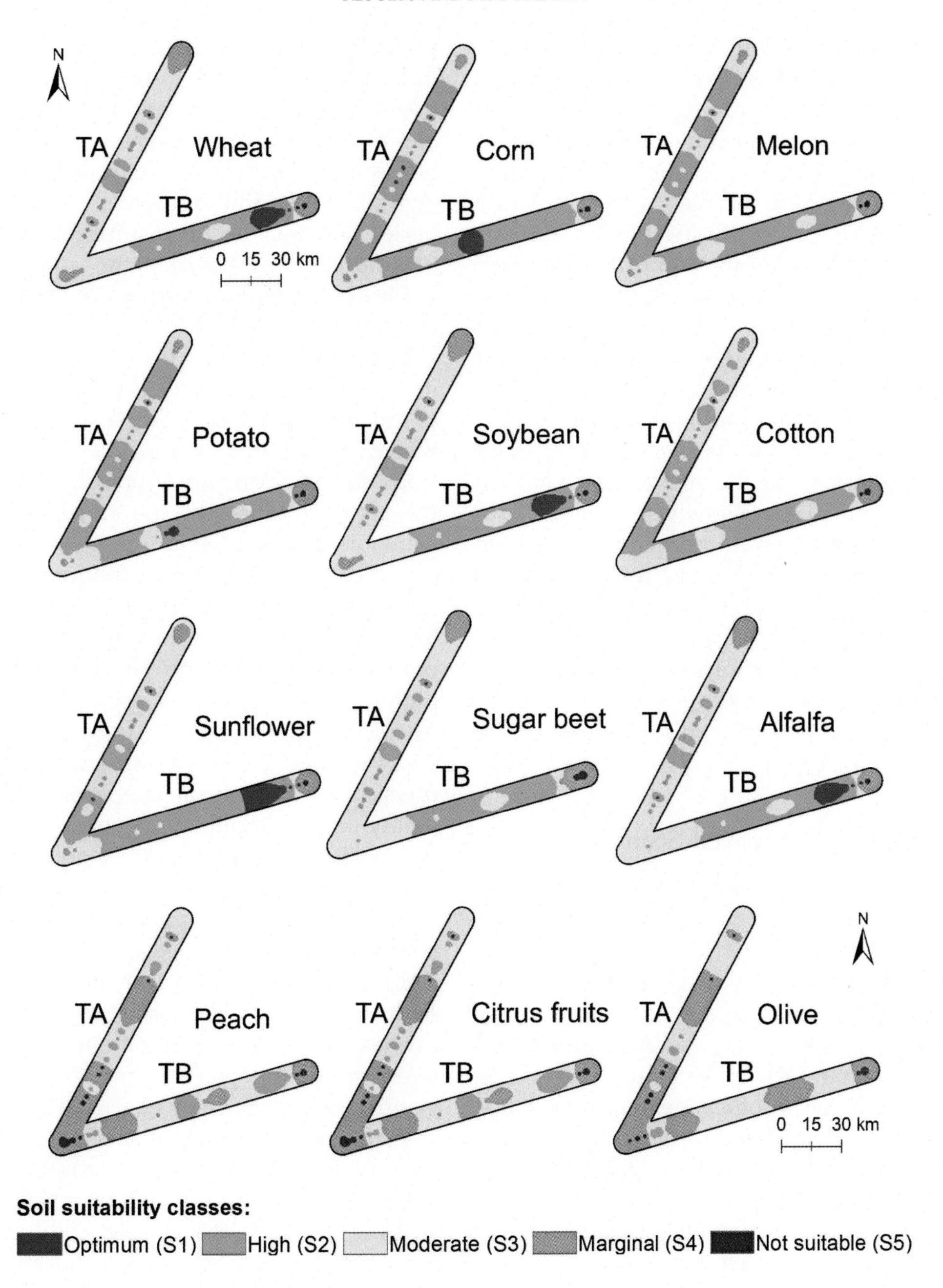

FIGURE 7.9 Spatial analysis of soil suitability classes under current situation of soil factors, according to the application of the Almagra model in TA and TB soil transects.

coarse texture, on the other hand, the final parts of transect TB was classified as an S5p subclass due to the shallow useful depth. The low content or the absence of calcium carbonate content became a limiting factor in some parts of TA, especially in the soil profiles with pH below than 7 SE1066 (pH 6.5), SE0404 (pH 5.5), SE0401 (pH 5.4), the acidity of the soil due to the parent material that contains acidic igneous rocks. In contrast the excessive concentrations of calcium carbonate appeared to be a limiting factor to crop suitability in some parts of the TB transect, especially in the profile locations SE1021 and SE0602, where the basic soils are formed from basic rocks and the pH values were 8.3 and 8.4, respectively. Moreover, the soil profiles that were located at lower elevations often had salinity problems, a heavy texture, and/or high values of ESP and/or very poor drainage and/or shallow soil depth, and/or incipient development of soil profiles. Consequently, the assessment results of these parts are found in marginal class or/and not suitable class for the 12 Mediterranean crops.

Soil Suitability Improved Scenario

The soil suitability was determined in case some of the improvement of soil limiting factors along transect TA and TB, proposed improvement scenario will be in the improbable soil factors as drainage or/and carbonate or/and salinity or/and sodium saturation. On the other hand the other limiting factors as depth, texture, and profile's degree of development were not considered in the improved scenario because it is difficult to accomplish. Fig. 7.10 represents the soil suitability for a hypothetical situation in which the soil factors (d, c, s, and a) would have been improved.

In both transects, there are some parts with a sandy texture and excessive drainage or with heavy clay texture and very poor drainage that can form marginal or not suitable soil. Therefore obtain a well-drained soil in such parts by adding organic matter and gypsum in both cases in order to improve soil drainage (Overstreet, 2009). In addition reducing tillage and adding sand in case of the clayey soil with very poor drainage its lead to increase the agriculture suitability under the improved scenario. Besides, the proper irrigation that aids to improve and modify the soil properties, while over-irrigation can lead to decreasing soil suitability.

Some segments of the transects especially in TB had a calcareous soil with a high content of calcium carbonate and therefore high pH, which will be not appropriate for plant nutrition and consequently crop suitability due to the transformation of the nutrients from an available to unavailable forms (Villen-Guzman et al., 2015). Consequently, the improvement in these parts will occur by adding organic matter and/or adding sulfur (Hu et al., 2005). The parts of transect TA and TB that had soils with high pH and high SAR it suggested to improve by adding gypsum, and according to (Anikwe and Ibudialo, 2016) gypsum can improve soil physical and chemical properties due to the ability of Ca^{2+} applied via gypsum to aggregate soil particles thereby creating an enabling soil physical condition for well-nutrient uptake.

Salinity has been appeared to be a limiting factor in few parts of TA and TB transects, i.e., TA01, TB01, and TB02 and the improved scenario recommend the leaching of salinity that can lead to increase soil suitability for the different studied crops. This improved scenario is in agreement with Letey et al. (2011).

Almagra model output subclasses have been codified form 1 that represents optimum evaluation class (S1) till 19 that represents the not suitable subclass ($S5_{(3)}$), the codification has been done for each soil profile in both transects and for the 12 Mediterranean crops evaluation (Fig. 7.11). The subscripted number in Fig. 7.11 represents the digit of limiting factors as useful depth and/or texture and/or drainage and/or calcium carbonate content and/or salinity

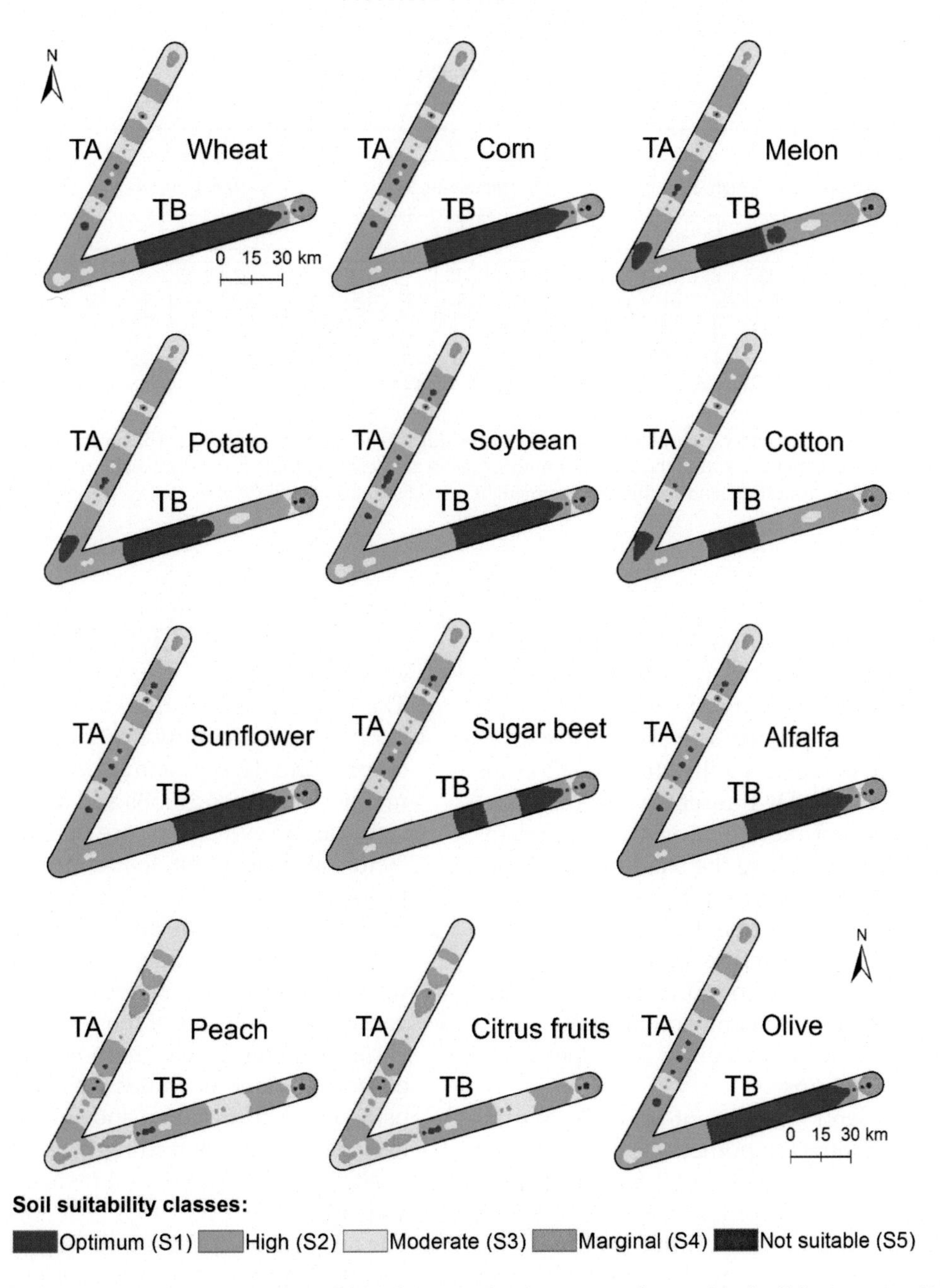

FIGURE 7.10 Spatial analysis of soil suitability classes under the improved scenario of soil factors, according to the application of the Almagra model in TA and TB soil transects.

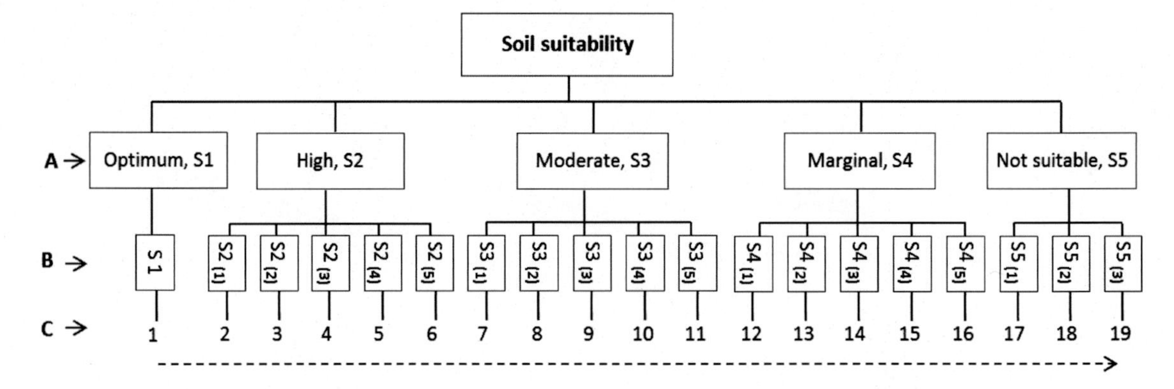

FIGURE 7.11 Soil suitability according to Almagra model outputs. (A) Suitability classes; (B) suitability subclasses with a subscripted number that represents the digit of limiting factors as useful depth, texture, drainage, calcium carbonate content, salinity, sodium saturation, and profile development. (C) Soil limitation degree.

and/or sodium saturation and/or profile development.

Decreasing the severity of some soil limiting factors (in the improved scenario) leads to an increase in the soil suitability for the 12 crops along of TA and TB transects. The only sections that did not receive an improved classification were those where the shallow soil depth and the very coarse texture where the limiting factors, as improvement of these factors, is not feasible, and therefore these factors were not included in this scenario (Fig. 7.12). In this hypothetical improvement scenario the highest improvement on suitability is for perennial crops in the TA transect.

Soil and land assessment models, spatial analyses and the hypothetical scenarios of agriculture management or climate (Muñoz-Rojas et al., 2015a, 2016) help to achieve a sustainable land management in the studied area, and to improve the soil characteristics and consequently increasing the soil suitability for diverse agriculture crops. Additionally the improvement of soil limitation it could be as adaptation strategies towards the long term of environmental changes as climate changes.

CONCLUSION

1. The purpose of our study was to determine agricultural suitability for 12 Mediterranean crops (annuals, semiannual, and perennials) under current conditions and an improved scenario in Seville province in relation to soil factors and elevation, to identify the limiting factors.
2. High variation in soil suitability classes was observed on both studied soil transects, where results varied between S1 and S5; this variability is a result of the soil properties variance along the elevation transects. This observation demonstrates the importance of using relief information and soil factors in decision-making regarding the formulation of site-specific soil use and management strategies.
3. The absence of calcium carbonate content became a limiting factor for the TA transect. In contrast the excessive content of calcium carbonate in TB transect appeared to be a limiting factor.
4. Similarities existed between the studied transects, where the forest parts are located

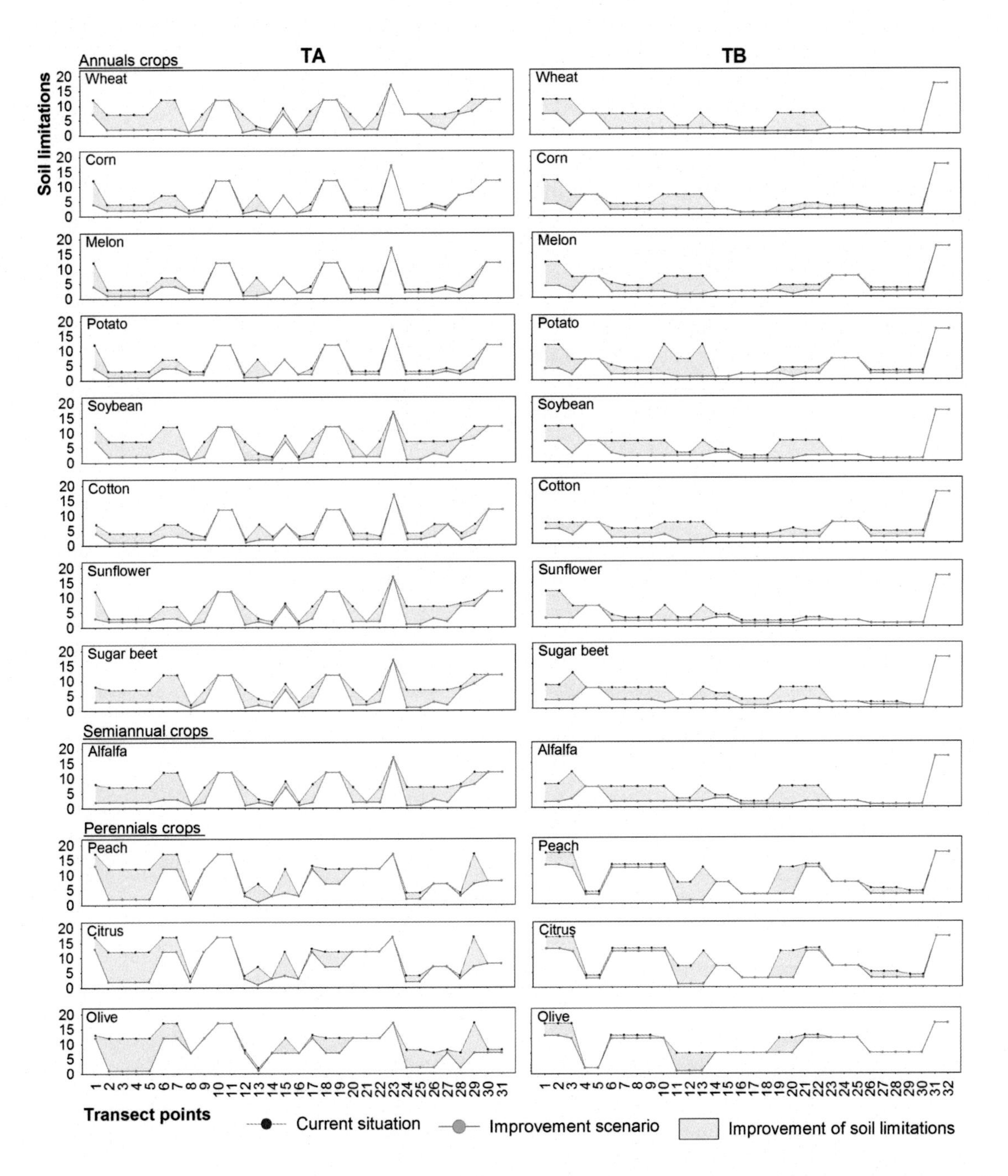

FIGURE 7.12 The results of the soil suitability evaluation in the current situation and the improvement scenario, for each point of transect TA and TB and for 12 Mediterranean crops. Lower values on the y-axis represent better suitability.

in the highest elevations. Crop suitability evaluation for the higher parts of Seville transect TA was not suitable due to the high content of gravels and coarse texture. For Seville transect, TB evaluation was not suitable due to shallow useful depth.
5. Soil salinity was the main limiting factor in the lowlands of both transects.
6. Leaching processes to remove the excess soluble salts from transects parts where the salinity is limiting factor to agriculture suitability, this can be accomplished by installing an effective drainage system before establishing the agricultural utilization projects. Also adding organic matter, gypsum, and lime it could improve soil suitability.
7. In the improvement scenario, all suitability crops have been improved along of the studied soil transects except in the locations that have shallow soil depth and very coarse of very fine texture.

References

Abd-Elmabod, S.K., 2014. Evaluation of soil degradation and land capability in Mediterranean areas, under climate and management change scenarios (Andalusia region, Spain and El-Fayoum province, Egypt). Ph.D. Thesis. University of Seville, Spain.

Abd-Elmabod, S.K., Ali, R.R., Anaya-Romero, M., Jordán, A., Muñoz-Rojas, M., Abdelmagged, T.A.L., et al., 2012. Evaluating soil degradation under different scenarios of agricultural land management in Mediterranean region. Nat. Sci 10, 103–116.

Albaladejo, J., Ortiz, R., Garcia-Franco, N., Ruiz-Navarro, A., Almagro, M., Garcia-Pintado, J., et al., 2013. Land use and climate change impacts on soil organic carbon stocks in semi-arid Spain. J. Soils Sediments 13, 265–277.

Anaya-Romero, M., Abd-Elmabod, S.K., De la Rosa, D., 2011. Agro-ecological soil evaluation for monitoring water quality using Microleis DSS. Ambientalia 2, 1–10.

Anaya-Romero, M., Abd-Elmabod, S.K., Muñoz-Rojas, M., Castellano, G., Ceacero, C.J., Alvarez, S., et al., 2015. Evaluating soil threats under climate change scenarios in the Andalusia region, Southern Spain. Land Degrad. Dev 26, 441–449.

Anikwe, M.A.N., Ibudialo, E.A.N., 2016. Influence of lime and gypsum application on soil properties and yield of cassava (Manihot esculenta Crantz.) in a degraded Ultisol in Agbani, Enugu Southeastern Nigeria. Soil Till Res 158, 32–38.

Antoine, J., Van Waveren, E.J., De la Rosa, D., Mayol, F., Moreno, J.A., 1995. FAO–ISRIC–CSIC multilingual soil database (SDBm) World Soil Resources Report 81. FAO, Rome.

Bekchanov, M., Ringler, C., Bhaduri, A., 2015. A water rights trading approach to increasing inflows to the Aral Sea. Land Degrad. Dev. http://dx.doi.org/10.1002/ldr.2394.

Beyene, F., 2015. Incentives and challenges in community-based rangeland management: evidence from eastern Ethiopia. Land Degrad. Dev 26, 502–509.

Branca, G., Lipper, L., McCarthy, N., Jolejole, M.C., 2013. Food security, climate change, and sustainable land management. A review. Agron. Sustain. Dev. 33, 635–650.

Brevik, E.C., 2013. The potential impact of climate change on soil properties and processes and corresponding influence on food security. Agriculture 3, 398–417.

Crescimanno, G., Lovino, M., Proven Zano, G., 1995. Influence of salinity and sodicity on soil structural and hydrolic characteristics. Soil Sci. Soc. Am. J. 59, 1701–1708.

Darwish, K.H.M., Wahba, M.M., Awad, F., 2006. Agricultural soil suitability of Haplo-Soils for some crops in newly reclamid areas of Egypt. J. Appl. Sci. Res. 12, 1235–1243.

De la Rosa, D., 1974. Soil survey and evaluation of Guadalquivir river terraces, in Sevilla province. Cent. Edaf. Cuarto Pub. Seville, Spain.

De la Rosa, D., 2005. Soil quality evaluation and monitoring based on land evaluation. Land Degrad. Dev 16, 551–559.

De la Rosa, D., Anaya-Romero, M., Diaz-Pereira, E., Heredia, N., Shahbazi, F., 2009. Soil-specific agro-ecological strategies for sustainable land use—a case study by using MicroIEIS DSS in Sevilla Province (Spain). Land Use Policy 26, 1055–1065.

De la Rosa, D., Cardona, F., Almorza, J., 1981. Crop yield predictions based on properties of soils in Sevilla, Spain. Geoderma 25, 267–274.

De la Rosa, D., Cardona, F., Paneque, G., 1977. Evaluacion de suelos para diferentes usos agricolas. Un sistema desarrollado para regiones mediterráneas. Anales de Edafologia y Agrobiologia 36, 1100–1112.

De la Rosa, D., Mayol, F., Diaz-Pereira, E., Fernandez, M., 2004. Aland evaluation decision support system (MicroLEIS DSS) for agricultural soil protection. Environ. Model. Softw 19, 929–942.

De la Rosa, D., Mayol, F., Moreno, J.A., Bonson, T., Lozano, S., 1999. An expert system/neural network model (ImpelERO) for evaluating agricultural soil erosion in Andalusia region. Agric. Ecosyst. Environ. 73, 211–226.

De la Rosa, D., Moreno, J.A., Garcia, L.V., 1993. Expert evaluation system for assessing field vulnerability to agrochemical compounds in Mediterranean region. J. Agric. Eng. Res. 56, 153–164.

De la Rosa, D., Moreno, J.A., Garcia, L.V., Almorza, J., 1992. MicroLEIS: a microcomputer-based Mediterranean land evaluation information system. Soil Use Manage 8, 89–96.

Diyabalanage, S., Navarathna, T., Abeysundara, H.T., Rajapakse, S., Chandrajith, R., 2016. Trace elements in native and improved paddy rice from different climatic regions of Sri Lanka: implications for public health. SpringerPlus 5, 1864.

Elsheikh, R., Shariff, A.R., Amiri, F., Ahmed, N.B., Balasundarm, S.K., Soom, M.A., 2013. Agriculture Land Suitability Evaluator (ALSE): a decision and planning support tool for tropical and subtropical crops. Comput. Electron. Agric. 93, 98–110.

Erdogan, H.E., Yüksel, M., De La Rosa, D., 2003. Evaluation of sustainable land management using agro-ecological evaluation approach in Ceylanpinar State Farm (Turkey). Turk. J. Agric. For. 27, 15–22.

ESRI, 2013. ArcGIS 10.2. Environmental Systems Research Institute. Redlands, CA.

FAO, 1993. Guidelines for land-use planning FAO Development Series 1. Food and Agriculture Organization of the United Nations, Rome, Italy.

FAO, 2007. Global Assessment of Land Degradation and Improvement GLADA Report 5. FAO, Rome, Italy.

Farroni, A., Magaldi, D., Tallini, M., 2002. Total sediment transport by the rivers of Abruzzi (Central Italy): prediction with the RAIZAL model. Bull. Eng. Geol. Environ. 61, 121–127.

Garcia, G.J., Antonello, S.L., Magalhaes, M.G.M., Fonseca Filho, H., 2003. SIAT-land evaluation system-V.2.0. Cientifica Jaboticabal 31, 19–30.

Gessesse, B., Bewket, W., Bräuning, A., 2015. Model-based characterization and monitoring of runoff and soil erosion in response to land use/land cover changes in the Modjo watershed, Ethiopia. Land Degrad. Dev 26, 711–724.

Giger, M., Liniger, H., Sauter, C., Schwilch, G., 2015. Economic benefits and costs of sustainable land management technologies: an analysis of WOCAT's global data. Land Degrad. Dev http://dx.doi.org/10.1002/ldr.2429.

Greve, M.H., Bou Kheir, R., Greve, M.B., Bøcher, P.K., 2012. Quantifying the ability of environmental parameters to predict soil texture fractions using regression-tree model with GIS and LIDAR data: the case study of Denmark. Ecol. Indic. 18, 1–10.

Hanh, H.Q., Azadi, H., Dogot, T., Ton, V.D., Lebailly, P., 2016. Dynamics of agrarian systems and land use change in North Vietnam. Land Degrad. Dev http://dx.doi.org/10.1002/ldr.2609.

Haroun, O.R.M., 2004. Soil evaluation systems as a guide to identify an economical feasibility study for agricultural purposes in El-Fayoum Province. Ph.D. Thesis. Faculty of Agriculture, Cairo University, Egypt.

Hilt, K., Harrison, J., Bowers, K., Stevens, R., Bary, A., Harrison, K., 2016. Agronomic response of crops fertilized with struvite derived from dairy manure. Water Air Soil Pollut 227, 388.

Horn, R., Simota, C., Fleige, H., Dexter, A.R., Rajkay, K., De la Rosa, D., 2002. Prediction of soil strength of arable soils and stress dependent changes in ecological properties based on soil maps. J. Plant Nutr. 165, 235–239.

Hu, Z.Y., Zaho, F.J., McGrath, S.P., 2005. Sulphur fractionation in calcareous soils and bioavailability to plants. Plant Soil 268, 103–109.

Jacobsen, S.E., Jensen, C.R., Liu, F., 2012. Improving crop production in the arid Mediterranean climate. Field Crops Res 128, 34–47.

John, R., Murari, S., Mustafa, P., 2008. Long-term cereal-based rotation trials in the Mediterranean region: implications for cropping sustainability. Adv. Agron. 97, 273–319.

Kabata-Pendias, A., 2001. Trace Elements in Soil and Plants, third ed. CRC Press, Boca Raton, FL, USA, ISBN: 0-8493-1575-1.

Kalema, V.N., Witkowski, E.T., Erasmus, B.F., Mwavu, E.N., 2015. The impacts of changes in land use on woodlands in an equatorial African savanna. Land Degrad. Dev. 26, 632–641.

Kelgenbaeva, K., Buchroithner, M., 2007. The use of GIS for grain suitability studies in intramontane continental basins. In: International Cartographic Conference, at Moscow, Russia, p. 23.

Kim, M.S., Min, H.G., Lee, S.H., Kim, J.G., 2016. The effects of various amendments on trace element stabilization in acidic, neutral, and alkali soil with similar pollution index. PLoS One 11, e0166335.

Lal, R., 2008. Soils and sustainable agriculture. A review. Agron. Sustain. Dev. 28, 57–64.

Lal, R., 1997. Soil degradative effects of slope length and tillage method on alfisols in western Nigeria II. Soil chemical properties, plant nutrient loss and water quality. Land Degrad. Dev. 8, 221–244.

Lal, R., 2013. Climate-strategic agriculture and the water-soil-waste nexus. J. Plant Nutr. Soil Sci. 176, 479–493.

Lehmann, J., Kleber, M., 2015. The contentious nature of soil organic matter. Nature 528, 60–68.

Letey, J., Hoffman, G.J., Hopmans, J.W., Grattan, S.R., Suarez, D., Corwin, D.L., et al., 2011. Evaluation of soil salinity leaching requirement guidelines. Agric. Water Manage 98, 502–506.

Levine, E.R., Knox, R.G., Lawrence, W.T., 1994. Relationships between soil properties and vegetation at the Northern Experimental Forest, Howland, Maine. Remote Sens. Environ. 47, 231–241.

Lindsay, W.L., 1991. Iron oxide solubilization by organic matter and its effect on iron availability. Plant Soil 130, 27–34.

López García, J., Acosta, R., Bojórquez Serrano, J.I., 2006. Aptitud relativa agrícola del municipio de Tuxpan, Nayarit, utilizando el modelo Almagra del Sistema MicroLEIS. Investigaciones Geográficas 59, 59–73.

Louis, P., 2011. Erosion Impacts on Soil and Environmental Quality: Vertisols in the Highlands Region of Ethiopia. University of Florida, Soil and Water Science Department, USA.

Lugo-Morin, D.R., 2006. Evaluación del riesgo agroambiental de los suelos de las comunidades indígenas del estado Anzoátegui, Venezuela. Revista Ecosistemas., 16.

Lugo-Morin, D.R., Rey, J.C., 2009. Evaluación de la vulnerabilidad a la degradación agroambiental a través del uso del sistema MicroLEIS en los suelos de los llanos centrales de Venezuela. Revista Internacional de Contaminación Ambiental 25, 43–60.

Matinfar, H.R., Sarmadian, F., Alavipanah, S.K., 2011. Use of DEM and ASTER sensor data for soil and agricultural characterizing. Int. Agrophys. 25, 37–46.

McCauley, A., Jones, C., Jacobsen, J., 2009. Soil pH and organic matter. Nutrient Management Module No. 8. Montana State University.

Moura, E.G.D., Gehring, C., Braun, H., Ferraz Junior, A.D.S.L., Reis, F.D.O., Aguiar, A.D.C.F., 2016. Improving farming practices for sustainable soil use in the humid tropics and rainforest ecosystem health. Sustainability 8, 841.

Muñoz-Rojas, M., Abd-Elmabod, S.K., Zavala, L.M., De la Rosa, D., Jordán, A., 2016. Climate change impacts on soil organic carbon stocks of Mediterranean agricultural areas: a case study in Northern Egypt. Agric. Ecosyst. Environ. http://dx.doi.org/10.1016/j.agee.2016.09.001.

Muñoz-Rojas, M., Doro, L., Ledda, L., Francaviglia, R., 2015a. Application of CarboSOIL model to predict the effects of climate change on soil organic carbon stocks in agro-silvo-pastoral Mediterranean management. Agric. Ecosyst. Environ. 202, 8–16.

Muñoz-Rojas, M., Jordán, A., Zavala, L.M., Abd-Elmabod, S.K., De la Rosa, D., Anaya-Romero, M., 2012. Organic carbon stocks in Mediterranean soil types under different land uses (Southern Spain). Solid Earth 3, 375–386.

Muñoz-Rojas, M., Jordán, A., Zavala, L.M., De la Rosa, D., Abd-Elmabod, S.K., Anaya-Romero, M., 2015b. Impact of land use and land cover changes on organic carbon stocks in Mediterranean soils (1956–2007). Land Degrad. Dev 26, 168–179.

Muñoz-Rojas, M., Jordán, A., Zavala, L.M., González-Peñaloza, F.A., De la Rosa, D., Pino-Mejias, R., et al., 2013. Modelling soil organic carbon stocks in global change scenarios: a CarboSOIL application. Biogeosciences 10, 8253–8268.

Nwer, B.A.B., 2006. The application of land evaluation technique in the north-east of Libya. Ph.D. Thesis. Chapter 6, 154–182.

Olson, G.W., 1981. Soils and the Environment. Chapman & Hall, New York, 178 pp.

Overstreet, L.F., DeJong-Huges, J., 2009. The Importance of Soil Organic Matter in Cropping Systems of the Northern Great Plains. Available at: <http://www.extension.umn.edu/distribution/cropsystems/M1273.html> (accessed 19.05.16).

Oyonarte, C., Aranda, V., Durante, P., 2008. Soil surface properties in Mediterranean mountain ecosystems: effects of environmental factors and implications of management. Forest Ecol. Manag. 254, 156–165.

Qadir, M., Schubert, S., 2002. Degradation processes and nutrient constraints in sodic soils. Land Degrad. Dev 13, 275–294.

Recha, C.W., Mukopi, M.N., Otieno, J.O., 2015. Socio-economic determinants of adoption of rainwater harvesting and conservation techniques in semi-arid Tharaka sub-county, Kenya. Land Degrad. Dev 26, 765–773.

Reijonen, I., Metzler, M., Hartikainen, H., 2016. Impact of soil pH and organic matter on the chemical bioavailability of vanadium species: the underlying basis for risk assessment. Environ. Pollut. 210, 371–379.

Rengasamy, P., 2006. World salinization with emphasis on Australia. J. Exp. Bot. 57, 1017–1023.

Rongjiang, Y., Jingsong, Y., 2010. Quantitative evaluation of soil salinity and its spatial distribution using electromagnetic induction method. Agric. Water Manag 97, 1961–1970.

Rossiter, D.G., 1996. A theoretical framework for land evaluation. Geoderma. 72, 165–202.

Safari, Y., Delavar, M.A., Zhang, C., Esfandiarpour-Boroujeni, I., Owliaie, H.R., 2015. The influences of selected soil properties on Pb availability and its transfer to wheat (Triticum aestivum L.) in a polluted calcareous soil. Environ. Monit. Assess. 187, 1–10.

Salas-Mélia, D., Chauvin, F., Déqué, M., Douville, H., Gueremy, J.F., Marquet, P., et al., 2005. Description and validation of the CNRM-CM3 global coupled model, Centre National de Recherches Meteorologiques, Meteo France, working note 103.

Sánchez, P.A., Couto, W., Buol, S.W., 1982. The fertility capability soil classification system. Interpretation, applicability and modifications. Geoderma 27, 283–309.

Schulte, R.P.O., Creamer, R.E., Donnellan, T., Farrelly, N., Fealy, R., O'Donoghue, C., et al., 2014. Functional land management: a framework for managing soil-based ecosystem services for the sustainable intensification of agriculture. Environ. Sci. Policy 38, 45–58.

Shahbazi, F., De la Rosa, D., Anaya-Romero, M., Jafarzade, A., Sarmadian, F., Neyshaboury, M., et al., 2010. Land use planning in Ahar area (Iran) using MicroLEIS DSS. Int. Agrophys. 22, 277–286.

Sonneveld, B.G.J.S., 2003. Formalizing expert judgements in land degradation assessment: a case study for Ethiopia. Land Degrad. Dev 14, 347–361.

Sterk, G., 2003. Causes, consequences and control of wind erosion in Sahelian Africa: a review. Land Degrad. Dev. 14, 95–108.

Syers, J., De Vries, F., Myamudeza, P., 2001. The Sustainable Management of Vertisols. CABI Publishing, New York.

Sys, C., 1980. Land Characteristics and Qualities and Methods of Rating Them. ITC Post-Graduate Lecture Note. Ghent, Belgium.

Triantafilis, J., Ward, W.T., McBratney, A.B., 2001. Land suitability assessment in the Namoi Valley of Australia, using a continuous model. Soil Res. 39, 273–289.

USDA, Soil Survey Staff, 2014. Keys to Soil Taxonomy, 12th ed. USDA-Natural Resources.

USDA, 2010. Keys to Soil Taxonomy, eleventh ed. United State Department of Agriculture. Natural Resources Conservation Service (NRCS), Washington, DC.

Van Diepen, C.A., Van Keulen, H., Wolf, J., Berkhout, J.A.A., 1991. Land evaluation: from intuition to quantification. In: Stewart, B.A. (Ed.), Advances in Soil Science. Springer, New York, pp. 139–204.

Villen-Guzman, M., Paz-Garcia, J.M., Amaya-Santos, G., Rodríguez-Maroto, J.M., Gomez-Lahoz, C., 2015. Effects of the buffering capacity of the soil on the mobilization of heavy metals Equilibrium and kinetics. Chemosphere 131, 78–84.

Vink, A.P.A., 1983. Landscape Ecology and Land Use. Longman, London.

Warrence, N.J., Bauder, J.W., Pearson, K.E., 2002. Basics of Salinity and Sodicity Effects on Soil Physical Properties. Department of Land Resources and Environmental Sciences: Montana State University-Bozeman, Bozeman, MT.

Wildemeersch, J.C., Timmerman, E., Mazijn, B., Sabiou, M., Ibro, G., Garba, M., et al., 2015. Assessing the constraints to adopt water and soil conservation techniques in Tillaberi, Niger. Land Degrad. Dev 26, 491–501.

Willaarts, B.A., Oyonarte, C., Muñoz-Rojas, M., Ibáñez, J.J., Aguilera, P.A., 2016. Environmental factors controlling soil organic carbon stocks in two contrasting Mediterranean climatic areas of southern Spain. Land Degrad. Dev 27, 603–611.

Wilson, G.A., Kelly, C.L., Briassoulis, H., Ferrara, A., Quaranta, G., Salvia, R., et al., 2016. Social memory and the resilience of communities affected by land degradation. Land Degrad. Dev http://dx.doi.org/10.1002/ldr.2669.

Wolfgramm, B., Shigaeva, J., Dear, C., 2015. The research–action interface in sustainable land management in Kyrgyzstan and Tajikistan: challenges and recommendations. Land Degrad. Dev 26, 480–490.

Yami, M., Snyder, K.A., 2015. After all, land belongs to the state: examining the benefits of land registration for smallholders in Ethiopia. Land Degrad. Dev http://dx.doi.org/10.1002/ldr.2371.

Yeshaneh, E., Salinas, J.L., Blöschl, G., 2015. Decadal trends of soil loss and runoff in the Koga catchment, northwestern Ethiopia. Land Degrad. Dev http://dx.doi.org/10.1002/ldr.2375.

Zeng, F., Ali, S., Zhang, H., Ouyang, Y., Qiu, B., Wu, F., et al., 2011. The influence of pH and organic matter content in paddy soil on heavy metal availability and their uptake by rice plants. Environ. Pollut. 159, 84–91.

CHAPTER

8

Soil and Land Use in the Alps—Challenges and Examples of Soil-Survey and Soil-Data Use to Support Sustainable Development

Clemens Geitner[1], *Jasmin Baruck*[1], *Michele Freppaz*[2], *Danilo Godone*[3], *Sven Grashey-Jansen*[4], *Fabian E. Gruber*[1], *Kati Heinrich*[5], *Andreas Papritz*[6], *Alois Simon*[7], *Silvia Stanchi*[2], *Robert Traidl*[8], *Nina von Albertini*[9] *and Borut Vrščaj*[10]

[1]University of Innsbruck, Innsbruck, Austria [2]University of Turin, Grugliasco, Italy [3]Geohazard Monitoring Group, CNR IRPI, Turin, Italy [4]University of Augsburg, Augsburg, Germany [5]Institute for Interdisciplinary Mountain Research, Austrian Academy of Sciences, Innsbruck, Austria [6]ETH Zurich, Zurich, Switzerland [7]Provincial Government of Tyrol, Innsbruck, Austria [8]Bavarian Environmental Agency, Marktredwitz, Germany [9]Umwelt Boden Bau, Paspels, Switzerland [10]Agricultural Institute of Slovenia, Ljubljana, Slovenia

INTRODUCTION

The present chapter broaches the issue of soil and land use in the Alps, focusing on their diversity and the related consequences for soil mapping, modeling, and managing. In recent decades the Alps as well as other mountain regions have been faced with both land-use and climate changes that significantly affect soils. Therefore soil protection and management strategies are in demand. Nevertheless, a precondition for this is that soil information is available and features both an adequate scale as well as an applicable form (Brevik et al., 2016). As this is not the case in many regions in the Alps, we first of all aim at providing insight into land-use-related soil issues. Second, we discuss which soil data are necessary to satisfy soil management requirements and how to obtain the relevant information in the Alps by mapping, measuring, and modeling soils and their properties.

Soil Mapping and Process Modeling for Sustainable Land Use Management.
DOI: http://dx.doi.org/10.1016/B978-0-12-805200-6.00008-6

The Alps and Their Soils

The Alps, the highest mountain range in Europe, extending over eight countries, are located in a transition zone from temperate to Mediterranean climates. In terms of soil formation the Alps hold a singular position as the intensity and relative importance of soil-forming factors differ considerably from those of other landscapes (Egli et al., 2006, 2008, 2009, 2010, 2014; Price and Harden, 2013; Stöhr, 2007). Alpine soils are prone to disturbances by natural processes to a high degree (Hagedorn et al., 2010) but are also strongly influenced by ancient and current human activities (Bätzing, 2015; FAO, 2015; Geitner, 2010; Hagedorn et al., 2010).

Recently, two comprehensive overviews have been published on mountain soils (FAO, 2015) and soils in the Alps (Baruck et al., 2016). Both reviews focus on the special characteristics of the development and pattern of Alpine soils, including available soil information, soil classification, and soil mapping in the Alpine area. Moreover, in the FAO contribution to the International Year of Soil 2015, mountain soils are discussed from a global perspective in terms of human activities, climate change, related threats, and cultural heritage (FAO, 2015).

Soil-forming conditions in the Alpine environment, including sites from the valleys up to the mountain peaks, are characterized by: (1) a wide range of climatic regimes from north to south and west to east (Schär et al., 1998); due to topography, altitudinal and aspect-related changes and variations in temperature, precipitation, and wind, (2) a climatic elevation gradient with distinct vegetation belts denoting nine different vegetation zones (Grabherr, 1997; Theurillat et al., 1998), (3) very high topographical variability at all scales (Egli et al., 2005, 2006; Geitner et al., 2011b), determining the meso- and microclimate as well as the local water budget, (4) the steep relief favors strong morphodynamics, in particular gravitational and fluvial processes, (5) great spatial variability of parent materials with a high proportion of young unconsolidated deposits, predominantly from the Pleistocene period with glacial, periglacial, and eolian deposits, and (6) highly diverse (historical) land-use practices with patterns that are variable over short spatial ranges.

Due to both the strong Pleistocene impact and the general exposure to morphodynamic processes, "time" must be considered as a special soil-forming condition in the Alps. The material removal, transport, and accumulation processes along the slopes determine soil genesis and soil depth, so that well-developed soils, even on the same parent material, may occur in the direct neighborhood of initial soils (Baize and Roque, 1998; Minghetti et al., 1997; Sartori et al., 2001).

Given this great variety of soil-forming conditions, the inherent properties and spatial distribution of Alpine soils are characterized by

- high variability over very short spatial scales, leading to complex patterns of soil characteristics (Egli et al., 2005; FAO, 2015; Geitner, 2007; Hagedorn et al., 2010; Theurillat et al., 1998; Veit, 2002);
- typical elevation gradients of some soil properties—by and large the portions of fine grain sizes, pH values, exchange capacity, clay minerals, the stability of aggregates, and the incorporation of organic matter decrease with altitude (Djukic et al., 2010; Egli et al., 2003; FAO, 2015; Veit, 2002; Zech et al., 2014);
- differences with regard to the aspect and the associated local climate conditions and water budgets (Egli et al., 2006; Sartori et al., 2005; Zech et al., 2014); and
- multilayered soil profiles and buried soils, which are quite common in this environment especially at geomorphologically active sites (Baruck et al., 2016; FAO, 2015; Geitner et al., 2011a; Veit et al., 2002).

Based on these conditions, specific problems arise when surveying but especially modeling and interpreting soils in the Alpine environment. The proper extrapolation from point to area (from the profile site to a cartographic unit) remains challenging (Baruck et al., 2016). However, knowledge of soil properties and soil patterns is an essential prerequisite for sustainable land-use management (Brevik et al., 2017; Pereira et al., 2017). The mosaic-like juxtaposition of soil types is a relevant issue at all scales as the high spatial variability also determines soil functions and therefore also soil-related services for society (FAO, 2015).

Land Use and Land-Use Changes in the Alps

Land use and land-use changes in the Alps are issues of great temporal and spatial complexity. Thus herein we can only present a rough overview of development trends and relevant regional differences. Thereby, we focus on the primary sector, as agriculture and forestry account for the largest areas used in the Alps. Furthermore, agriculture and forestry use the soil not only as a space but also in terms of its wealth of ecological functions. However, in examining agriculture and forestry in the Alps, we have to realize that both operate working under difficult and often limiting conditions due to the unfavorable climatic and topographical situation.

Agriculture in the Alps strongly depends on climatic conditions that are quite different due to the position of the Alps in the transition zone from temperate to Mediterranean and from oceanic to a more continental climate. In addition, some very dry inner-Alpine regions where irrigation measures are common must be considered (Grashey-Jansen, 2014). Furthermore the regional climates are modified by the elevation levels as well as by the position, in particular by aspect and inclination, within the pronounced topography. These climatic and topographical conditions resulted in specific local and regional patterns of land use, including a vertical stratification with the typical highest tier above the tree line that is only used for grazing seasonally. The traditional agricultural land use led to diversity as regards the cultural landscape that has been only partially preserved to the present day. However, land-use practices are determined not only by environmental conditions, but also by cultural roots, which in the Alps are just as diverse.

For centuries or even millennia, Alpine agriculture had to ensure the self-sufficiency of the local people. This meant that everything that was needed had to be produced on the farm, causing a highly diverse agrarian landscape with small units and a mosaic of fields, grassland, and woody plants. In most Alpine regions, this situation changed at the latest in the middle of the 20th century, because the development of the transportation sector now allows farmers to be market- and profit-oriented. Thus today most of the farmers in the Alps concentrate on one type of land use (e.g., grassland or arable farming or fruit growing). While monocultures are predominant in many areas, in rare cases there also exist mixed holdings with several products.

Forests are a formative landscape element in the Alps, as almost half (45.5%) of the whole area is covered by woodlands, whereas Slovenia (63.6%) and Austria (53.5%) have the highest values (von Andrian-Werburg et al., 2008). In recent decades, in many regions already starting shortly after World War II, forested areas have been steadily increasing, mainly due to retreating agricultural activities (Brändli, 2000; Tappeiner et al., 2008a,b; Veit, 2002). The tree line runs at altitudes between 1500 and 2300 m and is controlled mostly by human impacts but at its highest positions is limited by climatic conditions. Currently, in many regions the tree line is shifting upwards, causing a considerable change in soils and their organic matter. However, we must bear in mind that the Alpine woodlands were reduced significantly in previous centuries due to expanding settlements and mining activities with an enormous demand for wood. In these times, erosion and natural hazard events significantly increased. Even though the current percentage of woodlands is quite high again, the naturalness and age structure are often insufficient to ensure all forest functions (e.g., Grabherr et al., 1998; BMLFUW, 2015).

By means of two examples of the comprehensive map collection "Mapping the Alps," edited by Tappeiner et al. (2008a), we will illustrate the spatial heterogeneity of current agricultural use as well as the differences between general regional development trends.

In general, today Alpine agriculture is dominated by grassland and livestock farming (approx. 80%, all percent values regarding these types of farming refer to the total number of farms in the Alps) (Fig. 8.1), as under the widespread cold and wet conditions, especially at higher altitudes, this is the most suitable form of farming. The few arable fields left are mainly used to grow silage maize as feed. In a few parts of the Eastern and Western Alps the trend moves in the opposite direction, as there arable farming was intensified by specializing and optimizing production. However, purely arable farming accounts for less than 3%. In some of the inner-Alpine longitudinal valleys with drier and warmer conditions, as well as on the southern Alpine rim, farmers established permanent crops (10%, mainly grapes and apples). Mixed holdings (around 5%) as well as farms specialized in finishing animals (<2%) can be found near larger towns and agglomerations. In contrast to these intensification trends, in some regions of the Southern Alps many agricultural areas were abandoned completely and have already been reforested (Hoffmann et al., 2008). Across the Alpine space, 40% of agricultural enterprises were abandoned between 1980 and 2000. Accordingly, an average of 20% of the agricultural areas lay fallow (Streifeneder et al., 2007), mainly concerning sites with marginal yields as well as steep or inaccessible slopes (Tasser et al., 2007). Regarding this, regional differences in the Alps are also strongly influenced by national subsidies for mountain agriculture (Penz, 2005). In summary, this means that in recent years Alpine landscapes have been subject to greater intensification, on the one hand, as well as to abandonment and reforestation, on the other. This significant polarization, of course, also has strong effects on soils and their recent changes and threats.

Fig. 8.2 differentiates the Alpine municipalities with regard to their development, illustrating the results of a cluster analysis that considers selected indicators, covering all areas of sustainability (Gramm et al., 2008). The determination is based on statistical data on population, agriculture, tourism, transport infrastructure, and commuter balance. Without going into too much detail the spatial patterns of the map confirm a polarization between prosperous areas (cities and agglomerations as well as major tourism areas) and less favored areas (peripheral areas), which leads to significant land consumption for development, on the one hand, and extensification of land use and marginalization, on the other. The most sought-after land is located in the broad valley areas, causing land-use conflicts between agriculture and development for settlements, industrial parks, and transport infrastructure with extensive soil sealing. Land consumption at higher elevations is mainly concentrated in large tourism centers with extensive hotel complexes and, accordingly, a vast range of ski runs that have specific impacts on vegetation and soil (Freppaz et al., 2013; Roux-Fouillet et al., 2011). In terms of soil threats, summer tourism activities play a minor or only a very local role. Nevertheless Grabherr (1982) examined the impact of trampling by tourists on a high-altitude grassland and the various species thereof in the Tyrolean Alps. Originating from the frequent use of these pathways (increasingly also by bicycles), local erosion processes can be quite strong (Veit, 2002).

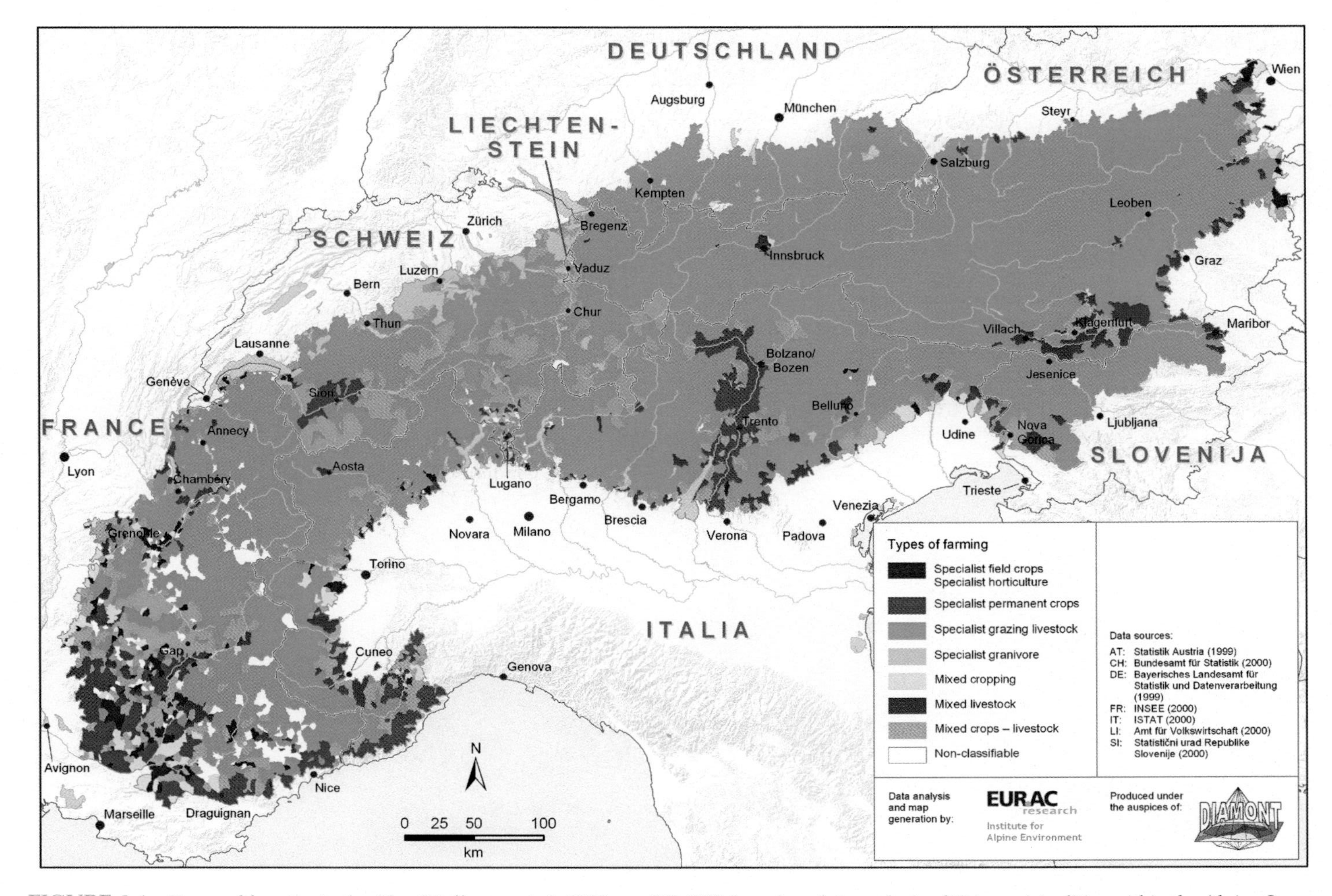

FIGURE 8.1 Types of farming in the Alps (Hoffmann et al., 2008, pp. 198–199); based on data analysis of 134 municipalities within the Alpine Space, delimited according to the Alpine Convention.

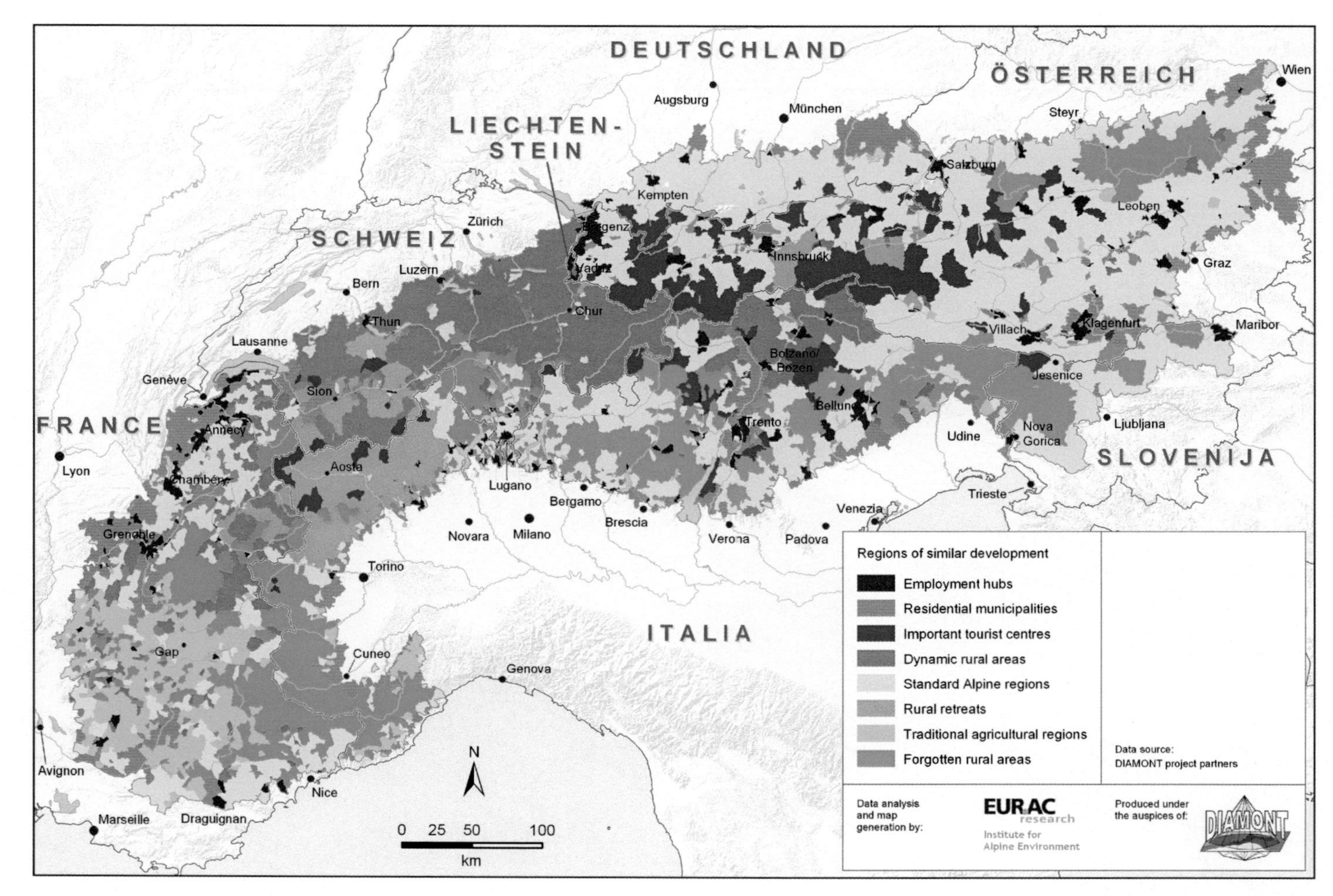

FIGURE 8.2 Regions of similar development in the Alps (Gramm et al., 2008, pp. 272–274); based on data analysis of 134 municipalities within the Alpine Space, delimited according to the Alpine Convention.

In spite of local and regional features, by focusing on land use and soils, the development types in Fig. 8.2 may help us assess potential soil changes and threats within different municipalities and regions. Furthermore, this map provides a link to the case studies presented and discussed in "Soils and Land Use in the Alps" and "Mapping and Modeling Soils in the Alps.".

A Conceptual Framework of Soil–Society Relations

With regard to the Alps, we have to realize that, except for the highest elevations, we are dealing with cultural landscapes. The soils in the Alps have experienced some form of human impact, even at sites that seem near-natural today. This is because human land use started thousands of years ago, usually increasing in the Middle Ages and the Modern Era. Nevertheless most of these traditional land-use activities came to an end within recent decades or centuries. However, soils often have a "long memory" and traces of former land-use practices can be found in soils, especially in their structure and morphology, but also in terms of their physical and chemical properties. Consequently, according to Richter and Tugel (2012, p. 38.1), we should consider soils, even in the Alps, as a "cultural-historic-natural system", and learn to look at them as a "human-natural body" (Richter and Yaalon, 2012, p. 766; see also Brevik and Arnold, 2015). With regard to such, McNeill and Winiwarter (2006, p. 4) point to a clear need for further research by emphasizing the following: "The history of soils is perhaps the most neglected subject within environmental history". That is just as true for the Alps.

Fig. 8.3 provides a conceptual framework to differentiate soil–society relations. As a common theme, it may provide orientation and help to better identify relevant aspects of soil–society

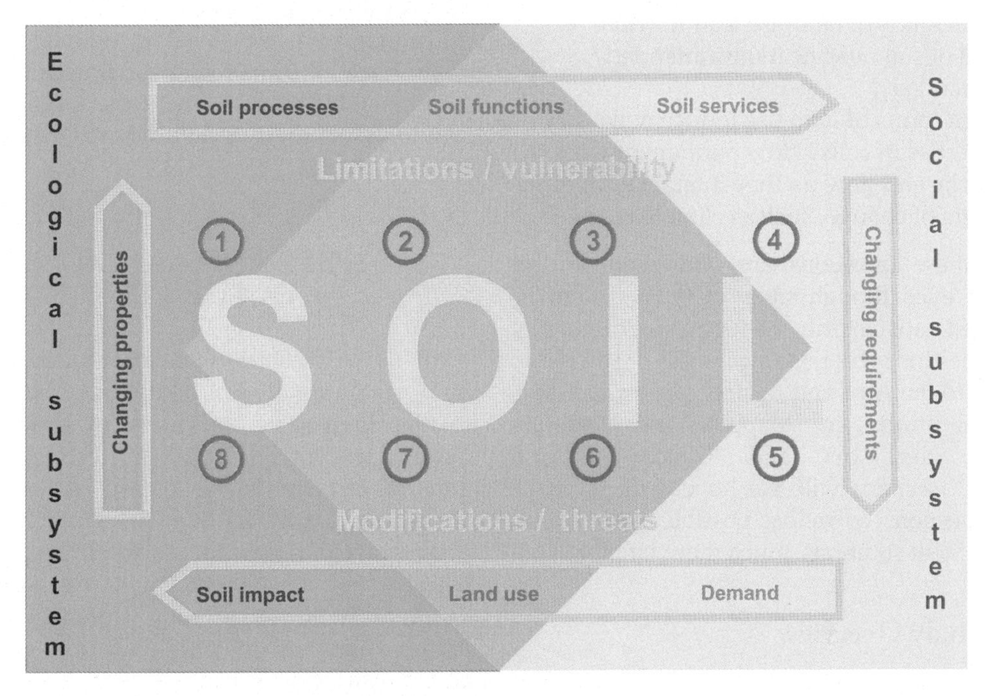

FIGURE 8.3 A conceptual framework of soil–society relations.

relations and to link them to the case studies in the present section. Basically, this concept can be applied from different perspectives, e.g., with regard to individual soil units, to a certain land-use type, to a specific region and elevation zone, or to a given (historical) period. By applying this framework we can formulate the following eight key questions in terms of Alpine soils and land-use practices (indicated in the figure by their number in the relevant sector):

1. What are the typical soil features in the Alps?
2. Which soil properties are limiting for certain forms of land use?
3. Which properties make the soils vulnerable to a particular type of land use?
4. What are the specific services that soils provide for society?
5. What are the relevant (historical or current) changes in societal needs in terms of soils and related land use?
6. Which kinds of land use lead to which specific threats for soils?
7. Which kinds of land use lead to what kinds of soil modifications (intended/ unintended)?
8. Which forms of land use induce new properties in soils? How permanent will these be and how do they interact with other drivers of change, such as climate change?

All these questions are addressed in this chapter, even though some of them cannot be answered satisfactorily. Nevertheless discussing these questions will raise awareness of soil issues specific to the Alps and related gaps in soil data. The case studies in "Mapping and Modeling Soils in the Alps" and "Conclusions and Outlook" sections will, we hope, help to make these questions, as well as possible answers and management solutions, much more tangible.

Case Study Overview

The value of the present chapter is that it provides a mixed bag of examples of how soils and land-use issues are closely interlinked in the Alps. These case studies are divided into two groups. Without claiming to be exhaustive, the examples in "Soils and Land Use in the Alps" section illustrate the relations between soils and land use in different landscape units. In doing this, we also include a historical perspective because soils often still reflect former impacts. In "Mapping and Modeling Soils in the Alps" section we present examples from scientific work, aiming to capture soil information in the Alps in order to make it available at a particular scale and for a specific purpose. In this sense the methodical approaches in "Mapping and Modeling Soils in the Alps" section should provide responses to some of the challenges that arise in "Soils and Land Use in the Alps" section.

In Fig. 8.4 the location of the case studies from both groups are marked in different colors; additional information about them is provided in the corresponding Table 8.1. The case studies come from five different countries and range from very local to regional and may even have a trans-regional dimension. They also consider several landscape units and elevations with their specific spatial pattern and land-use forms.

SOILS AND LAND USE IN THE ALPS

In the Alps various relations between the land use and soils can be found, which is due to the long-term traditions of the cultural landscape. Precisely because these relations in general, but even more so their regional influences, are barely known, the current section provides examples for those links between land use and soils, including case studies from different regions, reaching from the bottom of the valley up to high elevation grassland and considering recent as well as historical aspects.

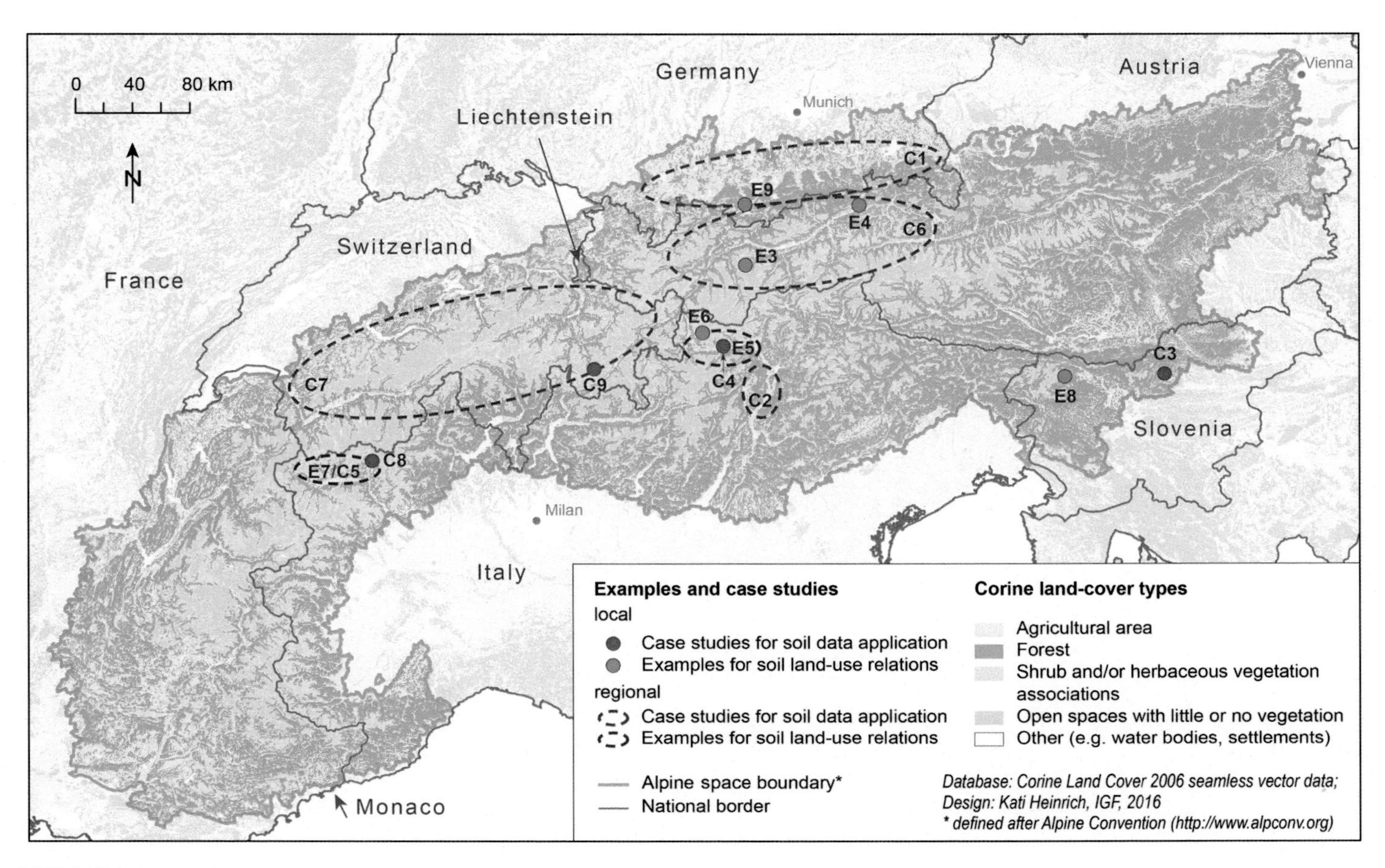

FIGURE 8.4 Localization of the examples and case studies of this chapter in the Alps.

TABLE 8.1 Brief Characteristics of the Examples and Case Studies of this Chapter

No./Page	Localization	Scale/Timeframe	Main Topic	Questions[a]
E1/235	*Not further localized*	–/past	Deforestation by mining	**6**, 7, 8
E2/235	*Not further localized*	–/past	Litter removal in forests	6, **7**, 8
E3/235	Niederthai/Umhausen, Oetz Valley, Austria (ca. 1600 m)	Local/past	Erosion/accumulation at steep arable lands, irrigation sediments	**5**, 6, **7**, **8**
E4/236	Municipality of Wörgl, Inn Valley, Austria (ca. 500 m)	Local/past–present	Spatial changes of forests, agriculture and settlements on valley floors	1, 2, 5, 7
E5/238	Vinschgau, Etsch-Valley, South Tyrol, Italy (ca. 600–1000 m)	Regional/present	Geomorphological structure, land-use change, dominance of vineyards and orchards	1, 2, 5, **7**, 8
E6/239	Laaser Leiten, Vinschgau, South Tyrol, Italy (ca. 1000–1500 m)	Local/past–present	Deforestation, grazing land, soil degradation, reforestation	3, 5, **6**, 8
E7/241	Aosta Valley Region, Italy (ca. 600–1200 m)	Regional/past–present	Impact of land management and reshaping in terraced agroecosystems	2, **5**, 7
E8/242	Pokljuka Plateau, Slovenia (ca. 900–1300 m)	Local/past–present	Soil diversity and its relation to land use (forests and pastures)	**1**, 6, **7**
E9/246	Zugspitz Region, Bavaria, Germany (ca. 2000–2500 m)	Local/present	Soil diversity due to parent material and grazing impacts	**1**, 2, 7, 8
C1/253	Alpine regions of Bavaria, Germany (ca. 800–2600 m)	Regional/present	The potential of soil overview maps as a basis for planning	2, **3**, **4**
C2/254	Study areas in South Tyrol, Italy (ca. 300–1400 m)	Regional/present	The potential of a layer-structured soil information system	2, 3, 4
C3/260	Municipality of Braslovče, Savinja Valley, Slovenia (ca. 300–900 m)	Local/present	Upgrading overview soil information for local spatial planning documents	2, 3, 4
C4/265	Test sites in Vinschgau, Etsch-Valley, South Tyrol, Italy (ca. 600–1000 m)	Local/present	Detailed mapping and measurements of soil properties for water management in orchards	**2**, 7
C5/269	Aosta Valley Region, Italy (ca. 600–1200 m)	Regional/present	Mapping land suitability for mountain viticulture	2, 3, 4
C6/271	North Tyrol, Austria (ca. 500–2300 m)	Regional/present	Considering soil characteristics in the designation of protection forests	2, 4
C7/274	Alpine regions of Switzerland (ca. 500–2300 m)	Regional-national/present	Geostatistical modeling of forest soil carbon stocks	**4**, 7, 8
C8/276	Monterosa Ski Resort, Aosta Valley Region, Italy (ca. 1500–3500 m)	Local-regional/present	Mapping susceptibility to snow gliding	6, 7
C9/278	Julier Pass, Graubünden, Switzerland (ca. 1900–2200 m)	Local/present	Using soil and ecological information to ensure soil protection within a construction project	1, 3, 4

E1-9 = examples of soil land-use relation; C1-9 = case studies of soil-data application.

[a]*Numbers related to the conceptual framework in "A Conceptual Framework of Soil–Society Relations" section bold numbers refer to the main question, Fig. 8.3.*

Traditional Land-Use Types and Related Long-Term Impacts on Soils

There are some historical land-use forms in the Alps that are almost forgotten today but which have left significant traces in the soils. The most important ones will be described and partly illustrated here, since this knowledge is crucial for understanding soil features and patterns in Alpine landscapes. Of the forested area in the Alps, we must bear in mind that nowadays it is significantly more widespread than in previous centuries. Notably, during the Middle Ages large-scale forest clearing took place in the Alps, mainly as a result of population growth and the related expansion of settlements. Moreover, the specific geological settings meant that in many regions of the Alps mining took place, in some cases since prehistoric times, which reached a peak during the Modern Period. Mining activities and related processing created an enormous demand for wood. Exploitative forest clearing increased the danger of natural hazards, such as erosion, flooding, debris flows, and avalanches. In terms of soils, these morphodynamic processes entail a translocation of soil material and unconsolidated sediments, resulting in a confusing pattern of buried, disturbed, and initial soils. The latter can also be found on the tailings within mining areas. Nevertheless the location and extent of these anthropogenic accumulation sites are often unknown today.

Apart from wood exploitation, some secondary forest uses were historically widespread in the Alps. In terms of soils, forest pasture activities as well as the practice of litter removal had the strongest impact (Schiechtl, 1969; Schiechtl and Neuwinger, 1980; Stuber and Bürgi, 2012). The purpose of the latter was to obtain organic material as livestock bedding, which was particularly common in those regions in the Alps where straw was not available. However, by raking the litter from the forest floor, all organic layers and even the topsoils were often removed and thus soils were greatly disturbed. Performing this every few years and over centuries degraded the soil humus severely and reduced the nutrient supply considerably (Bitterlich, 1991; Stuber and Bürgi, 2012). Even though this form of forest use was abandoned at the latest three or four decades ago, in many cases the traces of these continuous disturbances of forest soils are quite recognizable today and must be taken into consideration when mapping and evaluating soils.

In order to ensure self-sufficiency, arable farming in the Alps was expanded towards the altitudinal limit, frequently using the favorable radiation conditions of slopes with a southern aspect. However, most of the agricultural sites in the montane level are situated on steep slopes, where the plowed soils were strongly affected by erosion. With every heavy rainfall soil material was gradually displaced downhill and had to be carried back up the slope (Fig. 8.5). Despite these efforts, over time more and more soil material was accumulated at the foot of slopes and in sinks (Geitner, unpublished soil surveys in Niederthai, Oetz Valley; see also Geitner, 1999; Thalheimer, 2006). Nowadays, arable farming has normally been abandoned and replaced by grassland in such extreme positions in the Alps. However, these former accumulation areas with thick deposits and related soils rich in humus and nutrients can be verified easily in the field. According to the world reference base for soil resources (WRB), these anthropogenic soils can be classified as *Terric* or *Regic Anthrosols*.

Given its pronounced topography, it is characteristic of the Alps that the amount of annual precipitation ranges widely. Within the continental inner-Alpine zone annual precipitation is as low as 700–500mm in many places. In these regions irrigation measures have traditionally been widespread. For example, in the Ötz valley region intensive use of meadows with artificial irrigation is documented not only in the Middle Ages (Geitner, 1999), but as early as the Bronze Age (Oeggl et al.,

FIGURE 8.5 (A) At this south-facing slope in Niederthai (approx. 1550 m, Oetz Valley) the microrelief reveals the plots of former arable land. The area at the foot of the slope is dominated by soils developed from anthropogenic sediments, which are rich in organic matter and nutrients. (B) This photograph illustrates the arduous task of returning the soil to the upper part of the field ("Erd-Auftragn"), Juns in the Tux Valley (Tyrol) in 1943. *Source: (A) photo: C. Geitner; (B) photo: E. Hubatschek, ©Edition Hubatschek.*

1997). Today examples of such land-use measures in the Ötztal are only found in a handful of places (Fig. 8.6), while in drier Alpine regions (e.g., in South Tyrol) they are still common. Traditional irrigation practices are associated with the input of fine-grained mineral particles (Fig. 8.6). Their accumulation over a long period has improved the ecological conditions, especially the water regime, of these sites in the longer term, so that today agriculture continues to benefit therefrom. For mapping and understanding soil properties, we have to consider and possibly reconstruct these historical irrigation practices with the resultant sediments. According to the WRB, this anthropogenic soil can be referred to as *Irragric Anthrosol*.

Former and Recent Land-Use Changes on a Valley Floor (Inn Valley, Austria)

Today, most of the valley floors in the Alps are sought-after areas, not only for the agricultural sector but also as sites for settlements, industry, and infrastructure. Given this demand and the scarcity of flat areas, land-use planning faces special challenges in these regions. Nevertheless the properties and functions of such sites near rivers have changed fundamentally during the last 200 years, as can be illustrated by the small town of Wörgl in the Lower Inn Valley (Tyrol, Austria, see Fig. 8.4 and Table 8.1). Here, the Cadastral Map of Francis I (the land register, 1817–61) turns out to be an essential document with comprehensive plot-specific land-use maps on a scale of 1:2880. They indicate the environmental and economic conditions of the Habsburg monarchy in the first half of the 19th century. In Alpine regions most of the land-use patterns from around 150 years ago can be assumed to largely remain, in most instances since the medieval expansion of settlements. Thus these maps, which refer to long-term land use, can be used to assess past soil modifications as a result of specific land use in the past on a detailed scale.

FIGURE 8.6 (A) The position of the soil profile (C) in the upper part of a formerly irrigated meadow (Grastalfeld, Oetz Valley). (B) One of the last examples of traditional meadow irrigation in the Upper Oetz Valley near Obergurgl: where water is channeled into the meadow, the accumulation of suspended matter produces distinct ridges in the terrain. (C) Soil profile located on an irrigation ridge (Grastalfeld, 1720 m, Oetz Valley). As a result of artificial water discharge, a thick layer of fine-grained material has developed over the centuries. The lower, dark, humus-, and skeleton-rich fossil A-horizon reveals the original surface of this terrain. At its upper limit there are distinct concentrations of charcoal, which 14C dating has identified as medieval. The discernible current soil formation in the uppermost part of the profile suggests that irrigation was abandoned several decades ago. *Source: (A) photo: C. Geitner; (B) photo: D. Kreuzer; (C) photo: C. Geitner.*

Assessments of the land register map of Wörgl from 1855 by Geitner et al. (2007) reveal that before that time the floodplain was only used extensively for forests and meadows. This was on account of the dynamics of the Inn River, which caused perennial rises in the groundwater level and, more crucially, repeated extensive flooding. Consequently, arable land was mainly located around the settlement at the slightly elevated alluvial fans and thus restricted to rather unfavorable skeleton-rich soils. In Wörgl these areas around the town were traditionally used for ley farming, i.e., alternating between growing field crops and grass on a given plot.

However, the construction of the railway line (from Kufstein to Innsbruck) in this section of the valley between 1855 and 1858 had a strong environmental impact: In the course of this construction project, comprehensive flood protection and melioration measures in the floodplain were realized. This meant that arable land could be extended to floodplain areas. In terms of soils the farmers could now use the high potential of these riverside *Fluvisols* (from fine-grained and carbonate-rich fluvial sediments). For at least a century the floodplains provided the most important plots for arable farming. At the same time the population was growing fast and the settlement area began to cover more and more of the alluvial fan.

Nowadays the situation has again changed radically. In recent decades the importance of agriculture for food self-sufficiency has drastically declined while demand for building land has increased, giving rise to new pressure on these floodplain areas. In order to support and develop the local economy the municipalities encourage incoming industry and business. As a result, floodplain plots with the most productive soils are being sealed one by one. In this way, not only the agricultural potential, but also all other soil functions become irretrievably lost, which will cause new problems related to water and matter balance in the near future.

These fundamental land-use changes within the municipality of Wörgl represent a typical dynamic of the riverside regions of the Alps over the last 150 years. The local characteristics and specific features of these land-use changes and their impact on soils may, however, differ from one locality to another and have to be studied in detail in order to compare land-use impacts on the soils on valley floors in the Alps over time and across regions.

Geomorphic Structure and Land Use in an Inner-Alpine Dry Region (Etsch Valley, Italy)

Due to their Late- and Postglacial pedogenesis, all soils in the Alps are quite young. Additionally, annual precipitation of less than 500 mm in places also contributes to diminished soil development. The spatial distribution of soil types within the Upper and Middle Etsch Valley in South Tyrol (Fig. 8.7) is very heterogeneous and shows small-scale variations (Grashey-Jansen, 2014). Hillsides are mostly dominated by *Leptosols*, *Regosols*, and *Cambisols*. Due to land-use impacts at these sites, soils are influenced by erosion processes here and there (see "Soil degradation and reforestation on the slopes of an inner-Alpine dry region (Etsch Valley, Italy)" section). In some places, previous erosion caused the accumulation of colluvium on the lower slopes. Deep developed *Cambisols* with high nutrient content can be found on the Late- and Postglacial accumulations of the alluvial fans and cones of debris due to the well-weathered parent material. The landscape in the Upper Etsch Valley (Val Venosta) is especially characterized by numerous and voluminous debris cones. Well-stratified soils with hydromorphic gleyic properties are often located between or in front of these cones due to the tailback of the groundwater flow (Fig. 8.7; Grashey-Jansen and Schröder, 2009; Fischer, 1966, 1990). The hydromorphic soils at valley bottom sites tend to include paleo-residues of plants and fauna (Kleen et al., 2016), which can be used for reconstructing the development of the landscape.

Furthermore, it is important to note that a large part of the soils of the Etsch Valley are strongly influenced by land use, in particular agriculture. The present distribution of different soils in the valley floor has to be seen in the light of the straightening and regulation of the river and its tributaries during the 18th and 19th centuries. In addition, many sites were drained and ameliorated. Most soils have been modified profoundly through cultivation, above all by plugging and the addition of non-parent material. Thus many soils must be classified as *Anthrosols*.

For more than 1000 years wine grapes have been grown on the thermally favored hillsides of the debris cones. Before the regulation of the river, only these valley sites were protected against flooding. The dry and stone-rich soils provide the best growing conditions for the

FIGURE 8.7 A stony Regosol adjacent to a well-stratified Fluvisol in the Upper Etsch Valley. *Source: photos: S. Grashey-Jansen.*

modest grapevines. Furthermore, these sites are unsuitable for other agriculture use. However, these sites were usually strongly modified up to a soil depth of 50–100cm by deep plowing in order to improve soil water retention (Wittmann, 2006). Due to the higher aeration and warming the organic matter content is usually significantly lower than in undisturbed soils. High skeleton content reduces water-holding capacity, but enables deep rooting of grapevines. Furthermore, the stony substrate has a positive effect on the thermal balance due to its good thermal conductivity. Nevertheless, during some phenolical stages of the grapevines the edaphic dryness of these soils must be compensated for by irrigation.

In contrast the roots of the apple trees are more horizontally oriented and need less vertical root-space. Thus the soil properties of the upper soil horizons are more decisive. Most South Tyrolean orchards are located in the flat valley floor with well-stratified and deep soils where root penetration is no problem. However, these soils also have been strongly modified by agriculture and orcharding. But such changes are mostly limited to the first 40–50cm within the soil profiles (Grashey-Jansen, 2008). Due to the fact that these soils were not usable before the regulation of the Etsch River, the human impact is comparatively young and, in relation to soil structure changes, also less intensive than in the vineyards (Werth, 2003).

Soil Degradation and Reforestation on the Slopes of an Inner-Alpine Dry Region (Etsch Valley, Italy)

A characteristic feature of the Upper Etsch Valley (Vinschgau) is the extremely different hillsides due to aspect. In principle the northern exposed hillsides are more heavily vegetated—they are covered by coniferous forests with podzolic soils. The southern exposed slopes are overgrown with reforested woods and insular dry grasslands with undifferentiated soil development. The Vinschgau region was already populated 7000 years ago, thus changing from a natural to a cultural landscape. Besides the usual wood utilization the southern exposed hillsides have been intensively used as grazing land for sheep and goats. In combination with the sub-Mediterranean climate a kind of steppic vegetation has become dominant. Periods of intensive forest degradation and deforestation were followed by periods of natural regeneration and subsequently by different reforestation programs for some parts of the hillside. Reforestation activities started already in 1884, 2 years after devastating erosion and flooding events (Grashey-Jansen and Schroeder, 2011). These reforestation measures were carried out in order to counteract landslides and avalanches, which have repeatedly threatened the villages in the valley. The reforestation has caused a significant vegetation contrast between the reforested woodland and xeric grassland with steppic species on the same hillside.

The soil textures at these sites are dominated by coarse-grained or fine-grained unconsolidated substrates. However, all soils on the southern exposed hillside are characterized by weak development without any significant profile differentiation. Conversely, at less steep sites, some remarkable soil depths of 140 cm and more can be found, caused by colluvial processes (Grashey-Jansen and Schroeder, 2012). The darker soil color of the forest soil profiles (Fig. 8.8) is attributed to the retarded decomposition of coniferous litter and the formation of humic and fulvic acids. However, there are no significant differences in the pH values of the reforested and the open areas and podzolic processes could not be detected (Grashey-Jansen, 2012a). This may be explained by the regional climatic conditions with very low annual precipitation but also by the treetops, which shield the soil surface from precipitation. The high proportion of sandy fractions and the intensive root penetration offers good infiltration conditions. So the minor soil moisture of the sites restricts soil chemical weathering and chemical soil-forming processes. Finally the amount of time since former soil degradation is still too short to cause significant differences in soil development.

The long history of deforestation and the intensive grazing have resulted in soil compaction. Together with other unfavorable soil features, causing surface capping and water repellency, this led to increased surface runoff and intensive linear erosion with gullies in ancient times, as visible in Fig. 8.9. In principle the soil development on this south exposed hillside has been repeatedly interrupted by processes of soil erosion and the deposition of translocated material. Currently, soil erosion is minimal, because the topsoils in the reforested areas are protected by the canopy and those of the open grassland, with only extensive grazing use by sheep, are well fixed by the plaited root systems of the special steppic vegetation. In particular the dense stock of *Festuca valesiaca* stabilizes the soil surface. That means that the changeover to extensive grazing in the open areas has had a positive effect on topsoil stability (Schröder, 2010).

The Effect of Land-Use Changes on Soil Properties (Aosta Valley, Italy)

Intense land-use changes and consequent modifications of soil management have been observed over the last decades in the Alps

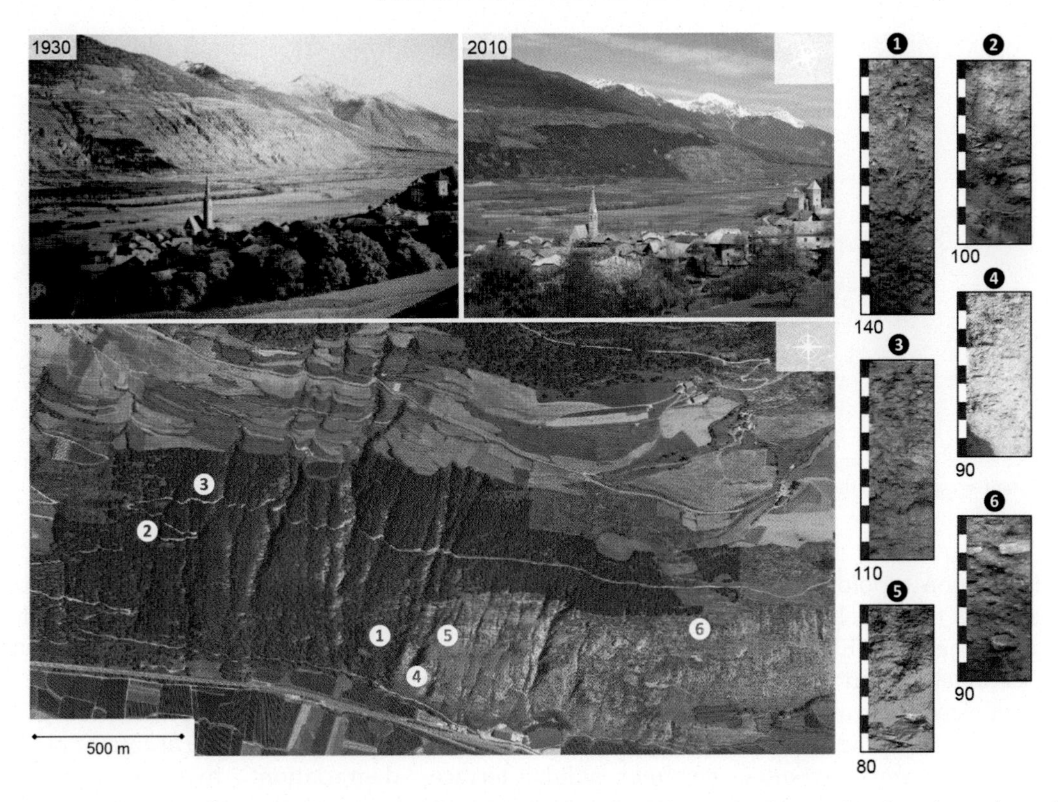

FIGURE 8.8 The progress of reforestation in 1930 and 2010 and soil profiles on the slope within the reforested area (1–3) and open grassland (4–6). Note the gullies inside and outside of the forest as a result of intensive erosion. *Source: photos: S. Grashey-Jansen; L. Schröder; Orthophoto 2011 by AGEA—agea.gov.it.*

FIGURE 8.9 Gullies created by running water as a consequence of deforestation and overgrazing. They can be clearly seen in the unwooded areas in Laaser Leiten, where they are still active in the event of heavy rainfall. *Source: photos: S. Grashey-Jansen.*

(MacDonald et al., 2000; Gellrich et al., 2007). Among the most relevant transformations the abandonment of terraced areas and land reshaping for agricultural and recreational use (e.g., ski slope construction) have affected large portions of mountain land, with consequent impacts on soil properties (Fig. 8.10). In the Aosta Valley (in the north-western Italian Alps) too, the abandonment of agricultural land is a common occurrence (Marguerettaz, 1983).

Terraced agroecosystems in marginal, steeply sloping areas cover large surfaces in this mountain region. They can be interesting for different land uses depending on climate and altitude, from extensive cattle grazing, cultivation of berries and officinal plants, chestnut forests, and vineyards (Stanchi et al., 2012). The

FIGURE 8.10 (A) Abandoned terraced vineyard Montjovet, Aosta Valley. (B) Land reshaping and new terraced fields (permanent grassland) in the Municipality of Gaby, Aosta Valley. *Source: (A) photo: S. Stanchi; (B) photo: M. Freppaz.*

most common typologies of terracing for this region and the whole Alpine area were collected in the ALPTER database (www.alpter.net, Scaramellini and Varotto, 2008). The database, collected after surveys in different study areas on the Alpine arc, includes: building typology (dry stone, wooden, concrete, solid wall), foundation type (on rock, sediments, terrain, mixed material), wall length (from approx. 1 m to hundreds of meters), height (from 0.2 to 0.3 m to several meters), wall and tread slope, tread width (very heterogeneous), terracing density, average terrain slope (often more than 100%). Soils of terraced sites are known to have improved properties in terms of fertility, organic matter, structure, and porosity (Sandor et al., 1990). They often have better agricultural quality when compared with the surrounding undisturbed soils, mostly due to rock removal, fertilization and manuring, i.e., constant organic matter inputs (Sandor and Eash, 1995).

The agricultural decline in this region has been widely documented, with the abandonment of the traditional agricultural practices, the diffusion of mechanization, and the "growth" of new land uses connected, for example, to touristic activities (Curtaz et al., 2014). Among the agricultural areas, terraced slopes were often the first to be abandoned, due to low accessibility and economic limitations. Abandonment is often followed by slope colonization by pioneer species, revegetation by natural processes (e.g., Lasanta et al., 2006), terrace degradation, hydrogeological hazards, and diffuse slope instability phenomena (Brancucci and Masetti, 2008; Freppaz et al., 2008). Sometimes, specific land-reshaping techniques have been proposed in this region to improve the accessibility and mechanization of mountain agricultural areas (including terraces), consequently increasing the economic profitability of crops. The interventions include a wide range of techniques with the long-term goal of recovering the original soil properties prior to disturbance in terms of quality and fertility (Schaffer et al., 2007; Curtaz et al., 2014).

These techniques are also applied in the construction and restoration of ski runs, with machine-grading, which is the process of smoothing slopes by the removal of topsoil, boulders, and vegetation, involving the use of heavy earth-moving machinery, also necessary for the construction of drainage and retaining walls. Due to the construction of ski runs the original soil thickness can be reduced with the loss of previous soil horizonation, generally resulting in altered topsoil. The perturbation of ski-run soils results in an almost complete loss of structure with subsequent problems of soil compaction and the reduction of water and air permeability (Freppaz et al., 2013). The rapid establishment of continuous plant cover after disturbance can protect and stabilize the substrate and hence minimize soil erosion, due to both its evident above-ground properties and its root systems (Barni et al., 2007; Rixen and Freppaz, 2015; Ola et al., 2015). Furthermore, ski-run management during the ski season can alter soil properties. In particular, groomed snow increases snow density and reduces snow depth compared to ungroomed areas, with a decrease in the insulation capacity of the snowpack. Consequently the soil is subjected to freezing temperatures and freeze–thaw cycles, with significant effects on soil nutrient cycling (Rixen et al., 2008).

Soil Diversity and Dynamics at Forest and Pasture Sites (Pokljuka Plateau, Slovenia)

The Slovenian Alps feature diverse soils that frequently change within very short distances. Such situations are well presented by the Pokljuka plateau case study, which is based on a soil survey carried out during summer 2013 and 2014. Pokljuka is an approx. 400 km^2 plateau west/north-west of Lake Bled in the Triglav National Park at altitudes mainly ranging between 900 and 1,300 m. The mean annual temperature is 2–1°C at altitudes above 1200 m and 1–0°C above 1400 m. The mean annual precipitation is between 2000 and 2600 mm and 2600 and 3200 mm at higher altitudes. The parent material of Pokljuka is generally carbonate, mainly Triassic dolomite and limestone (both with embedded chert), consolidated carbonate till, and, fragmentary, unconsolidated limestone-dolomite till, and a limited area of silicate parent material.

According to the Digital Soil Map of Slovenia at a scale of 1:25,000 (DSM25), the dominant soil types of the Pokljuka plateau are in particular *Rendzic Leptosol*, *Cambisols*, and *Lithosols*, while *Dystric Cambisol* is present to a limited extent. The geometry of the DSM25 shows that the areas of occurrence of these large groups of soil types are relatively homogeneous, according to the scale and purpose of this map (i.e., national and regional levels). However, at the local level at a scale of 1:5,000 the soil diversity is significantly larger, as very different soil types interlace within short distances. The combinations of soil-forming factors, here primarily the micromorphology and the land use, have created an interesting juxtaposition of diverse soils (*Lithosols*, *Rendzic* and *Lithic Leptosols*, shallow *Chromic Cambisols*, *Dystric Cambisols*, *Luvisols*, and *Podzolized Luvisols*) within an area that appears to be uniform at a first glance. In terms of land use, coniferous forests are most dominant on the Pokljuka plateau. The natural vegetation of mixed broad-leafed and coniferous forest (*Anemono-trifolio-Fagetum avr.geogr. Heleborus niger*) has largely been changed to almost pure spruce stands (*Picea abies*) over the previous 200 years (Arih, 2011), so that nowadays spruce forest dominates a significant part of the plateau with its typical surface cover of needles, branches, moss, and grass (Fig. 8.11).

The soft and heavily weathered limestone till, intersected by compact limestone rock outcrops, is the main parent material in some parts of the study area. In such places with undulating micromorphology, different soil types can develop over a very short distance (typically between 5 and 30 m), as the (micro)topography strongly determines the soil development (Kralj and Vrščaj, 2015). In microdepressions rainwater and also water from melting snow accumulate and therefore the quantity of percolating water is at least doubled if not tripled in larger microcatchment areas. Accordingly the soil leaching intensity is different. Preferential water flow, caused mainly by micromorphology, determines

FIGURE 8.11 (A) A spruce forest and its soils, Pokljuka plateau in north-west Slovenia. The micromorphology strongly determines the soil profile development. (B) A shallow Luvisol that developed on carbonate moraine. (C) The sidewalls of the same profile. Here the more concentrated water flow increasingly translocates the clay, humic substances, and free iron oxides through narrow preferential percolation areas and channels forming very initial stages of albeluvic glossae. In some parts of the profile reducing conditions appear and even stagnic properties develop. *Source: photos: B. Vrščaj.*

FIGURE 8.12 Podzolized soil, Pokljuka plateau. *Source: photo: B. Vrščaj.*

the variety of strongly leached to moderately leached areas. Fig. 8.11C shows an example of a shallow *Luvisol* profile developed in a small depression on typical gravelly limestone/dolomite till. *Luvisols* are common on lower slopes and level surfaces.

Podzolized soils in various stages of development can also be found in the small microdepressions. Fig. 8.12 presents a medium-deep podzolized soil that has developed on similar parent material at a distance of 1.5 km from the previous profile, at an altitude of 1,200 m and under the same land use—a spruce forest. Spruce has a very shallow fan-like shaped rooting system that does not penetrate deeply into the subsoil where leached cations accumulate or to saprolite depths where cations are released from weathering parent material. The litter of coniferous trees (spruce, pine, and fir) slowly decomposes mainly by means of acidophilus fungi and creates layers that maintain a fibrous structure (Kralj and Vrščaj, 2015). The typical $pHCaCl_2$ (measured in $CaCl_2$ solution) of the Oh horizon is 3.6–3.8. The material abundant in organic acids (including humic, oxalic, acetic, tannic acids) significantly accelerates the leaching process in topsoil mineral horizons. The organic matter, clay, and the iron oxides are prominently translocated into the deeper illuvial horizons. There, accordingly, the textural difference is abrupt. The tongue-like-shaped argic horizon follows the surface of the C horizon, which represents the soft, heavily weathered carbonate till (Fig. 8.12).

In general the canopy of coniferous forests returns many fewer base cations to the topsoil than species in deciduous and mixed forests, therefore the land-use change from mixed forest to pure spruce stands induces or contributes to soil degradation processes. Rapid soil leaching and accelerated podsolization are locally

FIGURE 8.13 (A) A Luvisol profile and (B) profile location in a depression used as seasonal pasture, Pokljuka plateau. *Source: photos: B. Vrščaj.*

FIGURE 8.14 (A) The flat bottom of a karst dolina with acidophilus vegetation, Rendzic Leptosols, and Chromic Cambisols, Pokljuka plateau. (B) Upper horizons of the podzolic soil profile, Pokljuka plateau. *Source: photos: B. Vrščaj.*

very prominent. According to some researchers the land-use change from mixed forest to pure coniferous stands in the past strongly promoted topsoil acidification, leaching, and induced podsolization processes (Kralj and Vrščaj, 2015). Nevertheless the whole area is not uniform as at close distances *Lithosols* develop on top of small ridges, boulders, and rock outcrops that are juxtaposed in the area, whereas *Rendzic Leptosols* are the characteristic soils that dominate short slopes.

In contrast to forest land use the soils in areas traditionally used for agriculture do not feature such strong leaching processes. For example, a typical *Luvisol* profile (Fig. 8.13) was excavated in a depression of a seasonal pasture. This location is 5.8 km to the west at the same altitude (1,220 m) and on the same parent material as in Fig. 8.12. The upper horizons have significantly lighter texture in comparison to the lower one. Podsolization processes have not developed.

The study area features karstic morphology where dolinas are frequent (Fig. 8.14). Partly the limestones are rich on black, gray, or pale-brown chert. In the course of carbonate weathering the undissolved quartz material accumulates on the bottom of the dolinas. From these beds of chert, featuring various thicknesses and grain sizes, podzolized soils have developed here and there. Fig. 8.14 shows a soil profile that has developed on such an accumulation of coarse quartz/chert sand and rubble on the bottom of a small dolina. The soil is acidic (the $pHCaCl_2$ of the A-horizon is 4.2). The transitional bleached CE horizon of irregular thickness consists of approx. 80% chert and 20% bleached eluvial material of finer texture. The underlying horizon is loamy, has a polyhedron structure, and contains approx. 15% chert skeleton. Consequently, this profile represents an early transitional soil development stage towards a Podzol. Nevertheless this soil is a very local special case, linked to an odd parent material, translocated from the slopes where *Rendzic Leptosols*, *Chromic Cambisols*, or *Lithosols* (rock outcrops) predominate.

According to this special soil type the flat or gently undulated depressions (Fig. 8.14) are characterized by acidophilus vegetation (i.e., *Nardus stricta*, *Luzula luzuloides*, and *Vaccinium myrtillus*), which is in strong contrast to the vegetation present in areas of *Rendzic Leptosols*, *Lithosols*, or *Chromic Cambisols*. That means that the soil diversity, here frequently within 10–30 m, determines the diversity of the vegetation cover and thus fundamentally contributes to the rich biodiversity of the area. This is also the case for peat soils (*Fibric Histosols*), which are dominated by *Sphagnum* spp. moss scattered in isolated depressions of the Pokljuka plateau.

Soil Pattern and Grazing Impacts at High Elevations (Zugspitz Region, Germany)

The Zugspitzplatt, with an altitude between 2000 and 2700 m, is a part of the Wetterstein Mountains (Northern Calcareous Alps) in southern Germany. It is built from very pure and karstified Triassic limestone (Fig. 8.15). Thus soil development is slow and influenced by high pH values. Due to the high altitude, physical weathering predominates in comparison to chemical weathering. Thus *Eutric Leptosols* and *Eutric Regosols* are the most frequent soil types between 2000 and about 2400 m. In contrast, allochthonous, mica-dominated brown deposits occur and the development of *Cambisols* with low pH values can also be observed there (Fig. 8.15). These brown deposits and their soils were described in detail by Küfmann (2002, 2003, 2006, 2008) as loess loam-like and "exotic" *Cambisols*. According to her studies, both the Late-Glacial and recent eolian deposition of mica from the crystalline Central Alps has strongly modified the basic soil development in patches.

Field studies by Grashey-Jansen et al. (2014) prove a clear pattern in the spatial soil distribution. Preferred areas of eolian deposition were mapped by the authors on the southern exposed karrenfields north of the so-called Gatterl. The Gatterl is a gap in the closed southern ridge of the Wetterstein Mountains where winds from the Central Alps can pass through (Fig. 8.16), becoming channelized by throughflowing this topographical bottleneck. Thereby, these winds and their turbulence cause a special pattern of reinforced deposition of mineral dust particles primarily at altitudes between 2100 and 2200 m (Grashey-Jansen et al., 2014, 2015).

The vegetation has adapted widely to the prevailing carbonatic soil conditions (e.g., *Carex firma*, *Saxifraga caesia*, *Androsace chamaejasme*; Fig. 8.15). Thus the current vegetation of the alpine zone of the Zugspitzplatt is dominated by different variations of the *Caricetum firmae* grassland. On the contrary, on mica-influenced acidified soil sites vegetation patches with acidophile species such as *Nardus stricta*, *Potentilla aurea*, and *Homogyne alpine* exist (Korch, 2014). Especially *Nardus stricta* also provides evidence of frequent grazing on these acidified soil formations. Accordingly the original plant

FIGURE 8.15 Slots and pits of limestone pavements on the karst landscape (A+D). Gaps are partially covered with dense-rooted Carex firma (B). Figure (E) depicts the silicate-enriched and cambic soil substrate. Figure (C) shows bare limestone pavement with karst solutional grooves and insular dots of Carex firma. The pavement of the karstified Wetterstein limestone submerges under the soil surface. The contrast between bare (C) and covered (F) karst is obvious. Rising water and the accumulation of mica depositions have caused processes of brunification within the soil substrate (F). *Source: photos: S. Grashey-Jansen (A–C); R. Rehm (D); R. Bernhard (E and F).*

FIGURE 8.16 The Gatterl Gap in the closed southern ridge of the Wetterstein Mountains, where southerly winds from the Central Alps can pass through. *Source: photo: S. Grashey-Jansen.*

community on such sites is often replaced by the *Geo montani-Nardetum strictae* (Korch, 2014).

Due to this traditional land use, various sites are influenced anthropo-zoogenically by grazing pressure (solely sheep). These sites are decisively influenced by soil compaction, nutrient inputs, and selective browsing due to minimal grazing management (Grashey-Jansen and Seipp, 2012). Furthermore, grazing is connected with the input of animal excrement (Fig. 8.17). Accordingly, studies by Korch and Friedmann (2012) show that nitrophilic species such as *Urtica dioica* occur even at altitudes of nearly 2400 m.

Summary of Soils and Land Use in the Alps

The examples of the previous sections, which concerned ancient agricultural techniques and forest use as well as river regulations and mining activities, illustrate the strong relation between land-use practices and soils in the Alps. Consequently, in order to characterize and understand Alpine soils we always have to reckon with significant, in some areas dominant anthropogenic impacts, often linked to land-use activities that are almost forgotten

FIGURE 8.17 Soils and vegetation dynamics are strongly influenced by sheep grazing at different topographical positions of the Zugspitzplatt (A and C), resulting in adapted vegetation at these sites (Alchemillo-Poetum supinae in (B) with Poa supina and Taraxacum alpinum on eutrophic grazing sites). (D) depicts Nardus stricta as an acidophilic plant species that has become established as a result of grazing. Concurrently, this plant is not eaten by the sheep. *Source: photos: S. Grashey-Jansen (A) and O. Korch 2012 (B–D).*

today. Focusing on agriculture, labor-intensive soil management activities in particular improved and maintained Alpine soil fertility. Although restricted mainly to manual work (e.g., soil returning, terracing, and irrigating), the sustainable land management practices significantly altered soils from a natural to a seminatural state and consequently ensured the prosperity and self-sufficiency of the Alpine people. Besides well-developed soil protection and sustainable management practices, it should be noted that some areas, mainly grazing land and to some degree also forest sites, not that small by extent, were overused for centuries, causing disproportionate erosion and nutrient removal and thereby significantly disturbed soils. With regard to forestry the change in the natural composition to coniferous-dominated or pure coniferous forests in some areas significantly changed the soils, with long-lasting consequences in the soils.

Land-use changes since as late as the middle of the 20th century decreased the spatial heterogeneity of agricultural use. On the one hand, market- and profit-orientations changed the traditional family-farms' production practices to the current dominant intensive livestock farming and monocultures. On the other hand, abandoned areas strongly increased in some regions, thus presenting new challenges (MacDonald et al., 2000). This polarization regarding the intensity of land use has naturally had a significant effect on the soils. Traditional land management practices as well as the current human influences must be recognized and acknowledged in order to properly adapt and fine-tune soil protection and management in the Alps, in particular as the environmental conditions in the Alps are often expected to be more or less natural and pristine.

The examples of soil land-use relations illustrate both the typical soil features and patterns in the Alps, as well as impacts to soil bodies due to land-use and tillage practices. Additionally, they raise overarching questions concerning the agricultural and forestry production capacities of soils and related limitations, soil vulnerability to different threats, and the current and potential ecosystem services the soils provide. Such topics are integrated in a conceptual framework of soil–society relations that considers soil as a "cultural-historical-natural system", as discussed by Richter and Tugel (2012, p. 38.1). In such context, it should be mentioned that in recent years, some small but interesting attempts to revitalize traditional agricultural landscapes have arisen (e.g., Varotto and Lodatti, 2014) and may grow and influence Alpine landscapes in the future.

With regard to soil information the small-scale relations between topography, meso- and microclimates, and water balance, followed by mosaic-like land-use patterns with small and highly diverse land units, play a crucial role in the resulting high variability of soils over very short spatial scales. This soil fragmentation and diversity require that soil maps applicable to protective and management purposes correspond to these conditions. Nowadays, we mostly lack large-scale soil maps (Brevik et al., 2017). In general, existing spatial information on soils is strongly generalized and thus provides a very rough overview of the spatial distribution of the main soil types. Nevertheless, in order to elaborate or improve soil maps, geological maps that consider Quaternary deposits in detail are needed (Brevik and Miller, 2015). Furthermore, historical land-use studies should definitely be integrated, as they can significantly contribute to identifying specific soils in the Alps.

Against this background of soil land-use relations, within the next section we will present and discuss different soil-survey approaches, dealing largely with the peculiarities of Alpine landscapes and their soils. Thereby, we will further scrutinize what type of soil information is suitable for application in planning and land-use management decisions.

MAPPING AND MODELING SOILS IN THE ALPS

Methods, Status, and Relevance

Traditional soil maps, designed as paper maps and often difficult to understand by nonexperts and experts from other disciplines, were mostly created in the precomputer era through fieldwork. Traditional soil mapping is a time- and resource-consuming process (Brevik et al., 2003). Soil maps, soil classifications, and the resulting soil-mapping units (SMUs) are representations of the structured expert knowledge of soil surveyors (Bui, 2004), and are based on field observations, knowledge of the functioning of dominant soil-forming factors, and available data. Detailed-scale, area-wide mapping is therefore difficult to implement using traditional methods. When GIS software and spatial data processing tools became widely available and affordable in the mid-nineties, SMUs and their attributes were digitized and soil profile databases were created.

Generally, the commonly available polygon-based soil maps are produced at different scales and using different methodologies (Brevik et al., 2017). For instance, Fig. 8.18 illustrates that the transfer from plot to landscape scale has become an important methodological issue. Accordingly the accuracy and semantics vary. In terms of Alpine soils, Baruck et al. (2016) described the national, and partly regional, differences concerning methodical approaches, scales, classification systems, and data completeness. They showed that soil surveying in the Alps is characterized by a strong emphasis on agricultural land, whereas soil information regarding forests and high mountain areas is rather limited due to the high costs of traditional soil surveying in remote areas. Furthermore, they discussed the basic challenges of and suggestions for soil surveying in the Alpine region caused by the mosaic-like juxtaposition of soil properties and soil-forming conditions (see "The Alps and Their soils" section).

While in the past spatial data on soil-forming factors were modest, data availability has improved significantly, especially over the last two decades. Earth observation, airborne, and other methods yield abundant digital information on soil-forming factors that can be utilized for spatial and semantic improvement of polygon soil maps to produce raster-based information on soil properties as well as soil maps at a more detailed scale (Escribano et al., 2017). In particular, high-resolution digital elevation (DEM) data and derived information are important for modeling and deriving soil parameters in mountainous areas where topography is frequently considered the most important soil-forming factor.

Digital soil mapping (DSM) as a survey method was introduced relatively early (e.g., Bui and Moran, 2001; Lathrop et al., 1995), alongside the computerization of soil science and the development of soil informatics. DSM, which according to Lagacherie and McBratney (2006), is the computer-assisted combination of field and laboratory work with spatial and nonspatial inference systems to create spatial soil information (maps of soil properties or soil types), can play a valuable role in making optimal use of existing soil information and closing information gaps. McBratney et al. (2003) and Behrens and Scholten (2006) provided overviews of the methods applied in DSM, and Grunwald (2009) reviewed multifactorial approaches in the recent literature. Methods to predict soil properties from topography include a wide range of methods, from geostatistical ones such as Kriging (Odeh et al., 1995), fuzzy logic (McBratney and Odeh, 1997), and other methods and techniques that have been exhaustively summarized (McBratney et al., 2003) and published (Lagacherie and McBratney, 2006) elsewhere. Of the soil-formation factors (climate, organisms, relief, parent

Soil landscape with several soil units

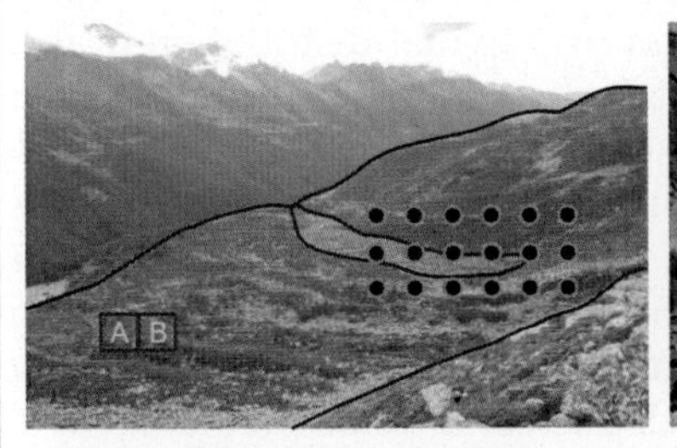

Various field survey methods are available, for instance:

a) Grid mapping ●
b) Transect ● and catena mapping ●

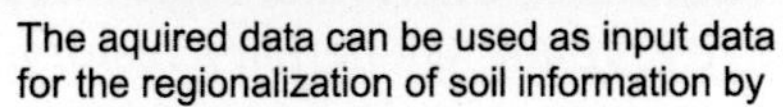

Considering the complex juxtaposition of soil bodies in the Alpine region, traditional soil surveying as well as digital soil mapping face specific challenges.

Soil bodies represented by soil profiles

Complex patterns

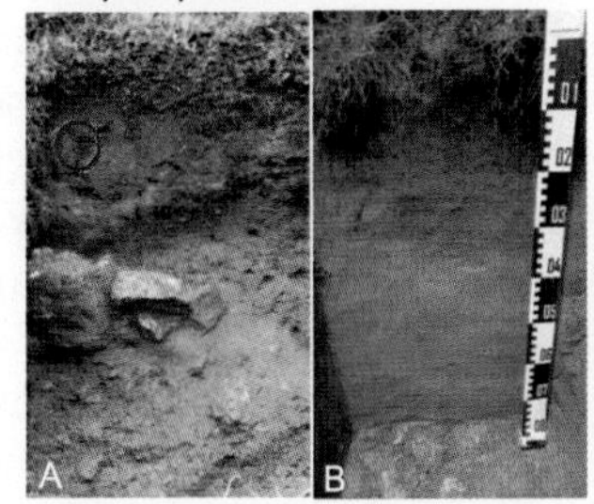

Multi layered profiles

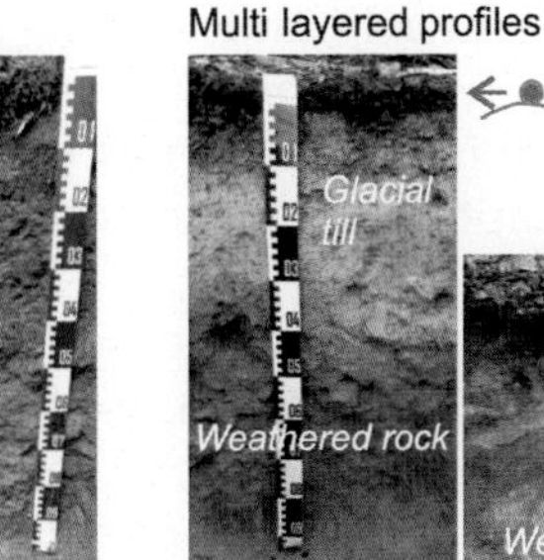

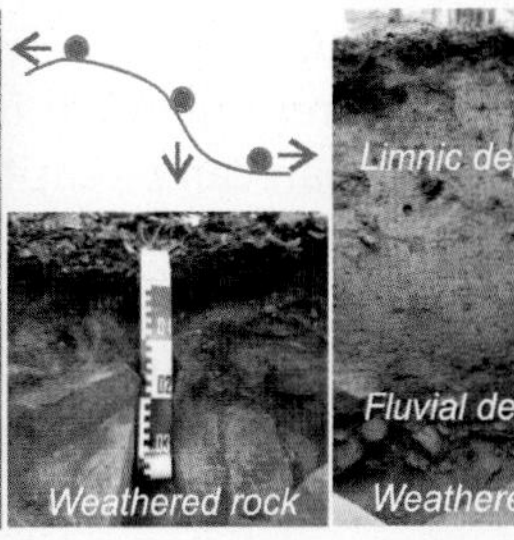

In the Alps, seven soil classification systems are presently in use: those of Austria (Nestroy et al., 2011), France (AFES, 2009; AFES, 1998; AFES, 2000), Germany (Ad-hoc-AG Boden, 2005), Slovenia (Škorič et al., 1973; Stritar, 1984; Kralj and Grčman, 2010), and Switzerland (BGS, 2010), as well as the World Reference Base (WRB) (IUSS, 2014) and the US Soil Taxonomy (Soil Survey Staff, 1999, 2014) (c.f. Baruck et al., 2016).

Soil sample

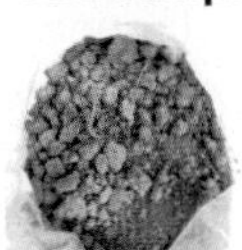

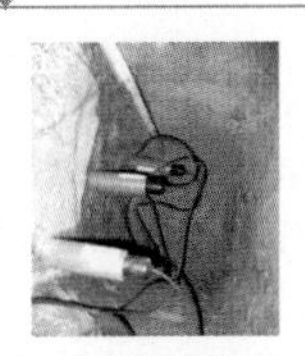

Soil sample analyses and field measurements at plot scale

Pedologic site assessment
Description, deduction, and classification of soil properties
Classification of soil types (and designation of the borders of soil units)

Regionalization, visualization and ecological evaluation
Soil maps and soil function maps at different scale

Soil information with landscape-related information
Soil information systems

FIGURE 8.18 Illustration of soil-survey approaches, considering different methods and scales. *Source: photos: J. Baruck and C. Geitner.*

material, and time) (McBratney et al., 2003), topography or relief is believed to play the dominant role in soil formation and, as a direct result, in DSM in the Alps (Geitner et al., 2011a; Herbst et al., 2012; Ballabio, 2009). However, despite the increasing availability of high-resolution digital elevation models, this soil-forming factor still requires increased attention and additional understanding in the Alpine environment, due to the high morphodynamic activity, young Quaternary history, and strong anthropogenic forces. Another important role regarding the results of DSM is the accuracy of thematic maps of environmental variables such as the soil parent material. This parent material at the surface is mainly formed as unconsolidated deposits by Pleistocene and Holocene processes. Due to these glacial, periglacial, gravitational, fluvial, and eolian processes within the highly variable Alpine topography, parent material is frequently much more diverse than can be captured in geological maps (Baruck et al., 2016).

Furthermore, to a certain degree, the soil information available from surveys not focused on soil, e.g., forestry or geology surveys, can be used for DSM. Examples where point databases from forest site surveys and/or forest soil monitoring have been used for DSM can be found in Bavaria (Germany) and Switzerland. Based on point data from forest soil profile sites in the Swiss Canton of Basel-Landschaft, Herbst and Mosimann (2010) compared a random forest approach to predict soil depth and stone content with an expert-based empirical-statistical model. They emphasized the influence of the quality of the test samples on the predictive accuracy of the random forest model. The same data points were used by Fracek and Mosimann (2013) to predict noncalcareous soil depth by applying a knowledge-based model, and by Herbst et al. (2012), who presented a geomorphographic terrain classification for a decision-based model for the prediction of forest soil properties such as soil depth, stone content, and acidity (topsoil and subsoil pH). In Germany, Häring et al. (2012) investigated the influence of topography on the spatial distribution of forest soils in Bavaria by applying a random forests classifier, based on forest soil profile data, to disaggregate complex SMUs into single soil types.

All in all, the trend in DSM in the Alps and similarly remote high mountain areas is moving away from the spatial delineation of soil types towards the continuous prediction of soil parameters, with a recent increase especially regarding soil organic carbon (SOC) due to its importance in climate change research (e.g., Ballabio, 2009; Dorji et al., 2014; Nussbaum et al., 2014; Yang et al., 2015).

Due to several reasons, information on soil quality and soil suitability, mainly soil maps, is seldom taken advantage of in spatial planning. The two main issues that should be resolved by soil experts and pedologists themselves are the following: first, the coarse scale, and second, the lack of semantic soil information specific to the planner's needs. Especially for vulnerable areas, more detailed soil maps are necessary to better reflect the spatial variability. This is particularly important in mountainous regions where key soil properties (i.e., the soil profile structure and total soil depth, the properties of topsoil horizons—depth, organic matter content, pH, texture, etc.) can vary significantly within very short distances. The semantic information of the SMUs, mainly limited to physical and chemical soil properties, should be presented in a manner so that the planner can directly present the suitability for a certain type of land use. In general the suitability of soil for agriculture, and, more recently, ecosystem soil quality are the two soil quality criteria that are used for soil protection and which steer the planning process. However, the land-use planning process could greatly benefit from additional important soil information. For instance, soil permeability, susceptibility to landsliding and erosion, buffering, and neutralization capacity are properties that can contribute to better decisions with regard to spatial planning for industrial, infrastructure, housing, and recreational areas as well as the preservation of agricultural areas or biodiversity hot spots in mountainous environments. The principles of ecological soil evaluation and the implementation thereof in specific planning cases in Austria are illustrated and discussed by Haslmayr et al. (2016).

Due to both the methodical challenges in terms of soil surveys in the Alps, as well as the application of soil information, the following sections provide a number of case studies from the Alps. Therein, not only different regions are presented, but also various land-use types and their soil information needs are discussed.

The Potential of Soil Overview Maps (Bavaria, Germany)

Soil surveying is a legal task of the Bavarian Environmental Agency and is divided into two categories: The "Soil Basis Inventory" deals just with punctiform locations, while the "Soil Area Inventory" records the spatial distribution of soils and their properties (soil mapping). Considering the large area of Bavaria and the extreme heterogeneity of the soil parent material, the development of a so-called "Overview map of the soil 1:25,000" (Übersichtsbodenkarte 1:25,000, ÜBK25) was initiated in the early 1990s to accelerate the soil cartographic work. This map series represents soil map overviews, verified by sampling selected sites in the field, which were finished in 2015.

Soil mapping in the Bavarian Alps started in the late 1990s. It was the last region of Bavaria to be examined in the course of mapping and therefore the smooth integration of the Alpine mapping units into the existing general legend of the soil map of Bavaria was required. That means that three fundamental features had to be included in every mapping unit: the soil types including their spatial shares, the texture, and the parent material. It was clear from the beginning that a stratigraphic structure of the mapping units, as in other Bavarian landscapes, was not applicable. As a consequence, the modification of previously used concepts with the aim of ensuring correct representation of the Alpine landscape had to be implemented.

Tectonics, relative relief, and both former and recent morphodynamics induce small-scale change in soil type and substrate conditions (Figs. 8.19 and 8.20). Thus it is difficult to define the legend units. A very important approach to overcoming those specific problems is to create hierarchical systems, enabling use of the fittest level, thus best reflecting reality. The most precise description of a mapping unit is the combination of soil type, texture, and lithology (Fig. 8.21, left and right). If the nature of rocks vary on a small scale within a sequence, attribution to a certain lithology is not possible, so the parent material is named on a higher level, for example carbonate rock (Fig. 8.21, middle).

Due to the strong morphodynamics, mapping units comprising the soil type and the related morphologically- or morphodynamically-structured parent material were established (Fig. 8.22A).

As far as the classification of soils is concerned, even by using a hierarchical classification system such as the German system (Ad-hoc-AG Boden, 2005), some of the Alpine soils cannot be described correctly (Fig. 8.22B). Regarding the small Alpine area of Germany, it is understandable that the classification system is not devoted to Alpine soils. This also implies a lack of knowledge with regard to interactions of soil-forming factors and soil-forming processes especially at higher altitudes. Therefore upgrading the knowledge of Alpine soils and classification systems is a necessary task at present (Baruck et al., 2016).

For some applications of the soil map it is possible to deduce necessary information directly from the map legend with the support of a soil-mapping expert, but this approach is not purposeful in light of the growing number of protection- and planning-related requirements

FIGURE 8.19 Cambisol (till with crystalline debris, left) and Regosol (till with carbonatic debris, right; see the dashed line in between), near Garmisch-Partenkirchen. *Source: photo: R. Traidl.*

(Traidl, 2005). Thus a data set referring to an appropriate map is necessary to provide suitable parameters for a wide range of applications. These parameters are offered by a special database that contains a data set illustrating every soil-mapping unit (SMU) with representative soil profiles. Various applications are possible in combination with the soil map (Fig. 8.23). These include, for instance, maps of soil functions, soil hazards, and prediction maps in the event of changing environmental conditions, natural as well as anthropogenic. Soil function and soil hazard maps demonstrate the potential of soil evaluation and can form a basis for planning and prediction.

Just a few soil function maps have been realized for parts of Bavaria. Hitherto, they have addressed the soil capacity as regards precipitation, the filtering and buffering of nitrates, buffering acidification, the bond strength regarding heavy metals (cadmium), and the habitat potential for natural vegetation (Bayerisches Geologisches Landesamt & Bayerisches Landesamt für Umweltschutz, 2003). Thematic gaps within these soil functional approaches will be filled in the future.

Soil hazard information and predictions deduced from soil maps can also have a wide range and have not yet been compiled in Bavaria. Besides global hazards such as soil erosion, other threats and hazards can be defined. Flooding, loss of forests due to storms, mass movements, drought stress, and organic matter decline are some of the subjects of possible evaluation by soil maps. Prediction refers to declined soil functions as well as to hazards resulting from changes in environmental conditions such as climate and land use. The latter in particular is significant for planning. Even though conflicts of interests between business and ecology will arise, they have to be faced and solved, if possible. One basis for this task is the collection and harmonization of sufficient data.

The Concept of a Digital, Layer-Structured Soil Information System (South Tyrol, Italy)

The overarching objective of the presented project ReBo, which was funded by the Autonomous Province of Bolzano-South Tyrol

FIGURE 8.20 Distinct soil development caused by recent morphodynamics, Reintal. *Source: photo: R. Traidl.*

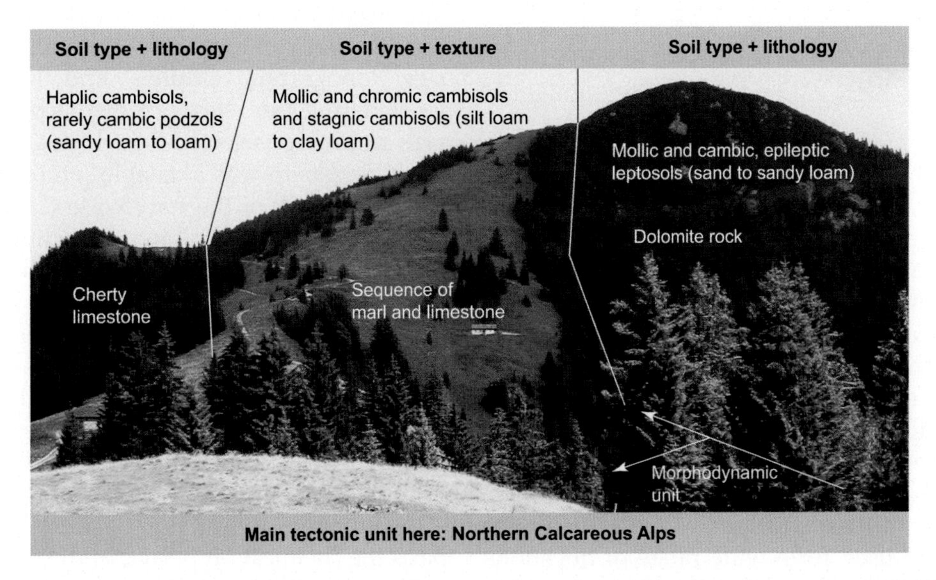

FIGURE 8.21 Mapping based on specific parent rock or on a rock sequence, Chiemgau. *Source: photo: R. Traidl.*

FIGURE 8.22 (A) Mapping units based on morphodynamic features at in the Chiemgau region. (B) Organic-rich debris below residual stone pavement at a periglacial zone, altitude 2500 m. *Source: photos: R. Traidl.*

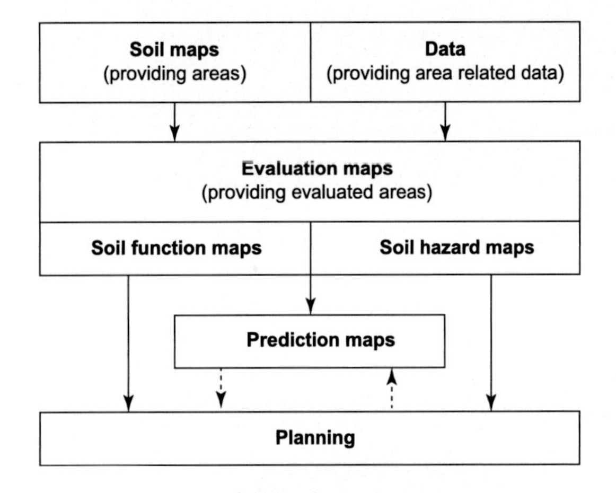

FIGURE 8.23 The potential of soil maps and information derived therefrom for planning issues.

(Italy), was to establish and examine a concept to create soil maps at a detailed scale. The procedure needed to be well adapted to the diverse Alpine landscape of South Tyrol (Italy), as well as to the specific availability of geodata. From a methodological perspective, some additional aims were paramount. Namely the approach and its results should (1) serve as a sound basis for further small-scale soil differentiation, (2) be used for a wide range of questions and land-use decisions, (3) be comprehensible by nonspecialists, (4) embed soil maps in a GIS with additional soil-relevant geodata layers that can be combined freely, emphasizing the landscape context, (5) consider legacy soil data, and (6) properly communicate uncertainties.

Based on these objectives a user-oriented methodical framework was developed to derive soil-relevant information in order to optimize and therefore minimize the pedological field survey. In the project, two test sites were investigated in order to cover a wide spectrum of soil-forming conditions (e.g., geology, topography, altitude, and land cover). Considering the relevance of the parent material, areas were selected for which a new digital geological map was available, prepared within the framework of the Italy-wide CARG project (APB, 2016a). The following section presents the workflow (Fig. 8.24) and the individual layers (Table 8.2) of the soil information system.

As the workflow demonstrates, the soil information system is based on various data. The parent-material base map is derived by combining two different geological databases. The new official geological map of South Tyrol turned out to be excellent regarding the spatial and genetic differentiation of Quaternary deposits. On this basis the chemical properties of the parent material were added, which was mapped within

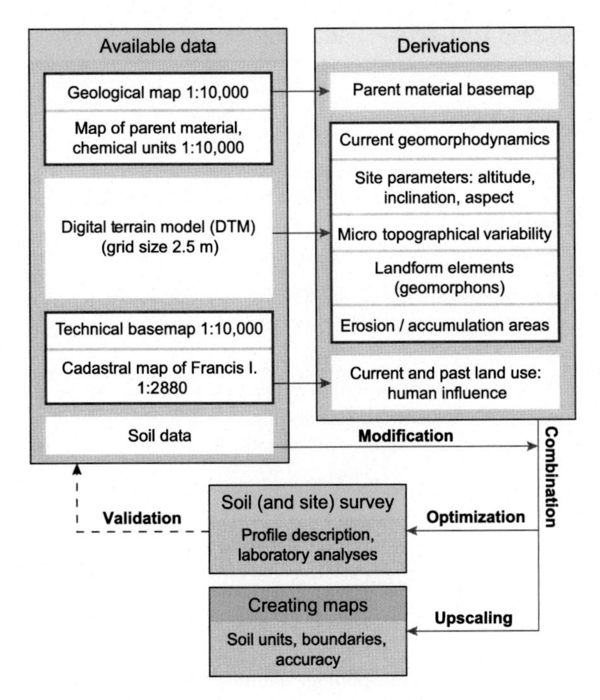

FIGURE 8.24 Workflow of the digital, layer-structured soil information system for South Tyrol.

the framework of the forest site survey (APB, 2010). Among other approaches and derivations a high-resolution digital terrain model (DTM) from airborne laser scanning was used to derive landform units with the open source GRASS GIS extension module r.geomorphon (Jasciewicz and Stepinski, 2013). As the model is based on visible neighborhood calculations, similarities with the site perception of the surveyor exist. The resulting landform element map was further differentiated regarding slope position by applying a statistical learning approach to the topographic site description of the ReBo project's soil profiles. In order to consider the anthropogenic soil modifications attributed to land use, human influence was estimated utilizing the information on current land use, adapted from the scale-variable (from 1:5,000 to 1:10,000) technical base map from 2006 (APB, 2006) as well as the Cadastral Map of Francis I (the land register) of the mid-19th century at a scale of 1:2,880 (APB, 2016b). Furthermore, numerous existing soil data were integrated (e.g., from forest sites and the agricultural soil survey). Supplementary and additional soil data were collected based on a field map at a scale of 1:10,000.

The presented Soil Basis Map at a scale of 1:10,000 (Fig. 8.25) is basically a conceptual derivation of potential soil units derived from the parent material. With very few exceptions the boundaries of the geological base map are not altered. Verified and supplemented by field surveys, each soil unit is represented by one or more reference soil profiles. Nevertheless the Soil Basis Map must be understood as an overview map (see "The potential of soil overview maps (Bavaria, Germany)" section) for orientation purposes, i.e., it can be the basis for other detailed mapping approaches. The designated soil units are an adaptation of the concept of the Bavarian "Overview map of the soil 1:25,000" (see "The potential of soil overview maps (Bavaria, Germany)" section). This approach, proposed by Traidl (2008), considers the Alpine peculiarities by creating SMUs that take into account both the parent material and the morphodynamic background. The soil types themselves are classified according to the Austrian Soil Classification System (Nestroy et al., 2011).

In order to identify further differences within the soil units, they can be combined with 14 geoecological layers (see Table 8.2) that are relevant to the genesis and distribution of soil properties and types. Fig. 8.25 illustrates such a combination. In addition to information on the parent material, topography, and land use, new ideas were implemented concerning the reasons for, and the distinctness of, soil unit boundaries, as well as information on the positional accuracy of the profile sites.

In order to demonstrate the benefits of these combinations, one reference profile site from the agricultural survey presented in Fig. 8.25 (the red points), with is located near to and south-southwest of the ReBo project reference

TABLE 8.2 Additional Geoecological Layers of the Soil Basis Map 1:10,000 for South Tyrol

Layer	Data Source, Scale, Methods	Content
Parent Material: The Basis for the Direction and Pace of Soil Development		
Geological parent material	Geological map (1:10,000), Map of Parent Material Chemical Units (1:10,000) → expert-based, soil-related modification and combination	Units with lithological, chemical, and geomorphological (genesis) information
Topography: Morphodynamics Limit Soil Formation. Soil-Forming Processes are Determined by the Position in the Landscape. Microtopographical Variability Increases the Small-Scale Diversity of Soils. Erosion and Accumulation Control Soil Depth and the Nutrient and Water Balance		
Current geomorphodynamics	Geological map (1:10,000), DTM (grid size 10 m), own mapping (from case to case)	Differentiation of processes and materials; identifying areas of prevented or at least disturbed soil formation
Location parameters	Digital terrain model (grid size 10 m) → derivation of slope i`nclination and aspect	Slope classes according to Blum et al. (1996); climatic altitudinal classes according to the South Tyrol forest survey (APB, 2010)
Microtopographical variability	Digital terrain model (grid size 2.5 m) → derivation of roughness (Terrain Ruggedness Index)	Threshold values based on legacy soil data
Landform elements	Digital terrain model (grid size 10 m) → computation and classification with r.geomorphons (GRASS GIS)	Geomorphological units: flat, peak, ridge, shoulder, spur, slope, hollow, footslope, valley, pit (Jasciewicz and Stepinski, 2013); the unit slope was further differentiated based on profile curvature and mid-slope position
Erosion/accumulation areas	Geomorphons, topographic index	Landform elements reclassified with regard to erosion and accumulation
Land Use and the Related Degree of Human Influence: Land Use, Land-Use Duration, and Land-Use Changes Modify Soils		
Current land use	Based on the technical base map (1:10,000)	The most relevant categories: forest, vineyard, apple orchard
Past land use	Cadastral Map of Francis I. (1851–1861) (1:2,880) → comparison with recent land use	This land-use pattern represents the traditional and typically long-term land use
Boundaries Between Soil Units: Provide Additional Information Regarding Landscape–Soil Relations		
Reason for soil unit boundaries	Different databases and expert knowledge	Classes: not specified, geology, topography, water regime, land use
Distinctness of soil unit boundaries	Different databases, expert knowledge, and own mapping	Differentiation: sharp (<10 m), intermediate, diffuse (>50 m)
Positional Accuracy: Provides Information to Estimate Positional Uncertainties and Thus Helps to Prevent Misinterpretations		
Positional accuracy of profile sites	Subject to different measuring procedures	Circles used to illustrate possible deviations from the central point

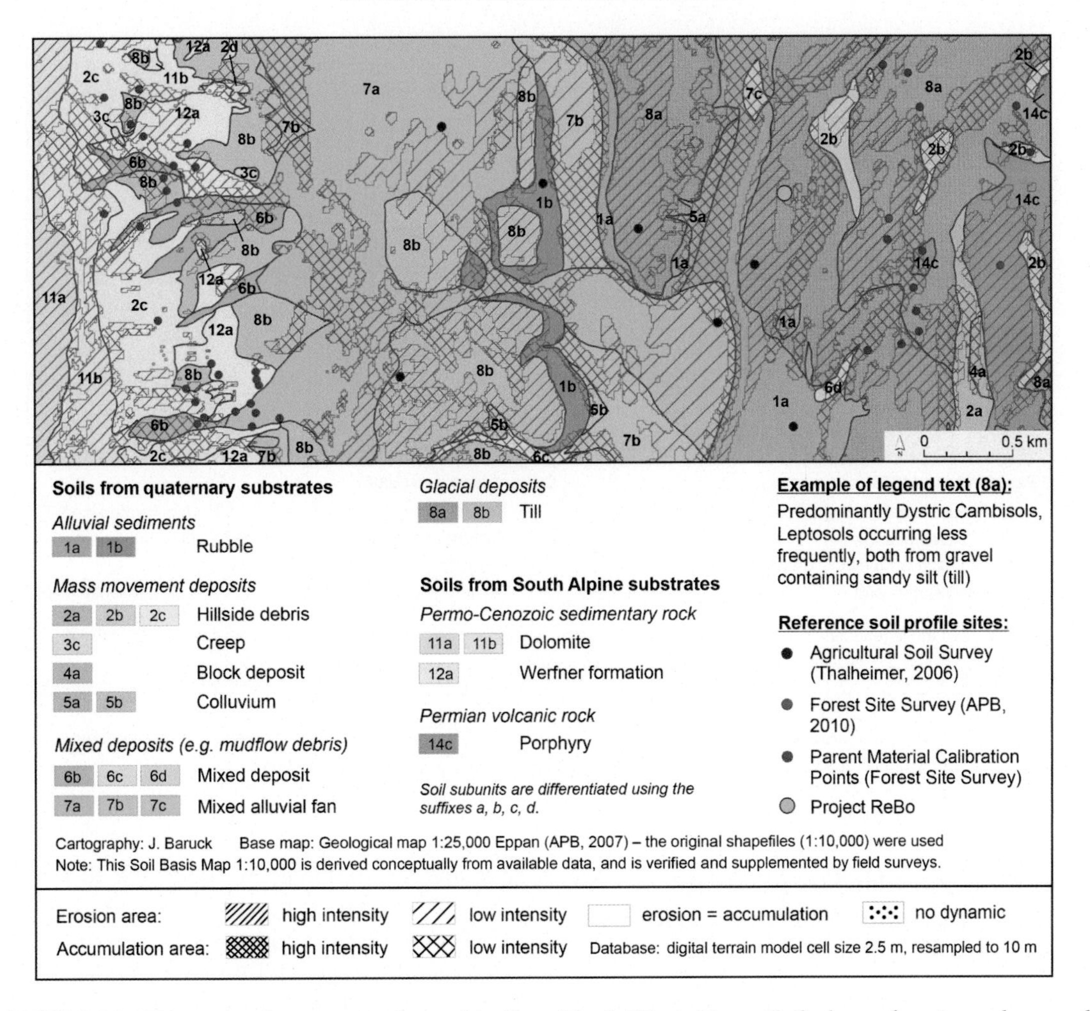

FIGURE 8.25 This section demonstrates the combination of the Soil Basis Map with the layer of erosion and accumulation areas that originate from a test site that covers the area from the Kalterer Lake in the South to the western slope above the alluvial fan of Andrian (South Tyrol, Italy).

profile (the green point), was examined more closely. For this site the presented maps (Figs. 8.25 and 8.26) provide the following information: soils from glacial deposits, characterized by gravel containing sandy silt (till)—(i.e., predominantly Dystric Cambisols, with Leptosols occurring less frequently). The site is situated in a potential erosion area. Furthermore, it is characterized by land-use change from forest to currently intensive vineyard cultivation (Fig. 8.26), meaning that erosion can indeed occur. Combinations like this can be carried out for each site and the expert-based conclusions will help to both estimate and upscale soil properties and soil types as well as to guide more detailed future fieldwork.

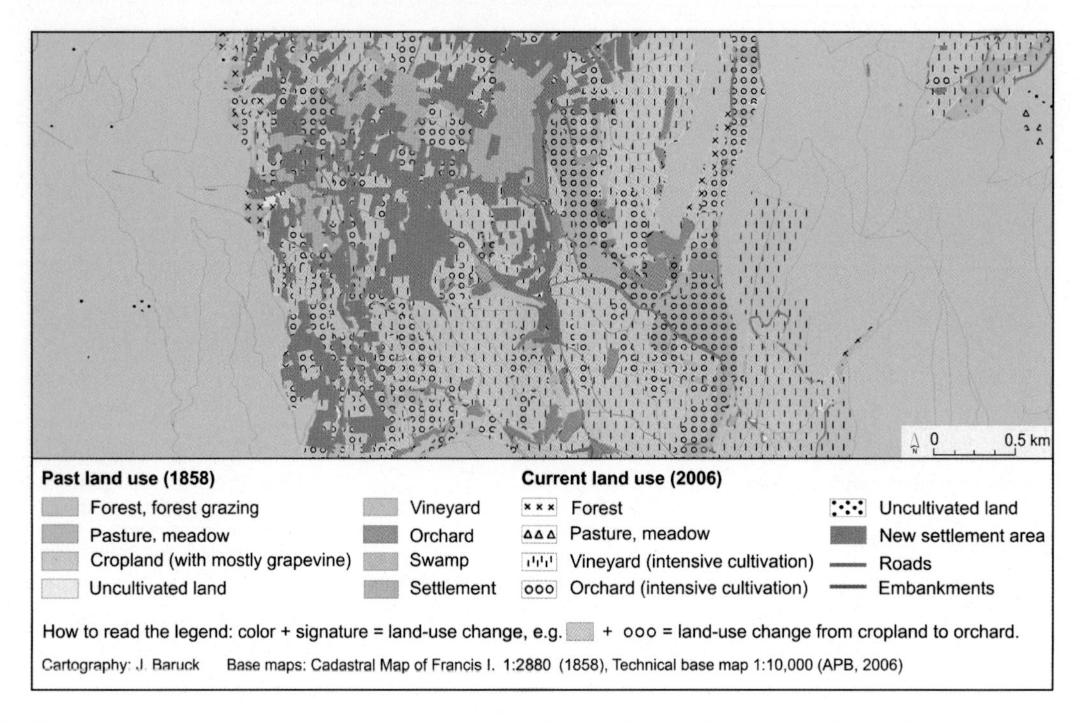

FIGURE 8.26 Land use and land-use changes, which fundamentally modify soil properties, are derived by combining the layer regarding current land use and the layer regarding past land use (South Tyrol, Italy).

Improving Soil Maps to Assist the Land-Use Planning Needs of a Local Community (Braslovče, Slovenia)

For Slovenia the soil information in digital form is relatively rich (Vrščaj et al., 2005). The vector-based 1:25,000 Digital Soil Map of Slovenia (DSM25) covers the entire country. Nevertheless, scale and accuracy are not suitable for modeling at municipality levels. Individual SMU polygons are defined by three predominant soil typological units (STU) and the proportion of the SMU they occupy (Vrščaj and Lobnik, 1999). The spatial location of STUs within SMUs is not defined. The SMU geometry (shape and size) and the STU composition were defined by field survey and expert knowledge assessments and are for that reason subjective to different extents. Nevertheless, on the country level, the 1:25,000 polygon soil maps represent detailed and valuable information that should be spatially improved and semantically upgraded in order to be utilized on a local/municipality level (Vrščaj, 2007a). The DSM25 was completed in its digital form in 1999. Newer digital spatial data of the main pedogenetic factors, which were not available at the time of the soil mapping, enable the spatial improvement of the SMU geometry, and the inclusion of semantic information on soil properties.

Slovenia underwent significant soil loss after becoming an independent state. Soil urbanization and soil sealing were especially intensive after the year 2000, mainly between 2002 and

2007, before the first signs of the global economic crisis appeared in 2007 (Vrščaj, 2007b, 2011). The construction of motorways and other infrastructure, intensive housing development, and the expansion of new commercial areas in the suburbs of cities raised concern among experts already at that time, especially since primarily the best lowland agricultural soils were sealed (Vrščaj, 2008).

The Municipality of Braslovče measures 54.9 km^2 and is located in the Alpine Savinja Valley. Its territory is composed of three main geomorphologic areas: (1) the bottom of the valley, approximately 280 m, where shallow *Fluvisols* and *Eutric Cambisols* dominate, (2) flat or gently undulating Pleistocene terraces with *Planosols* on very mild slopes (e.g., 3°), *Stagnosols* and *Gleysols* in depressions, and (3) the mountain area of Dobrovlje (917 m), which features steep (e.g., 41°) to moderate (e.g., 22°) slopes, plateaus, and karst dolinas on the top. The municipality has faced some difficulties in the land-use planning process in recent years. Namely the valley is famous for highly valued Savinja Golding hops production, a valuable exportable good of the local agricultural sector. Thus the agricultural production in the valley is important and profitable. At the same time the valley is burdened by a fringe of settlements and, most importantly, a motorway that is planned to follow a path along the bottom of the valley. The adaptation and modifications of the municipal spatial plan and design of new settlement areas required additional information on soil quality and soil suitability. The upgraded soil information should serve spatial planers to better adjust and justify planning decisions and to adapt the land-use planning documents of the municipality. Following the decision of the local authorities to acquire more accurate soil information than available by the DSM25, a more detailed study on the soil resources of the municipal area was completed (Radišek et al., 2014a, 2014b). The main purpose of the study was to describe the main soil types present in the municipality, to assess the soil quality, soil capacities, and suitability for agriculture, possible threats that may occur due to future land-use changes, and to elaborate a soil map on a more detailed scale. Due to the very limited time and resources available the tasks were carried out using a combination of a field survey and computer modeling. In order to create a soil map at a 1:10,000 scale, five soil-type prediction models for four different uniform lithological test areas were developed and tested. The 12.5 m DEM was used to determine relief derivatives (e.g., slope, curvature, aspect). The uniform lithological areas were vectorized from the 1:25,000 raster geological map and the land use was utilized from the digital 1:5,000 scale land-use database. The modeling was performed in four main stages (Fig. 8.27). Fig. 8.28 shows a section of the initial polygon soil map at 1:25,000.

1. The first stage was the definition of four main areas with uniform parent material: (1) compact Triassic Limestone, (2) crystalline rocks, (3) carbonate gravel deposits, and (4) areas of clayey Pleistocene terraces.
2. The second stage was the definition of the principal soil-forming factors for each lithologic area: DEM and four DEM-derived datasets, land-use, and climate data. Individual soil-forming factors were related to the existence/depth of individual diagnostic horizons that develop within catenas of uniform lithological areas. The surveyors' mental model that developed during surveying and field observations was used to elaborate a series of expert knowledge-based pedotransfer functions that predict the occurrence (presence/thickness) of individual diagnostic soil horizons in relation to pedogenetic factors

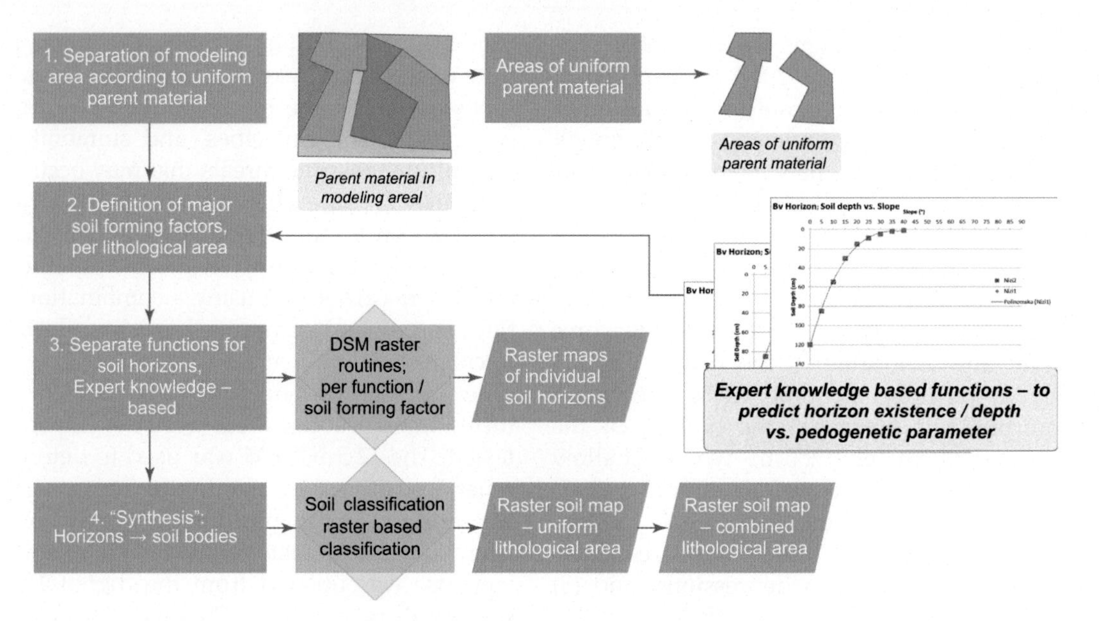

FIGURE 8.27 The main stages of the 1:10,000 soil-map modeling in which expert knowledge was utilized, Municipality of Braslovče (B. Vrščaj).

dominant in an individual lithological area (e.g., the thickness of the Bv horizon in the area of siliceous parent materials, or the Brz horizon in limestone areas). Pedotransfer functions were elaborated for all diagnostic horizons that define the soil types of individual soil series. Legacy soil profile data combined with limited fieldwork—the data from approximately 100 descriptions of soil augering locations—were used to calibrate the functions.

3. The main task in the third stage was to program raster computer-assisted mapping routines. For this the semantics of pedotransfer functions were implemented in GIS software scripting routines, which were used to build horizon depth-predicting models. The results of the routines were individual raster layers presenting the distribution/thickness of individual diagnostic soil horizons (Fig. 8.29). The GIS and relational database systems were used to link average soil horizon properties data to raster cells and, thereby, to elaborate continuous raster maps of soil properties—i.e., information important for end users.
4. In the fourth step the soil classification algorithm was programmed in raster GIS script to synthesize the raster layers of individual horizons and to predict the distribution of individual soil types (STUs). The decision trees were elaborated for each STU represented in the catenas of individual lithological areas. The algorithms were run to produce raster maps representing the distribution of individual soil types, which in the final operation were spatially merged with the raster soil map of the area

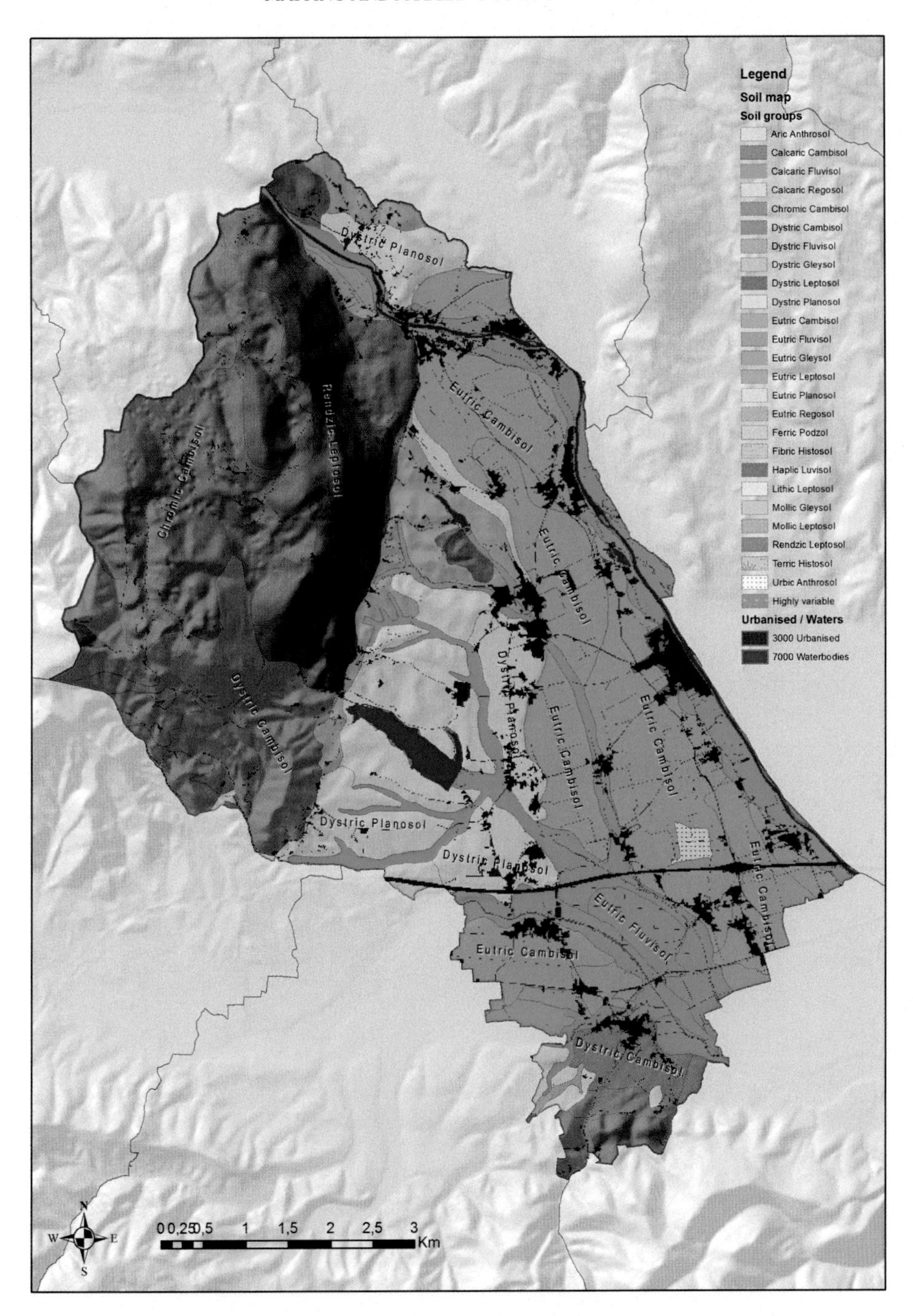

FIGURE 8.28 The initial polygon soil map at a scale of 1:25,000, Municipality of Braslovče (Radišek et al., 2014b).

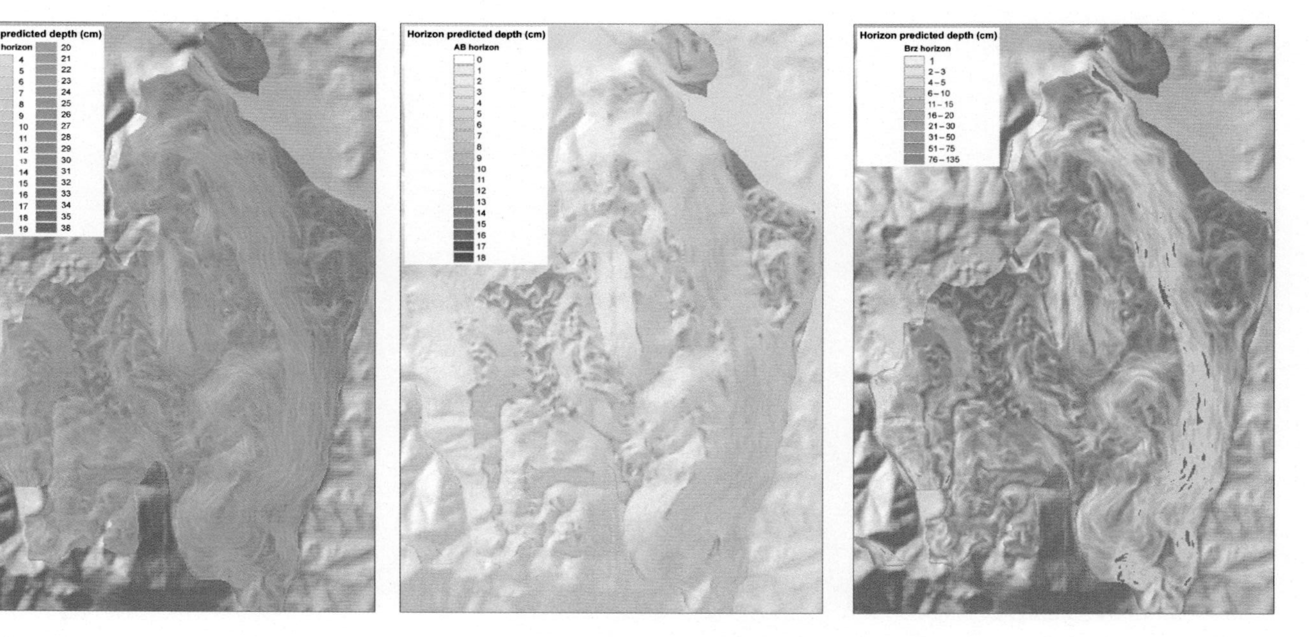

FIGURE 8.29 Raster maps of predicted individual soil horizon depths (example—a Tertiary Limestone area, modeled depth of horizons A, AB, and Brz, from left to right), Municipality of Braslovče (B. Vrščaj).

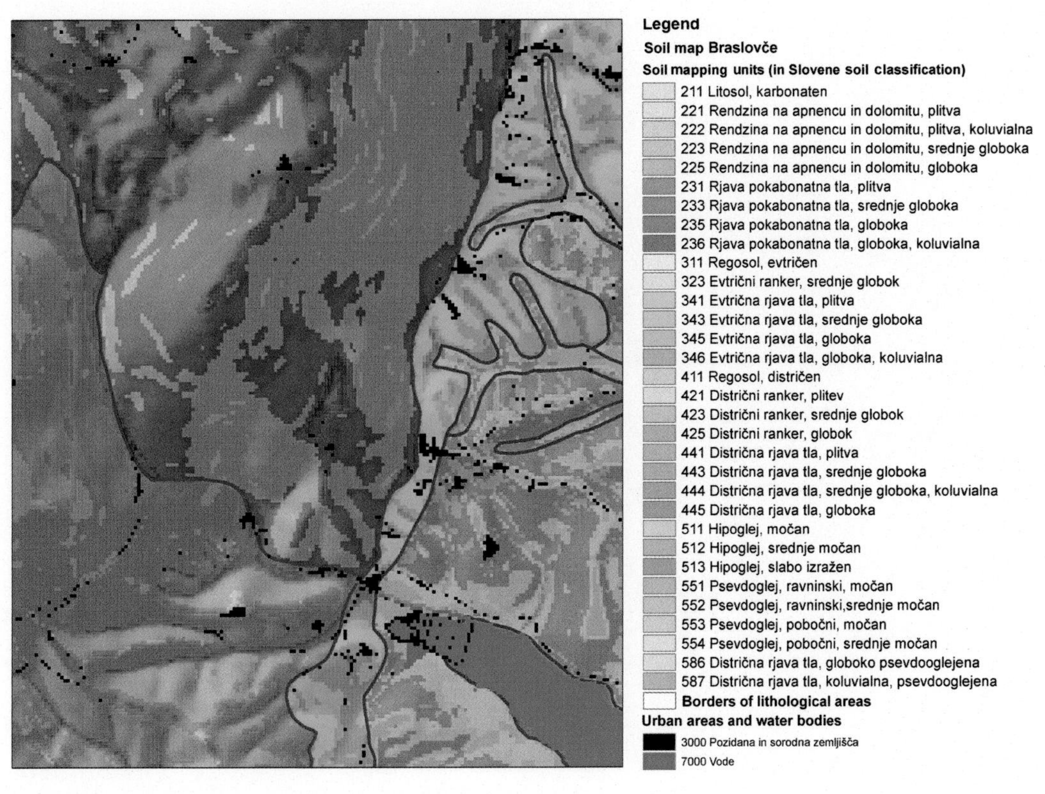

FIGURE 8.30 Final raster soil map composed of separate soil maps of lithologically different areas, Municipality of Braslovče (B. Vrščaj).

at a resolution of 12.5 m per cell size (Fig. 8.30). The last operation—an automatic vectorization of the raster soil map to produce a vector, polygon-based soil map—was used to draw a soil map that would corresponds to a 1:10,000 scale (Fig. 8.31) and to compare it to the initial polygon-based DSM25.

The procedure presented was developed to: (1) test the potential of embedding the expert knowledge of soil surveyors in DSM procedures; (2) develop a method to improve the geometry of the existing 1:25,000 vector-based soil map, and (3) to develop spatial information regarding individual soil properties at better resolution.

The result is better spatial resolution and significantly less subjective data due to the utilization of measured digital information on soil-forming factors. The outputs of this method can replace and significantly improve coarse DSM25 soil information and is better suited to soil quality evaluation and planning in local communities. The expert knowledge-based soil inference maps can be further utilized to produce relevant covariate information for other statistical and geostatistical-based soil inference/DSM methods.

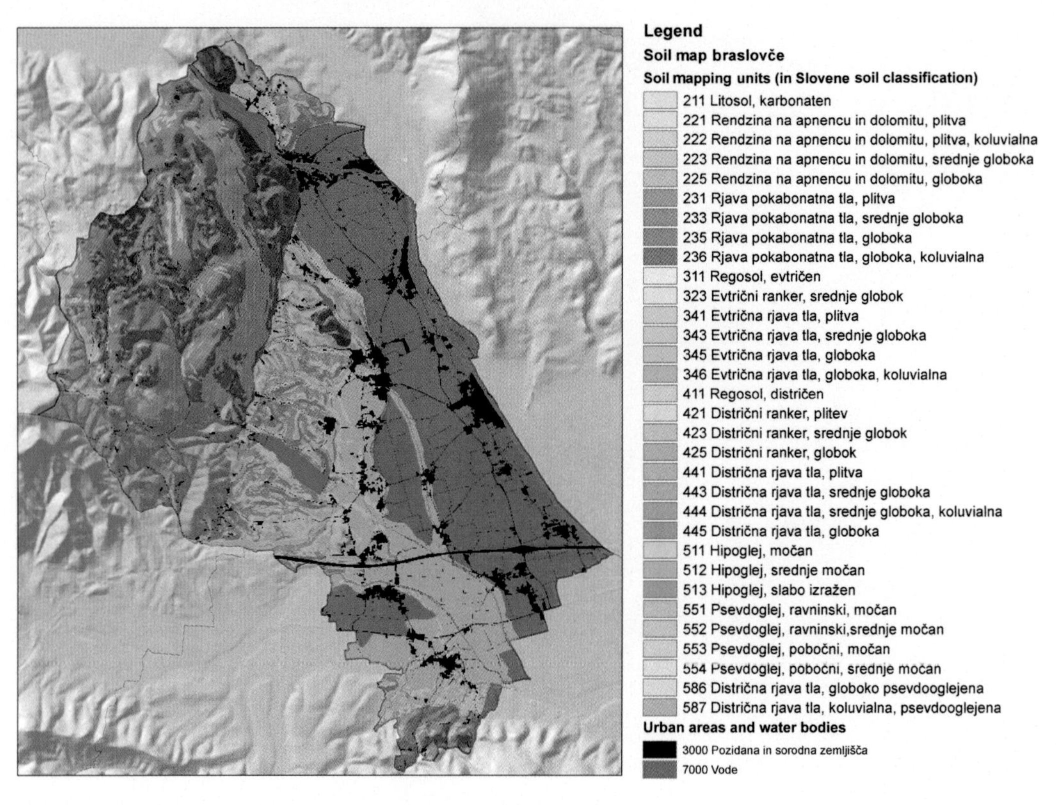

FIGURE 8.31 Final DSM polygon soil map (Slovenian Soil Classification), Municipality of Braslovče (Radišek et al., 2014b).

Soil Properties and Water Management (South Tyrol, Italy)

South Tyrol in Northern Italy is the largest coherent apple-growing area in Europe (18,541 ha and an annual production of 1,130,000 tons in 2015) (APB, 2015). Orcharding in South Tyrol covers a wide land area. Due to the mix of Mediterranean and middle-European climatic conditions in this region, profitable orcharding is possible up to 1100 m. However, the average annual precipitation in the upper Etsch Valley (Val Venosta) is only about 500 mm per annum. Thus irrigation is a necessary factor of production for intensive orcharding and more than 85% of the orchards in South Tyrol are equipped with overhead irrigation systems. These irrigation systems cause a very unequal distribution of water on the soil surface. Nevertheless soil properties have not been considered in irrigation thus far. Furthermore, the water losses of these systems by evaporation and wind drift can be very high and decrease the efficiency of irrigation significantly. Thus the plant roots (main root zone in 40 cm soil depth) are supplied by uneven amounts of water (Grashey-Jansen, 2008; Stimpfl et al., 2006).

Mostly the irrigation follows subjective criteria and at many locations much more water is used for irrigation than the plants actually need (Grashey-Jansen, 2014). The current haphazard system of such irrigation wastes great quantities of fresh water and energy (according to APB (2016c) the current agricultural consumption of water in South Tyrol of 150 mL m^3 per

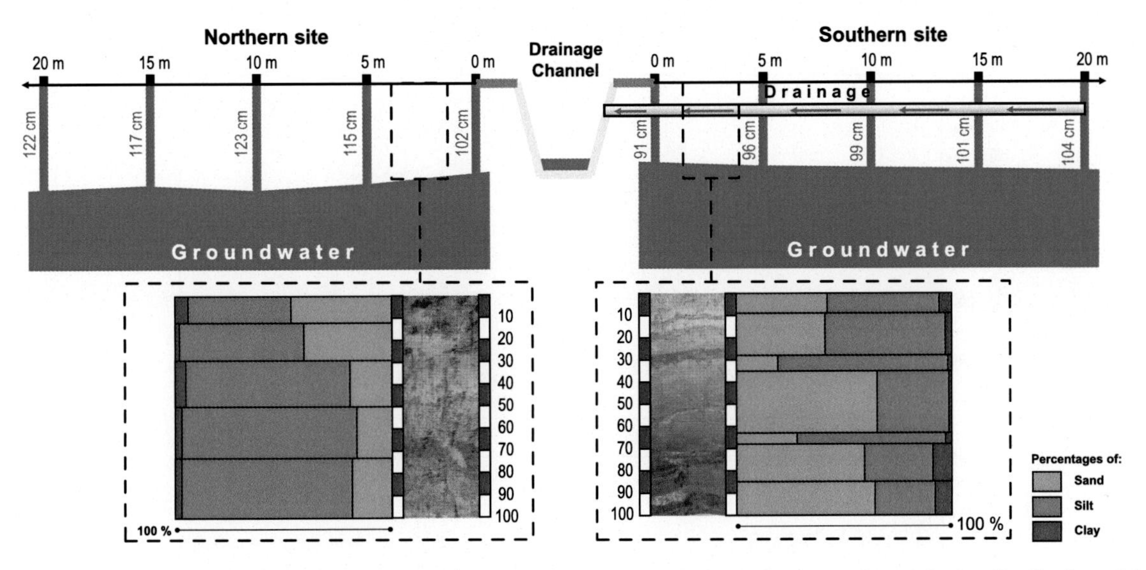

FIGURE 8.32 A comparative depiction of the measured depths of the groundwater and particle size distribution of the two study sites.

annum is three times as high as the industrial water consumption of the region).

The distribution of soil types with different characteristics of *Cambisols*, *Fluvisols*, and *Gleysols* is very heterogeneous and the spatial variability of physical soil properties can vary significantly within small distances across irrigated fields. Furthermore, large areas of the irrigated orchards are very close to the groundwater. Thus reductive pedogenetic features can often be observed, pointing to *Gleysols* and other soil types with gleyic properties. However, paradoxically, these locations are also still irrigated. Furthermore at these sites the groundwater may rise capillarily and supply the tree roots with sufficient water (Grashey-Jansen, 2010, 2012b). To counteract this type of water surplus in the soils and in order to achieve amelioration, some locations are fitted with drainage systems.

Fig. 8.32 shows two neighboring locations in the Etsch-Valley of the Vinschgau (South Tyrol) near the Municipality of Kastelbell, which are in the immediate vicinity of each other. The southern site is equipped with a drainage system, the other one not. Both orchards are fitted with an irrigation system, but due to the rainy weather during the measured period in 2008 (Fig. 8.32), the locations were not irrigated during the observation period.

According to Grashey-Jansen (2010) and Grashey-Jansen and Eben (2009) the soils at both sites can be classified as *Calcaric Gleysols* (with 2–24% $CaCO_3$) and with typical gleyic horizons, which emphasize the influence of groundwater. The soil texture of both sites is similar, but the different particle size distributions in the root zone are crucial: sandy silt at the northern site (sand: 19.2%; silt: 76.1%; clay: 4.7%) and silty sand in the south (sand: 64.9%; silt: 34.3%; clay: 0.8%). The content of organic matter in the main root zone of both soil sites is similar, so this factor is negligible (1.8% at the northern site; 1.1% at the southern site). Due to the grain sizes, soil-hydrological properties could be expected, which has, in turn, consequences for the water supply of the tree roots.

According to Grashey-Jansen (2010), Fig. 8.33 shows the progress of soil water tension

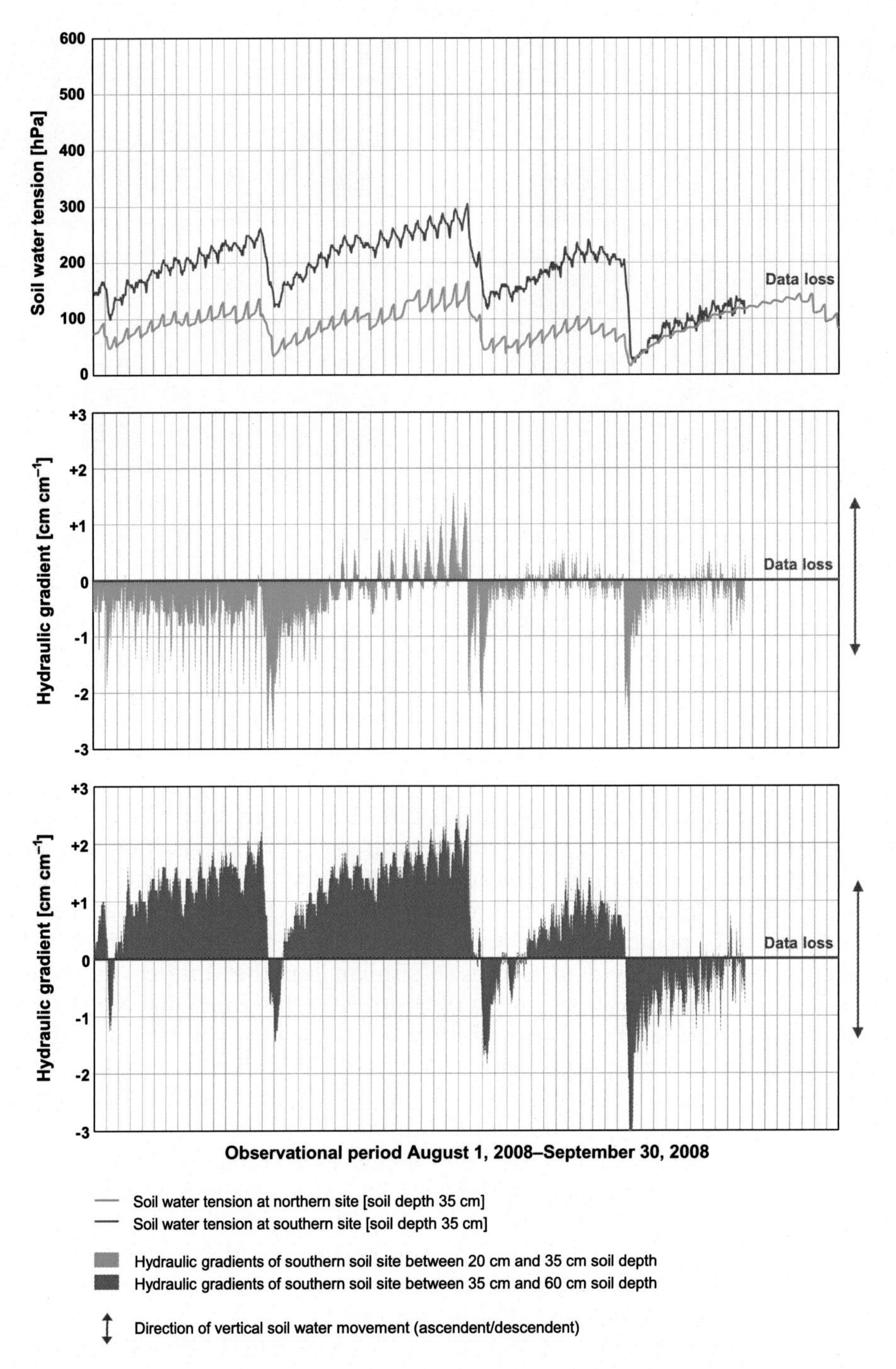

FIGURE 8.33 The soil water tension at both locations (see Fig. 8.32) and the hydraulic gradients of the southern soil site.

measured at a soil depth of 35 cm at both locations. Soil water tension mostly proceeds synchronously and shows an almost identical pattern. The three regressions are caused by areal precipitation. However, for most of the time a tension-difference of about 100 hPa can be noted. This indicates principal differences in the level of soil water tension.

The values of the northern soil site vary at a very low level, which is critical from the perspective of fruit growing. The soil water tension barely exceeds 100 hPa and accordingly suggests saturated conditions. In contrast the soil water tension at the southern soil site ranges between 150 and 250 hPa (much more suitable for apple trees) and therefore is on average 100 hPa higher than at northern soil site. The differences between the water tension cannot only be caused by the mentioned drainage system. The groundwater level is, in spite of the drainage system, about 10–20 cm higher than the level at the other site. The higher soil water tension at the southern site is definitely caused by the significantly higher content of sand.

The hydraulic gradients in the lower part of Fig. 8.33 show the dominance of an ascendant soil water movement between 60 and 40 cm of soil depth. However, a descendent soil water movement predominates between 40 and 20 cm. Thus a capillary rising process from the groundwater table (which, on average, is located 90 cm under the surface) into the upper horizons might exist. This relates to the increase in clay content with soil depth: 0.8% at about 40 cm, 2.4% at about 60 cm, 5.5% at about 80 cm, and 7.9% at about 100 cm. Thus it might be possible to cultivate this site without any additional irrigation (Grashey-Jansen, 2010).

This case study is representative of many locations featuring South Tyrolean orcharding, showing that even between two orchards in close proximity there can be big differences in the soil-physical parameters. These differences are typically neglected in irrigation practice, but have a direct influence on the soil–water-balance and therefore also on the natural water supply of these orchards in the Tyrolese fruit-growing areas. In order to optimize irrigation strategies in an objective way it is very important to pay appropriate attention to the pedological conditions and the substrate-specific differences. Regarding the aspects of climatic change and the scarcity of water resources, the term "precision irrigation" is increasingly being discussed in this region. Thus one can meet the requirements only by considering and mapping the small-scale heterogeneities of the soils.

Mapping Land Suitability to Mountain Viticulture (Aosta Valley, Italy)

The Aosta Valley, as with many other mountain regions in Europe, has a large surface area occupied by vineyards. According to the Centre for Research, Environmental Sustainability, and Advancement of Mountain Viticulture (CERVIM, 2016), the start of viticulture in the area dates back to the Roman age, or even before. The diffusion of autochthonous vines was promoted during the Middle Ages, and the maximum vineyard surface was recorded in the 19th century (around 3000 ha). Since then, the Aosta Valley has experienced a progressive reduction of vineyards, with a minimum of 500 ha in 2000, followed by a steady increase in recent years. At present the estimated vineyard extent is around 520 ha (CERVIM, 2016, the data refer to 2006). The majority of these vineyards suffer from intrinsic limitations related to extreme land fragmentation, as most of the farms have less than 1 ha, are located at high altitude (up to 1100 m) on steep slopes, and are often on man-made terraces (Freppaz et al., 2008). Fig. 8.34 shows two examples of vineyards in the Aosta Valley.

CERVIM has defined such viticulture systems as "heroic viticulture", due to having to cope with extreme environmental conditions,

FIGURE 8.34 (A) A terraced pergola vineyard in Arnad, Valle d'Aosta Region. (B) A terraced vineyard in Montjovet, Valle d'Aosta Region. *Source: photos: S. Stanchi.*

i.e.: (1) an altitude exceeding 500m, (2) a slope >30%, (3) vines planted on terraces or embankments, and (4) vines planted in difficult growing conditions. Additional problems can result from intense water erosion processes, related to severe slopes, shallow soils with limited profile development, and limited organic matter content. Despite these intrinsic limitations, many small-scale wineries can produce high-quality wines, and contribute to preserving the landscape value of heroic vineyards. Moreover, "extreme" environmental conditions (e.g., a water deficit) may determine moderate stress on plants that can be beneficial for the organoleptic properties of mountain wines. In addition, the marginal site conditions are not suited to mechanization and often do not warrant the economic sustainability of vine growing. Considering the potential production and existing limitations, careful planning is needed in order to optimize production, and possibly to find new areas for high-quality wine production.

Land evaluation is a branch of soil science that can support sustainable land use and management. In particular, land suitability evaluation (Calzolari et al., 2006) enables defining the suitability of a site to a specific use (e.g., vine growing). Assessing the suitability to mountain viticulture in marginal areas may have productive, aesthetic, socio-economic, and landscape effects. Site and soil variables need to be mapped and managed in a GIS environment at the municipal or regional scale. Considering the extreme land fragmentation, an ideal sampling strategy for soil should be at plot scale, but very often the soil and environmental data needed for modeling are not available at such detail.

Stanchi et al. (2013b) presented a case study located in the Aosta Valley. The land suitability to mountain viticulture was assessed at the 1:10,000 scale in an 85 km^2 watershed (7°30′33″E; 45°48′01″N) with a prevailing south-facing exposure. Soils in the study area were mainly represented by *Regosols* and *Cambisols* (unpublished data) according to WRB (World Reference Base for Soil Resources; IUSS Working Group, 2006). All soils showed limited total depth (with a range of 35–75cm), high amounts of skeleton, limited chemical fertility, low organic matter content, and potential summer drought due to quick drainage and dry microclimatic conditions.

The main site characteristics affecting wine quality and production were considered, i.e., altitude, slope, aspect, and the main soil chemical and physical properties. According to the FAO land suitability classification, the suitable classes were subdivided into: S1 (suitable, with no significant limitations); S2 (suitable, with moderate limitations); S3 (marginally suitable, with severe limitations). The unsuitable ones were instead identified as N1 (marginally not suitable) and N2 (permanently not suitable due to intrinsic limitations). With regard to the soil variable the effects thereof on vineyard suitability are rather complex, however some general considerations can be drawn: shallow soils, rich in skeleton, with slight to moderate water stress, can significantly improve wine quality, and enhance the production of secondary metabolites; an excess of N has a negative effect on grape quality; organic matter has a positive effect on soils with limited fertility.

In the case study the following parameters were included in the automated GIS procedure (the details are described in Stanchi et al., 2013b): the Winkler index—i.e., the effects of the thermal gradient (Orlandini et al., 2005), slope classes, aspect, and soil type (at the level of soil associations). The result was a 1:10,000 suitability map for vine growing in the mountain study area, using as input the topographic and soil data (Fig. 8.35). An automatic classification and overlay procedure was carried out, which can be implemented for other study areas. The most limiting factor was altitude, which influences climate, followed by slope and aspect. The approach developed for the study area can be transferred to similar mountain sites, eventually adding further parameters. Of course, a basic set of soil properties should be mapped, i.e., pH, texture, C and N content, and carbonate content.

Considering Soil Characteristics in the Designation of Protection Forests (Tyrol, Austria)

The tasks of forest planning include proactive spatial planning of the protection service of forests as an important forest ecosystem function in mountain areas. Thereby the designation of protection forests is an important part of the Austrian Forest Development Plan, which is prescribed by forest law (Forest Act, 1975, § 8) and serves as a basis for political and legal decisions of relevance to forestry.

The two main groups of protection forests in Austria are object-protecting and site-protecting forests. The former ones are directly connected with the protection of a certain object of high public interest, e.g., settlements, infrastructure, or agricultural land. In contrast, site-protecting forests serve for the maintenance of forests, soil, and reforestation when these goals are threatened by environmental factors (e.g., water, gravity) or management. For the present considerations the latter ones are more relevant as they are mainly defined by site and soil characteristics. More specifically, these are sites on eolian soils, shallow soils, sites with erosion risk, or potential exposure to karstification (Forest Act, 1975) that qualify for site-protecting forests.

The level of public interest in the protection service and therefore in the maintenance of site-protecting forests is assessed in terms of three categories (BMLFUW, 2012). These categories represent the value of the site-protecting forest in question and range from S1 (little public interest), and S2 (increased public interest), to S3 (special public interest in the protection service). Since hitherto no spatial soil data on Tyrolean forests are available, soil characteristics had only a minor influence. The designation of the protection category is mainly based on natural hazards such as avalanches, rock falls, or landslides, slope inclination, slope aspect, and altitudes close to the timberline. Although some of these factors no doubt influence soil characteristics, their importance for the designation of site-protecting forests should be explained with the case study below.

In order to investigate the influence of soil characteristics on the designation of protection forests, 1,058 soil profiles from different categories of site-protecting forests were analyzed. The assigned category was taken from the current Forest Development Plan. The focus was on soil depth in general and terrestrial soils with high organic carbon content (*Histosol* and *Leptosol*) as soils vulnerable to erosion, SOC (Soil Organic Carbon) reduction, and exposure to karstification.

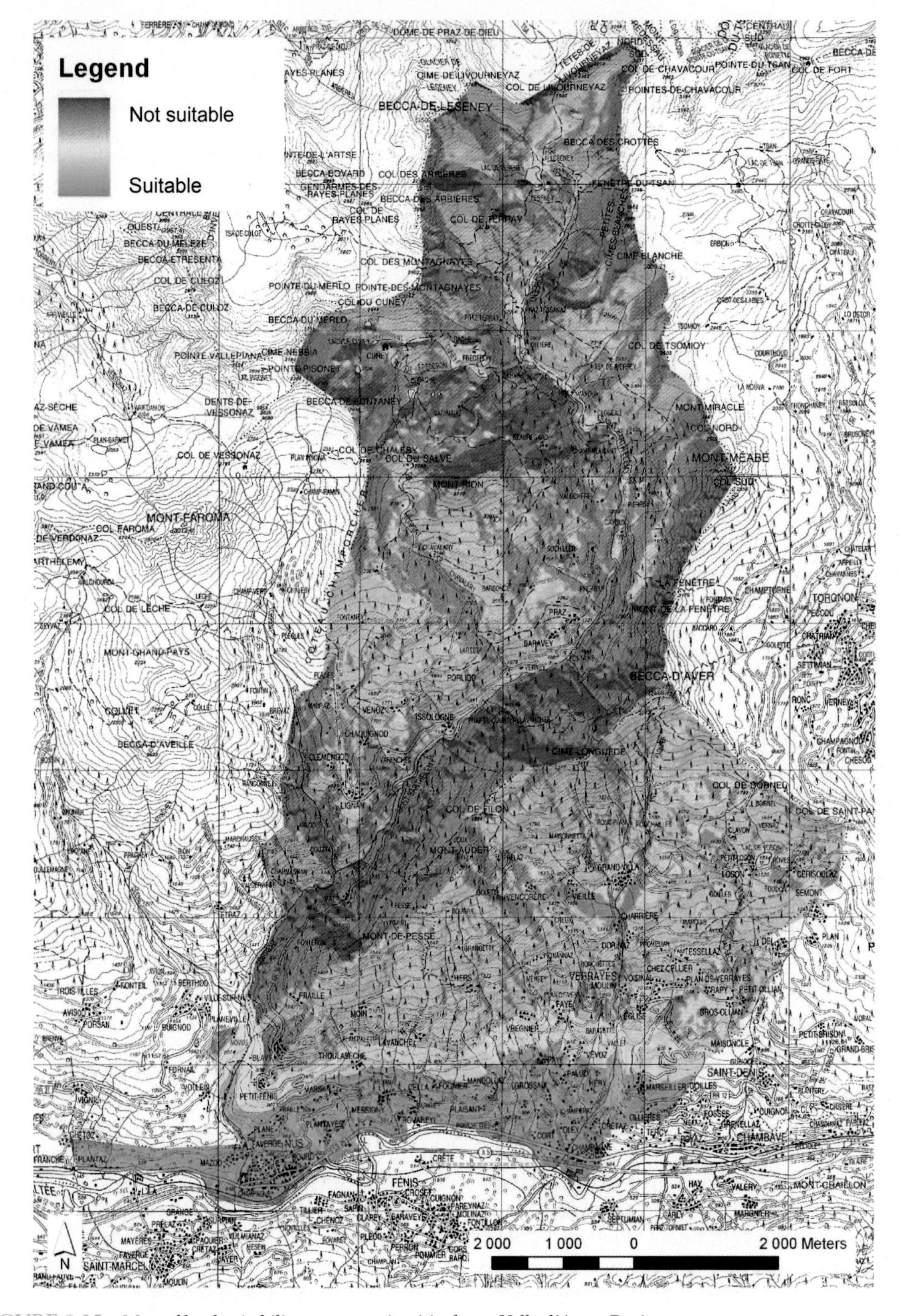

FIGURE 8.35 Map of land suitability to mountain viticulture, Valle d'Aosta Region.

In order to account for soil depth, at each profile, for up to 100 cm of depth or until bedrock was reached, the amount of fine earth was calculated as

$$\text{Fine soil}\,[\text{dm}^3 * \text{m}^{-2}] = \sum_{i=1}^{n} \mathbf{D}[\text{dm}] * ((100\text{-}\mathbf{CG}[\%])/100) * 100$$

where D is the soil horizon depth and CF is the coarse fraction (>2 mm).

The distribution of the amount of fine earth (<2 mm) for different protection categories is displayed in Fig. 8.36. Forests with higher public interest in the protection service have lower amounts of fine earth, S3 with 314 ± 235 $dm^3 m^{-2}$, S2 with 450 ± 280 $dm^3 m^{-2}$, and S1 with 531 ± 282 $dm^3 m^{-2}$. These differences are highly significant (Kruskal–Wallis test, $\chi^2 = 126.6$, $P < .0001$), confirmed by a post-hoc test (Nemenyi test) with significant differences between all categories ($P < .005$). These results emphasize that soil depth can be seen as an important factor in the designation of protection forests. Nevertheless it demonstrates that extremely shallow soils are partly assigned to less vulnerable categories. The areas of shallow soils are difficult to determine, particularly when access is difficult or there is closed ground vegetation cover, and a lack of soil augering, regular soil outcrops, or spatial soil information.

In order to examine another example of the influence of soil characteristics the designated protection category at terrestrial sites with high organic carbon content (*Histosol* and *Leptosol*) should be investigated. As Fig. 8.37 shows, most of the 252 sites are in category S3 (78%), while category S2 accounts for 12%, and category S1 for 9%. Thereby, terrestrial *Histosols* and *Leptosols*, especially at low slope inclination, need to be discussed from the perspective of organic carbon reduction. After disturbances in the forest ecosystem (e.g., due to windthrow), this soil type turned out to be highly vulnerable to organic carbon losses and exposure of bedrock (Göttlein et al., 2014). Hence, it is important to consider them in the designation of site-protecting forests even at flat sites. In order to tackle this challenge, raising awareness of the importance of organic

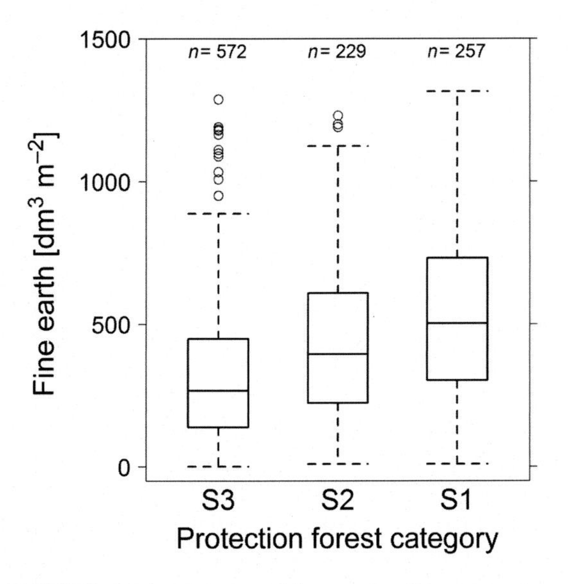

FIGURE 8.36 Amount of fine earth of the protection-forest categories (line = median, n = number of soil profiles).

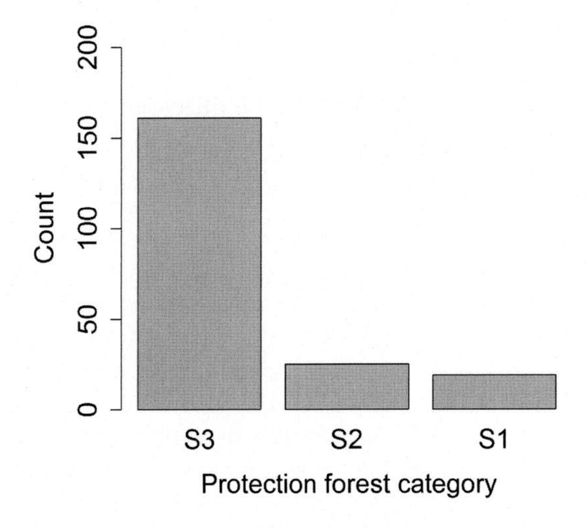

FIGURE 8.37 Protection-forest categories at Histosol and Leptosol sites (total n = 205)

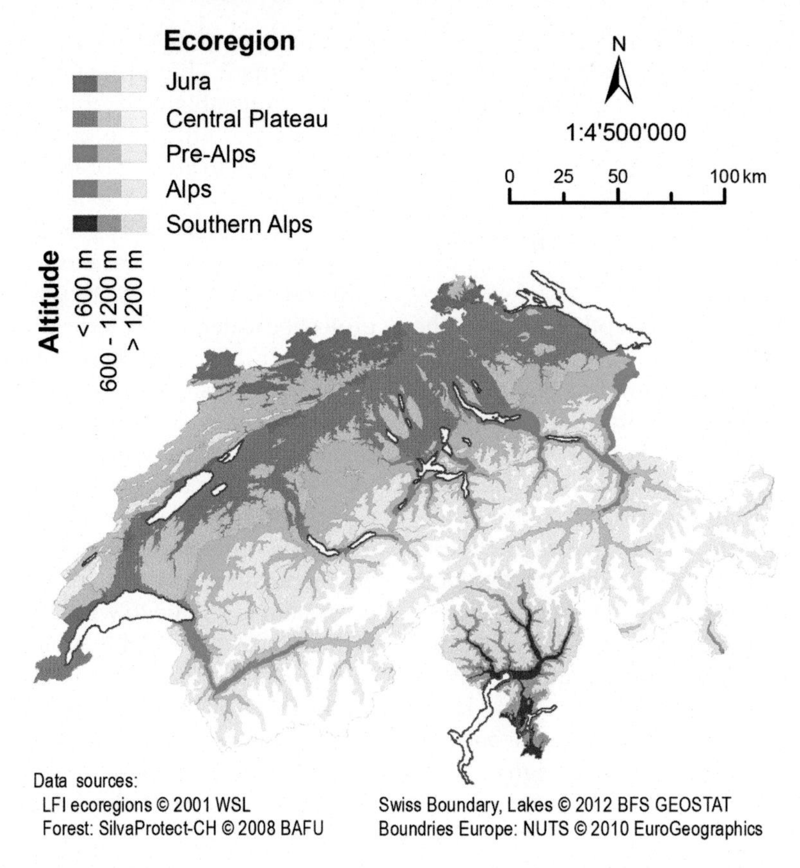

FIGURE 8.38 Map of the ecoregions of Switzerland (Brassel and Lischke, 2001), stratified by altitudinal class (© M. Nussbaum et al., 2014).

carbon (humus layers) as the rhizosphere and its spatial distribution is needed.

The examples presented display only a small section of the basic principles for the designation of protection forests. Nevertheless it supports the importance of considering soil characteristics, as is intended in the legal framework (Forest Act, 1975; BMLFUW, 2012). Furthermore, the examples show how relevant soil information can be in order to further improve and objectify the designation of protection forests. By providing spatial soil information for forest planning tools, a contribution towards sustainable land-use management can be achieved.

Modeling Soil Carbon Stocks in Forests (Switzerland)

Estimates of SOC (Soil Organic Carbon) stocks are required for national greenhouse gas (GHG) inventories in the frame of the United Nations Framework Convention on Climate Change and the Kyoto Protocol. SOC stocks are used to estimate carbon (C) sources and sinks caused by land-use change. In its GHG inventory (FOEN, 2016), Switzerland reports on SOC stocks for the whole country as well as for its five ecoregions, i.e., Jura, the Central Plateau, the Pre-Alps, the Alps, and the Southern Alps, further stratified by altitude (Fig. 8.38).

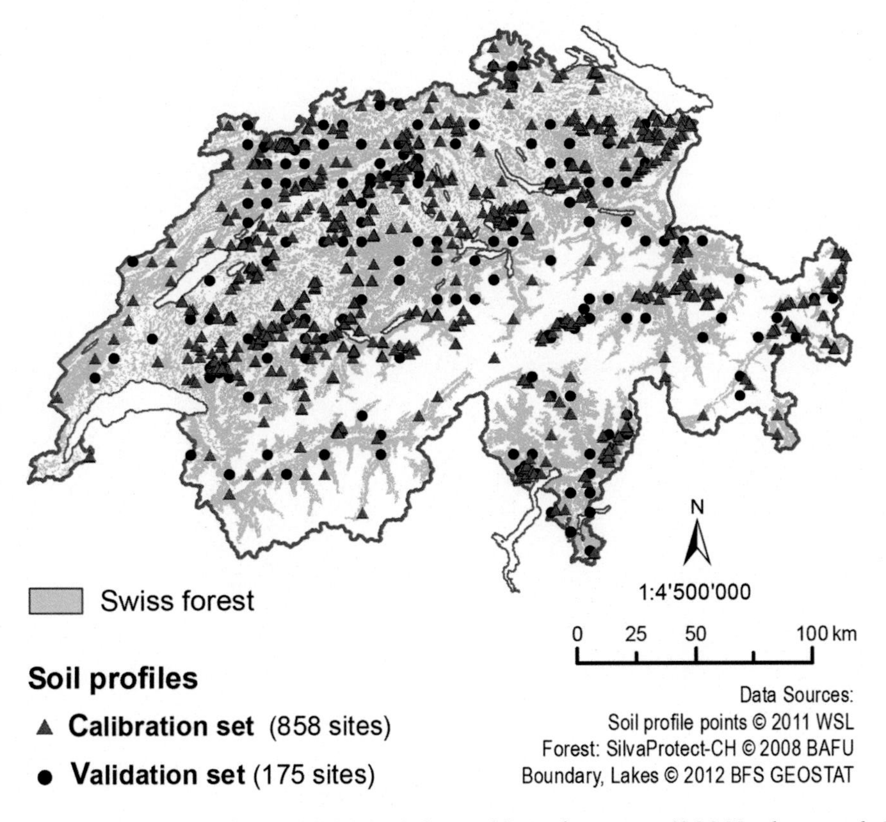

FIGURE 8.39 Locations of soil profiles with SOC stock data and Swiss forest areas (© M. Nussbaum et al., 2014).

In a recent study, Nussbaum et al. (2014) used a geostatistical DSM approach to map forest soil SOC stocks across the country and to estimate the national mean SOC stock and regional means for the ecoregions. The study was limited to forest soils because harmonized and geo-referenced soil data was only available for this land use. Forest SOC stocks are important for the Swiss GHG inventory because forests cover about 45% of the vegetated area of Switzerland—and have an even larger share in the Alps—and per area store more organic C than arable and grasslands soils (FOEN, 2016). This section summarizes the main findings from the study by Nussbaum et al. (2014) and discusses some aspects relating to the Alps.

The study used data from 1,033 forest soil profiles (Fig. 8.39) provided by the Swiss Federal Institute for Forest, Snow, and Landscape Research WSL. Data on SOC stocks stored in the top 1 m of the mineral soil of these profiles were related by a robust geostatistical regression analysis to environmental covariates that characterize the soil-formation conditions (geologic parent material, climate, vegetation, and topography) at the profile sites. The fitted model was then used to compute predictions of SOC stocks at 0–100 cm soil depth by robust external-drift Kriging (Fig. 8.40).

Mapping units of the 1:200,000 soil suitability map (Swiss Federal Statistical Office, 2000), precipitation, reflectance in the near-infrared (NIR) band of the SPOT5 mosaic (Mathys and

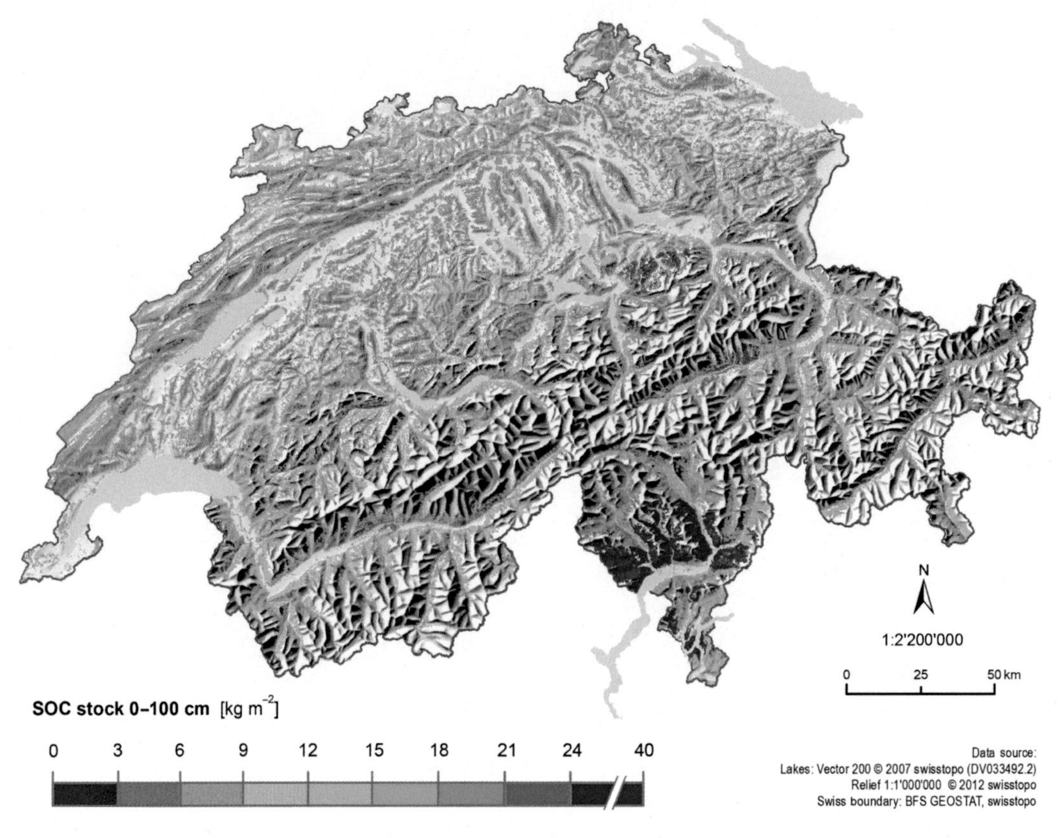

FIGURE 8.40 Predictions of SOC stocks at 0–100 cm depth of the mineral soil of Swiss forests (© M. Nussbaum et al., 2014).

Kellenberger, 2009), and slope were selected from a very large set of 360 potential covariates as explanatory variables for the regression model. The units of the soil suitability map accounted for differences in carbon stocks between the ecoregions and regions with special parent material or climate (Nussbaum et al., 2014). Generally, small stocks were predicted for lower altitudes in the central Alpine valleys (Valais, Lower Engadin) and on the Central Plateau (Fig. 8.40), but the predicted stocks were the smallest for areas in the eastern Pre-Alps and Alps where Permian Verrucano sand stone forms the bedrock. At these sites, SOC accumulates in the forest floor, which was not considered when computing mineral soil stocks. Fig. 8.40 further shows large stocks up to $40\,kg\,C\,m^{-2}$ in parts of the eastern Pre-Alps that get a lot of precipitation and where waterlogged soils on Flysch prevail. Very large SOC stocks were also predicted for the Southern Alps, where a combination of forest fires and Al-rich soil on metamorphic parent material led to an accumulation of organic matter.

Stock increased with increasing precipitation and decreased with increasing NIR reflectance. These findings agree with the common understanding that a wet and cool climate and the predominance of conifer trees (small NIR reflectance) favor SOC accumulation. However, the interpretation of the positive dependence of SOC stocks on slopes is less clear: One would rather expect decreasing stocks with increasing slope due to more severe erosion on steep terrain. Furthermore, cosine (northerness) and sine (easterness) of the aspect angle had unexpectedly no influence on SOC stock.

A discrete approximation of external-drift block Kriging was used to predict the mean SOC stocks for the whole country and the ecoregions (Fig. 8.41). In the Pre-Alps—as wells as on the Central Plateau and in the Jura mountains—predicted SOC stocks increased significantly with altitude, likely due to increasing precipitation and decreasing temperature. In the Southern Alps, stocks did not depend on altitude. Sites at lower altitude in Ticino are influenced by forest fires (leading to the accumulation of black carbon in the soils), and a large amount of aluminum and iron weathered from silicate-rich bedrock stabilize SOC there (Blaser et al., 1997).

A comparison with earlier studies (Perruchoud et al., 2000; Bolliger et al., 2008) revealed that previous estimates had underestimated the mean national forest SOC stocks for Switzerland. Furthermore, the current estimate of 12.58 kg C m^{-2} (standard error 0.24 kg C m^{-2}) is more precise than the previous estimates. Geostatistical DSM (Digital Soil Mapping) techniques thus not only enable concise and interpretable summarization of the structure of the spatial distribution of a soil property and the mapping thereof over an area of interest. In addition, these methods enable the prediction of spatial means of a response variable over arbitrary regions of interest.

Mapping Susceptibility to Snow Gliding in the Monterosa Ski Resort (Aosta Valley, Italy)

Despite involving slow movements, snow-gliding processes can be responsible for significant damage to buildings and structures (e.g., Höller et al., 2009; Margreth, 2007a) and significant soil erosion (e.g., Ceaglio et al., 2012; Stanchi et al., 2014; see Fig. 8.42). The intensity of the gliding process is known to be strongly affected by the conditions at the soil/snow interface (e.g., temperature and water content of the snow, temperature at the snow/soil interface, surface roughness), besides the topographically predisposing variables (slope, aspect).

Ski area managers need to consider snow-gliding processes in their planning actions,

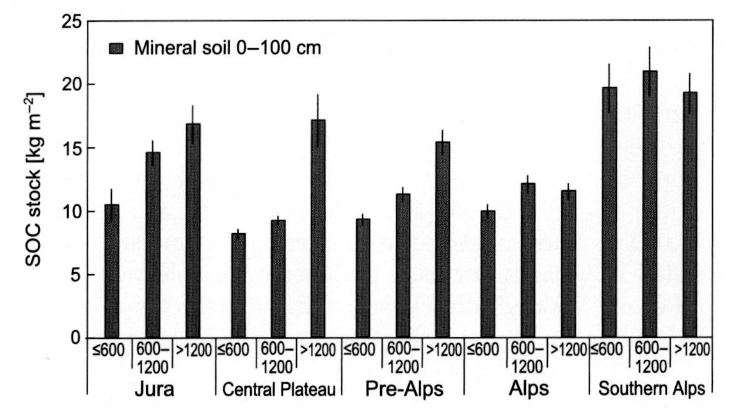

FIGURE 8.41 Block Kriging predictions of mean SOC stocks at 0–100 cm soil depth for Swiss ecoregions stratified by altitudinal class (vertical lines: 95% prediction intervals for regional mean stocks, © M. Nussbaum et al., 2014).

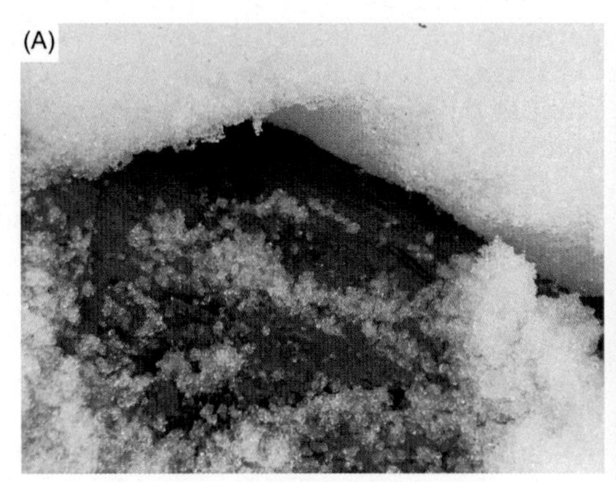

FIGURE 8.42 (A) Soil surface after a snow-gliding process. It is possible to observe a thin layer of soil (through liquefaction and loss of consistency) which may contribute to the snow-gliding process. (B) Soil erosion patches in areas prone to snow gliding and glide-snow avalanches. *Source: photos: M. Freppaz.*

TABLE 8.3 Classification and Weighting Factors of the Different Input Parameters for the Susceptibility Index

Slope Angle	Weight	Land Use	Weight	Roughness	Weight	Aspect	Weight
0–15°	0.0	Grassland	10.0	Low	10.0	South	10.0
15–25°	5.3	Dwarf schrubs	5.2	Medium	2.5	East, West	6.0
25–35°	7.8	Bushes	2.4	High	1.0	North	2.0
35–60°	10.0	Coarse scree	0.8				
60–90°	0.0	Forests	0.3				
		Other	0.0				

as the damage caused by snow gliding may be greater than that produced by avalanches. Although some guidelines for building and engineering purposes in areas subject to snow gliding already exist (e.g., Margreth, 2007b), no homogeneous mapping criteria have been proposed yet. Maggioni et al. (2016) proposed a model to assess the susceptibility of snow to gliding using a GIS-based procedure (at 2 m and 10 m grid sizes) in a ski resort located in the NW Italian Alps (Monterosa Ski Resort, between Piedmont and the Aosta Valley). The study area (~140 km^2) ranges from 1500 to 3500 m and the average annual accumulated snowfall is 631 cm (time series 1928–2001) with 49 days of snowfall and 228 days of snow cover (Mercalli et al., 2003). A southern aspect prevails (48% on SE–S–SW aspects), and the vegetation cover is characterized mostly by grassland (33%), and secondly by forest (20%).

The factors considered were: slope angle, land cover, roughness, and aspect. These factors can deeply affect the conditions at the snow/soil interface (e.g., altering soil moisture). The range for each parameter variable was subdivided into classes, and for each class a weight in the range 0–10 was assigned, based on literature data (Table 8.3).

A map was obtained (Fig. 8.43) by processing the weighted rasters and making a

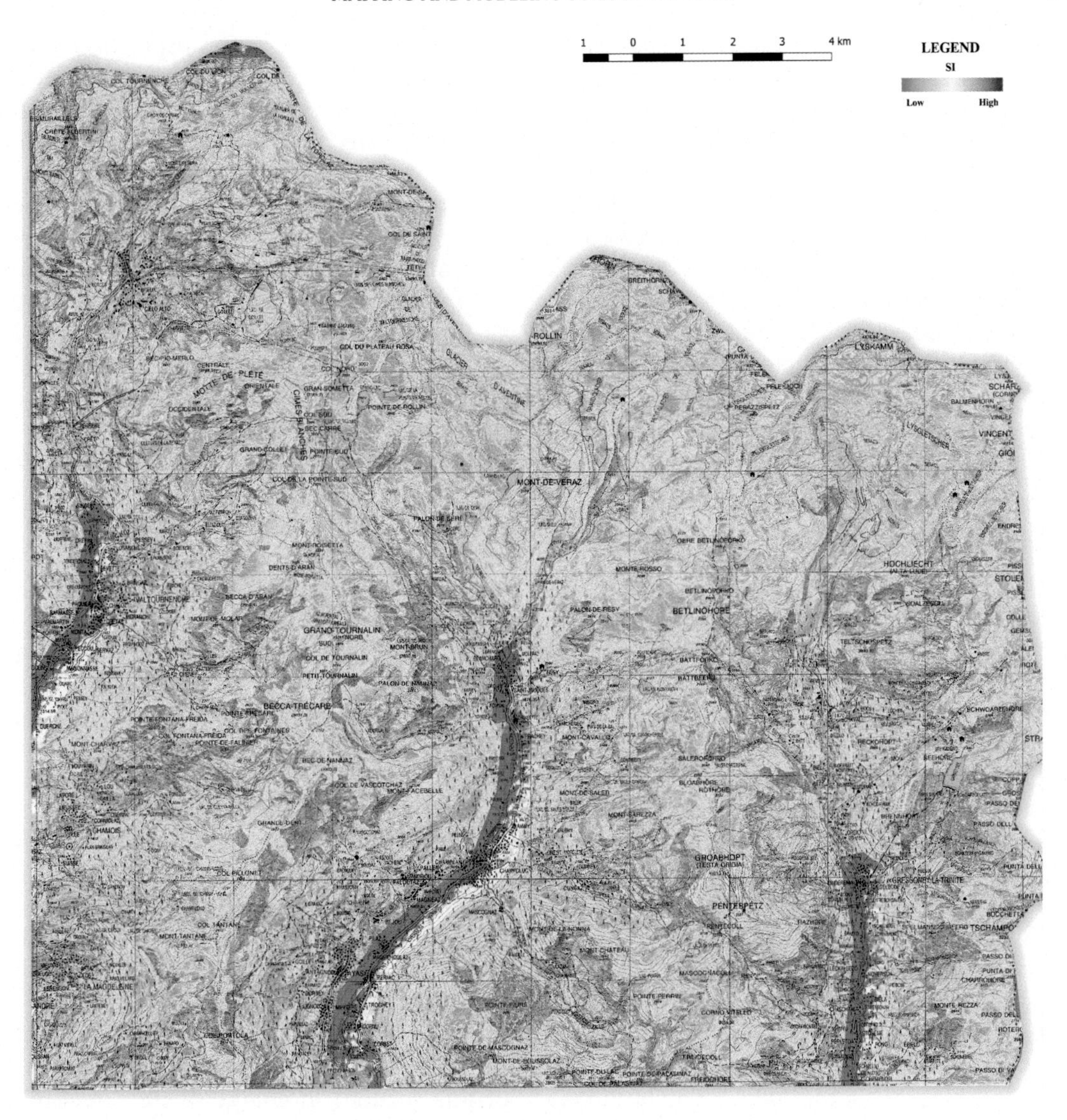

FIGURE 8.43 Extract from the snow-gliding susceptibility map.

comparison with (1) qualitative survey information provided by the Monterosa Ski Resort personnel (expert-based information), and (2) available photographs. The overall correspondence between the susceptibility map and the field observations was satisfactory. The presence of an extensive and complete soil database would help improve the model, enabling the use of relevant input data such as soil texture, soil erodibility, and Atterberg limits (Stanchi et al., 2013a). Furthermore, continuously mapping the preparatory and triggering components (Heckmann and Becht, 2006), such as snow temperature, total snow depth, and soil and air temperature, would be beneficial. Also the soil moisture at the soil/snow interface and the soil strength properties might provide useful insight in terms of the preparatory and triggering conditions in soil, but in order to do so, focus on sites equipped for snow-gliding assessment is needed.

In the future, snow-gliding susceptibility maps could be used to identify the areas most prone to snow-glide avalanches on a large scale, while statistical models might provide threshold values for the most relevant soil, snow, and weather parameters, to be used as an early warning monitoring system for the most dangerous sites shown in the map.

Sustainable Land Use in Construction Projects (Switzerland)

In Switzerland there currently exist two important instruments for environmental and soil protection within construction projects outside official construction zones. First the Environmental Impact Assessment, second the instrument of the Environmental Protection Expert for Construction Projects (EPECP), and for soil-related construction projects, the Soil Protection Expert for Construction Projects (SPECP, recognized by the Swiss Soil Science Society). The expert's duties are prescribed in the Swiss Standards (SN 640583) and he or she has discretionary power (Bono et al., 2014). In Switzerland the necessity of soil protection on construction projects is widely accepted. As the quality check of the Swiss Soil Science Society revealed (BGS, 2015), the SPECP has become a successful and effective instrument for chemical and physical soil protection since its introduction in 2001. The SPECP's tasks are to accompany building projects from the planning and authorization phase to the submission and construction phases, and finally to the follow-up phase.

Although various laws, regulations, and guidelines clarify soil protection in general, practical guidelines that include an integrated approach for the mountainous and alpine region are still missing. Protection methods adapted to lower altitudes are often not applicable in alpine areas. Due to enhanced construction activities, the alpine environment is under increasing pressure. Different sites illustrate that destructive construction methods lead to long-term damage. Therefore preemptive and protective methods are necessary. It is of utmost importance that the heterogeneity of the alpine nature (see "The Alps and Their soils" section) is incorporated within the solution-finding process in construction projects.

In the Swiss alpine territory little basic information exists for the work of EPECP/SPECPs. Large-scale geological and hydrogeological maps, occasionally vegetation maps and maps of protected areas, might be available. High-resolution relief plans are a valuable and oftentimes accessible tool. Other information, such as geomorphological observations or estimations of water balance or land use, as well as the overlapping of different factors, is obtained through on-site assessment by the EPECP/SPECP. Usually, additional soil core samples or soil profiles help to characterize soil formation and soil properties, as well as the related ecosystem functions. Oftentimes, detailed soil maps are unavailable. The financial input needed to prepare high-resolution soil maps for construction sites is generally not deemed to be justifiable.

These were the basic conditions that the EPECP/SPECP (Nina von Albertini, the author of this section) in charge of the street-renewal project at the Julier Pass of the Cantonal Civil Engineering Office encountered. The project entailed the significant enlargement and a partly new path of the pass road. During the execution phase from 2008 to 2013, important new embankments (fillings and cuttings), road dismantling, protection dams against avalanches, and renaturations of stretches of water were executed (Fig. 8.44). 230,000 m^3 of material was moved and 18 ha affected. The vegetation plans showed agricultural land, different alpine

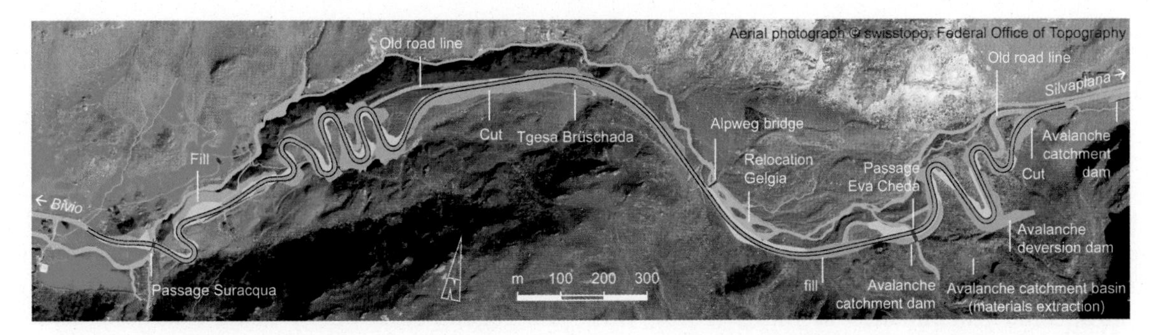

FIGURE 8.44 Part of the construction project at the Julier Pass with the new alignment of the road, filling and cutting constructions, avalanche barriers, road dismantling, and further elaborated constructions. Builder and plan: Cantonal Civil Engineering Department Graubünden.

grasslands, alpine heaths, scree vegetation, avalanche deposit zones, spring vegetation, and fens, as well as watercourses with riparian zones. Based on point-based soil samples the EPECP/SPECP elaborated a soil protection concept. This phase enabled knowledge of the territory to be gathered and the development of practicable ideas.

At the Julier Pass the thickness of soil layers varied from raw soils with a minimal soil cover to depths of up to 1 m. The heterogeneity, not only of the soils (Fig. 8.45), but also of the developed ecosystems, called for interdisciplinary approaches and protection measures in order to ensure that the complex key conditions were respected. Conventional alpine greeneries of the last 30 years show that usually the site-specific vegetation with its original heterogeneity is lost. As Curtaz et al. (2014) demonstrated, rebuilding soils often leads to the nonreversible homogenization of the soil's composition and horizons as well as a simplification of the relief (Fig. 8.46A). In addition, soil losses and work delays resulting from bad weather conditions can occur as a logical consequence of the customary method of loading, transporting, depositing, reloading, and, finally, positioning the soil.

Therefore in this special case, it was essential to develop new methods that would enable the preservation or rebuilding of the heterogeneous and interconnected structures. Consequently, for the elaboration of adequate soil protection measures the SPECP not only had to deal with little existing data and a degree of high complexity, but also to make allowance for the necessity of an interdisciplinary overview and the extensive

FIGURE 8.45 Soil profile at the Julier Pass at a hillside location rich in organic matter and with a high presence of boulders. *Source: photo: N. v. Albertini.*

applicability of the chosen methods in various ecosystems. Additionally the methods needed to be economically acceptable and easy to integrate into the construction process.

The direct-shifting of soil and its vegetation, as realized and further developed at the Julier Pass, is the ideal solution to the aforementioned challenges in the context of alpine conditions. As shown for the Julier Pass in the analysis of site-specific resemblance in Marti (2015), this way of adapting the landscape enables the integration of technical construction in a plausible way without destroying the character of the landscape. The restoration of a natural, well-adapted relief by means of direct-shifting is feasible even with low labor input (Fig. 8.46B).

In order to successfully plan the direct-shifting method and to determine its costs the thickness of the soil and its layers and the presence of soil skeleton on the surface and within the soil profile is the most important information. In direct-shifting the soil is lifted up together with its vegetation cover by using a large excavator shovel and redeposited within the reach of the excavator on newly prepared subsoil in one single movement (Figs. 8.47 A and B and 8.48).

FIGURE 8.46 (A) A conventionally constructed embankment with technical drainage. The forming of the relief, the soil, and the vegetation are not adapted to the environment of the site. (B) A direct-shifted, completely new embankment is shown below the red line; above is the natural terrain. Compare to Fig. 8.49A, where the slope is being shifted. *Source: photos: N. v. Albertini.*

FIGURE 8.47 (A) Extraction of fen sods (Flachmoor-Soden) in the area of the future alignment of the road. (B) Direct placement of the fen sods in the newly prepared area. *Source: photos: N. v. Albertini.*

Thereby, the complex is loosened and slightly disrupted. Compared to conventional handling the soil experiences only minimal physical interference. As an ecosystem with its specific components, the soil remains comparably functional and heterogenic.

Therefore the method can be applied even under nonoptimal weather conditions. Regarding the short vegetation and construction season in the Alps, this is a very positive characteristic in terms of project costs. Thanks to a reduction in deposit surfaces, temporary access roads or technical protection against erosion and seeding, direct-shifting allows savings as regards time and the cost of transport, machines, and labor.

FIGURE 8.48 Newly built fen 10 months after direct-shifting. *Source: photo: N. v. Albertini.*

The preservation of the autochthonous ecosystems enables the continued existence of soil material in its typical layering, including root horizons, soil organisms such as mycorrhiza (Graf and Frei, 2013), the seed pool, and the growing vegetation. Consequently, erosion risk is kept low without any high-input technical methods. Marti (2015) demonstrated this even in the case of slopes with more than 30° inclination and with almost immediate effect (Fig. 8.49A). Only minimal erosion occurs in the variable spaces between the sods with accumulation in the close proximity (Fig. 8.49B).

As noted by von Albertini and Regli (2012), about 65% of the newly built embankments (mostly alpine grasslands) as well as about 2,000 m^2 of fens and riparian zones were successfully compiled by direct-shifting of sods, including their root horizons (Figs. 8.47A–8.51A). The direct-shifting even included the transfer of boulders with their proper lichen

FIGURE 8.49 (A) Advanced filling construction with direct-shifted sods integrating structural elements such as boulders and dwarf shrubs. The sods and their root horizons are removed from behind the excavator as the filling gains height. (B) Sod mosaic with irregular gaps 1 year after direct-shifting. Before the emergence of complete vegetation cover through the expansion of the autochthonous flora, small-scale erosion and accumulation of soil material within the gaps may occur. These processes can be accepted as they contribute to site-specific heterogeneity. *Source: photos: N. v. Albertini.*

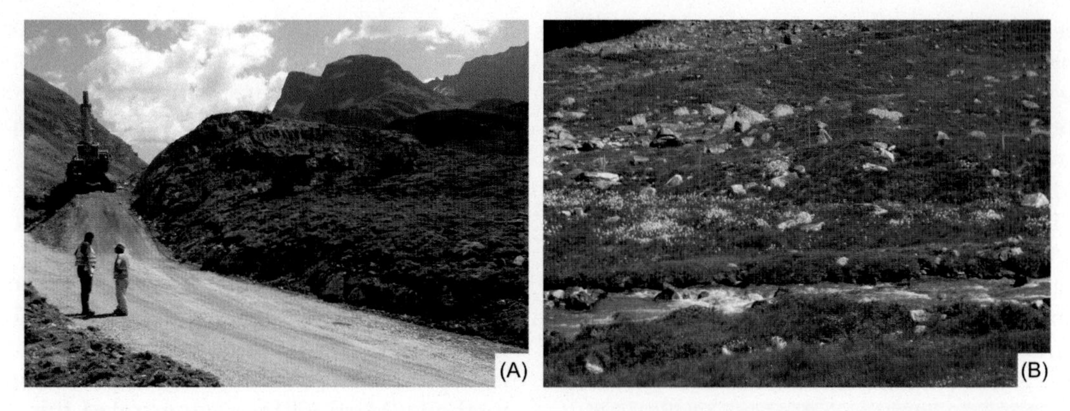

FIGURE 8.50 (A) A recently constructed, direct-shifted slope adapted to its natural surroundings. The terrain was lowered up to 7m. (B) The newly built fen from Fig. 8.47, 4 years after direct-shifting (below the fence, shifted alpine grassland followed by fen vegetation). *Source: (A) photo: Tiefbauamt Graubünden; (B) photo: N. v. Albertini.*

FIGURE 8.51 (A) The same slope as in Fig. 8.49A less than 2 years after completion. Right of the newly built creek is the natural terrain, on the left the direct-shifted slope. (B) The same surface as shown in Fig. 8.49B 3 years after the shifting. The autochthonous sod vegetation has colonized most of the gaps. *Source: photos: N. v. Albertini.*

or dwarf shrubs. Floristic monitoring (von Albertini, 2014) was able to show that 3 years after construction 80% of nearly 200 mechanically shifted dwarf shrubs had survived (Fig. 8.46B). Furthermore, the number of species, the composition of vegetation, and the coverage of vegetation in direct-shifted areas and in natural surroundings corresponded highly. Moreover, there was no import of alien species and varieties caused by seeding, which reduced the settling of problematic species or neophytes. To conclude, direct-shifting leads to a reduction in the overall impact on soil, flora, and fauna in the construction area over space and time.

In the context of the Julier Pass road project the method of direct-shifting was further developed and broadly implemented. This becomes evident following a review of the literature, such as the guidelines for the revegetation of high-altitude areas (Locher-Oberholzer et al.,

2008) and the guide for natural revegetation in Switzerland (Bosshard et al., 2013) or Krautzer et al. (2012). In these publications, sod transplantation usually gets marginal attention and is often only mentioned in combination with seeding, and with a feasible realization limited to spring and autumn. Unfortunately, in the literature, the assumption that sod transplantation is very laborious and feasible only with manual labor, thus quickly leading to high costs and time overruns, is incorrect and widespread.

In Wittmann and Rücker (2012) a technically and floristically successful implementation seems to have been carried out. The consideration of the landscape aesthetic, though, seems to have the potential to be further developed, as was shown by Marti (2015) and von Albertini (2014) with regard to the Julier Pass. Furthermore, in Wittmann and Rücker (2012), sod transplantation was always realized in combination with the previous removal, the temporary deposit, and the subsequent placement of a so-called "inter soil". On the contrary, on the Julier Pass, the direct rearrangement of sods including their entire root zone was realized without any separation and with intermediate positioning.

Just as important as the direct-shifting of the soil is the building of a rough subsoil as an appropriate base for the sods to be imported. Thereby the heterogeneity of the macro- and microrelief has to be rearranged also in the terrain's subsoil. In this context, it has been observed that heterogeneous transfer, including the structural elements, leads to structural and material interdigitation. Consequently, a positive effect on infiltration combined with a reduction in erosion and landslide risks results.

Depending on the availability of sods, it is possible to restore a tighter or looser mosaic, integrating the structural elements as blocks or shrubs (Fig. 8.51A). Thanks to the existing seed pool in the soil material, as well as to the vegetative and seed spread parting from the sods, the in-between spaces will be colonized in the short term (about 3 years) with site-specific vegetation (Fig. 8.51B). Therefore the combination of direct-shifting with spontaneous greening represents a low-priced and effective solution. It not only enables the achievement of vegetation cover that sufficiently protects against erosion, but it also eliminates the necessity to import seeds.

Thanks to direct-shifting the stabilizing function of the vegetation and its root activity is reestablished within a short time, also demonstrating the importance of keeping the disturbance of soil organisms and the concomitant microbial activity at a minimum. Transplanting purely peeled vegetation sods to newly arranged soil layers may facilitate drifting and concomitantly enhance the possibility of landsliding. To lower such risk, Wittmann and Rücker (2012) propose securing the sods using ground anchors or nails and to additionally cover the surface with an erosion blanket. Nevertheless, such bioengineering methods lead to higher costs, are less natural, and due to the clean layering the erosion risk remains high for a longer period. Therefore an experienced SPECP and careful execution can be as important as the technique itself. The goal should be a holistic approach and low input solutions. In order to achieve successful implementation of this complex task, it is essential that the SPECP communicates in a competent manner with the various partners, especially with the permission-granting authorities, the project planners, the construction management, and the construction workers. A well-informed machine operator handles soil with care.

Due to the low accessibility of high-resolution soil maps for alpine areas the SPECP has to provide project-related soil data based on soil profiles or soil core samples for the environmental impact study and for the planning phase. Of further importance is indirect information obtained through geology, topography, and related geomorphodynamics, aspect, vegetation and its cover rate, as well as knowledge of the water balance in the surroundings of the project. The combination of indirect and directly obtained information helps decision-making with regard to protection measures and efficient construction methods. Considering the difficult conditions, it is absolutely necessary for the SPECP to be equipped with quality site-specific knowledge. Within the context of construction projects, not only time and money are scarce resources, but also good soil specialists. Therefore ensuring methods for work with DTMs and aerial images would be a good addition to work with soil data (see "Methods, Status, and Relevance" section). As the construction proceeds, the SPECP will obtain more detailed soil information on the site. Unfortunately the data so collected is often not integrated into existing data pools and therefore this information subsequently becomes inaccessible or even lost.

Soil type, water dynamics, soil horizons, grain size distribution, soil structure, and the presence of soil skeleton are the necessary soil data within alpine construction projects. In general, maps can provide basic information, but decisions on handling methods should be based on holistic comprehension of the site, and its ecosystems and landscape. Whether soil data can act as a restrictive factor and therefore be integrated into guidelines for soil protection and soil handling is difficult to generalize. With regard to rare ecosystems, such as very dry areas or wetlands, soil data could arguably determine protection measures. Regarding the use of soil data in alpine construction projects a few questions remain open and thereby signify the need for further research in this area:

1. Is the time and financial expenditure needed for area-wide soil maps too high compared to their utility and possible impact on decision-making procedures?
2. Does the value of soil maps at a scale of 1:25,000 depend on the possibilities of combining them with DTMs, microclimate and water balance conditions, existing soil data, and risk maps, or are topography and the water balance more significant variables than the soil type when seeking to identify adequate protection solutions?
3. Should the data obtained by soil experts at construction projects be collected? How could its quality be guaranteed or made uniform?

The presented method of direct-shifting achieves one major goal: to minimize the environmental impact and destroy as little of the environment as possible. The formerly accepted destruction, "desertification", and homogenization of surfaces accompanying construction projects can nowadays be avoided. The method as proposed in this section combines the process of direct-shifting with controlled spontaneous autochthonous greenery and represents a practicable and low-priced solution to maintaining site-specific biodiversity. However, the success of environmental and soil protection on construction sites is based on the EPECP/SPECP's ability to constructively and positively communicate with the construction personnel.

The specific know-how gained in the process of adapting such new methods should be further developed, disseminated, and also be made accessible across national borders. In addition, binding guidelines adapted to the alpine environment should be provided in addition to basic soil data on a scale of 1:10,000. Fortunately, in some EU countries, the introduction of EPECP/SPECPs is already proceeding. Switzerland should act preemptively in this sector and implement corresponding guidelines as soon as possible.

The experience obtained within the framework of the Julier Pass project shows that the proposed direct-shifting is not only an efficient greenery method, but also represents an extensive construction technique with far-reaching positive impact. Soil protection, environmental and nature conservation, landscape protection, and time and cost savings can be covered by a single technique. Direct-shifting therefore can significantly contribute to sustainable land-use within the context of construction projects in heterogeneous areas such as the Alps, a requirement postulated by the Alpine Convention (Art. 2, 2.d "Soil conservation", 1995).

CONCLUSIONS AND OUTLOOK

Against the background of soil land-use examples in "Traditional Land-Use Types and Related Long-Term Impacts on Soils" section, the case studies in "Mapping and Modeling Soils in the Alps" section should provide answers related to soil-survey and soil-information use in the Alps. They all illustrate a high degree of diversity regarding aims, scales, and methodological approaches. Nevertheless, all of them try to meet the specific requirements of Alpine landscapes. Furthermore, most of the case studies aim to use soil information to support land-use planning. On the basis of these examples and with regard to the need for future improvement by integrating soil information into land management decisions, the following issues should be emphasized:

1. Soil information is vital for a wide range of land-use planning and management activities that aim to be sustainable and resource efficient.
2. Soil maps for the Alps should not be less detailed than they are at a scale of 1:25,000.
3. Besides soil types, soil maps should include information on the most relevant physical and chemical soil properties in order to meet the demands of site-specific land use and management.
4. For planning issues, soil maps should be interpreted in terms of soil potential and vulnerabilities, considering ecological processes and functions as well as susceptibility to various threats.
5. For special planning cases, soil information must be spatially detailed and more specific.

Taking this into consideration and also realizing the high costs of soil surveys in the field, the following challenges are evident:

1. Soil surveys should provide data that can be used flexibly in GIS systems.
2. Soil information should be a constitutive part of any ecosystem-oriented landscape research or management; data should always be combined with other geoecological information.
3. Soil classification systems should be more flexible (e.g., within the hierarchy) to better integrate the soil information into other classification or evaluation systems.
4. Soil surveys should be more focused on soil properties in order to obtain data that are more applicable and free the user from soil classification. Soil property data that can also be complemented by measurements provide important input for the calculation of pedotransfer functions.
5. In order to reduce time and costs, soil-survey fieldwork should be supported, combined, and improved by the integration of DSM techniques. The DSM approaches can also be used to further differentiate already existing soil maps/information. Nevertheless, ground-truth soil surveys are not deemed to be replaceable by methods providing purely digital and remote sensing data.
6. DSM, using different geo-informatics approaches, in particular helps to extrapolate soil property point data to wider areas, independent of SMUs. At the same time, we should bear in mind that the prediction of

soil-forming factors largely depends on the availability and accuracy of data.

7. Currently, new soil data collection methods, e.g., remote sensing and geophysics, provide new possibilities to support soil surveys. For instance, demand-oriented UAV-flights (UAV = Unmanned Aerial Vehicle) can be performed in order to produce both detailed aerial images as well as digital elevation models, which, in combination, are perfectly suitable for identifying soil-vegetation complexes above the timberline. In this context, intensive collaboration between methods and disciplines is necessary.
8. Considering the wide variety of soil-survey methods, scales, and systematics, cross-border soil data and knowledge transfer in the Alps should be expanded, for instance by using some type of cross-border, sustainable soil-management platform.

Ranging from carbon stock estimates to the management of construction sites, the diversity of the presented case studies shows how different soil data applications can be in the Alpine environment. Nevertheless, this short overview cannot replace intensive cross-border communication and the exchange of best practices regarding soil data exploitation. International communication amongst the Alpine countries and regions is also needed in order to overcome data incompatibility (e.g., soil classification differences) and to promote cross-border cooperation in the management of the Alpine territory.

In any case the scale of soil information seems to be crucial and should be addressed through the intensive collaboration of the soil scientists of the Alpine countries. An optimal framework can be achieved by combining soil overview maps at a scale of 1:25,000 with application-specific soil maps, e.g., at a scale of 1:5,000. Such detailed soil maps and data can be prepared when (1) general soil maps are available, (2) contemporary soil data collection methods are utilized, and (3) data processing methods are applied.

As the parent material presents the basis for soil development, the detailed genetic and spatial differentiation of quaternary deposits is particularly important. Unfortunately, data on these relatively thin sediment strata are often not represented in geological maps. Consequently the utilization of geological maps with their units may lead to insufficient or, in the worst case, incorrect conclusions and misleading soil information.

Nevertheless, all these scientific efforts will be in vain if the stakeholders are not convinced how, which, and to what extent soil information should be integrated into land-use decision-making processes in order to ensure sustainable development. Regrettably, soils are often only regarded as a homogeneous surface for agricultural production and are used according to subjective criteria and tradition. In order to strengthen a new level of soil awareness, more intensive communication between the end users of soil information (decision-makers, administrations) and soil experts is needed. Focusing on soil properties, on the one hand, and embedding soil information in a landscape context may contribute to a better understanding of soil and consequently more sustainable and rational planning and management decisions. Nevertheless, in order to promote soil-related issues in practice, also the legal frameworks must be improved.

The chapter presented contributes to creating focused, purpose-driven, and applicable soil information and thereby better exploitation of soil information in support of sustainable development in the Alps. It highlights specific challenges that need approaches and solutions tailored to the specifics of the Alpine landscape. Furthermore, the variety of examples and methods from these five Alpine countries evidences the existence of an interdisciplinary and transdisciplinary scientific and lay community that aims to meet soil-related challenges in the Alps on an international level.

Acknowledgments

Most of the case studies reported in this chapter formed part of research projects and were financially supported by national or international institutions, to which we would like to take this opportunity to express our gratitude:

The German Research Foundation
Case Study E3 ("Traditional Land-Use Types and Related Long-Term Impacts on Soils" section), Project: Sedimentologische und vegetationsgeschichtliche Untersuchungen im Horlachtal
European Union
Case Study E4 ("Former and Recent Land-Use Changes on a Valley Floor (Inn Valley, Austria)" section), Alpine Space Project TUSEC-IP: Techniques of Urban Soil Evaluation in City Regions—Implementation in Planning Procedures
Case Study E7 ("The Effect of Land-Use Changes on Soil Properties (Aosta Valley, Italy)" section), Project: Effect of land-use changes on soil properties (Aosta Valley, Italy)
Case Study E8 ("Soil Diversity and Dynamics at Forest and Pasture Sites (Pokljuka Plateau, Slovenia)" section), Project recharge.green: Balancing Alpine Energy and Nature
Case Study C5 ("Mapping Land Suitability to Mountain Viticulture (Aosta Valley, Italy)" section), Project Alpter: Terraced landscapes of the alpine arc
Case Study C6 ("Considering Soil Characteristics in the Designation of Protection Forests (Tyrol, Austria)" section), VOLE Project: Innovative Waldtypisierung—Grundlage und Maßnahmenkatalog zur Prävention von Naturgefahren und den Auswirkungen des Klimawandels
Autonomous Province of Bolzano-South Tyrol (Italy), Department of Education, University and Research
Case Study C2 ("The Concept of a Digital, Layer-Structured Soil Information System (South Tyrol, Italy)" section), Project ReBo: Acquisition and Regionalization of Soil Data
Provincial Government of Tyrol (Austria), Forest Administration Tyrol
Case Study C6 ("Considering Soil Characteristics in the Designation of Protection Forests (Tyrol, Austria)" section), Project: Waldtypisierung Tirol
The Austrian Federal Ministry of Agriculture, Forestry, Environment and Water Management
Case Study C6 ("Considering Soil Characteristics in the Designation of Protection Forests (Tyrol, Austria)" section), VOLE Project: Innovative Waldtypisierung—Grundlage und Maßnahmenkatalog zur Prävention von Naturgefahren und den Auswirkungen des Klimawandels

Furthermore, we would like to thank the following institutions and persons for excellent cooperation and support:

Regione Autonoma Valle d'Aosta FSE and FESR
Monterosa Ski Resort
Case Study C8 ("Mapping Susceptibility to Snow Gliding in the Monterosa Ski Resort (Aosta Valley, Italy)" section)
Research Centre for Agriculture and Forestry Laimburg, South Tyrol (Italy)
Bewässerungskonsortium Galsaun
VI.P Verband der Vinschgauer Produzenten für Obst und Gemüse
Case Studies E5 ("Geomorphic Structure and Land Use in an Inner-Alpine Dry Region (Etsch Valley, Italy)" section) and E6 ("Soil Degradation and Reforestation on the Slopes of an Inner-Alpine Dry Region (Etsch Valley, Italy)" section)
Tiefbauamt Graubünden (Switzerland)
Case Study C9 ("Sustainable Land Use in Construction Projects (Switzerland)" section), Projekt Julierstrasse: Strassenkorrektion Mot–Sur Gonda
Dr. Odoardo Zecca (Institut Agricole Regional—Aosta)

We would also like to express our appreciation to Dean J. DeVos for his thorough and patient English-language editing and proofreading of this work, which included numerous authors from a number of countries.

References

Ad-hoc-AG Boden, 2005. Bodenkundliche Kartieranleitung, fifth ed. Schweizerbart, Hannover.

Alpine Convention, 1995. Frame Convention. Article 2 General obligations, 2.d 'soil conservation'. <http://www.alpconv.org/en/convention/framework/default.html> (accessed 10.01.16).

APB, 2006. Autonome Provinz Bozen – Südtirol, Abteilung Natur, Landschaft und Raumentwicklung. Technische Grundkarte. <http://www.provinz.bz.it/natur-raum/themen/landeskartografie-technische-grundkarte.asp> (accessed 03.02.16).

APB, 2010. Autonome Provinz Bozen – Südtirol, Abteilung Forstwirtschaft. Waldtypisierung Südtirol. Band 1-2. <http://www.provinz.bz.it/forst/studien-projekte/waldtypisierung.asp#anc940> (accessed 03.02.16).

APB, 2015. Autonome Provinz Bozen – Südtirol, Abteilungen Landwirtschaft und Forstwirtschaft sowie Land- und forstwirtschaftliches Versuchszentrum Laimburg. Agrar- & Forstbericht 2015. <http://www.provinz.bz.it/landwirtschaft/flip/afb2015/default.htm> (accessed 25.10.16).

APB, 2016a. Autonome Provinz Bozen – Südtirol, Abteilung Hochbau und Technischer Dienst. Projekt CARG. <http://www.provinz.bz.it/hochbau/projektierung/812.asp> (accessed 03.02.16).

APB, 2016b. Autonome Provinz Bozen – Südtirol, Südtiroler Landesarchiv. Katastralmappen (Franziszeischer

Kataster). <http://www.provinz.bz.it/kunst-kultur/landesarchiv/katastralmappen.asp> (accessed 25.10.16).

APB, 2016c. Autonome Provinz Bozen – Südtirol, Landesagentur für Umwelt. Wassernutzung. <http://umwelt.provinz.bz.it/wasser/wassernutzung.asp> (accessed 25.10.16).

Arih, A., 2011. Analiza stanja gozdarstva. In: Izhodišča za Načrt upravljanja Triglavskega narodnega parka 2012-2022, 37. <http://www.tnp.si/nacrt_upravljanja_tnp/C262/> (accessed 31.10.16).

Baize, D., Roque, J., 1998. Les sols des alpages du Beaufortin (Alpes françaises). Cartographie et particularités. Écologie 29 (1), 147–152.

Ballabio, C., 2009. Spatial prediction of soil properties in temperate mountain regions using support vector regression. Geoderma 151, 338–350.

Barni, E., Freppaz, M., Siniscalco, C., 2007. Interactions between vegetation, roots and soil stability in restored high altitude ski runs on the Western Alps, Italy. Arc. Antar. Alp. Res 39, 25–33.

Baruck, J., Nestroy, O., Sartori, G., Baize, D., Traidl, R., Vrščaj, B., et al., 2016. Soil classification and mapping in the Alps: the current state and future challenges. Geoderma 264 (Part B), 312–331.

Bätzing, W., 2015. Die Alpen. Geschichte und Zukunft einer europäischen Kulturlandschaft, third ed. C.H. Beck, München.

Bayerisches Geologisches Landesamt & Bayerisches Landesamt für Umweltschutz, 2003. Das Schutzgut Boden in der Planung. Bewertung natürlicher Bodenfunktionen und Umsetzung in Planungs- und Genehmigungsverfahren. Augsburg.

Behrens, T., Scholten, T., 2006. A comparison of data-mining techniques in predictive soil mapping In: Lagacherie, P. McBradney, A. Voltz, M. (Eds.), Digital Soil Mapping. An Introductory Perspective. Developments in Soil Science, 31. Elsevier, Amsterdam, pp. 353–617.

BGS, 2015. Bodenkundliche Gesellschaft Schweiz. Qualitätskontrolle 2014/2015 Bodenkundliche Baubegleiter, 14–19. <http://www.soil.ch/cms/fileadmin/Medien/BBB/Qualitaetsbericht/Bericht_QK2014_BBB_D.pdf> (accessed 25.01.16).

Bitterlich, W., 1991. Der Einfluss der historischen Landnutzung (Streunutzung und Schneitelung) auf einen Gebirgswald in Finkenberg im Zillertal. Unpublished Dimploma Thesis, BOKU University, Wien.

Blaser, P., Kernebeek, P., Tebbens, L., van Breemen, N., Luster, J., 1997. Cryptopodzolic soils in Switzerland. Eur. J. Soil Sci 48, 411–423.

Blum, W.E.H., Spiegel, H., Wenzel, W.W., 1996. Bodenzustandsinventur–Konzeption, Durchführung und Bewertung; Empfehlungen zur Vereinheitlichung der Vorgangsweise in Österreich. Überarbeitete Neuauflage. BMFL und BMWVK, Wien.

BMLFUW, 2012. Bundesministerium für Land- und Forstwirtschaft, Umwelt und Wasserwirtschaft (Ed.). Waldentwicklungsplan, Richtlinien über Inhalte und Ausgestaltung. Wien.

BMLFUW, 2015. Federal Ministry of Agriculture, Forestry, Environment and Water Management. Sustainable Forest Management in Austria. Austrian Forest Report 2015. Vienna.

Bolliger, J., Hagedorn, F., Leifeld, J., Böhl, J., Zimmermann, S., Soliva, R., et al., 2008. Effects of land-use change on carbon stocks in Switzerland. Ecosystems 11, 895–907.

Bono, R., von Albertini, N., Clement, J.-P., Klaus, G., Vogt, M., 2014. Bodenkundliche Baubegleitung: Der Schweizer Weg. Bodenschutz 01/14, 6–14.

Bosshard, A., Mayer, PH., Mosimann, A., 2013. (Ed. Ö + L Ökologie und Landschaft GmbH) Leitfaden für naturgemässe Begrünungen in der Schweiz. Mit besonderer Berücksichtigung der Biodiversität. Oberwil-Liel.

Brancucci, G., Masetti, M., 2008. Terraced systems: heritage and risk. In: Scaramellini, G., Varotto, M. (Eds.), Terraced Landscapes of the Alps: Atlas. Marsilio, Venice, pp. 46–53.

Brändli, U.-B., 2000. Waldzunahme in der Schweiz – gestern und morgen. Informationsblatt Forschungsbereich Landschaft 45, 1–4.

Brassel, P., Lischke, H. (Eds.), 2001. Swiss National Forest Inventory: Methods and Models of the Second Assessment. Swiss Federal Institute for Forest, Snow and Landscape Research WSL, Birmensdorf.

Brevik, E.C., Arnold, R.W., 2015. Is the traditional pedologic definition of soil meaningful in the modern context? Soil Horizons 56 (3). http://dx.doi.org/10.2136/sh15-01-0002.

Brevik, E.C., Calzolari, C., Miller, B.A., Pereira, P., Kabala, C., Baumgarten, A., et al., 2016. Soil mapping, classification, and modeling: history and future directions. Geoderma 264, 256–274.

Brevik, E.C., Fenton, T.E., Jaynes, D.B., 2003. Evaluation of the accuracy of a central iowa soil survey and implications for precision soil management. Precis. Agric. 4, 331–342.

Brevik, E.C., Miller, B.A., 2015. The use of soil surveys to aid in geologic mapping with an emphasis on the Eastern and Midwestern United States. Soil Horizons 56 (4). http://dx.doi.org/10.2136/sh15-01-0001.

Brevik, E.C., Pereira, P., Munoz-Rojas, M., Miller, B.A., Cerdà, A., Parras-Alcántara, L., et al., 2017. Historical perspectives on soil mapping and process modeling for sustainable land use management. In: Pereira, P., Brevik, E.C., Munoz-Rojas, M., Miller, B. (Eds.), Soil Mapping and Process Modelling for Sustainable Land Use Management. Elsevier, Amsterdam, pp. 3–30.

Bui, E.N., 2004. Soil survey as a knowledge system. Geoderma 120 (1–2), 17–26.

Bui, E.N., Moran, C.J., 2001. Disaggregation of polygons of surficial geology and soil maps using spatial modelling and legacy data. Geoderma 103 (1–2), 79–94.

Calzolari, C., Costantini, E., Venuti, L., 2006. La valutazione dei suoli e delle terre: storia, definizioni e concetti. In: Cantagalli, E. (Ed.), Metodi di valutazione dei suolie delle terre. Cantagalli, Siena, pp. 2–45.

Ceaglio, E., Meusburger, K., Freppaz, M., Zanini, E., Alewell, C., 2012. Estimation of soil redistribution rates due to snow cover related processes in a mountainous area (Valle d'Aosta, NW Italy). Hydrol. Earth Syst. Sci. 16, 517–528.

CERVIM, 2016. Centro di Ricerche, Studi e Valorizzazione per la Viticulture Montana. Viticulture de Montagne. Aosta, Italy. <http://www.cervim.org/en/mountain-and-steep-slope-vitigulture.aspx> (accessed 27.05.16).

Curtaz, F., Stanchi, S., D'Amico, M., Filippa, G., Zanini, E., Freppaz, M., 2014. Soil evolution after land-reshaping in mountains areas (Aosta Valley, NW Italy). Agric. Ecosyst. Environ 199, 238–248.

Djukic, I., Zehetner, F., Tatzber, M., Gerzabek, M.H., 2010. Soil organic-matter stocks and characteristics along an Alpine elevation gradient. J. Plant Nutr. Soil Sci. 173, 30–38.

Dorji, T., Odeh, I.O., Field, D.J., Baillie, I.C., 2014. Digital soil mapping of soil organic carbon stocks under different land use and land cover types in montane ecosystems, Eastern Himalayas. For. Ecol. Manag 318, 91–102.

Egli, M., Dahms, D., Norton, K., 2014. Soil formation rates on silicate parent material in alpine environments: different approaches—different results? Review article. Geoderma 213, 320–333.

Egli, M., Margreth, M., Vökt, U., Fitze, P., Tognina, G., Keller, F., 2005. Modellierung von Bodentypen und Bodeneigenschaften im Oberengadin (Schweiz) mit Hilfe eines Geographischen Informationssystems (GIS). Geogr. Helv. 60 (2), 87–96.

Egli, M., Mirabella, A., Sartori, G., 2008. The role of climate and vegetation in weathering and clay mineral formation in late Quaternary soils of the Swiss and Italian Alps. Geomorphology 102, 307–324.

Egli, M., Mirabella, A., Sartori, G., Fitze, P., 2003. Weathering rates as a function of climate: results from a climosequence of the Val Genova (Trentino, Italian Alps). Geoderma 111, 99–121.

Egli, M., Mirabella, A., Sartori, G., Zanelli, R., Bischof, S., 2006. Effect of north and south exposition on weathering and clay mineral formation in Alpine soils. Catena 67, 155–174.

Egli, M., Sartori, G., Mirabella, A., Favilli, F., Giaccai, D., Delbos, E., 2009. Effect of north and south exposure on organic matter in high Alpine soils. Geoderma 149, 124–136.

Egli, M., Sartori, G., Mirabella, A., Giaccai, D., 2010. The effects of exposure and climate on the weathering of late Pleistocene and Holocene Alpine soils. Geomorphology 114, 466–482.

Escribano, P., Schmid, T., Chabrillat, S., Rodríguez-Caballero, E., García, M., 2017. Optical remote sensing for soil mapping and monitoring. In: Pereira, P., Brevik, E.C., Munoz-Rojas, M., Miller, B. (Eds.), Soil Mapping and Process Modelling for Sustainable Land Use Management. Elsevier, Amsterdam, pp. 91–128.

FAO, 2015. Understanding Mountain Soils. A contribution from mountain areas to the International Year of Soils 2015. Rome.

Fischer, K., 1966. Die Murkegel des Vinschgaus. Der Schlern 40 (4), 24–34.

Fischer, K., 1990. Entwicklungsgeschichte der Murkegel im Vinschgau. Der Schlern 64 (2), 93–97.

FOEN, 2016. Switzerland's Greenhouse Gas Inventory 1990–2014. National inventory report, including reporting elements under the Kyoto Protocol, submission of 15 April 2016 under the United Nations Framework Convention on Climate Change and under the Kyoto Protocol, Federal Office for the Environment FOEN, Climate Division, Bern. <http://www.bafu.admin.ch/klima/13879/13880/14577/index.html?lang=en> (accessed 04.05.16).

Forest Act, 1975. Austrian Forest Act. in the valid version at 01/2016. StF: BGBl. Nr. 440/1975; NR: GP XIII RV 1266 AB 1677S. 150. BR: 1392 AB 1425, p. 344.

Fracek, K., Mosimann, T., 2013. Wissensbasierte Modellierung der Mächtigkeit des kalkfreien Bodenbereiches in den Waldböden des Kantons Basel-Landschaft (Nordwestschweiz). Waldökologie, Landschaftsforschung und Naturschutz 13, 5–16.

Freppaz, M., Agnelli, A., Drusi, B., Stanchi, S., Galliani, C., Revel Chion, V., et al., 2008. Soil quality and fertility: studies in the Valle d'Aosta. In: Fontanari, E., Patassini, D. (Eds.), Terraced Landscapes of the Alps. Projects in Progress. Marsilio, Venice, pp. 37–39.

Freppaz, M., Filippa, G., Corti, G., Cocco, S., Williams, M.W., Zanini, E., 2013. Soil properties on ski runs. In: Rixen, C., Rolando, A. (Eds.), The Impacts of Skiing and Related Winter Recreational Activities on Mountain Environments. Bentham Science Publishers, Sharjah, United Arab Emirates, pp. 45–64.

Geitner, C., 1999. Sedimentologische und vegetationsgeschichtliche Untersuchungen an fluvialen Sedimenten in den Hochlagen des Horlachtales (Stubaier Alpen/Tirol). Ein Beitrag zur zeitlichen Differenzierung der fluvialen Dynamik im Holozän. Münchener Geographische Abhandlungen, Reihe B, 31, München, 247 pp.

Geitner, C., 2007. Böden in den Alpen – Ausgewählte Aspekte zur Vielfalt und Bedeutung einer wenig beachteten Ressource. In: Borsdorf, A., Grabherr, G. (Eds.), Internationale Gebirgsforschung, IGF-Forschungsberichte 1, Innsbruck, Wien, pp. 56–67 and 82–83.

Geitner, C., 2010. Soils as archives of natural and cultural history; examples from the Eastern Alps. In: Borsdorf, A., Grabherr, G., Heinrich, K., Scott, B., Stötter, J. (Eds.), Challenges for Mountain Regions–Tackling Complexity. Böhlau, Vienna, pp. 68–75. Wien.

Geitner, C., Bussemer, S., Ehrmann, O., Ikinger, A., Schäfer, D., Traidl, R., et al., 2011. Bodenkundlich-stratigraphische Befunde am Ullafelsen im hinteren Fotschertal sowie ihre landschaftsgeschichtliche Interpretation. In: Schäfer, D. (Ed.), Das Mesolithikum-Projekt Ullafelsen (Teil 1). Mensch und Umwelt im Holozän Tirols 1. Philipp von Zabern, Innsbruck, pp. 109–151.

Geitner, C., Moser, Y., Wanker, C., Tusch, M., 2007. Aspekte historischer und aktueller Wechselbeziehungen von Boden und Landnutzung in den Gemeinden Wörgl und Bruneck. In: Innsbrucker Geographische Gesellschaft, Alpine Kulturlandschaft im Wandel, Festschrift für Hugo Penz zum, 65. Geburtstag, Innsbruck, pp. 124–145.

Geitner, C., Tusch, M., Meißl, G., Kringer, K., Wiegand, C., 2011a. Der Einfluss des Reliefs auf die räumliche Verteilung der Böden: Grundsätzliche Überlegungen zu Wirkungszusammenhängen und Datengrundlagen anhand von Beispielen aus den Ostalpen. Z. Geomorphol. 55 (Suppl. 3), 127–146.

Gellrich, M., Baur, P., Kock, B., Zimmermann, N.E., 2007. Agricultural land abandonment and natural forest re-growth in the Swiss mountains: a spatially explicit economic analysis. Agric. Ecosyst. Environ 118, 93–108.

Göttlein, A., Katzensteiner, K., Rothe, A., 2014. Standortssicherung im Kalkalpin – SicALP. Abschlussbericht zum Forschungsprojekt INTERREG BY/Ö J00183, 212. Forstliche Forschungsberichte, München. 1–172.

Grabherr, G., 1982. The impact of trampling by tourists on a high altitudinal grassland in the Tyrolean Alps, Austria. Vegetatio 48 (3), 209–217.

Grabherr, G., 1997. The high-mountain ecosystems of the Alps. In: Wielgolaski, F.E. (Ed.), Polar and Alpine Tundra. Ecosystems of the World, 3. Elsevier, Amsterdam, pp. 97–121.

Grabherr, G., Koch, G., Kirchmeir, H., Reiter, K., 1998. Hemerobie österreichischer Waldökosysteme. Veröffentlichungen des Österreichischen MaB-Programmes, Österreichische Akademie der Wissenschaften 17, 493 pp.

Graf, F., Frei, M., 2013. Soil aggregate stability related to soil density, root length, and mycorrhiza using site-specific Alnus incana and Melanogaster variegatus s.l. Ecol. Eng. 57, 314–323.

Gramm, D., Tasser, E., Tappeiner, U., 2008. Regions of similar development. In: Tappeiner, U., Borsdorf, A., Tasser, E. (Eds.), Alpenatlas–Atlas des Alpes–Atlante delle Alpi–Atlas Alp–Mapping the Alps. Spektrum Akademischer Verlag, Heidelberg, pp. 272–274.

Grashey-Jansen, S., 2008. Raum-zeitliche Differenzierung der Bodenwasserdynamik auf obstbaulich genutzten Standorten in Südtirol unter Bewässerungseinfluss Geographica Augustana, 3. University of Augsburg., 271 pp.

Grashey-Jansen, S., 2010. Pedohydrological case study of two apple-growing locations in South Tyrol (Italy). Agric. Water Manag. 98, 234–240.

Grashey-Jansen, S., 2012a. Soil hydrology and soil properties at a partially reforested hillside in the Central Alps. J. Forest Sci. 58 (8), 363–371.

Grashey-Jansen, S., 2012b. The influence of physical soil properties on the water supply of irrigated orchards – some examples from Val Venosta (South Tyrol/Northern Italy). J. Environ. Biol. 33, 417–424. (Special Issue: Environment and Geography in Mediterranean).

Grashey-Jansen, S., 2014. Optimizing irrigation efficiency through the consideration of soil hydrological properties—examples and simulation approaches. Erdkunde 68 (1), 33–48.

Grashey-Jansen, S., Eben, B., 2009. Bodenphysikalische Untersuchungen über den Wasserhaushalt von zwei Standorten. Obstbau/Weinbau 46 (4), 144–149.

Grashey-Jansen, S., Korch, O., Beck, C., Friedmann, A., Bernhard, R., Dubitzky, C., 2014. Distributional patterns of mica influenced soil sites in consideration of atmospheric circulation types—a case study on the alpine belt of the Zugspitzplatt (Northern Limestone Alps, Germany). Int. J. Geol. Agric. Environ. Sci. 2 (4), 11–19.

Grashey-Jansen, S., Korch, O., Beck, C., Friedmann, A., Bernhard, R., Dubitzky, C., 2015. Äolisch beeinflusste Bodenentwicklung in der alpinen Zone des Zugspitzplatts unter pedogenetischer Berücksichtigung lokaler Windströmungen und Großwetterlagen. UFS-Scientific Results 2013/2014, 7–11.

Grashey-Jansen, S., Schröder, L., 2009. Zur Physiogeographie des Vinschgaus. Der Schlern 83 (6), 52–65.

Grashey-Jansen, S., Schroeder, L., 2011. Geschichte der Aufforstungsversuche in den Vinschgauer Leiten. Allg. Forst Z. Waldwirtsch. Umweltvorsorge 22, 12–13.

Grashey-Jansen, S., Schroeder, L., 2012. Gebirgswaldböden in einem Aufforstungsbereich der südlichen Ötztaler Alpen. Allg. Forst Z. Waldwirtsch. Umweltvorsorge 5, 25–27.

Grashey-Jansen, S., Seipp, S., 2012. Azonale Boden- und Vegetationsformationen in der subalpinen und alpinen Stufe des Zugspitzplatts. UFS-Scientific Results 2011/2012, 41–43.

Grunwald, S., 2009. Multi-criteria characterization of recent digital soil mapping and modeling approaches. Geoderma 152, 195–207.

Hagedorn, F., Mulder, J., Jandl, R., 2010. Mountain soils under a changing climate and land-use. Biogeochemistry 97, 1–5.

Häring, T., Dietz, E., Osenstetter, S., Koschitzki, T., Schröder, B., 2012. Spatial disaggregation of complex soil map units: a decision-tree based approach in Bavarian forest soils. Geoderma 185–186, 37–47.

Haslmayr, H.-P., Geitner, C., Sutor, G., Knoll, A., Baumgarten, A., 2016. Soil function evaluation in Austria—development, concepts and examples. Geoderma 264 (Part B), 379–387.

Heckmann, T., Becht, M., 2006. Statistical disposition modelling of mass movements In: Böhner, J. McCloy, K.R. Strobl, J. (Eds.), SAGA-Analysis and Modelling Applications. Göttinger Geogr. Abh., 115. Selbstverlag des Geographischen Instituts, Göttingen, pp. 61–73.

Herbst, P., Gross, J., Meer, U., Mosimann, T., 2012. Geomorphographic terrain classification for predicting forest soil properties in Northwestern Switzerland. Z. Geomorphol 56, 1–22.

Herbst, P., Mosimann, T., 2010. Prognose ökologisch wichtiger Waldbodeneigenschaften mit Random Forest in der Nordwestschweiz – Vergleich der Vorhersagen mit wissensbasierter empirisch-statistischer Modellierung. Geomatik Schweiz 4, 140–144.

Hoffmann, C., Streifeneder, T., Ruffini, F.V., Franz, A., Tröbinger, H., 2008. Types of farming. In: Tappeiner, U., Borsdorf, A., Tasser, E. (Eds.), Alpenatlas–Atlas des Alpes–Atlante delle Alpi–Atlas Alp–Mapping the Alps. Spektrum Akademischer Verlag, Heidelberg, pp. 198–199.

Höller, P., Fromm, R., Leitinger, G., 2009. Snow forces on forest plants due to creep and glide. For. Ecol. Manag. 257, 546–552.

IUSS Working Group, 2006. World Reference Base for soil resources 2006 World Soil Resources Reports 103, second ed. FAO, Rome.

Jasciewicz, J., Stepinski, T.F., 2013. Geomorphons—a pattern recognition approach to classification and mapping of landforms. Geomorphology 182, 147–156.

Kleen, C., Grashey-Jansen, S., Stojakowits, P., 2016. Die Vinschgauer Talböden als Archiv der Landschaftsgeschichte. Obstbau/Weinbau, Obstbau/Weinbau 53 (5), 12–17.

Korch, O., 2014. Untersuchungen zu Flora und Vegetation des Zugspitzplatts (Wettersteingebirge, Bayerische Alpen)—Rezente Vegetationsdynamik unter besonderer Berücksichtigung klimatischer und anthropozoogener Prozesse. Unpublished Doctoral Thesis, University Augsburg.

Korch, O., Friedmann, A., 2012. Phytodiversität und Dynamik der Flora und Vegetation des Zugspitzplatts. Jahrbuch des Vereins zum Schutz der Bergwelt 2011 (12), 217–234.

Kralj, T., Vrščaj, B., 2015. Pestrost tal na območju Pokljuke, Pršivca in Lopučniške doline [Soil diversity of Pokluka plateau], Pršivec and Lopučniška valley. Acta Triglavensia 3, 40–59.

Krautzer, B., Uhlig, C., Wittmann, H., 2012. Restoration of Arctic-Alpine ecosystems. In: van Andel, J., Aronson, J. (Eds.), Restauration Ecology, The New Frontier, second ed. Blackwell, Oxford, UK, pp. 189–202.

Küfmann, C., 2002. Erste Ergebnisse zur qualitativen Untersuchung und Quantifizierung rezenter Flugstäube in den Nördlichen Kalkalpen (Wettersteingebirge). Mitt. Geogr. Ges. München 86, 59–84.

Küfmann, C., 2003. Soil types and eolian dust in high-mountainous karst of the Northern Calcareous Alps (Zugspitzplatt, Wetterstein Mountains, Germany). Catena 53, 211–227.

Küfmann, C., 2006. Quantifizierung und klimatische Steuerung von rezenten Flugstaubeinträgen auf Schneeoberflächen in den Nördlichen Kalkalpen (Wetterstein-, Karwendelgebirge, Berchtesgadener Alpen, Deutschland). Z. Geomorphol 50 (2), 245–268.

Küfmann, C., 2008. Are Cambisols in Alpine Karst Autochthonous or Eolian in origin? Arc. Antar. Alp. Res. 40 (3), 506–518.

Lagacherie, P., McBratney, A.B., 2006. Spatial soil information systems and spatial soil inference systems: perspectives for digital soil mapping In: Lagacherie, P. McBradney, A.B. Voltz, M. (Eds.), Digital Soil Mapping. An Introductory Perspective. Developments in Soil Science, 31. Elsevier, Amsterdam, pp. 3–22.

Lasanta, T., González-Hidalgo, J.C., Vicente-Serrano, S., Sferi, E., 2006. Using landscape ecology to evaluate an alternative management scenario in abandoned Mediterranean mountainous areas. Landsc. Urban Plan 78, 101–114.

Lathrop Jr, R.G., Aber, J.D., Bognar, J.A., 1995. Spatial variability of digital soil maps and its impact on regional ecosystem modeling. Ecol. Model 82 (1), 1–10.

Locher-Oberholzer, N., Streit, M., Frei, M., Andrey, C., Blaser, R., Meyer, J., et al., 2008. Richtlinien Hochlagenbegrünung, 2. Mitteilungsblatt Verein Ingenieurbiologie, Hochlagenbegrünung. 3–33.

MacDonald, D., Crabtree, J.R., Wiesinger, G., Dax, T., Stamou, N., Fleury, P., et al., 2000. Agricultural abandonment in mountain areas of Europe: environmental consequences and policy response. J. Environ. Manag. 59 (1), 47–69.

Maggioni, M., Godone, D., Oppi, L., Freppaz, M., Stanchi, S., Höller, P., 2016. Snow gliding susceptibility maps for ski resort management: the case study of the Monterosa Ski in the NW Italian Alps. J. Maps 12 (S1), 115–121.

Margreth, S., 2007a. Snow pressure on cableway masts: analysis of damages and design approach. Cold Reg. Sci. Technol 47, 4–15.

Margreth, S., 2007b. Defense structures in avalanche starting zones. Technical guideline as an aid to enforcement. Environment in Practice no. 0704. Federal Office for the Environment, Bern; WSL Swiss Federal Institute for Snow and Avalanche Research SLF, Davos.

Marguerettaz, O., 1983. I terreni incolti: cause e conseguenze dell'abbandono e possibilità di ricupero. Revue Valdôtaine d'Histoire Naturelle 36/37, 263–267.

Marti, N., 2015. Direktumlagerung, Erfolgskontrolle einer neuartigen Wiederbegrünungsmethode an der Julierpassstrasse. Unpublished Master Thesis, Zurich University of Applied Sciences.

Mathys, L., Kellenberger, T., 2009. Spot5 RadcorMosaic of Switzerland. Technical report, National Point of Contact for Satellite Images NPOC: Swisstopo; Remote Sensing Laboratories, University of Zurich, Zurich.

McBratney, A.B., Mendonça Santos, M., de, L., Minasny, B., 2003. On digital soil mapping. Geoderma 117 (1), 3–52.

McBratney, A.B., Odeh, I.O.A., 1997. Application of fuzzy sets in soil science: fuzzy logic, fuzzy measurements and fuzzy decisions. Geoderma 77, 85–113.

McNeill, J.R., Winiwarter, V., 2006. Soils, soil knowledge and environmental history: an introduction. In: McNeill, J.R., Winiwarter, V. (Eds.), Soils and Societies. Perspectives from Environmental History, Isle of Harris. The White Horse Press, Isle of Harris, U.K., pp. 1–6.

Mercalli, L., Cat Berro, D., Montuschi, S., Castellano, C., Ratti, M., Di Napoli, G., et al., 2003. Atlante climatico della Valle d'Aosta. SMI.

Minghetti, P., Sartori, G., Lambert, K., 1997. Relations sol-végétation dans les pinèdes a Pinus mugo Turra du Trentin (Italie). Rev. Ecol. Alp. 4, 23–34.

Nestroy, O., Aust, G., Blum, W.E.H., Englisch, M., Hager, H., Herzberger, E., et al., 2011. Systematische Gliederung der Böden Österreichs. Österreichische Bodensystematik 2000 in der revidierten Fassung von 2011. Mitt. Österr. Bodenkdl. Ges. 79 Wien, 98 pp.

Nussbaum, M., Papritz, A., Baltensweiler, A., Walthert, L., 2014. Estimating soil organic carbon stocks of Swiss forest soils by robust external-drift kriging. Geosci. Model Dev 7, 1197–1210.

Odeh, I.O.A., McBratney, A.B., Chittleborough, D.J., 1995. Further results on prediction of soil properties from terrain attributes: heterotopic cokriging and regression-kriging. Geoderma 67 (3–4), 215–226.

Oeggl, K., Patzelt, G., Schäfer, D. (Eds.), 1997. Alpine Vorzeit in Tirol, Begleitheft zur Ausstellung. Innsbruck.

Ola, A., Dodd, I.C., Quinton, J.N., 2015. Can we manipulate root system architecture to control soil erosion? Soil 1, 603–612.

Orlandini, S., Grifoni, D., Mancini, M., Barcaioli, G., Crisci, A., 2005. Analisi degli effetti della variabilità meteo climatica sulla qualità del Brunello di Montalcino. Ital. J. Agrometeorol 2, 37–44.

Penz, H., 2005. Vom Vollerwerb zur Nebenbeschäftigung. In: Borsdorf, A. (Ed.), Das neue Bild Österreichs. Verlag ÖAW, Wien, pp. 82–83.

Pereira, P., Brevik, E.C., Oliva, M., Estebaranz, F., Deppellegrin, D., Novara, A., et al., 2017. Goal oriented soil mapping: applying modern methods supported by local knowledge. In: Pereira, P., Brevik, E.C., Munoz-Rojas, M., Miller, B. (Eds.), Soil Mapping and Process Modelling for Sustainable Land Use Management. Elsevier, Amsterdam, pp. 65–88.

Perruchoud, D., Walthert, L., Zimmermann, S., Lüscher, P., 2000. Contemporay carbon stocks of mineral forest soils in the Swiss Alps. Biogeochemistry 50, 111–136.

Price, M.F., Harden, C.P., 2013. Mountain soils. In: Price, M.F., Byers, A.C., Friend, D.A., Kohler, T., Price, L.W. (Eds.), Mountain Geography. University of California Press, Berkley, Los Angeles, London, pp. 167–182.

Radišek, J., Huršec, M., Vrščaj, B. 2014a. Digitalna pedološka karta Občine Braslovče 1:10.000 [The Soil Map of Braslovče County 1:10,000].

Radišek, J., Huršec, M., Vrščaj, B., 2014b. Izdelava pedološke karte in karte bonitetnih vrednosti ter interpretacija pedoloških podatkov za potrebe pridobivanja bonitetnih ocen kmetijskih, gozdnih in drugih zemljišč v občini Braslovče [Elaboration of soil map, soil data interpretation and land evaluation in the Municipality Braslovče]. Braslovče, Slovenija [Slovenia], ProTellus d.o.o, Kmetijski inštitut Slovenije [ProTellus d.o.o. and Agricultural Institute of Slovenia].

Richter, D.D., Tugel, A.J., 2012. Soil change in the anthropocene: bridging pedology, land use and soil management In: Huang, P.M. Li, Y. Sumner, M.E. (Eds.), Properties and Processes. Handbook of Soil Science, 38. CRC Press, Boca Raton, pp. 31–61.

Richter, D.D., Yaloon, D.H., 2012. "The Changing Model of Soil" Revisited. Soil Sci. Soc. Am. J 76, 766–778.

Rixen, C., Freppaz, M., 2015. Winter sports: the influence of ski piste construction and management on soil and plant characteristics Understanding Mountain Soils: A Contribution from mountain areas to the International Year of Soils 2015. FAO, Rome. 81–83.

Rixen, C., Freppaz, M., Stoeckli, V., Huovinen, C., Huovinen, K., Wipf, S., 2008. Altered snow density and chemistry change soil nitrogen mineralization and plant growth. Arc. Antar. Alp. Res. 40 (3), 568–575.

Roux-Fouillet, P.S., Wipf, S., Rixen, C., 2011. Long-term impacts of ski piste management on alpine vegetation and soils. J. Appl. Ecol. 48, 906–915.

Sandor, J.A., Eash, N.S., 1995. Ancient agricultural soils in the Andes of southern Peru. Soil Sci. Soc. Am. J. 59, 170–179.

Sandor, J.A., Gersper, P.L., Hawley, J.W., 1990. Prehistoric agricultural terraces and soils in the Mimbres area, New Mexico. World Archeol 22, 70–86.

Sartori, G., Mancabelli, A., Corradini, F., Wolf, U., 2001. Verso un catalogo dei suoli del Trentino: 3. Rendzina (*Rendzic Leptosols*) e suoli rendziniformi. Studi Trent. Sci. Nat. Acta Geol. 76, 43–70.

Sartori, G., Mancabelli, A., Wolf, U., Corradini, F., 2005. Atlante dei suoli del Parco Naturale Adamello-Brenta. Suoli e paesaggi. Museo Tridentino di Scienze Naturali—Monografie II, Trento.

Scaramellini, G., Varotto, M., 2008. Introduction. In: Scaramellini, G., Varotto, M. (Eds.), Terraced Landscapes of the Alps: Atlas. Marsilio, Venice, pp. 7.

Schaffer, B., Attinger, W., Schulin, R., 2007. Compaction of restored agricultural soil by heavy agricultural machinery—soil physical and chemical aspects. Soil Till. Res 93, 28–43.

Schär, C., Davies, T.D., Frei, C., Wanner, H., Widmann, M., Wild, M., et al., 1998. Current alpine climate. In: Cebon, P., Dahinden, U., Davies, H., Imboden, D.M., Jaeger, C.C. (Eds.), Views from the Alps. Regional Perspectives on Climate Change. The MIT Press, Cambridge, London, pp. 21–72.

Schiechtl, H.M., 1969. Auswirkungen der Ast- und Bodenstreunutzung in Fichtenwäldern des Zillertales. Mitteilungen der ostalpin-dinarischen pflanzensoziologischen Arbeitsgemeinschaft 6, 40–41.

Schiechtl, H.M., Neuwinger, I., 1980. Regeneration von Vegetation und Boden nach Einstellung der Beweidung und Bodenstreunutzung in einem zentralalpinen Hochlagen-Aufforstungsgebiet. In: Bundesversuchsanstalt, F. (Ed.), Beiträge zur subalpinen Waldforschung. Österreichischer Agrarverlag, Wien, pp. 63–80.

Schröder, L., 2010. Geoökologische Untersuchungen in den Laaser Leiten (Südtirol/Italien) unter besonderer Berücksichtigung der Bodenentwicklung. Unpublished Diploma Thesis, University of Augsburg.

Stanchi, S., Freppaz, M., Agnelli, A., Reinsch, T., Zanini, E., 2012. Properties, best management practices and conservation of terraced soils in Southern Europe (from Mediterranean areas to the Alps). A review. Quat. Int 265, 90–100.

Stanchi, S., Freppaz, M., Ceaglio, E., Maggioni, M., Meusburger, K., Alewell, C., et al., 2014. Soil erosion in an avalanche release site (Valle d'Aosta: Italy): towards a winter factor for RUSLE in the Alps. Nat. Hazards Earth Syst. Sci 14, 1761–1771.

Stanchi, S., Freppaz, M., Godone, D., Zanini, E., 2013a. Assessing the susceptibility of alpine soils to erosion using soil physical and site indicators. Soil Use Man 29, 586–596.

Stanchi, S., Godone, D., Belmonte, S., Freppaz, M., Galliani, C., Zanini, E., 2013b. Land suitability map for mountain. viticulture: a case study in Aosta Valley (NW Italy). J. Maps, 586–596.

Stimpfl, E., Aichner, M., Cassar, A., Thaler, C., Andreaus, O., Matteazzi, A., 2006. Zustandserhebung der Südtiroler Böden im Obstbau. Laimburg J. 3 (1), 74–134.

Stöhr, D., 2007. Soils—heterogeneous at a microscale In: Wieser, G. Tausz, M. (Eds.), Trees at their Upper Limit. Treelife Limitation at the Alpine Timberline. Plant Ecophysiology, 5. Springer, Dordrecht, pp. 37–56.

Streifeneder, T., Tappeiner, U., Ruffini, F.V., Tappeiner, G., Hoffmann, C., 2007. Selected aspects of agrostructural change within the Alps—a comparison of harmonised agro-structural indicators on a municipal level in the alpine convention area. La revue de géographie alpine. J. Alp. Res. 95 (3), 27–52.

Stuber, M., Bürgi, M., 2012. Hüeterbueb und Heitisrähl: Traditionelle Formen der Waldnutzung in der Schweiz 1800–2000 Bristol-Stiftung, second ed. Haupt, Bern.

Swiss Federal Statistical Office, 2000. Swiss Soil Suitability Map. BFS GEOSTAT. <http://www.bfs.admin.ch/bfs/portal/de/index/dienstleistungen/geostat/datenbeschreibung/digitale_bodeneignungskarte.html> (accessed 04.05.16).

Tappeiner, U., Borsdorf, A., Tasser, E. (Eds.), 2008a. Alpenatlas–Atlas des Alpes–Atlante delle Alpi–Atlas Alp–Mapping the Alps. Spektrum Akademischer Verlag, Heidelberg.

Tappeiner, U., Tasser, E., Leitinger, G., Cernusca, A., Tappeiner, G., 2008b. Effects of historical and likely future scenarios of land use on above- and belowground vegetation carbon stocks of an Alpine Valley. Ecosystems 11 (8), 1383–1400.

Tasser, E., Walde, J., Tappeiner, U., Teutsch, A., Noggler, W., 2007. Land-use changes and natural reforestation in the Eastern Central Alps. Agric. Ecosyst. Environ. 118, 115–129.

Thalheimer, M., 2006. Kartierung der landwirtschaftlich genutzten Böden des Überetsch in Südtirol (Italien). Laimburg J. 3 (1), 135–177.

Theurillat, J.-P., Felber, F., Geissler, P., Gobat, J.-M., Fierz, M., Fischlin, A., et al., 1998. Sensitivity of plant and soil ecosystems of the Alps to climate change. In: Cebon, P., Dahinden, U., Davies, H., Imboden, D.M., Jaeger, C.C. (Eds.), Views from the Alps. Regional Perspectives on Climate Change. The MIT Press, Cambridge, London, pp. 225–308.

Traidl, R., 2005. Die Bodenkundliche Landesaufnahme in Bayern–Stand und Perspektiven. Mitt. Österr. Bodenkdl. Ges 72, 87–91.

Traidl, R., 2008. Large-scale soil maps as basis for soil protection in Bavaria—with special reference to the Bavarian

Alps. Poster at Congress on "Soil Protection: Towards the integration among regional thematic strategies" Bologna, November 27–28, 2008.

Varotto, M., Lodatti, L., 2014. New family farmers for abandoned lands—the adoption of terraces in the Italian Alps (Brenta Valley). Mt. Res. Dev 34 (4), 315–325.

Veit, H., 2002. Die Alpen - Geoökologie und Landschaftsentwicklung. Stuttgart., UTB.

Veit, H., Mailänder, R., Vonlanthen, C., 2002. Periglaziale Deckschichten im Alpenraum: bodenkundliche und landschaftsgeschichtliche Bedeutung. Peterm. Geogr. Mitt. 146 (4), 6–14.

von Albertini, N., 2014. Begrünungsmethoden und Monitoring, H3a Julierstrasse, Strassenkorrektion Mot-Sur Gonda. Tiefbauamt Graubünden, Chur unpublished.

von Albertini, N., Regli, L., 2012. Erfolgreiche Begrünungsmethode beim Bau der Julierpassstrasse, 3. Mitteilungsblatt Verein Ingenieurbiologie, Hochlagenbegrünung. 12–22.

von Andrian-Werburg, S., Fischer, S., Schönthaler, K., 2008. Forest areas. In: Tappeiner, U., Borsdorf, A., Tasser, E. (Eds.), Alpenatlas–Atlas des Alpes–Atlante delle Alpi–Atlas Alp–Mapping the Alps. Spektrum Akademischer Verlag, Heidelberg, pp. 216–217.

Vrščaj, B., 2007a. Towards soil infomation for local scale soil protection in Slovenia. In: Hengl, T., Panagos, P., Jones, A., Tòth, G. (Eds.), Status and Prospects of Soil Information in South-Eastern Europe. Soil databases, projects and applications. ESBN Research Reports 21, 165–177.

Vrščaj, B., 2007b. Urbanizacija tal v Sloveniji [Soil urbanisation in Slovenia] Strategija Varovanja Tal v Sloveniji [Strategy of Soil Protection in Slovenia]. Presented at the Strategija varovanja tal v Sloveniji [Soil Protection Strategy in Slovenia]. Pedološko društvo Slovenije [Slovenian Soil Science Society], Ljubljana, Slovenia. 6.

Vrščaj, B., 2008. The structural changes of agricultural land, their quality and urbanization between 2002 and 2007. Hmelj. Bilt 15, 73.

Vrščaj, B., 2011. The urbanization of agricultural land in Slovenia between 2002-2007, land use changes and soil quality Soil Protection Activities and Soil Quality Monitoring in South Eastern Europe. EU JRC Scientific and Technical Reports, Ispra, Italy. 171–181.

Vrščaj, B., Lobnik F., 1999. Talni informacijski sistem [Soil Information System]. Zbornik Biotehniške fakultete Univerze v Ljubljani 23.

Vrščaj, B., Prus, T., Lobnik, F., 2005. Soil information and soil data use in Slovenia, second ed. In: Jones, R.J.A. Houšková, B. Bullock, P. Montanarella, L. (Eds.), Soil Resources of Europe, 9 European Soil Bureau Research Report, pp. 331–344.

Werth, K., 2003. Die Geschichte der Etsch zwischen Meran und San Michele: Flussregulierung, Trockenlegung der Möser, Hochwasserschutz. Athesia Tappeiner Verlag, Lana.

Wittmann, H., Rücker, T., 2012. Standortgerechte Hochlagenbegrünung in Österreich – ein Bericht aus der Praxis, 3. Mitteilungsblatt Verein Ingenieurbiologie, Hochlagenbegrünung. 23–33.

Wittmann, O., 2006. In: Blume, H.-P. Henningsen, P. Frede, H.-G. Guggenberger, G. Horn, R. Stahr, K. (Eds.), Deutsche Weinbaustandorte, 4. Handbuch der Bodenkunde, pp. 1–46.

Yang, R., Rossiter, D.G., Liu, F., Lu, Y., Yang, F., Yang, F., et al., 2015. Predictive mapping of topsoil organic carbon in an alpine environment aided by landsat TM. PLoS ONE 10 (10), e0139042.

Zech, W., Schad, P., Hintermayer-Erhard, G., 2014. Böden der Welt. Ein Bildatlas, second ed. Springer, Heidelberg.

CHAPTER

9

Compilation of Functional Soil Maps for the Support of Spatial Planning and Land Management in Hungary

László Pásztor[1], *Annamária Laborczi*[1], *Katalin Takács*[1], *Gábor Szatmári*[1], *Nándor Fodor*[2], *Gábor Illés*[3], *Kinga Farkas-Iványi*[1], *Zsófia Bakacsi*[1] and *József Szabó*[1]

[1]Hungarian Academy of Sciences, Budapest, Hungary [2]Hungarian Academy of Sciences, Martonvásár, Hungary [3]National Agricultural Research and Innovation Centre, Sárvár, Hungary

INTRODUCTION

National level spatial planning requires adequate, preferably timely, detailed spatial knowledge about soil cover. Hungary has long and rich traditions in soil survey and mapping. Large amounts of soil information are available in various dimensions and generally presented in maps, serving different purposes as to spatial and/or thematic aspects (Várallyay, 2012). An increasing proportion of soil related data has been digitally processed and organized into various spatial soil information systems (Pásztor et al., 2013). The existing maps, data, and systems served the society for many years, but the available data are no longer fully satisfactory for the recent needs of policy making, since (1) the original data collection did not and could not target the present data demands and (2) the produced datasets cannot be considered omnipotent. In addition, the demands on spatial soil information more and more frequently do not refer to primary soil properties, but to more complex or derived soil information, functions, processes, or services (Blum, 2005; Omuto et al., 2013; Panagos et al., 2012). The new information requirements, however, generally cannot be fulfilled with specifically targeted data collections (Montanarella, 2010). Traditional soil surveys, being time consuming and expensive, are very unlike to be carried out in the near future.

In Hungary spatial soil data requirements have been fulfilled so far with formerly elaborated, legacy map-based datasets either by their direct usage or after certain specific and

Soil Mapping and Process Modeling for Sustainable Land Use Management.
DOI: http://dx.doi.org/10.1016/B978-0-12-805200-6.00009-8

generally conditional, thematic, and/or spatial inference (Dobos et al., 2010; Pásztor et al., 2013; Sisák and Benő, 2014; Szabó et al., 2007; Szatmári et al., 2013; Waltner et al., 2014). Several national programs have emerged recently, whose successful completion necessitates the application of suitable data. That data needs to be spatially exhaustive and consistent as well as both globally and locally reliable. Also, accurate soil information is needed not only for primary soil properties, but dominantly for specific basic and or higher level soil features, which formerly had not even been mapped.

The Hungarian National Spatial Development Plan uses zones. Croplands with "Excellent" or "Good Productivity" are protected against being built on by industrial and commercial investments. Since these zones are not mathematically defined categories, the accurate identification and delineation of these types of areas has been a great challenge, which was earlier approximated by different approaches. First, conditions for agricultural production were defined in terms of the one-and-a-half century old "Golden Crown Standard"(Dömsödi, 2006) using its averaged land value index reflecting the area's potential productivity. In the next turn, related to the National Rural Development Plan (2004), areas characterized by excellent productivity and those affected by specific handicaps (see later) were identified and spatially delineated jointly (Ángyán et al., 1998) using, among other environmental data, the AGROTOPO (1994) spatial soil information system, whose spatial resolution corresponds to about 1:500,000–1:250,000 scale. The revision of old, and the preparation of a new development plan required novel, more spatially detailed thematic inputs together with revised, upgraded methods for their application.

The European Union's (EU) Common Agricultural Policy encourages maintaining agricultural production in less favored areas to secure both stable production and income to farmers and to protect the environment. Recently the delimitation of Areas with Natural Constraints (ANCs) is suggested to be carried out using common biophysical diagnostic criteria on low soil productivity and poor climate conditions all over Europe. The criterion system was elaborated by the European Commission's Joint Research Center (Van Orshoven et al., 2014) but its operational implementation was directed to member state authorities. The objective, transparent, and science-based common criteria system includes eight criteria related to climate, soil, and topography. The soil criteria consist of an additional 11 subcriteria, whose regionalization requires proper spatial information on several basic and derived soil properties.

The land use management and forestry sector requires information on the feasibility of areas for different land use purposes. Suitability of land parcels for both agriculture and forestry (from fuel wood production to lumber, or even soil protection) is site dependent. Decisions on the establishment of new forests or woodlands requires detailed spatial data about site characteristics according to a specific forestry criteria system. The needed information covers soil typology, soil hydrology, and soil physical properties such as depth and texture classes. Furthermore, specialized soil information is needed, which is appropriate for choosing species and the main objectives of forest management. On the basis of this information, it is possible to assess future yield classes of new forests that help to set up a priority order among forestry or other land use forms. The Agroclimate. 2 project aims to develop a fine resolution, countrywide decision support system (DSS) for farmers, foresters, landowners, agricultural, and forestry companies. The core of the DSS is the projected climate change impact on forests, croplands, and grasslands. The targeted system evaluates changes in site characteristics according to climate change models. Based on the knowledge of site characteristics, yield potentials of different crops are estimated. For forestry, the third most important site element following climate and hydrology is soil and its properties.

The National Adaptation Geo-information System (NAGIS, 2016) is a multipurpose geographic information system that can facilitate the policy-making, strategy-building, and decision-making process related to the impact assessment of climate change and initiating necessary adaptation measures in Hungary. AGRAGIS (2016) is the extension of NAGiS into the agrarian sector, established in order to estimate agri-environmental related impacts of climate change and support the associated vulnerability assessment. Both NAGiS and AGRAGiS heavily rely on adequate spatial information related to soil, which is incorporated into complex environmental modeling.

Due to the above outlined challenges the Hungarian spatial soil data infrastructure is significantly changing. Based on widely used, but almost forgotten, legacy data sources originating from past surveys, new map products have been created targeting the fulfilment of the recent demands.

A World Reference Base (WRB) based, harmonized digital soil map and database for the Danube Basin of the Danube-region (encompassing Hungary) have been compiled very recently by Dobos et al. (2016). Its legend uses the Reference Soil Group level with a spatial resolution of 463 m. The applied, slightly modified e-SOTER methodology used automated classification algorithms and soil diagnostic property maps as regionalized qualifiers, which were elaborated upon based on proper reference data (Dobos et al., 2013, 2011).

A unified, national soil type map with spatially consistent predictive capabilities that unifies expert inputs and databases from both agricultural farmland and forested sectors was compiled (Pásztor et al., 2017) by applying traditional and newly tested digital soil mapping (DSM) classification methods (i.e., segmentation of a synthesized image consisting of the predictor variables, multistage classification by Classification and Regression Trees, Random Forests and Artificial Neural Networks). Classifications were carried out on two levels—main soil type group and soil type—to achieve better results. The soil type map was supplemented by some basic soil property maps, also based on integrated and harmonized data sources of the two areas. These products can be equally used for agricultural or forestry oriented purposes, providing interoperability between sectors. The newly developed, fine resolution maps are very important for both forest management and forest research; they make it possible to prepare realistic spatial predictions for the expected area of species or their potential future growth rates. Additionally, it is also possible by these maps that the management plan of a single forestry unit can be prepared or even supervised. Because of the robustness and huge data background the map set is suitable for supporting nationwide programs.

The National Pedological and Crop Production Database (NPCPD) was rediscovered after some years of being lost for utilization (Kocsis et al., 2014). It provides valuable legacy data for the upper, cultivable layer of soils for arable lands. The data collection dates back to the '80s. Tóth et al. (2015) integrated NPCPD and the data provided by Land Use/Land Cover Area Frame Survey (LUCAS) (Eurostat, 2015) for the croplands of Hungary to produce soil property maps. The map products provide primary soil feature predictions for the uppermost 0–20 cm layer of topsoil for the agricultural land of the country at 250 m pixel size (NÉBIH, 2016).

The DOSoReMI.hu (Digital, Optimized, Soil Related Maps and Information in Hungary; Pásztor et al., 2014, 2015, 2016) project aims to significantly extend the potential for how soil information requirements could be satisfied in Hungary. In the framework of DOSoReMI.hu, numerous soil property maps have been compiled so far with proper DSM techniques. The resulting maps are partly in accordance with GlobalSoilMap.net specifications (Arrouays et al., 2015) and partly by slightly changing or more strictly defining some of the predefined parameters (depth intervals, pixel size, soil property, etc.). The elaborated soil property maps have been further utilized as DOSoReMI.hu was originally intended to take steps toward the mapping of higher level, more complex soil features. The digital soil maps presented in our paper were elaborated in the frame of DOSoReMI.hu.

MAPPING SOIL FEATURES BEYOND PRIMARY PROPERTIES

Quantification of soil functions and services is a great challenge in itself, even if the spatial relevance is supposed to be identified and regionalized. Proxies and indicators are widely used in ecosystem service mapping (Staub et al., 2011). Soil functions and services could also be approximated by elementary soil features (Adhikari and Hartemink, 2016; Calzolari et al., 2016). One solution is the association of soil types with services as basic principle. Soil property maps, however, provide quantified spatial information, which can be utilized more versatilely for the spatial inference of soil functions and services. The goal of the soil mapping is to reveal and visualize the spatial relationships of the thematic knowledge related to soil cover. Soil maps are thematic maps, where theme is determined by some specific information related to soils. This can be a primary or secondary (derived) soil property or class as well as any knowledge characterizing functions, processes, or services of soils. The greatest and inevitable challenge of the compilation of soil maps is the regionalization of the local knowledge, its spatial inference (Miller and Schaetzl, 2014). Reconnaissance of specific soil properties is carried out by sampling, which provides definite, point-like information. To create maps the data related to locations should be spatially inferred using appropriate methods. From a certain point of view the development of soil mapping (Brevik et al., 2016) is the conscious expansion of the repository of these methods: from mental space usage, along (base)map delimitation based on soil-landscape models till the various (mechanical, geometrical, geostatistical) interpolation methods and further until the introduction of ancillary, environmental, spatial data as auxiliary covariables related to various components of soil forming processes. Essentially, this latter is the base of DSM.

A soil map is an object specific spatial model of the soil cover, whose compilation is dominated by the consideration of soil forming processes (Böhner et al., 2001). There have been significant and essentially concurrent changes concerning three central elements of this definition. The expansion of DSM in the last decade can be attributed to the effects of these changes (Boettinger et al., 2010; Dobos et al., 2006; Lagacherie, 2008; Lagacherie et al., 2007). The framework of DSM (Hartemink et al., 2008; Lagacherie and McBratney, 2007; McBratney et al., 2003) involves spatial inference of the information collected at sampled points based on ancillary environmental variables related to soil forming processes. The concept of DSM also provides opportunity for the elaboration of goal specific soil maps, since the parameters characterizing a map product (thematic, resolution, accuracy, reliability, etc.) may be predefined. The activity of DSM goes beyond mapping purely primary and secondary soil properties, the regionalization of further levels of soil related features (processes, functions, and services) is also targeted (Minasny et al., 2012). These initiatives may be considered as new directions for soil interpretation.

Reconnaissance of higher level soil features (functions, services, processes) starts with the characterization, identification, and measurement of primary soil properties. Basically, there

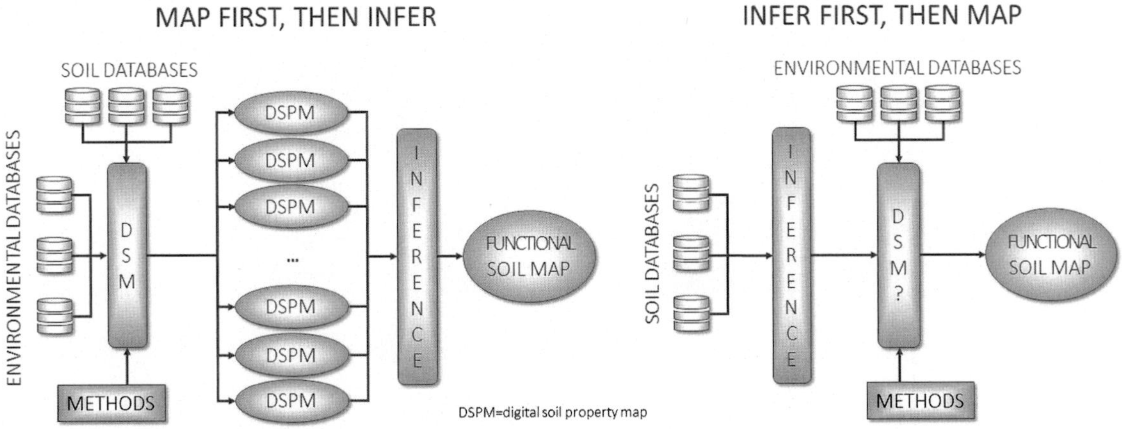

FIGURE 9.1 The two alternatives for the compilation of functional soil maps.

are two approaches for the mapping of these advanced soil (related) features (Fig. 9.1). The "infer first, then map" starts with the inference of higher level features from basic soil properties at the field-plot, observation level using appropriate methods and models. The thematically inferred values are then spatially extended with a specific mapping procedure. There might be two difficulties in the feasibility: data shortage and the lack of a suitable mapping method. Complex and accurate inference methods may require the availability of various parameters, which are not necessary accessible for identical or at least (cor)related locations even if specific component information is obtainable for numerous locations. This may involve the failure of inference for the majority of the affordable reference sites, which also makes the mapping procedure troublesome.

The "map first, then infer" starts from existing soil property maps, which provide spatially exhaustive raw data, and thematic inference is carried out on the spatial object (soil mapping unit or rather pixel) level. This approach is based on available map products and optimizes data usage, since the applied maps can be produced from different sources, using various base datasets of observations. Furthermore, error originating from the spatial inference of raw data as well as its propagation can be accounted for in a more straightforward manner taking into consideration the inherently produced predictions on the spatial accuracy of the input maps.

In the present chapter a simplified framework is presented how digital soil map products can support national level spatial planning and land management. Two distinguished topics are unfolded in more detail to present actual examples. (1) We summarize the steps for how the productivity function of soils was mapped to be utilized in various programs. (2) We present some aspects of the most recent delineation of ANCs according to a common European biophysical criterion system.

VARIATIONS FOR THE APPLICATION OF DIGITAL SOIL MAP PRODUCTS IN SPATIAL PLANNING AND LAND MANAGEMENT

In Fig. 9.2 we outlined three basic approaches for how digital soil map products

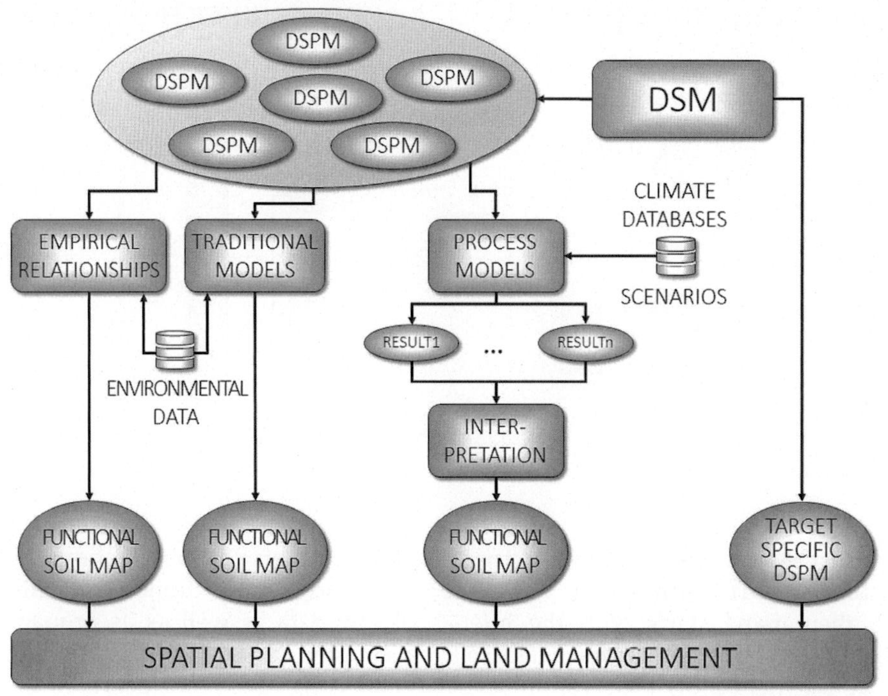

FIGURE 9.2 Involvement of digital soil map products in the process of spatial planning and land management.

can be involved in the process of spatial planning and land management.

1. There are various traditional models, empirical relationships established either in agricultural or forestry practice, which use basic soil parameters in the assessment of specific land related properties and processes. These models do not necessarily work in space, however, they can be transformed into spatial form. Once the input is spatialized, the output is also produced in map format.
2. Similar to traditional models, digital process or crop models can also be built in as engines into spatial inference systems. Their flexibility however can only be maintained in the geographical domain, if all the input data could be elaborated in map form. This is a great challenge in the case where the various input (e.g., soil vs. climate) are available at a significantly different resolution. Another problem may emerge if numerous spatial outputs are processed; for example, due to the usage of scenarios. In this case, suitable aggregation and/ or interpretation of the results should be elaborated too.
3. In some cases, spatial information from relatively simply defined soil properties is able to answer the demands, but the feature was never mapped, or there is not even direct data about them in the available data sources. In this case the elaboration of target-specific digital soil maps can solve the problem, where suitable thematic and spatial inference is carried out in a properly harmonized approach.

APPLICATION OF DIGITAL SOIL PROPERTY MAPS IN CONVENTIONAL MODELS

Both agriculture and forestry have utilized empirical relationships and traditional models to support their decision-making processes, which use basic soil parameters—mainly primary properties—assessed during standard surveys. One of the strongest examples comes from the great relevance soil information has in forestry. What makes reliable soil information valuable for forestry (except for intensive round-wood and biomass plantations) is the simple fact that there is no chance to alter site characteristics significantly in order to fit them to the requirements of grown trees (Szodfridt, 1993). In other words, agro-technological solutions are not used for forestry purposes by default (e.g., watering, fertilizers, pesticides). There are several reasons for this from the limited accessibility of sites, through the high expenses of applied technology. The expense of applying agro-technology shows no return until the ecosystem services of forests, which are tightly bound to diverse site conditions, are accounted for (Chen et al., 1997; Lal, 2005). These ecosystem services include the natural carbon dynamics of natural forest sites. The diversity of forests is reflected in their soils (Osman, 2013). This diversity is based mainly on the incredible genetic resource of trees (Kremer, 1994) by what they had been able to adapt to their environment over the millions of years.

Because of the reasons stated above and the multifunctionality of forests (Führer, 2000), instead of making pointless efforts to convert various forest sites into a generally good growing environment, foresters are to establish a management structure that fits the best practices for the existing forest ecosystem (Tonteri et al., 1990). This can only happen via the most reliable knowledge on site conditions and the demands of tree species. Foresters in Hungary developed the site evaluation and mapping methodology in order to ensure the most reasonable forest management practice (Babos et al., 1966; Járó, 1963). In spite of the continuous improvement of the system (Führer et al., 2011; Szodfridt, 1993), the main pillars of forest site evaluation remained:

- information on forest climate (Araújo and Rahbek, 2006);
- information on the hydrological conditions of forest sites (Csáki et al., 2014);
- knowledge of basic soil characteristics such as soil type, texture class, the depth of soil layers suitable for rooting.

These are the basic elements of site evaluation that make the assessment of forest type, the appropriate species, and their potential growing capacity or possible yield class. Both statistically developed yield models, such as yield tables, and tree growth simulation-based yield models (Assmann, 1970; Petritsch et al., 2007; Pretzsch, 2009) use these inputs. Yield class and expected timber volume by the end of the rotation period can be estimated only if the site properties are known. Therefore detailed soil type maps, texture class maps, and classified rooting depth maps are necessary for proper assessments and decision support.

Evaluation of the agricultural land requires the assessment of the productivity of soils. A traditional (Mueller et al., 2010) and in Hungary heavily embedded approach for this purpose is the application of properly developed scoring systems (Ángyán et al., 1998; Máté, 1960; Podmaniczky et al., 2007; Tóth et al., 2014; Várallyay et al., 1985). The translation of expert knowledge manifests in the assignment of small natural numbers to the categories of primary properties. In our opinion the spatialization of expert knowledge in this manner may have some drawbacks. (1) We see risk in the selection of the scoring values. The simplified projection of the realty onto a short range of discrete values may ignore numerous important effects and relations. Its effect on the thematic accuracy is generally unconsidered. (2) The aggregation of scored variables is generally carried out by rather simple operations too. The weights used in linear approximations mean similar problems as those caused by the above-mentioned scores. (3) Finally the effect of thematic vagueness on the overall accuracy is very difficult to predict and rarely assessed.

There was also a trial of DOSoReMI.hu provided digital soil property maps for application in expert scoring, similar to former approaches, which have been frequently used previously for the regionalization of soil productivity. Nevertheless, we initiated a new concept, which also relies on the produced digital soil property maps, but their quantified spatial information was utilized more versatilely for the spatial inference of soil productivity. The basic idea is, if the productivity function of soil is to be identified, it is a straightforward idea to "interview" the plants about their circumstances. The following section discusses how recent technology gives sense and background for a similar approach.

APPLICATION OF DIGITAL SOIL PROPERTY MAPS IN PROCESS MODELS

Nowadays crop models properly simulate the plant environment conditioned by various factors (characteristically pedologic and climatic) based on actual, predicted, or presumed data. The majority of soil properties are much less variable in time than climatic features. This is especially valid for the set of the soil parameters used in crop simulation models. Consequently, these can be represented by static data in modeling and there is no need for new data collection; input requirements can be fulfilled by available datasets. Applying rational meteorological and management scenarios, modeling yields for different crops produce multiple results, whose proper aggregation provides an appropriate approximation of land productivity. Basically crop models work on single plots, but if all the input is available in map form, the results will be also spatialized.

The Applied 4M Crop Model

Crop simulation models are often used for estimating the possible effects of climate change either on local or global scale. Theoretically system models are the only scientific tools with which we can look into the future and assess the impact of climate change. The 4M crop model (Fodor et al., 2014) was calibrated and then coupled with two of the latest climate change scenarios as well as with a high-resolution national geodatabase of soil and land use in order to create an impact projection on the biomass production of five major arable crops (Fig. 9.3).

4M is basically a crop environment resource synthesis (CERES) (Godwin and Singh, 1998; Ritchie, 1998; Ritchie et al., 1998) clone (with some minor modifications), rewritten in Delphi and extended with a graphical user interface that provides an access to the model parameters as well as to the input data and presents the output data via tables and graphs. Consequently, 4M is a daily time step, deterministic model that simulates the following main processes of the soil–plant system: soil water balance including plant water uptake, soil heat balance, soil nitrogen balance including nitrate leaching and plant N uptake, plant development, growth and senescence. The 4M model requires daily meteorological, static soil, and plant specific input data, as well as some basic information about the agro-technical operations (i.e., timing and applied amount of materials). Starting from the initial conditions set by the user the model simulates the evolution of the soil–plant system driven by

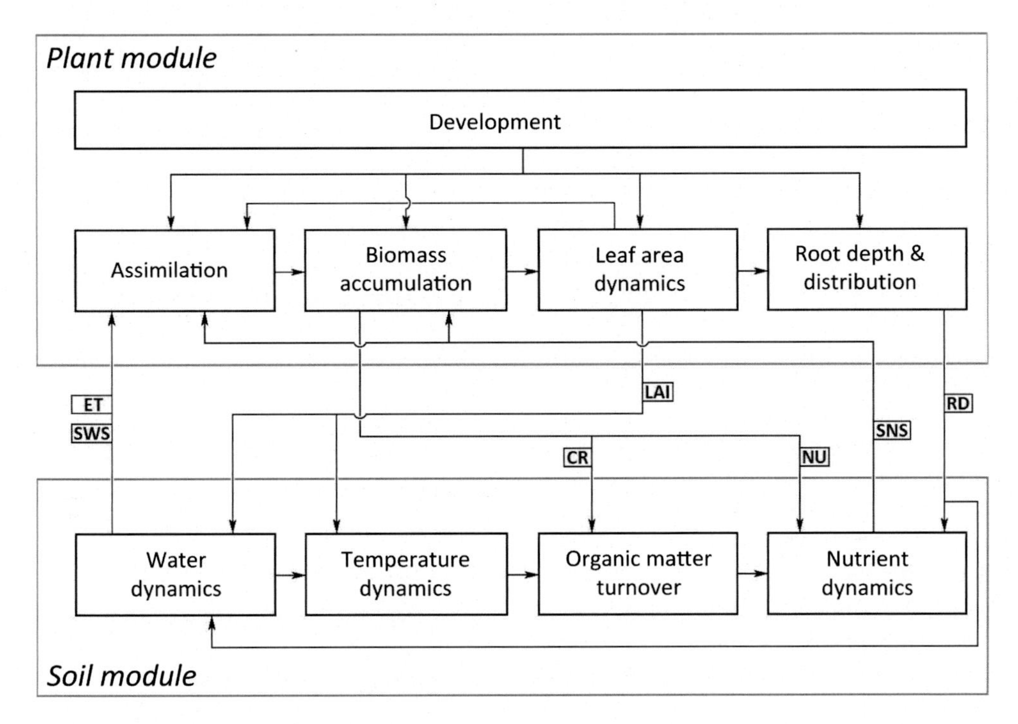

FIGURE 9.3 Flow chart of the 4M crop model: main processes and key variables. *ET*, evapotranspiration; *SWS*, soil water status; *LAI*, leaf area index; *CR*, crop residue; *NU*, nutrient uptake; *SNS*, soil nutrient status; *RD*, rooting depth.

daily climatic data and governed by the system parameters and agro-management options that are also provided by the users as inputs. The model outputs (yield, biomass, nitrate leaching, etc.) could be used in (1) research: the results of observations as well as of experiments could be extrapolated in time and space; thus, for example, the expected impacts of climate change could be projected; and in (2) practice: model calculations could be used in intelligent irrigation control and DSSs as well as for providing scientific background for policy makers.

For the model calibration, meteorological inputs (global radiation, minimum and maximum temperature, and precipitation) were retrieved from the 10 × 10km spatial resolution CarpatClim database (Spinoni et al., 2015), covering the area of Hungary with 1104 cells and providing the necessary climatic data for the 1961–2010 period. Primary soil property maps (i.e., texture of topsoil and subsoil, organic matter content, and rootable depth) were provided by DOSoReMI.hu. Derived soil hydrophysical properties were estimated by using pedotransfer functions (PDF; Fodor and Rajkai, 2011) based on the available clay, sand, and organic matter content data by the following PDF: bulk density (Rawls, 1983); field capacity, wilting point (Rajkai et al., 2004); and saturated hydraulic conductivity (Campbell, 1985). Farm accountancy data network (FADN) data of 294 representatively selected (by location and size) Hungarian agricultural enterprises were used from the period of 2001–10. The available data included the geographical location of the parcels with maize, sunflower, wheat, barley, and rapeseed, time of the agro-technical operations, amount of the applied fertilizers and the final yields.

Spatial Prediction of Soil Productivity

Model calculations were carried out for the five most important crops whose summarized territorial representation in Hungarian agriculture reaches 80–85%, namely: winter wheat, barley, maize, sunflower, and rapeseed. Each crop was grown for 30 years applying daily, actual as well as generated climate data. In the former case, data collected in the period between 1961 and 1990 were provided by the CarpatClim database. In the latter case a weather generator was used for the generation of adequate daily estimations conditioned by shorter range, weekly and monthly time series available in more spatially detailed form. In addition to the application of various climatic data inputs, further scenarios were tested by changing agro-management parameters. Water and nutrient supplies were altered in three distinct grades: optimal, normal, and poor supply categories.

Soil input for the model was available with the highest spatial resolution. Originally, soil parameters were produced quantitatively using continuous values. As a consequence, the calculations should have been carried out for each pixel within the country's border. Even excluding the irrelevant areas (sealed soils, water surfaces) this would have required enormous computation capacity to execute on more than 50 million entities. To decrease the number of model runs, soil parameters were categorized. Soil texture was taken into consideration in the form provided by the USDA categorization (12 classes), rooting depth was classified into three (<50 cm, 50–100 cm, and >100 cm); soil organic matter into six categories (<1%, 1–1.5%, 1.5–2%, 2–3%, 3–5%, and >5%). Finally, instead of the theoretical $12 \times 12 \times 3 \times 6 \times 1104 = 2{,}861{,}568$ soil–climate combinations, only 66,834 different, existing types resulted, which were used as input in the model calculation. The results of model runs were projected back to the geographical space with the aid of these joining entities.

Yields predicted along crop, water, and nutrient supply scenarios resulted in numerous maps, modeling specific components of productivity functions of soils. To produce a unique, easily communicable product, we aggregated the results. Crop yields were standardized using wheat equivalent yield, then the scenario results were weighted according to their estimated representativeness and then summarized. The result characterizes the land's productivity in the case of the most rational practices and most frequent environmental conditions (Fig. 9.4). The variation of the results was considered as a proxy for the vulnerability of the productivity.

Beyond Soil Productivity Assessment

In addition to spatially modeling soil productivity the elaborated environment provided two further opportunities:

1. Applications of climate data provided by down-scaled global climate scenarios provide a framework for the estimation of agro-environmental related impacts of climate change and support the associated vulnerability assessment. The model calculation carried out on 1 ha pixels was aggregated to the much rougher CarpatClim grid in order to support the thematically more extended investigation proceeding within NAGiS and AGRAGiS. An example of the base results is displayed in Fig. 9.5.
2. 4M output is not confined to yield prediction. Biomass production, evapotranspiration, and N-leaching are calculated, whose parameters can be also spatialized based on the introduced soil–climate combinations. The resulting maps then can be applied in the mapping of further soil functions like N-leaching in the approximation of attenuation (Fig. 9.6).

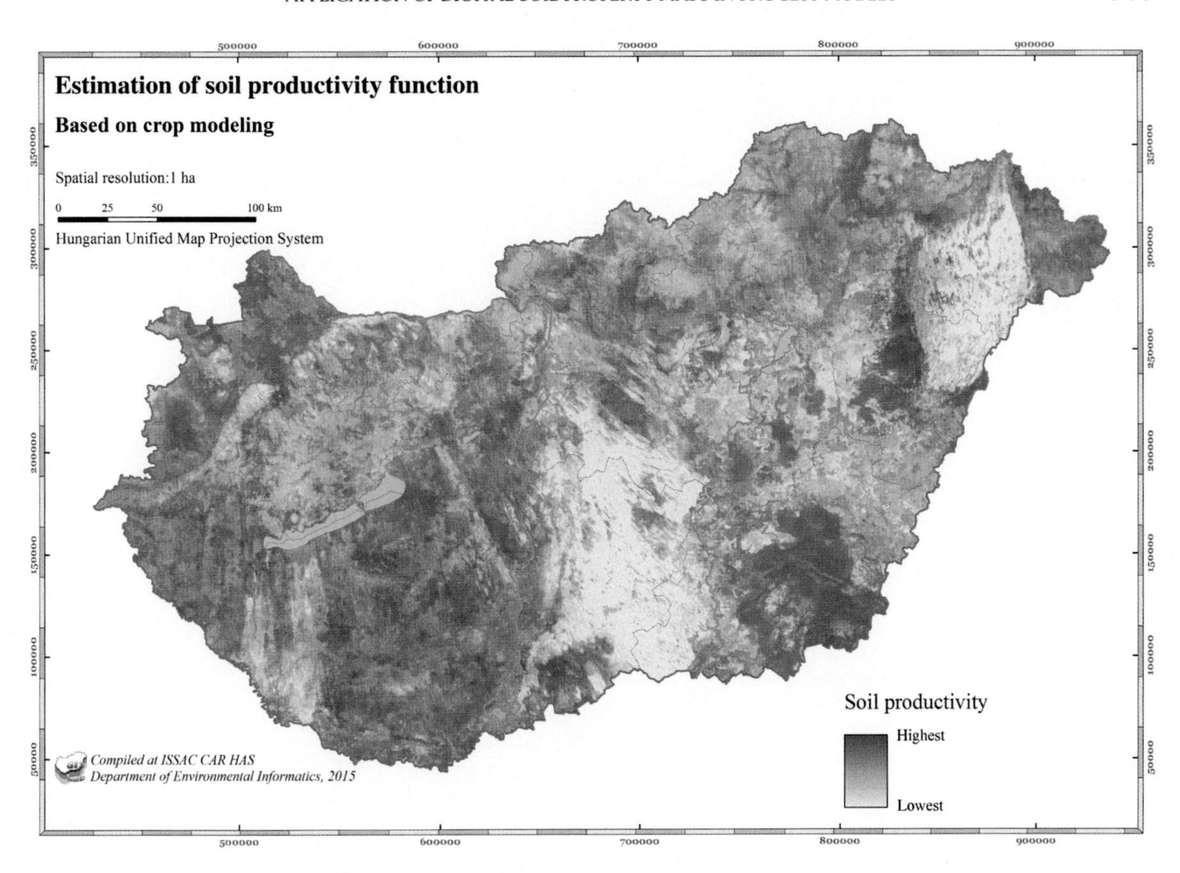

FIGURE 9.4 Predicted soil productivity map based on the aggregation of the crop modeling results representing the most rational practices and most frequent environmental conditions.

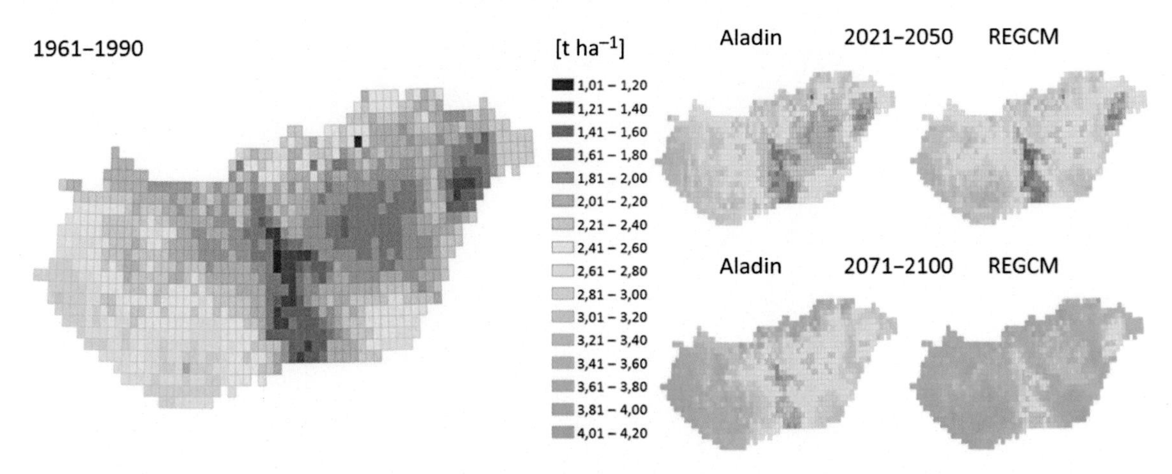

FIGURE 9.5 Potential impact of climate change on RAPESEED yields. The 30-year average yields of the 2021–50 and 2071–2100 periods are compared to that of the reference period under INTENSIVE crop production.

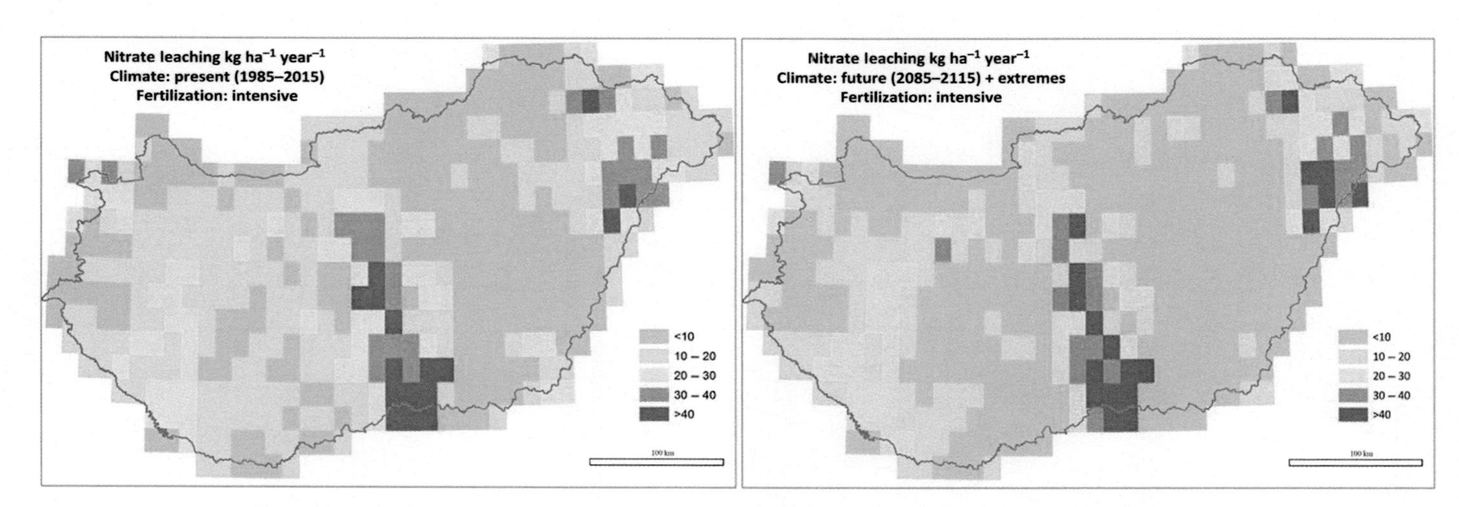

FIGURE 9.6 Assessment of specific attenuation function of soils: estimation of N-leaching; and potential impact of climate change on this function. The 30-year average nitrate leaching of the 1985–2015 and 2085–2115 periods under INTENSIVE crop production.

ELABORATION OF TARGET-SPECIFIC DIGITAL SOIL MAPS

Designation of ANCs according to the common European regulation requires information on several soil properties (Table 9.1). The criteria, which refer to basic soil properties—like rooting depth or pH—can be relatively easily mapped by the data of the most commonly used soil information systems. But the other criteria, which have not been directly observed in Hungary or refer to a more complex soil property, need more data sources or the reprocessing of existing soil survey data. In this section, we summarize how the requirements of delineation of ANC according to the common European biophysical criteria were fulfilled by specific DSM products based on the reinterpretation of former soil survey information.

Soil Data

The Digital Kreybig Soil Information System (DKSIS; Pásztor et al., 2010) was compiled based on the most detailed nationwide soil survey (Kreybig, 1937), and it covers the whole area of Hungary. DKSIS consists of two types of soil information. Soil mapping units are defined by the physical and chemical properties of the rooting zone, but it is only a rough categorization. The soil profile dataset contains many measured records about the physical and chemical soil properties on a layer level. Detailed profile descriptions are available for about 22,000 sampling sites, which is spatially transferred to an additional 250,000 locations. The Hungarian Soil Information and Monitoring System (SIMS) consists of 1234 observation locations, which have been selected to represent

TABLE 9.1 Soil Criteria and Thresholds for Delineation of ANCs According to EU Regulation (Van Orshoven et al., 2014)

Criterion	Definition	Severe Threshold
Limited soil drainage	Areas which are water logged for a significant duration of the year	Wet 80 cm >6 months, or 40 cm >11 months
		Poorly or very poorly drained
		Gleyic color pattern within 40 cm
		≥15% of topsoil volume is coarse material, rock outcrop, and boulder
Unfavorable texture and stoniness	Relative abundance of clay, silt, sand, organic matter (wt%), and coarse material (vol%) fractions	Texture class in half or more (cumulatively) of the 100 cm soil surface is sand and loamy sand
		Topsoil texture class is heavy clay (≥60% clay)
		Organic soil (organic matter ≥30%) of at least 40 cm
		Topsoil contains 30% or more clay and there are vertic properties within 100 cm of the soil surface
Shallow rooting depth	Depth (cm) from soil surface to coherent hard rock or hard pan	Rooting depth ≤30 cm
Poor chemical properties	Presence of salts, exchangeable sodium, excessive acidity	Salinity ≥4 dS/m in topsoil
		Sodicity ≥6 ESP in half or more of the 100 cm surface layer
		Soil acidity topsoil pH (H_20) ≤5

physiographical-soil-ecological units. SIMS contains detailed and up-to-date quantitative soil information about physical and chemical properties at the layer level (Várallyay, 2002). The Hungarian Detailed Soil Hydrophysical Database (MARTHA) contains harmonized soil hydrophysical and chemical information collected from different databases. In MARTHA the soil information is available for 3937 profiles, but they are representative mainly for the cultivated area (Makó et al., 2010). Three soil related criteria were selected to present as representative examples for the feasibility of their designation considering data availability, thematic, and spatial inference.

The criterion defined for unfavorable texture due to significant amounts of sand prescribes that the texture class in half or more of the 100cm soil surface is sand to loamy sand (Van Orshoven et al., 2014). The designation of this criterion requires the knowledge of the amount of sand and silt fractions of the 100cm soil surface. The data from SIMS were used since it contains measurements on seven particle size fractions, from which the percentage of sand, silt, and clay can be calculated. Based on Food and Agriculture Organization (FAO) particle size classes the limit between silt and sand is 0.063mm (FAO, 2006), but in SIMS it is 0.05mm. SIMS data were converted into FAO particle size classes by log linear interpolation (Nemes et al., 1999). The criterion of sandiness was tested by layers, and the layer level results were summarized for the 100cm of soil surface.

The criterion defined for unfavorable texture due to vertic properties prescribes the presence of vertic features within 100cm of the soil surface. Vertic properties emerge in the form of wedge-shaped soil aggregates with a longitudinal axis, slickensides, and shrink-swell cracks, which are typical for soils with high clay content (IUSS Working Group WRB, 2014). Vertic properties cannot be directly measured, but they can be observed in the field. Soil profile descriptions from DKSIS contain indications in the form of notes, such as shrink-swell cracks or slickensides. A binary parameter was created for the indication of the occurrence of a vertic property. If the profile description includes notes about vertic properties, the value of the parameter was set to 1, in all other cases it was coded with 0. Based on this reference the probability of occurrence was spatially inferred, resulting in a continuous map, with probability values for the presence of vertic properties. The presence of vertic properties was considered to be verified when above 66% probability.

The criterion defined for poor chemical properties due to salinity is determined by measuring the electrical conductivity of a solution extracted from a water-saturated soil paste (EC_e). The prediction of salinity requires the knowledge of EC_e in topsoil. None of the Hungarian soil information systems contains data on directly measured EC_e; therefore this parameter had to be estimated by other basic soil properties. EC_e can be calculated by the liquid limit and the total salt content of the soil (Filep and Wafi, 1993). MARTHA was used as the data source in the case of EC_e estimation. MARTHA was created primarily not for mapping purposes; therefore the spatial coverage is not consistent enough for the whole country and not ideal for countrywide prediction. Therefore the data of MARTHA was completed from another soil information system to achieve sufficient spatial coverage for the spatial prediction. In DKSIS, locations where salt content (and EC_e as well) is certainly zero can be identified. By the aid of these auxiliary points the spatial delineation of saline soils with higher EC_e could be refined.

Auxiliary Data

For the mapping of soil properties, relevant auxiliary environmental variables were used. The variables were selected to characterize the soil forming factors, which determine the predicted soil properties. Furthermore, multitemporal and multispectral remotely sensed images were used, which provide direct information on the surface and certain indirect information on subsurface conditions ruled by soil features.

For the characterization of terrain we used the relevant part of EU-digital elevation model (DEM), which is a 25 m spatial resolution DEM compiled for Europe (EU-DEM, 2015). Numerous morphometric derivatives were calculated using SAGA GIS software (Conrad et al., 2015), which provide information not only on pure terrain properties, but also on other environmental parameters. Flow line curvature, general curvature, real surface area, and vector ruggedness characterize the morphometry. Relative slope position index and topographic position index are in connection with topographic situation. Channel network base level, mass balance index, multiresolution index of valley bottom flatness, multiresolution index of the ridge top flatness, stream power index, and vertical distance to channel network are in connection with the hydrological and run-off properties of the area. Diurnal anisotropic heating and SAGA wetness index are in relation with microclimate.

Imagery provided by satellite remote sensing provides direct information on land cover, land use, vegetation condition, and bare soils (Escribano and Garcia, 2016). For the modeling, satellite images from the moderate-resolution imaging spectroradiometer (MODIS) sensor (on board the Terra and Aqua satellites) were used, which have a 250 m spatial resolution. The applied MODIS images were acquired in spring and fall (March 16, 2012 and September 7, 2013) in line with plant phenology. Data from the red (R) and the near infra-red (NIR) bands were used and the Normalized Difference Vegetation Index (NDVI) (Rouse et al., 1973) was calculated. These remotely sensed variables were selected because these bands and the calculated index have a strong relationship with the state of vegetation and biomass, which reflects certain soil features.

The climatic properties of the country were represented by four parameters. Spatial layers of average annual evapotranspiration, average annual precipitation, average annual temperature, and annual evaporation were interpolated applying the Meteorological Interpolation based on Surface Homogenized Data Basis (Szentimrey and Bihari, 2007) method for gridding from hourly station data. It was developed at the Hungarian Meteorological Service specifically for the interpolation of near-surface meteorological data and is based on the idea that the highest quality interpolation formula can be obtained when certain statistical parameters are known. These parameters are derived by modeling, using long term homogenized data of neighboring stations.

Lithology was derived from the Geological Map of Hungary 1:100,000 (Gyalog and Síkhegyi, 2005). In order to simplify the large number of lithology and facies categories, units were correlated with the nomenclature of parent material defined in the FAO Guidelines for soil description (Bakacsi et al., 2014; FAO, 2006). The groundwater level was taken from the Geological Atlas of Hungary (Pentelényi and Scharek, 2006). This polygon-based map displays rather broad interval categories, not continuous depth.

Land use was taken from the CORINE Land Cover 1:50,000 (CLC50; Büttner et al., 2004), which is a national land cover database elaborated on the basis of the CORINE nomenclature of the European Environment Agency (EEA), and adapted to fit the characteristics of Hungary. In order to stratify regions with different land cover, merged categories of CLC50 were used, maintaining its spatially fine resolution. The main objective of this approach is to improve the predictive applicability of remotely sensed information.

The use of legacy soil data supports the applicability of DSM and improves the accuracy of DSM products (Pásztor et al., 2016, 2013). In the present work we also used the soil mapping units of DKSIS.

In addition, a set of 70 environmental variables were compiled. For the mapping of each target soil property the suitable covariates were selected. The auxiliary dataset needed some preprocessing before the spatial inference. The maps of covariables were converted to rasters with a 100 m spatial resolution. The DEM and derivatives and the satellite images and meteorological layers were resampled, the vector data were rasterized to the standardized 100 m resolution grid. The categorical data (land cover, geology, groundwater and soil mapping units by properties) were split into separate layers by attribute and given a binary code for presence or absence, which were used in indicator form.

Spatial Inference

The target-specific maps were created by intentionally selected DSM methods. The final set of environmental covariables in the case of each target variable consisted of at least 40 layers. To reduce the number of predictor variables and to avoid their multicollinearity, principal component analysis was carried out. In the further analysis the first principal components, which explain together 99% of the variance, were used.

Regression Kriging (RK; De Carvalho et al., 2014; Dobos et al., 2010; Hengl et al., 2004; Illés et al., 2011; Marchetti et al., 2010) was used for spatial prediction, which is widely used in DSM (Miller, 2017). RK combines the regression of the dependent variable by auxiliary variables with kriging of the regression residuals (Hengl, 2007). First the target soil property was modeled by multiple linear regression analysis (MLRA) of the auxiliary variables with stepwise selection method (5% significance level). The interpolation of the differences between the predicted and observed values was carried out by ordinary kriging. The result of the estimation is the sum of the regression model and the interpolated residuals. RK was carried out in SAGA GIS environment.

The overall accuracy of the predicted maps was checked by Leave-One-Out-Cross Validation (LOOCV; Stone, 1974). By LOOCV the estimation of the target soil property is carried out $(n - 1)$ times, leaving out each time one of the samples. Then the predicted and the measured values of the left-out sample are compared. The estimation of the overall accuracy was tested by the following parameters: mean error (ME), mean absolute error (MAE), root mean square error (RMSE), and root mean normalized square error (RMNSE). The expected value of the ME and RMNSE are 0 and 1, respectively. MAE and RMSE refer to the accuracy of the estimation, the lower the value of the MAE and the RMSE, the better is the prediction accuracy. Because the occurrence of vertic properties was a binary classification, it was validated in a different manner, using coincidence matrix.

Compilation of the Criterion Maps

The spatial inference of the cumulative thickness of sand to loamy sand layers of the 100 cm soil surface used the following environmental auxiliary variables, which were selected by the stepwise method of the MLRA:

- relief: elevation, slope, multiresolution ridge top flatness, multiresolution valley bottom flatness;
- climate: yearly mean precipitation, actual evaporation, evapotranspiration
- geology: blown sand;
- satellite imagery: spring-NDVI, fall-NDVI, spring-NIR;
- land cover: arable land, forest;
- legacy soil map: physical properties: poor water retention and very high permeability, and infiltration rate.

On the predicted sandiness criterion map (Fig. 9.7) the areas with coarser particle size are well-delineated and coincide with the location of the main sand ridges in Hungary.

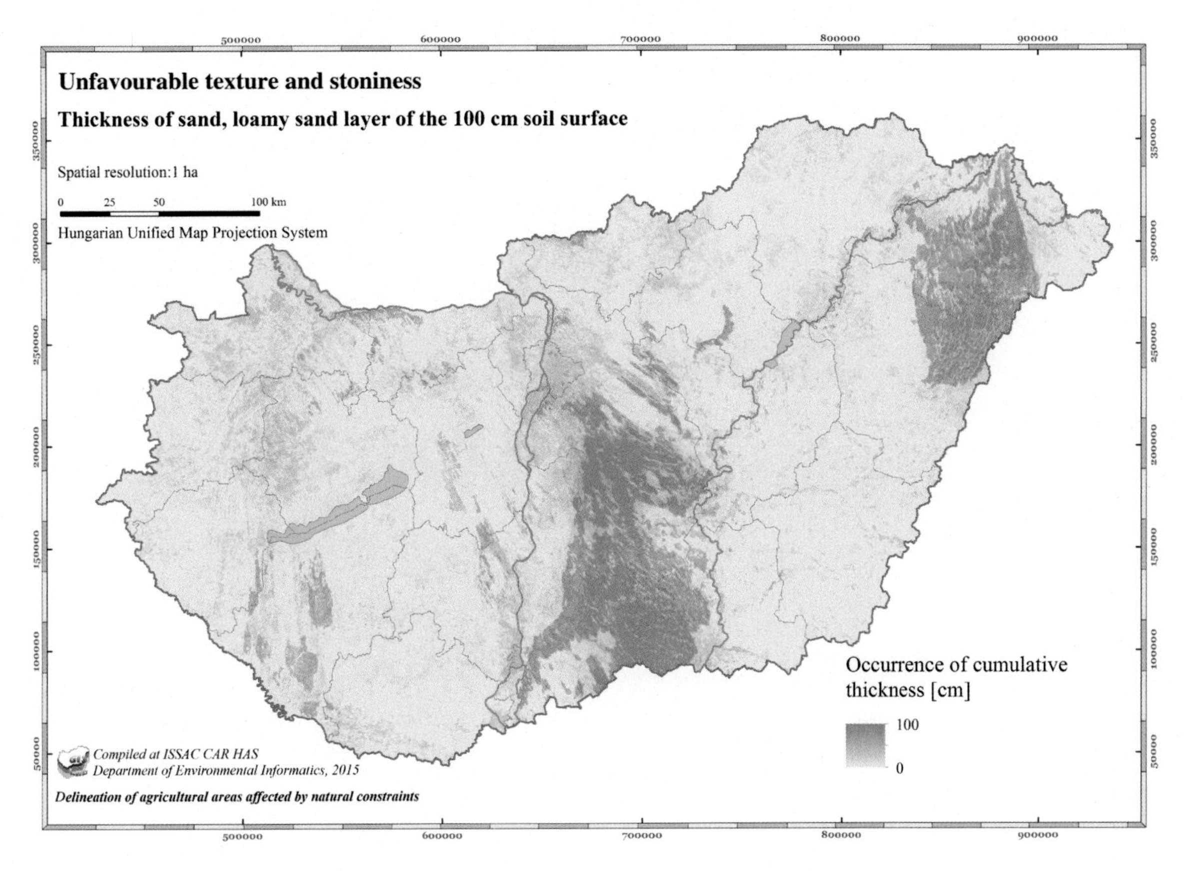

FIGURE 9.7 Thickness of sand to loamy sand layers of the 100 cm soil surface in Hungary.

TABLE 9.2 Results of the LOOCV Validation for the Prediction of Sandiness

Validation Method	Validation Result
ME	−0.003
MAE	12.478
RMSE	19.853
RMNSE	1.094

The result of the LOOCV is summarized in Table 9.2. ME and RMNSE are close to their expected value which is 0 and 1, respectively. MAE and RMSE are below the 20th percentile of the value set. These measures of accuracy suggest the prediction acceptable.

The spatial inference of the probability for the occurrence of vertic properties used the following environmental auxiliary variables, which were selected by the stepwise method of the MLRA:

- relief: aspect, topographic wetness index, SAGA wetness index, multiresolution ridge top flatness, multiresolution valley bottom flatness, diurnal anisotropic heating;
- climate: yearly mean temperature, precipitation, actual evaporation, evapotranspiration;
- satellite imagery: spring-NDVI, fall-NDVI, spring-NIR, fall-NIR, spring-R, fall-R;
- legacy soil map: physical properties: good and high water retention, saline soils, peaty soils.

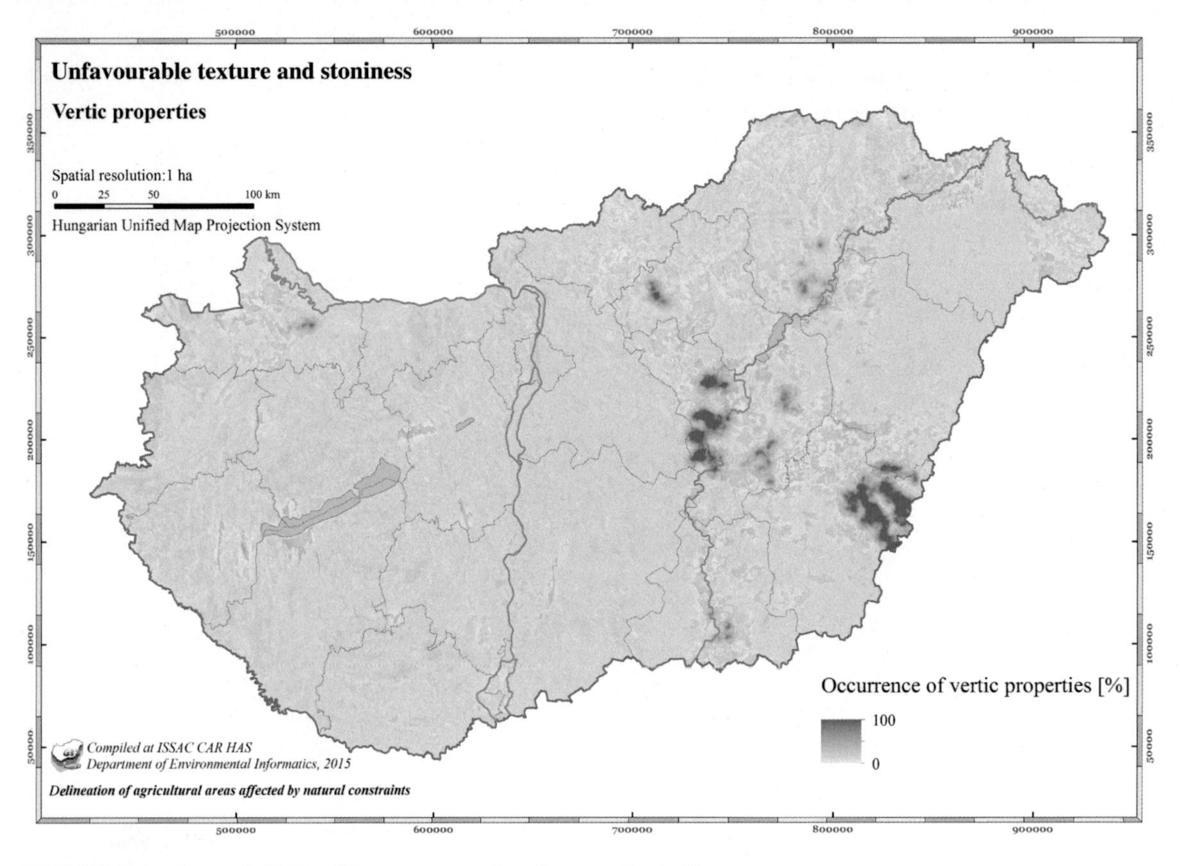

FIGURE 9.8 The probability of the occurrence of vertic properties in Hungary.

On the predicted vertic property criterion map (Fig. 9.8), where the probability values were converted to percentage, the occurrence of vertic properties coincides with lowlands covered with fluvial sediment and where the inland excess water inundation is frequent in Hungary.

For validation the predicted probability values (0–1) and the observed parameters (0 or 1) were compared; the predicted values above 0.66 were considered as 1, while under 0.66 as 0. The results referring to vertic property map are shown in Table 9.3. The overall accuracy of the map is rather good, since misclassification occurs only in 3% of cases.

The spatial inference of salt content used the following environmental auxiliary variables, which were selected by the stepwise method of the MLRA:

TABLE 9.3 Result of the Validation in the Case of Being Vertic

		Predicted	
Fulfilled Criterion		**Yes (1)**	**No (0)**
Observed	Yes (1)	34.8%	1.8%
	No (0)	1.2%	62.2%

- relief: SAGA wetness index, multiresolution ridge top flatness, multiresolution valley bottom flatness;
- climate: yearly mean precipitation, actual evaporation;

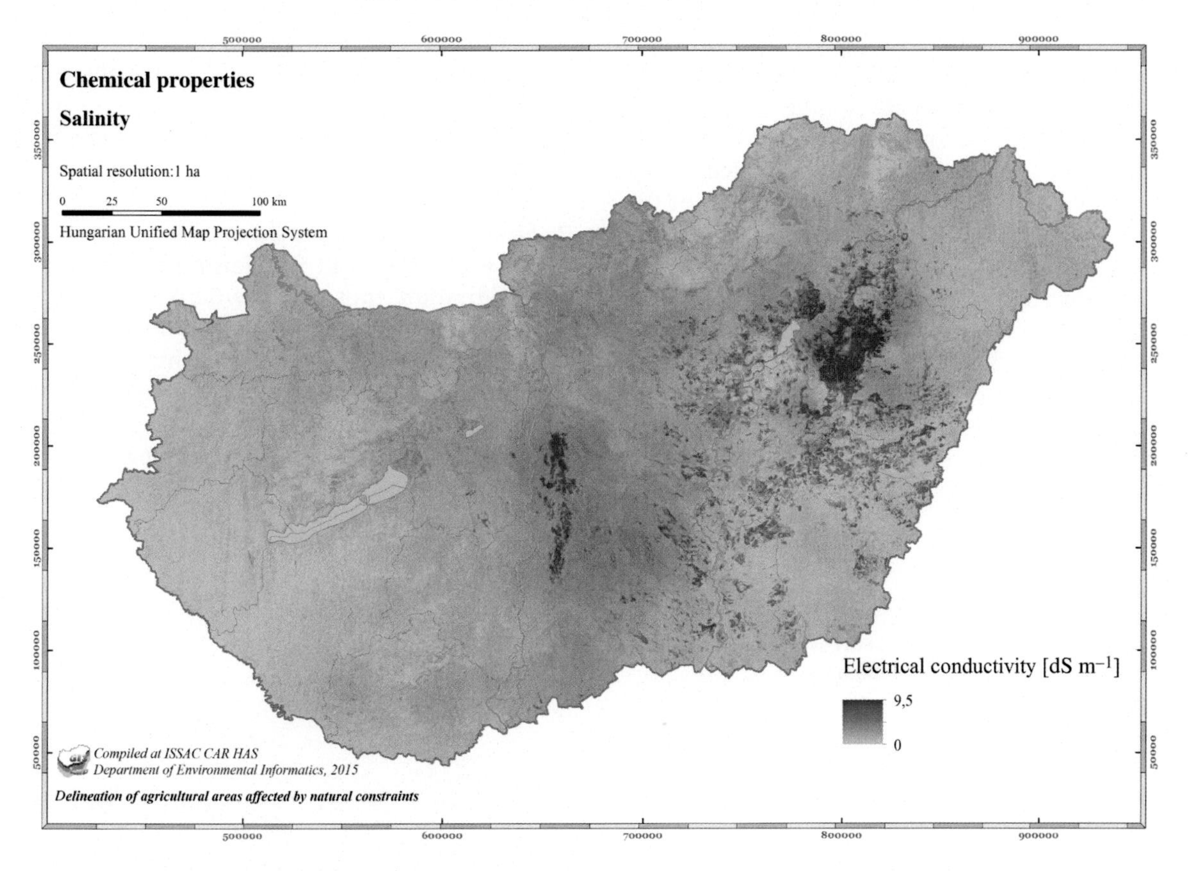

FIGURE 9.9 Soil salinity in Hungary.

- satellite imagery: fall-NDVI;
- land cover: grassland, sparse vegetation;
- legacy soil map: chemical properties: saline soils, neutral, and calcic soils.

On the predicted salinity criterion map (Fig. 9.9) the areas with higher EC and higher salt content are in good agreement with the areas covered by saline soils in Hungary.

The result of the LOOCV is summarized in Table 9.4. Based on the ME result the model somewhat overestimates. RMNSE is also close to its expected value. Based on MAE and RMSE the overall accuracy of the prediction is acceptable. MAE and RMSE have relatively low values, they are lower than the 10% of the value set.

TABLE 9.4 Results of the LOOCV Validation According to the Prediction of Salinity

Validation Method	Validation Result
ME	0.010
MAE	0.850
RMSE	1.430
RMNSE	0.935

Further Processing of the Elaborated Maps to Support Agricultural Policy

Semantic and spatial inferences of specific limiting factors resulted in a digital map series. Each map was queried according to a severe and a so-called subsevere (generally 20% less strict) threshold value. Results of the subsevere queries were evaluated further. Cooccurrence of two biophysical criteria in negative synergy, as a consequence of their joint presence was considered equally as that of a single severe criterion. The cumulative existence of both severe and subsevere soil related biophysical criteria is presented in Fig. 9.10. Biophysical evaluation ends with the nationwide integration of the areas with natural constraints. However, two further steps are expected for the direct support of agricultural policy:

1. Completion of biophysical criteria has to be aggregated at the LAU2 level and evaluated according to a so-called 66% rule related to utilized agricultural area (UAA). Settlements with affected UAA over 66% are designated as ANCs.
2. Fine tuning is also to be carried out for the exclusion from the Natural Handicap designation, those areas where the handicap has been overcome and where the agricultural production achieves, on average, outputs and results comparable to the average in the Member State and where there is no risk of land abandonment. Fine tuning is supposed to also be based

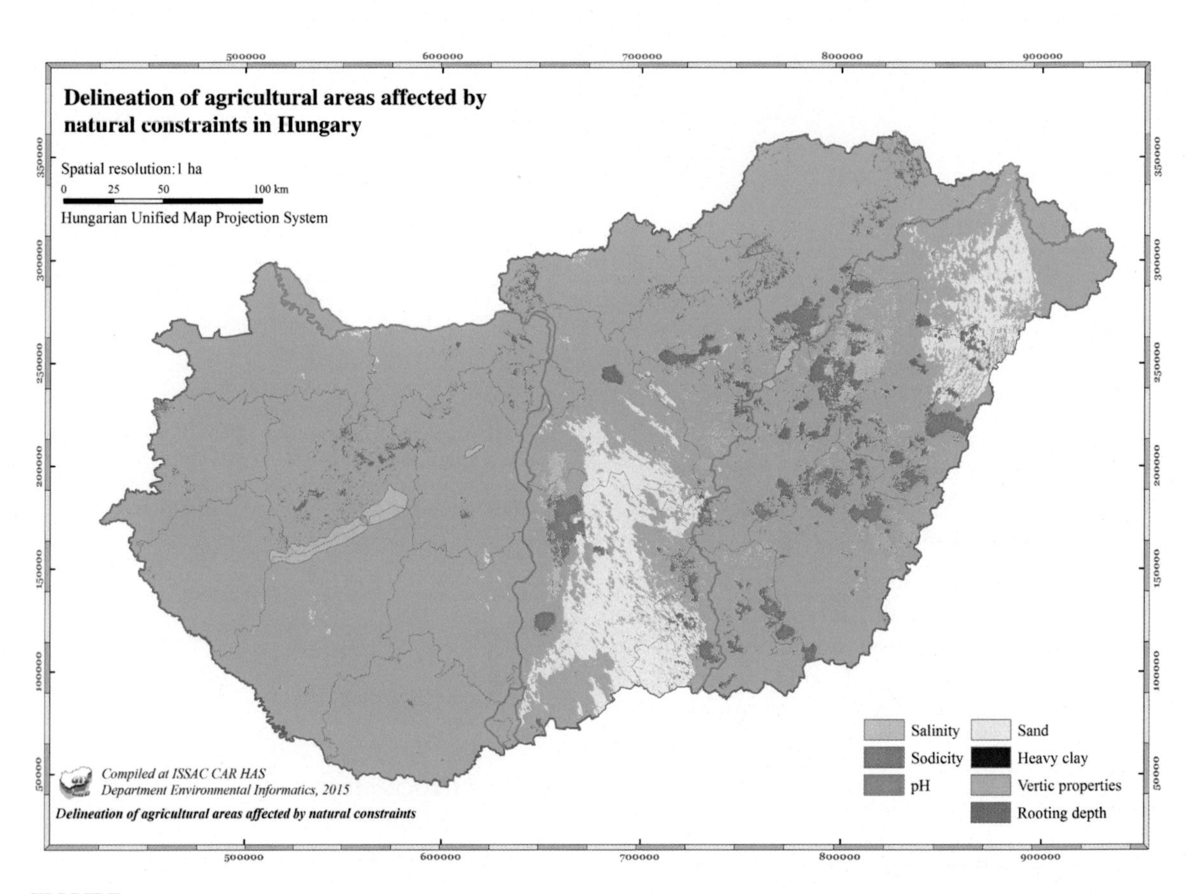

FIGURE 9.10 Result of the nationwide integration of severe and subsevere soil related biophysical criteria in Hungary.

on objective criteria, with the purpose of excluding areas in which significant natural constraints have been documented but have been overcome by investments, or by economic activity, or by evidence of normal land productivity, or in which production methods or farming systems have offset the income loss or added costs.

DISCUSSION

In the present chapter some ideas were outlined and examples were demonstrated for the application of digital soil property maps for the support of spatial planning and land management. The ideas concerned general issues, even if the examples themselves represented recent challenges in Hungary. However, they are not confined to our country. As a consequence, the discussed concepts and the results can be to the interest of the wider international community.

Some aspects of mapping higher level, complex soil features (derived properties, functions, services) were addressed, presenting variations for their spatial assessment. In our opinion, relation of thematic and spatial inference has great impact on the feasibility and the result of the regionalization process. Spatialization of empirical formulas, expert knowledge, criteria, and model results can be carried out in various manners; the final products should be evaluated according to their reliability, global, and local accuracy.

Two more extended case studies were presented on how digital soil property maps work in action. (1) Possibilities provided by process modeling for the spatial inference of soils' productivity function were demonstrated using the 4M crop simulation model. (2) Target-specific DSM was introduced for the designation of ANCs according to the common European regulation, which uses standard biophysical criteria for the objective and transparent identification and delineation of less favored areas.

We sincerely hope that we could corroborate trustfully, that:

- digital soil property maps provide great versatility for mapping higher level soil features;
- many opportunities are promised by their implementation in process modeling;
- target-specific DSM can produce directly applicable products.

Nevertheless we think general and direct mapping of soil functions is not definitively solved; it is still a promising challenge. Finally, we keep on producing proper spatial soil information in the service of the continuously renewing societal demands in the International Decade of Soils (2015–24).

Acknowledgements

Our work has been supported by the Hungarian National Scientific Research Foundation (OTKA, Grant No. K105167), Iceland, Liechtenstein, and Norway through the EEA Grants and the REC (Project No: EEA C12-12), AGRARKLÍMA.2 VKSZ_12-1-2013-0034. Authors thank J. Matus for her indispensable contribution.

References

Adhikari, K., Hartemink, A.E., 2016. Linking soils to ecosystem services—a global review. Geoderma 262, 101–111.

AGRAGiS, 2016. AGRATéR [WWW document]. <http://agrater.hu/>.

AGROTOPO, 1994. AGROTOPO Spatial Soil Information System. RISSAC HAS, Budapest.

Ángyán, J., Dorgai, L., Halász, T., Janowszky, J., Makovényi, F., Ónodi, G., et al., 1998. Az országos területrendezési terv agrárvonatkozásainak megalapozása. Agrárgazdasági Kutató és Informatikai Intézet, Budapest.

Araújo, M.B., Rahbek, C., 2006. How does climate change affect biodiversity? Science 313, 1396–1397.

Arrouays, D., McBratney, A.B., Minasny, B., Hempel, J.W., Heuvelink, G.B.M., MacMillan, R.A., et al., 2015. Specifications Tiered GlobalSoilMap Products, Release 2.4.

Assmann, E., 1970. The Principles of Forest Yield Study. Studies in the Organic Production, Structure, Increment and Yield of Forest Stands. Pergamon Press, Oxford.

Babos, I., Horváthné Proszt, S., Járó, Z., Király, L., Szodfridt, I., Tóth, B., 1966. Erdészeti term-helyfeltárás és térképezés. Akadémiai Kiadó, Budapest.

Bakacsi, Z., Laborczi, A., Szabó, J., Takács, K., Pásztor, L., 2014. Proposed correlation between the legend of the 1:100.000 scale geological map and the FAO code system for soil parent material. Agrokémia és Talajt 63, 189–202. (in Hungarian).

Blum, W.E.H., 2005. Functions of soil for society and the environment. Rev. Environ. Sci. Biotechnol. 4, 75–79.

Boettinger, J.L., Howell, D.W., Moore, A.C., Hartemink, A.S., Kienast-Brown, S., 2010. Digital soil mapping; bridging research, environmental application and operation Progress in Soil Science. Springer, Dordrecht.

Bőhner, J., Kőthe, R., Conrad, O., Gross, J., Ringeler, A., Selige, T., 2001. Soil regionalisation by means of terrain analysis and process parameterisation. Eur. Soil Bur. 7, 213–222. Research report no.

Brevik, E.C., Calzolari, C., Miller, B.A., Pereira, P., Kabala, C., Baumgarten, A., et al., 2016. Soil mapping, classification, and pedologic modeling: history and future directions. Geoderma 264, 256–274.

Büttner, G., Maucha, G., Bíró, M., Kosztra, B., Pataki, R., Petrik, O., 2004. National land cover database at scale 1:50000 in Hungary. EARSeL eProc. 3, 323–330.

Calzolari, C., Ungaro, F., Filippi, N., Guermandi, M., Malucelli, F., Marchi, N., et al., 2016. A methodological framework to assess the multiple contributions of soils to ecosystem services delivery at regional scale. Geoderma 261, 190–203.

Campbell, G.S., 1985. Soil Physics with BASIC: Transport Models for Soil-Plant Systems. Elsevier BV, Amsterdam.

Chen, Z.S., Hsieh, C.F., Jiang, F.Y., Hsieh, T.H., Sun, I.F., 1997. Relations of soil properties to topography and vegetation in a subtropical rain forest in southern Taiwan. Plant Ecol 132, 229–241.

Conrad, O., Bechtel, B., Bock, M., Dietrich, H., Fischer, E., Gerlitz, L., et al., 2015. System for automated geoscientific analyses (SAGA) v. 2.1.4. Geosci. Model Dev. 8, 1991–2007.

Csáki, P., Kalicz, P., Brolly, G.B., Csóka, G., Czimber, K., Gribovszki, Z., 2014. Hydrological impacts of various land cover types in the context of climate change for Zala County. Acta Silv. Lignaria Hungarica 10, 117–131.

De Carvalho, W., Da Silva Chagas, C., Lagacherie, P., Calderano Filho, B., Bhering, S.B., 2014. Evaluation of statistical and geostatistical models of digital soil properties mapping in tropical mountain regions. Rev. Bras. Cienc. do Solo 38, 706–717.

Dobos, E., Bialkó, T., Micheli, E., Kobza, J., 2010. Legacy soil data harmonization and database development. In: Boettinger, J.L., Howell, D.W., Moore, A.C., Hartemink, A.E., Kienast-Brown, S. (Eds.), Digital Soil Mapping: Bridging Research, Environmental Application, and Operation. Springer, Heidelberg, pp. 309–323.

Dobos, E., Hengl, T., Reuter, H., 2006. Digital Soil Mapping as a Support to Production of Functional Maps. Office for Official Publications of the European Communities, Luxembourg.

Dobos, E., Seres, A., Vadnai, P., Michéli, E., Fuchs, M., Láng, V., et al., 2013. Soil parent material delineation using MODIS and SRTM data. Hungarian Geogr. Bull. 62, 133–156.

Dobos, E., Vadnai, P., Michéli, E., Láng, V., Fuchs, M., Seres, A., 2011. Új generációs nemzetközi talajtérképek készítése, az e-SOTER módszertan. In: Az Elmélet És a Gyakorlat Találkozása a Térinformatikában II. pp. 53–60.

Dobos, E., Vadnai, P., Pásztor, L., Micheli, E., Kovács, K., Bertóti, R.D., 2016. A WRB based harmonized digital soil map of the Carpathian-basin. In: Geophysical Research Abstracts. pp. EGU2016–12592.

Dömsödi, J., 2006. Földhasználat (Land Use). Dialóg Campus Kiadó, Budapest-Pécs.

Escribano, P., Garcia, M., 2016. Remote and proximal sensing techniques: importance to soil mapping. In: Pereira, P., Brevik, E.C., Munoz-Rojas, M., Miller, B.A. (Eds.), Soil Mapping and Process Modeling for Sustainable Land Use Management. Elsevier this volume.

EU-DEM, 2015. Digital Elevation Model over Europe [WWW document]. <http://www.eea.europa.eu/data-and-maps/data/eu-dem>.

Eurostat, 2015. LUCAS—A Multi-Purpose Land Use Survey [WWW document]. <http://epp.eurostat.ec.europa.eu/>.

FAO, 2006. Guidelines for Soil Description. FAO, Rome.

Filep, G., Wafi, M., 1993. A talajoldat sókoncentrációjának és a talaj nátriumtelítettségének (ESP) számítása a telítési kivonat jellemzőiből. Agrokémia és Talajt 42, 245–256.

Fodor, N., Pásztor, L., Németh, T., 2014. Coupling the 4M crop model with national geo-databases for assessing the effects of climate change on agroecological characteristics of Hungary. Int. J. Digit. Earth 7, 391–410.

Fodor, N., Rajkai, K., 2011. Computer program (SOILarium 1.0) for estimating the physical and hydrophysical properties of soils from other soil characteristics. Agrokémia és Talajt 60, 27–40.

Führer, E., 2000. Forest functions, ecosystem stability and management. For. Ecol. Manage. 132, 29–38.

Führer, E., Horváth, L., Jagodics, A., Machon, A., Szabados, I., 2011. Application of a new aridity index in Hungarian forestry practice. Időjárás 115, 205–216.

Godwin, D.C., Singh, U., 1998. Nitrogen balance and crop response to nitrogen in upland and lowland cropping systems. In: Tsuji, G.Y., Hoogenboom, G., Thornton, P.K. (Eds.), Understanding Options for Agricultural Production. Kluwer Academic Publishers, Dordrecht, pp. 55–77.

Gyalog, L., Síkhegyi, F., 2005. Geological Map of Hungary, 1:100,000. Geological Institute of Hungary, Budapest [in Hungarian]. [WWW document]. <http://loczy.mfgi.hu/fdt100/>.

Hartemink, A.E., McBratney, A.B., Mendonça-Santos, Mde L., 2008. Digital Soil Mapping with Limited Data. Springer, Dordrecht.

Hengl, T., 2007. A Practical Guide to Geostatistical Mapping of Environmental Variables. JRC, Ispra.

Hengl, T., Heuvelink, G.B.M., Stein, A., 2004. A generic framework for spatial prediction of soil variables based on regression-kriging. Geoderma 120, 75–93.

Illés, G., Kovács, G., Heil, B., 2011. Comparing and evaluating digital soil mapping methods in a Hungarian forest reserve. Can. J. Soil Sci. 91, 615–626.

IUSS Working Group WRB, 2014. World reference base for soil resources 2014. International soil classification system for naming soils and creating legends for soil maps. World Soil Resources Reports No. 106.

Járó, Z., 1963. Talajtípusok. Országos Erdészeti Főigazgatóság, Budapest.

Kocsis, M., Tóth, G., Berényi Üveges, J., Makó, A., 2014. Presentation of soil data from the National Pedological and Crop Production Database (NPCPD) and investigations on spatial representativeness. Agrokémia és Talajt 63, 223–248.

Kremer, A., 1994. Genetic diversity and phenotypic variability of forest trees. Genet. Sel. Evol. 26, S105–S123.

Kreybig, L., 1937. Magyar Királyi Földtani Intézet talajfelvételi, vizsgálati és térképezési módszere (The survey, analytical and mapping method of the Hungarian Royal Institute of Geology). Magy. Királyi Földtani Intézet Évkönyve 31, 147–244.

Lagacherie, P., 2008. Digital soil mapping: a state of the art. In: Hartemink, A.E., McBratney, A.B., Lourdes Mendonça-Santos, Mde (Eds.), Digital Soil Mapping with Limited Data. Springer, Dordrecht, pp. 3–14.

Lagacherie, P., McBratney, A.B., 2007. Spatial soil information systems and spatial soil inference systems: perspectives for digital soil mapping. In: Lagacherie, P., McBratney, A.B., Voltz, M. (Eds.), Digital Soil Mapping—An Introductory Perspective. Elsevier, Amsterdam, pp. 3–22.

Lagacherie, P., McBratney, A.B., Voltz, M., 2007. Digital Soil Mapping: An Introductory Perspective. Elsevier, Amsterdam.

Lal, R., 2005. Forest soils and carbon sequestration. For. Ecol. Manage. 220, 242–258.

Makó, A., Tóth, B., Hernádi, H., Farkas, C., Marth, P., 2010. Introduction of the Hungarian Detailed Soil Hydrophysical Database (MARTHA) and its use to test external pedotransfer functions. Agrokémia és Talajt 59, 29–38.

Marchetti, A., Piccini, C., Francaviglia, R., Santucci, S., Chiuchiarelli, I., 2010. Estimating soil organic matter content by regression kriging In: Boettinger, J.L. Howell, D.W. Moore, A.C. Hartemink, A.E. Kienast-Brown, S. (Eds.), Digital Soil Mapping, Progress in Soil Science, 2. Springer Science + Business Media B.V, Dordrecht, pp. 241–254.

Máté, F., 1960. Megjegyzések a talajok termékenységük szerinti osztályozásához. Agrokémia és Talajt 9, 419–426.

McBratney, A., Mendonça Santos, M., Minasny, B., 2003. On digital soil mapping. Geoderma 117, 3–52.

Miller, B.A., 2017. GIS and spatial statistics applied for soil mapping: a contribution to land use management. In: Pereira, P., Brevik, E.C., Munoz-Rojas, M., Miller, B.A. (Eds.), Soil Mapping and Process Modeling for Sustainable Land Use Management. Elsevier, pp. 129–152.

Miller, B.A., Schaetzl, R.J., 2014. The historical role of base maps in soil geography. Geoderma 230-231, 329–339.

Minasny, B., Malone, B., McBratney, A.B., 2012. Digital Soil Assessments and Beyond. Taylor and Francis Group, London.

Montanarella, L., 2010. Need for interpreted soil information for policy making. In: 19th World Congress of Soil Science, Soil Solutions for a Changing World, 1–6 August 2010, Brisbane, Australia. p. DVD.

Mueller, L., Schindler, U., Mirschel, W., Shepherd, T.G., Ball, B.C., Helming, K., et al., 2010. Assessing the productivity function of soils. A review. Agron. Sustain. Dev. 30, 601–614.

NAGis, 2016. National Adaptation Geo-information System (NAGiS) [WWW document]. <http://nagis.hu/en>.

National Rural Development Plan, 2004. Nemzeti Vidékfejlesztési Terv az EMOGA Garanciarészleg Intézkedéseire.

NÉBIH, 2016. Magyarország mezőgazdasági területeinek talajtulajdonság-térképei [WWW document]. <http://airterkep.nebih.gov.hu/gis_portal/talajvedelem/kiadv.htm>.

Nemes, A., Wösten, J.H.M., Lilly, A., Oude Voshaar, J.H., 1999. Evaluation of different procedures to interpolate particle-size distributions to achieve compatibility within soil databases. Geoderma 90, 187–202.

Omuto, C., Nachtergaele, F., Rojas, R.V., 2013. State of the art report on global and regional soil information: Where are we? Where to go? Global Soil Partnership Technical Report. FAO, Rome.

Osman, K.T., 2013. Forest Soils: Properties and Management. Springer, Dordrecht.

Panagos, P., Van Liedekerke, M., Jones, A., Montanarella, L., 2012. European Soil Data Centre: response to European policy support and public data requirements. Land use policy 29, 329–338.

Pásztor, L., Laborczi, A., Bakacsi, Z., Szabó, J., Illés, G., 2017. Compilation of a unified, national soil-type map for Hungary by integrated, 1 object-based and 2 multi stage classification methods. Geoderma (in press), submitted for publication.

Pásztor, L., Laborczi, A., Takács, K., Szatmári, G., Bakacsi, Z., Szabó, J., 2016. Variations for the implementation of SCORPAN's "S.". In: Zhang, G.-L., Brus, D.J., Liu, F., Song, X.-D., Lagacherie, P. (Eds.), Digital Soil Mapping Across Paradigms, Scales and Boundaries. Springer Science+ Business Media, Singapore, pp. 331–342.

Pásztor, L., Laborczi, A., Takács, K., Szatmári, G., Dobos, E., Illés, G., et al., 2015. Compilation of novel and renewed, goal oriented digital soil maps using geostatistical and data mining tools. Hungarian Geogr. Bull. 64, 49–64.

Pásztor, L., Szabó, J., Bakacsi, Z., 2010. Digital processing and upgrading of legacy data collected during the 1:25,000 scale Kreybig soil survey. Acta Geod. Geophys. Hungarica 45, 127–136.

Pásztor, L., Szabó, J., Bakacsi, Z., Laborczi, A., 2013. Elaboration and applications of spatial soil information systems and digital soil mapping at the Research Institute for Soil Science and Agricultural Chemistry of the Hungarian Academy of Sciences. Geocarto Int. 28, 13–27.

Pásztor, L., Szabó, J., Bakacsi, Z., Laborczi, A., Dobos, E., Illés, G., et al., 2014. Elaboration of novel, countrywide maps for the satisfaction of recent demands on spatial, soil related information in Hungary. In: Arrouays, D., McKenzie, N., Hempel, J., DeForges, A., McBratney, A.B. (Eds.), GlobalSoilMap: Basis of the Global Spatial Soil Information System. Taylor and Francis, Dordrecht, pp. 207–212.

Pentelényi, A., Scharek, P., 2006. A talajvízszint mélysége a felszín alatt, 1:500.000. (Groundwater level map of Hungary). [WWW document]. <http://map.mfgi.hu/tvz_251020/>.

Petritsch, R., Hasenauer, H., Pietsch, S.A., 2007. Incorporating forest growth response to thinning within biome-BGC. For. Ecol. Manag. 242, 324–336.

Podmaniczky, L., Vogt, J., Schneller, K., Ángyán, J., 2007. Land suitability assessment methods for developing a European Land Information System for Agriculture and Environment (ELISA). Multifunctional Land Use: Meeting Future Demands for Landscape Goods and Services. Springer, Heidelberg.225–250

Pretzsch, H., 2009. Forest Dynamics, Growth and Yield, Forest Research. Springer, Heidelberg.

Rajkai, K., Kabos, S., van Genuchten, M.T., 2004. Estimating the water retention curve from soil properties: comparison of linear, nonlinear and concomitant variable methods. Soil Tillage Res 79, 145–152.

Rawls, W.J., 1983. Estimating soil bulk density from particle size analysis and organic matter content. J. Soil Sci. 135, 123–125.

Ritchie, J.T., 1998. Soil water balance and plant water stress. In: Tsuji, G.Y., Hoogenboom, G., Thornton, P.K. (Eds.), Understanding Options for Agricultural Production. Kluwer Academic Publishers, Dordrecht, pp. 41–54.

Ritchie, J.T., Singh, U., Godwin, D.C., Bowen, W.T., 1998. Cereal growth, development and yield. In: Tsuji, G.Y., Hoogenboom, G., Thornton, P.K. (Eds.), Understanding Options for Agricultural Production. Kluwer Academic Publishers, Dordrecht, pp. 79–98.

Rouse, J.W., Haas, R.H., Schell, J.A., Deering, D.W., 1973. Monitoring vegetation systems in the Great Plains with ERTS. In: Third ERTS Symposium, NASA SP-351 I. pp. 309–317.

Sisák, I., Benő, A., 2014. Probability-based harmonization of digital maps to produce conceptual soil maps. Agrokémia és Talajt 63, 89–98.

Spinoni, J., Szalai, S., Szentimrey, T., Lakatos, M., Bihari, Z., Nagy, A., et al., 2015. Climate of the Carpathian Region in the period 1961–2010: Climatologies and trends of 10 variables. Int. J. Climatol. 35, 1322–1341.

Staub, C., Ott, W., Heusi, F., Klingler, G., Jenny, A., Häcki, M., et al., 2011. Indicators for Ecosystem Goods and Services: Framework, methodology and recommendations for a welfare-related environmental reporting, Environmental studies no. 1102.

Stone, M., 1974. Cross-Validatory Choice and Assessment of Statistical Predictions. J. R. Stat. Soc 36, 111–147.

Szabó, J., Pásztor, L., Bakacsi, Z., László, P., Laborczi, A., 2007. A Kreybig Digitális Talajinformációs Rendszer alkalmazása térségi szintű földhasználati kérdések megoldásában (Application of the Kreybig Digital Soil Information System to solve land use problems at regional level). Agrokémia és Talajt 56, 5–20.

Szatmári, G., Laborczi, A., Illés, G., Pásztor, L., 2013. A talajok szervesanyag-készletének nagyléptékű térképezése regresszió krigeléssel Zala megye példáján (Large-scale mapping of soil organic matter content by regression kriging in Zala County). Agrokémia és Talajt 62, 219–234.

Szentimrey, T., Bihari, Z., 2007. Mathematical background of the spatial interpolation methods and the software MISH (Meteorological Interpolation based on Surface Homogenized Data Basis). In: Proceedings from the Conference on Spatial Interpolation in Climatology and Meteorology. Budapest, pp. 17–27.

Szodfridt, I., 1993. The Discipline of Forest Landsites (in Hungarian: Erdészeti termőhelyismeret-tan). Mezőgazda Kiadó, Budapest.

Tonteri, T., Hotanen, J.P., Kuusipalo, J., 1990. The Finnish forest site type approach: ordination and classification studies of mesic forest sites in southern Finland. Vegetatio 87, 85–98.

Tóth, G., Hengl, T., Hermann, T., Makó, A., Kocsis, M., Tóth, B., et al., 2015. Magyarország mezőgazdasági területeinek talajtulajdonság-térképei (Soil property maps of the agricultural land of Hungary) EUR 27539.

Tóth, G., Rajkai, K., Máté, F., Bódis, K., 2014. Magyarországi kistájak szántóföldjeinek minősége. Tájökológiai Lapok 12, 183–195.

Van Orshoven, J., Terres, J.-M., Tóth, T., 2014. Updated Common Bio-Physical Criteria to Define Natural Constraints for Agriculture in Europe.

Várallyay, G., 2002. Soil survey and soil monitoring in Hungary. Eur. Soil Bur. Res. Report, ESB, Ispra 9, 139–149.

Várallyay, G., 2012. Talajtérképezés, talajtani adatbázisok. Agrokémia és Talajt 61, 249–267.

Várallyay, G., Szűcs, L., Zilahy, P., Rajkai, K., Murányi, A., 1985. Soil factors determining the agro-ecological potential of Hungary. Agrokémia és Talajt 34, 90–94.

Waltner, I., Michéli, E., Fuchs, M., Láng, V., Pásztor, L., Bakacsi, Z., et al., 2014. Digital mapping of selected WRB units based on vast and diverse legacy data. In: Arrouays, D., McKenzie, N., Hempel, J., Richer de Forges, A., McBratney, A.B. (Eds.), GlobalSoilMap: Basis of the Global Spatial Soil Information System. Taylor and Francis Group, London, pp. 313–317.

CHAPTER

10

Mapping Ash $CaCO_3$, pH, and Extractable Elements Using Principal Component Analysis

Paulo Pereira[1], *Eric C. Brevik*[2], *Artemi Cerdà*[3], *Xavier Úbeda*[4], *Agata Novara*[5], *Marcos Francos*[4], *Jesus Rodrigo Comino*[6,7], *Igor Bogunovic*[8] *and Yones Khaledian*[9]

[1]Mykolas Romeris University, Vilnius, Lithuania [2]Dickinson State University, Dickinson, ND, United States [3]University of Valencia, Valencia, Spain [4]University of Barcelona, Barcelona, Spain [5]University of Palermo, Palermo, Italy [6]Trier University, Trier, Germany [7]Málaga University, Málaga, Spain [8]The University of Zagreb, Zagreb, Croatia [9]Iowa State University, Ames, IA, United States

INTRODUCTION

Ash is the most common residue left on the soil surface after a fire, draping it with black, gray, and white colors (Pereira et al., 2015). It is the most valuable soil protection in the immediate period after a fire, reducing the impact of erosive agents by acting as a mulch, facilitating water retention and infiltration, and increasing the quantity of available nutrients in the soil profile (Hosseini et al., 2016). These nutrients are fundamental for vegetative recuperation (Bodi et al., 2014; Campos et al., 2016). The type and amount of nutrients released by ash depend on the temperature of combustion and burned species (Brook and Wittenberg, 2016). Ash is also a very mobile material, especially in the period immediately after a fire and before the first rainfall. This highly dynamic nature makes ash very difficult to map because in very short time periods (seconds, minutes, or hours) spatial distribution can be totally different. After the first rainfalls, ash binds onto the soil surface and ash transport becomes more difficult. Mapping ash supplied nutrients is very important in fire-affected areas since ash is one of the main nutrient sources in the period immediately after a fire.

Despite the fact that there are enormous advanced statistical methods to assess soil characteristics, principal component analysis (PCA) is one of the most common methods for the

Soil Mapping and Process Modeling for Sustainable Land Use Management.
DOI: http://dx.doi.org/10.1016/B978-0-12-805200-6.00010-4

analysis of chemical elements (Bro and Smilde, 2014). This nonparametric method extracts important information from complex datasets facilitating their interpretation and reveals hidden relations between data. PCA reduces the set of original data and extracts the small number of latent factors, known as principal components (PCs), and evaluates the relationships between the variables. These PCs are composed of samples with the same variation. The basis of PCA is the correlation or covariance matrix which calculates the interrelationships between the variables. Normally the first factor explains the largest proportion of the variation and the subsequent PC explain progressively less of the variance. Depending on the data properties a multivariate dataset can be reduced to one or two factors that account for the majority of the variance. PCA is commonly applied to identify relationships between variables and samples, identify outliers, generate new hypothesis, and find and quantify patterns (Li et al., 2013; Bro and Smilde, 2014). It has been extensively used for mapping proposes (Abson et al., 2012; Hengl et al. 2014) in order to identify spatial patterns in the retained PCs (Henriksson et al., 2013). The objective of this work was to map ash $CaCO_3$, pH, and selected extractable elements using a PCA.

MATERIALS AND METHODS

Study Area, Experimental Design, and Sampling Methods

The data used for this work are from a fire that occurred at the end of July 2008 south of Lisbon (Portugal) at 38°33′N and 09°03′W at 55 m above sea level. The parent material is composed of Plio-pleistocene dunes and the soils are classified as Podzols (FAO, 2006). The soils have a loamy sand texture (83 ± 2.19% sand, 10 ± 2.31% silt, and 7 ± 2.97 clay) with low organic matter content and cation exchange capacity. The mean annual rainfall is 639.20 mm and the mean annual temperature is 14.8°C. The vegetation was mainly composed of *Pinus pinaster* and *Quercus suber*. The fire had a medium to high severity (Pereira et al., 2012).

Four days after the fire we designed a grid across a 9 × 27 m area on a west facing slope with an average inclination of 11%; the constant aspect elimated possible aspect affects from the data. Ash samples were collected every 3 m for a total of 40 sampling points. Ash was collected carefully with a brush and spoon in order to discard the soil mineral particles. Samples were stored in plastic bags and taken to the laboratory for chemical analysis. Ash chemical analyses are described in Pereira et al. (2010). In this work we analyzed ash $CaCO_3$, pH, electrical conductivity (EC), and extractable calcium (Ca), magnesium (Mg), sodium (Na), potassium (K), aluminum (Al), manganese (Mn), iron (Fe), zinc (Zn), total phosphorous (TP), and silica (Si).

Statistical and Spatial Analysis

Prior to data analysis the Grubbs test was carried out to identify data outliers and the Shapiro-Wilktest to identify data normality. For both tests, if the P level is lower than 0.05 there is the probability of the existence of an outlier in the distribution (Grubbs test) and data did not follow the Gaussian distribution (Shapiro–Wilk). The data normality assumption is required to conduct Pearson correlation coefficient and PCA analysis, however, not all data follows the Gaussian distribution (Pernet et al., 2013). In our case the distribution of the majority of the data did not respect the normality assumption; thus we tested several data transformation methods, such as logarithm (Log), square root (SQR), and Box-Cox in order to see which method was the most appropriate for data modeling. The criteria to choose the best method was the one that normalized the most data distributions according to the Shapiro–Wilk test. Significant correlations were considered at $P < 0.05$.

PCA was carried out based on the correlation matrix. In order to simplify the interpretation of the loadings the Varimax rotation was applied. After the application of this rotation technique, each original variable is more associated with one factor. A factor represents a certain number of variables (Abdi, 2003). Factors were retained according to the Kaiser criteria. Only the ones that had an eigenvalue greater than 1 were retained and used for spatial analysis (Pereira et al., 2009).

The values obtained from the factor scores were used for variogram modeling. In the present work the omni-direction alvariogram was used, which assumes that the variability of the variable is equal in all directions. Variable spatial dependence was calculated by using the nugget/sill ratio. A ratio between 0% and 25% shows that the variable has a strong spatial dependency, between 25% and 75% a moderate spatial dependency and between 75% and 100% a weak spatial dependency (Chien et al., 1997). In this work, we used ordinary kriging to interpolate each factor. The performance of ordinary kriging to interpolate the data of the factors was assessed using the leave one out cross validation method. This technique estimates the values of certain grid points from the remaining ones in the sampling pool. The errors produced (observed-estimated) allowed calculation of the root mean square error (RMSE), used to measure the accuracy of ordinary kriging to interpolate the different factors obtained from the PCA. The lower the RMSE, the better the prediction accuracy of ordinary kriging (Stevens et al., 2012; Martin et al., 2014). In addition, we measured the relationships between the observed and estimated values. Statistical analysis was carried out with Statistica 10.0 (StatSoft Inc., 2006) and spatial analysis with Surfer 9.0 (Golden Software).

RESULTS AND DISCUSSION

Descriptive Statistics

The descriptive statistics for the studied elements is shown in Table 10.1. The transformations carried out increased data symmetry and eliminated the presence of outliers in some cases ($CaCO_3$, pH, EC, Ca, Mg, K, Fe, Zn, and TF) and decreased them in others (Na and Al). This was due to the characteristics of the data distribution. Logarithmic and SQR data transformation were effective when the data were positively skewed, and the Box-Cox method performed well with positively and negatively skewed data but was less effective than log and SQR (Table 10.1). In relation to Na and Al data the original data were normally distributed, and transformations had a negative impact on data normality. In total, log transformation normalized eight variables, SQR 7 and Box-Cox 6. Thus we used log-transformed data for correlation, PCA, and mapping the factors identified by the PCA. Data transformation is a common practice previously used for data analysis in PCA studies (Olsen et al., 2012; Demsar et al., 2013) and mapping (Henriksson et al., 2013; Yuan et al., 2014) when data were too skewed and did not meet normality requirements. Data transformation generally increases the accuracy of the models and the ordinary kriging method requires normality of the data (Yao et al., 2013).

Correlations and Principal Component Analysis

Ash pH had a significant positive correlation with $CaCO_3$, while EC had a significant positive correlation with Ca, Mg, and K and negative with Al. Calcium had a very strong positive correlation with Mg and negative correlation with Al and Fe. This was observed with Mg as well. Aluminum and Mn had a significant positive

TABLE 10.1 Descriptive Statistics for the Soil Properties Studied

Variable		Mean	SD	CV	Min	Q1	Median	Q3	Max	Skew	Kur	GT *P* Value	SW *P* Value
$CaCO_3$	Original	19.16	8.83	46.09	5.30	13.91	17.23	21.11	48.89	1.36	2.62	0.0107	0.0015
	Log	1.24	0.20	15.93	0.72	1.14	1.24	1.32	1.69	−0.30	0.90	0.2594	0.2670
	SQR	4.27	0.96	22.48	2.30	3.73	4.15	4.59	6.99	0.57	1.04	0.1156	0.1517
	Box-Cox	6.97	1.03	14.75	4.62	6.42	6.89	7.36	9.64	0.21	0.78	0.2756	0.3525
pH	Original	7.95	0.22	2.74	7.52	7.85	7.96	8.07	8.64	0.55	1.86	0.0300	0.0842
	Log	0.90	0.01	1.31	0.88	0.89	0.90	0.91	0.94	0.42	1.59	0.0462	0.1258
	SQR	2.82	0.04	1.36	2.74	2.80	2.82	2.84	2.94	0.49	1.72	0.0373	0.1042
	Box-Cox	5.32	0.05	0.93	5.22	5.30	5.32	5.35	5.47	0.46	1.66	0.0409	0.1132
EC	Original	1.30	0.45	34.62	0.63	1.06	1.22	1.40	2.76	1.50	2.73	0.0194	0.0003
	Log	0.09	0.14	145.95	−0.20	0.03	0.09	0.15	0.44	0.44	0.77	0.3134	0.2663
	SQR	1.13	0.18	16.31	0.80	1.03	1.10	1.18	1.66	0.99	1.48	0.0855	0.0158
	Box-Cox	2.72	0.34	12.55	2.06	2.55	2.69	2.84	3.66	0.76	1.10	0.1524	0.0661
Ca	Original	4371.01	2184.68	54.05	1246.97	2138.97	3458.52	3676.61	11355.00	2.30	5.42	0.0047	0.0000
	Log	3.59	0.19	5.34	3.26	3.48	3.54	3.72	4.10	0.76	0.67	0.2260	0.1035
	SQR	64.30	15.58	24.23	42.58	54.79	58.81	72.72	112.67	1.44	2.57	0.0370	0.0006
	Box-Cox	36.76	5.04	13.70	29.00	33.67	35.10	39.78	51.43	1.14	1.63	0.0834	0.0061
Mg	Original	1196.01	626.18	52.36	468.12	834.54	1087.79	1478.65	4036.32	2.15	5.41	0.0000	0.0000
	Log	3.03	0.19	6.43	2.67	2.92	3.04	3.17	3.61	0.30	0.77	0.0748	0.3004
	SQR	33.68	7.97	23.67	21.64	28.89	32.98	38.45	63.53	1.29	3.81	0.0013	0.0039
	Box-Cox	25.06	3.44	13.73	19.34	23.01	24.91	27.29	36.74	0.83	2.14	0.0093	0.0460
Na	Original	1737.42	686.58	39.52	286.72	1446.53	1753.49	2178.64	3071.12	−0.19	−0.12	1.0000	0.4025
	Log	3.19	0.23	7.29	2.46	3.16	3.24	3.34	3.49	−1.54	2.34	0.0293	0.0008
	SQR	40.69	9.15	22.49	16.93	38.03	41.87	46.68	55.42	−0.84	0.60	0.2732	0.0214
	Box-Cox	28.02	4.01	14.32	16.67	27.11	28.71	30.62	33.89	−1.13	1.20	0.1169	0.0023

K	Original	3394.29	2184.68	64.36	1246.97	2138.97	2876.53	3676.61	11355.00	2.30	5.42	0.0024	0.0000
	Log	3.47	0.22	6.27	3.10	3.33	3.46	3.57	4.06	0.84	1.01	0.1953	0.0298
	SQR	56.14	15.79	28.12	35.31	46.25	53.63	60.63	106.56	1.63	2.93	0.0247	0.0007
	Box-Cox	33.86	5.33	15.75	25.94	30.45	33.25	35.74	49.78	1.30	2.01	0.0624	0.0010
Al	Original	12.51	7.01	56.05	0.61	7.79	12.70	16.48	31.24	0.62	0.92	0.2088	0.1237
	Log	0.99	0.37	37.40	−0.21	0.89	1.10	1.22	1.49	−1.73	3.23	0.0188	0.0003
	SQR	3.37	1.09	32.32	0.78	2.79	3.56	4.06	5.59	−0.47	0.48	0.5612	0.1992
	Box-Cox	5.89	1.41	24.00	2.02	5.28	6.23	6.79	8.36	−1.00	1.22	0.1660	0.0096
Mn	Original	4.25	1.76	41.37	0.39	3.06	4.14	4.95	10.12	0.76	2.22	0.0126	0.1515
	Log	0.58	0.23	39.70	−0.41	0.48	0.62	0.69	1.01	−2.07	7.87	0.0000	0.0006
	SQR	2.01	0.45	22.14	0.62	1.75	2.03	2.22	3.18	−0.40	2.02	0.0352	0.2755
	Box-Cox	4.18	0.70	16.73	1.66	3.80	4.24	4.51	5.77	−1.04	3.52	0.0031	0.0198
Fe	Original	9.38	9.44	100.63	1.63	4.53	6.36	10.76	53.55	3.44	13.50	0.0000	0.0000
	Log	0.86	0.29	34.07	0.21	0.66	0.80	1.03	1.73	0.78	1.54	0.0620	0.0729
	SQR	2.85	1.14	39.94	1.28	2.13	2.52	3.28	7.32	2.19	6.33	0.0004	0.0001
	Box-Cox	5.26	1.32	25.07	3.01	4.38	4.92	5.89	9.92	1.59	3.80	0.0044	0.0004
Zn	Original	1.25	1.11	88.24	0.08	0.49	1.07	1.61	5.33	2.04	5.32	0.0018	0.0001
	Log	−0.07	0.42	619.75	−1.10	−0.31	0.03	0.21	0.73	−0.57	−0.05	0.4402	0.1270
	SQR	1.03	0.45	44.07	0.28	0.70	1.03	1.27	2.31	0.69	0.95	0.1135	0.0886
	Box-Cox	2.45	0.90	36.61	0.69	1.84	2.55	2.99	4.63	0.12	0.05	0.4809	0.5010
TP	Original	92.42	77.35	83.68	4.68	31.41	59.88	135.92	314	1.07	0.65	0.1011	0.0108
	Log	1.79	0.44	24.85	0.67	1.49	1.78	2.13	2.50	−0.55	−0.17	0.3617	0.1551
	SQR	8.77	4.00	45.60	2.16	5.59	7.74	11.66	17.72	0.36	−0.73	0.8489	0.1843
	Box-Cox	10.80	3.23	29.87	4.43	8.35	10.27	13.26	17.14	0.02	−0.83	1.0000	0.4912
Si	Original	1050.56	384.26	36.58	224.76	848.84	1032.44	1311.18	1890.11	−0.11	−0.11	0.9955	0.0898
	Log	2.98	0.20	6.69	2.35	2.93	3.01	3.12	3.28	−1.35	1.97	0.0280	0.0010
	SQR	31.79	6.41	20.17	14.99	29.13	32.13	36.21	43.48	−0.70	0.46	0.2514	0.1476
	Box-Cox	24.23	3.08	12.69	15.48	23.13	24.52	26.33	29.36	−0.97	0.97	0.1094	0.0222

Standard deviation (SD), minimum (Min), quartil 1 (Q1), quartil 3 (Q3), maximum (Max), skewness (Skew), kurtosis (Kur), Grubbs test (GT), Shapiro–Wilk test (SW), logarithmic (Log), and square root (SQR).

correlation with Fe and TP was significantly positively correlated with Si (Table 10.2).

PCA identified five different factors. In total these five factors explained almost 73% of the variability in the soil properties (Table 10.3). Factor 1 showed high positive loadings in EC, Ca, and Mg and negative in Al and Fe. This shows that at the grid points where EC, Ca, and Mg had high contents, Al and Fe were present in small amounts. Factor 2 had high positive loadings in TP and Si and factor 3 in Mn and Fe. Factor 4 had high negative loadings in $CaCO_3$ and pH and factor 5 had high positive loadings in Na and K (Table 10.4).

Spatial Structure Analysis

The results of variograms modeled with the loadings from the different factors are given in Table 10.5 and Fig. 10.1. The best-fit model for factor 1 was the Gaussian. For factors 2, 3, and 5 wave (hole effect) was the most suitable and for factor 4 linear (Table 10.4). This shows that the factors extracted from the PCA analysis had a different spatial pattern and the ash extractable nutrients distribution is different in the immediate postfire period.

The factor 1 variogram had a good spatial structure and the variables explained (EC, Ca, Al, and Fe) had a specific spatial pattern, with a low small-scale variation (high values in some parts of the plot and low in others). Factors 2, 3, and 5 variograms showed that the distribution of these variables had a cyclical pattern, especially for factors 3 and 5. A wave hole-effect variogram is evidence of the presence of cyclical or periodic effects (Mosammam, 2015). Normally, this type of spatial structure is observed in mountainous areas where orography imposes cyclic changes in environmental variables (Diodato and Bellocchi, 2016), geological features (Chihi et al., 2013), groundwater levels (Triki et al., 2014), agricultural fields (Jemo et al., 2014), and other areas of strong human impact and induce strong diversity in landscape mosaics (Balaguer-Beser et al., 2013). In this case, this periodic pattern was induced by the different type, distribution, and conditions of the fuel (e.g., moisture) previous to and during the fire. It is well known that different types of vegetation respond differently to fire (Pausas et al., 2016). Plant chemical composition has impacts on the vulnerability to flames. *Pinus pinaster* fuel is rich in oils and resins and more flammable than *Quercus suber* fuel, known to be very resistant to fire impact (Dimitrakopoulos and Papaioannou, 2001; Ganteaume et al., 2009). The differences in moisture conditions and vegetation type had impacts on the type of ash produced and therefore on the heterogeneity of TP, SI, Mn, Zn, Na, and K. The factor 4 variogram did not reach the sill and this showed that the spatial correlation of $CaCO_3$ and pH increases in all the area of interest.

The nugget effect was high in factors 2 and 4 and low in factor 1 (Fig. 10.1 and Table 10.4). Nugget effect represents the nondetectable error and the spatial variation within the minimum sampling spacing and can be attributed to the limited number of samples and the existence of outliers (Rodriguez-Martin et al., 2013). Factors 2 and 4 had the highest nugget effect (0.25), which means that the small-scale variability (contiguous grid points with very different values) was higher comparing to the other factors. The error variance was not high and the factors extracted from the PCA have a spatial dependence, as was observed in the results of the nugget/sill ratio. Factor 1 had a high spatial dependence, while the other factors showed a moderate spatial dependence (Table 10.4).

The range was high on factor 4, and the spatial correlation of this factor was observed in all the study area. The lowest range was observed in factor 3 (Table 10.4). The range in a variogram model corresponds to the spacing

TABLE 10.2 Correlation Coefficients Between the Studied Variables

	$CaCO_3$	pH	EC	Ca	Mg	Na	K	Al	Mn	Fe	Zn	TP	Si
$CaCO_3$	–												
pH	0.53***	–											
EC	−0.06	−0.09	–										
Ca	−0.05	−0.06	0.53***	–									
Mg	−0.04	−0.00	0.54***	0.90***	–								
Na	−0.06	0.11	0.22	0.17	0.23	–							
K	−0.01	−0.10	0.47**	0.16	0.22	0.28	–						
Al	0.05	0.24	−0.32*	−0.38*	−0.42**	0.02	0.02	–					
Mn	−0.20	−0.16	0.11	0.29	0.24	0.13	0.13	0.24	–				
Fe	−0.13	−0.06	−0.27	−0.37*	−0.56***	0.13	−0.02	0.44**	0.36*	–			
Zn	0.17	−0.04	0.20	0.18	0.03	0.07	−0.04	0.20	0.28	0.22	–		
TF	−0.10	−0.18	0.06	0.09	0.10	0.20	0.29	−0.00	−0.07	0.02	0.15	–	
Si	−0.12	−0.23	0.22	−0.00	−0.01	0.24	0.20	−0.22	−0.09	0.17	−0.01	0.41**	–

Significant correlations (in red) at $^*P < 0.05$, $^{**}P < 0.01$, and $^{***}P < 0.001$.

TABLE 10.3 Eigenvalues Results

	Eigenvalue	%Total Variance	Cum. Eignevalue	Cum. (%)
Factor 1	3.16	24.33	3.16	24.33
Factor 2	2.09	16.06	5.25	40.38
Factor 3	1.72	13.26	6.97	53.64
Factor 4	1.48	11.42	8.46	65.06
Factor 5	1.03	7.91	9.49	72.96

TABLE 10.4 Loadings Results. Loadings Marked in Red are Explained by the Factor

	Factor 1	Factor 2	Factor 3	Factor 4	Factor 5
$CaCO_3$	0.02	0.01	0.07	−0.87	−0.10
pH	−0.09	−0.30	−0.07	−0.80	0.19
EC	0.66	0.20	0.12	0.03	0.35
Ca	0.88	−0.07	0.24	0.07	0.14
Mg	0.92	−0.10	0.06	0.03	0.22
Na	0.06	0.15	0.07	−0.06	0.74
K	0.17	0.22	−0.04	0.04	0.72
Al	−0.63	−0.25	0.39	−0.16	0.24
Mn	0.08	−0.36	0.67	0.37	0.30
Fe	−0.68	0.05	0.46	0.22	0.20
Zn	0.09	0.24	0.86	−0.19	−0.15
TP	0.03	0.74	0.12	0.04	0.19
Si	−0.02	0.77	−0.07	0.17	0.22
Expl. Var.	2.95	1.60	1.66	1.68	1.58
Prp. Totl	0.23	0.12	0.13	0.13	0.12

TABLE 10.5 Variogram Properties of the Retained Factors

	Model	Nugget Effect	Slope/Sill	Range (m)	Nug/Sill Ratio × 100
Factor 1	Gaussian	0.10	0.95	4.40	10.52
Factor 2	Wave (hole effect)	0.25	0.69	5.32	36.23
Factor 3	Wave (hole effect)	0.15	0.79	3.82	18.98
Factor 4	Linear	0.25	0.11	10	–
Factor 5	Wave (hole effect)	0.20	0.96	4.10	20.83

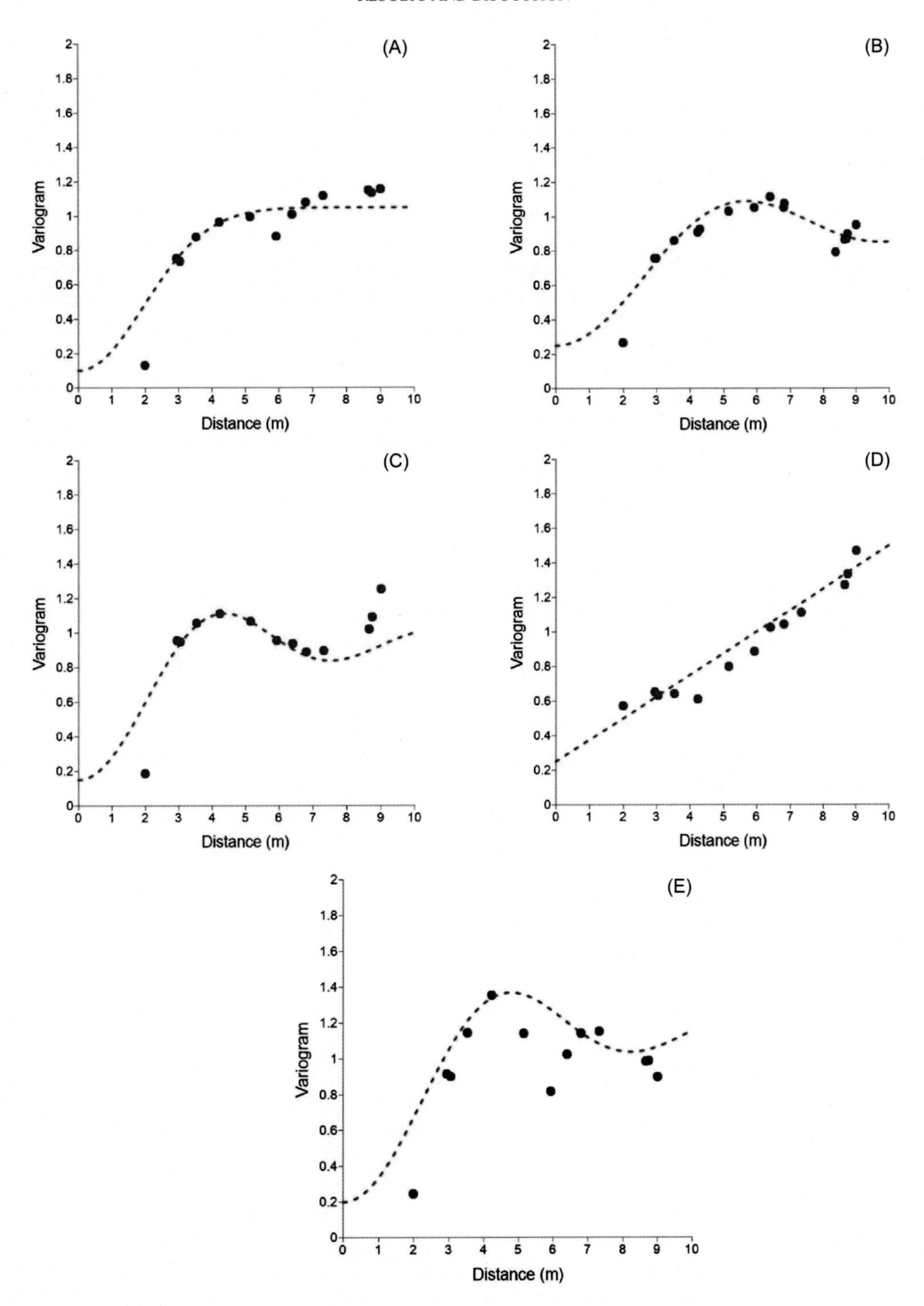

FIGURE 10.1 Omni-directional variogram calculated factor scores results: (A) factor 1, (B) factor 2, (C) factor 3, (D) factor 4, (E) factor 5.

between samples, beyond which samples are not spatially correlated (Chung et al., 2014). Thus, with the exception of factor 4, the variograms modeled reached the sill, or the point beyond which there is no spatial correlation, at 4.40m (factor 1), 5.32m (factor 2), 3.82m (factor 3), and 4.10m (factor 5). In all the cases the variogram range was higher than the sample density (3m), showing that the sampling design was appropriate to describe the factors extracted from the PCA.

Factor Maps and Interpolation Performance Analysis

The maps of the factors are shown in Fig. 10.2. The spatial structure of factor 1 had a better spatial structure than the remaining ones and this is visible in the map (Fig. 10.2A). The areas in red (high positive loadings in the south of the plot) had a high content of EC, Ca, and Mg and low content of Al and Fe. In the blue areas (high negative loadings), located mainly in the north of the plot, the opposite was observed. In the factor 2 map, it is possible to observe some cycling patterns identified in the variogram. High contents of TP and Si (red areas with high positive loadings) at to the east, low in the center (blue areas with high negative loadings) and again high in the west. Some close grid points can also be observed in the west part of the plot showing the short-scale variation (Fig. 10.2B). In the factor 3 map, some periodic patterns can be identified as well. Low values of Mn and Zn (blue areas with high negative loadings) are seen at some pointsalong the west and east of the plot with high values (red areas with high positive loadings) in the central part of the area of interest (Fig. 10.2C). In the map of factor 4, high values of $CaCO_3$ and pH are located in the western and central parts of the plot (blue areas with high negative loadings), while the low values (red areas with high positive loadings) are especially observed in the eastern part of the plot. There are two different patterns in these areas of the plot and there are not sudden changes, just a spatial trend from an area with low values to an area with high values, as shown by the modeled variogram (Fig. 10.2D). Finally the factor 5 map (Fig. 10.2E) showed a cyclic pattern (as identified in the variogram) with low values of Na and K in the western part of the plot (blue areas with high negative loadings), high values in the center-west (red areas with high positive loadings), low values in the center-east, and high values again in the east part of the area of interest. Some contiguous grid points with high positive and high negative loadings were also observed, showing highsmall-scale variability in the distribution of the factor 5 loadings (the nugget effect was 0.20, Table 10.5).

The ordinary Kriging method performance was acceptable in interpolating the loadings from the factors 1, 2, and 4, where the correlation between the observed and estimated values was above 0.40 and significant at $P < 0.05$ (Fig. 10.3A, B, and D). On the other hand the accuracy of this method to interpolate the loadings from factors 3 and 5 was low since the relationships between the observed and estimated values were poor (Fig. 10.3C and E). The results of the RMSE confirmed this. RMSE was lowest in factors 1, 2, and 4, with values of 0.889, 0.892, and 0.875, respectively. The errors of interpolation were higher in the factors 3 and 5, with an RMSE of 1.079 and 1.091, respectively. In the variograms and maps produced by these factors we observed some cyclic patterns (more evident than in factor 2) and the short spatial correlation (Table 10.5) may reduce the capacity of ordinary kriging to estimate the observed values with accuracy compared to the other factors (Yao et al., 2013). In this case some other methods such as inverse distance to a weight may be more advantageous and interpolate the dataset more accurately, as previous works have observed (Shahbazi et al., 2013;

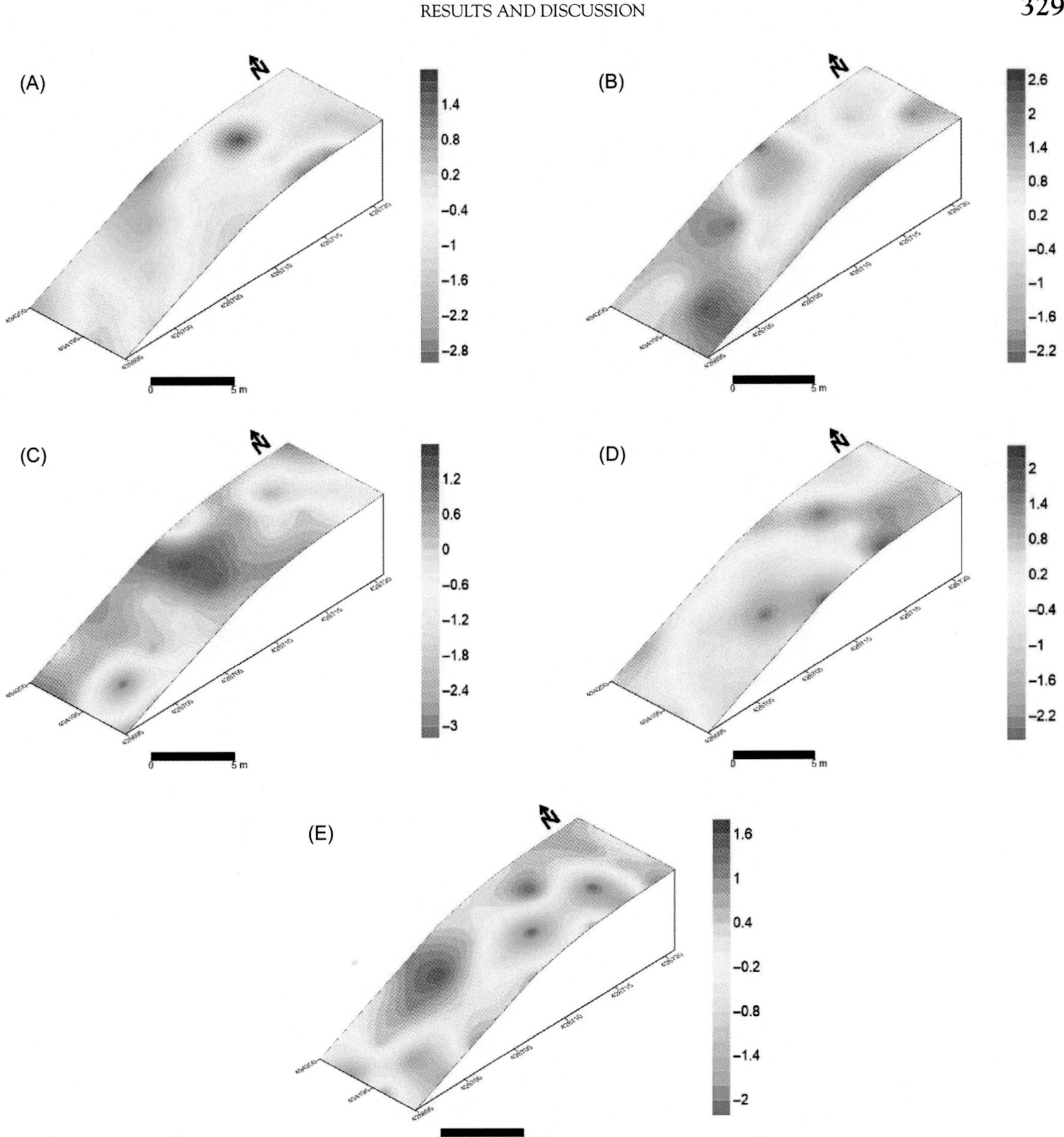

FIGURE 10.2 Interpolated maps with the results of the factor scores: (A) factor 1, (B) factor 2, (C) factor 3, (D) factor 4, (E) factor 5.

Gong et al., 2014). Kriging methods are good predictors when the data have a good spatial structure and are spatially homogeneous without the presence of small-scale variation and/or periodic phenomena (Lu and Wong, 2008), as we observed in the cases of factors 1, 2, and 4 where the periodic patterns and/or small-scale variation was not so marked.

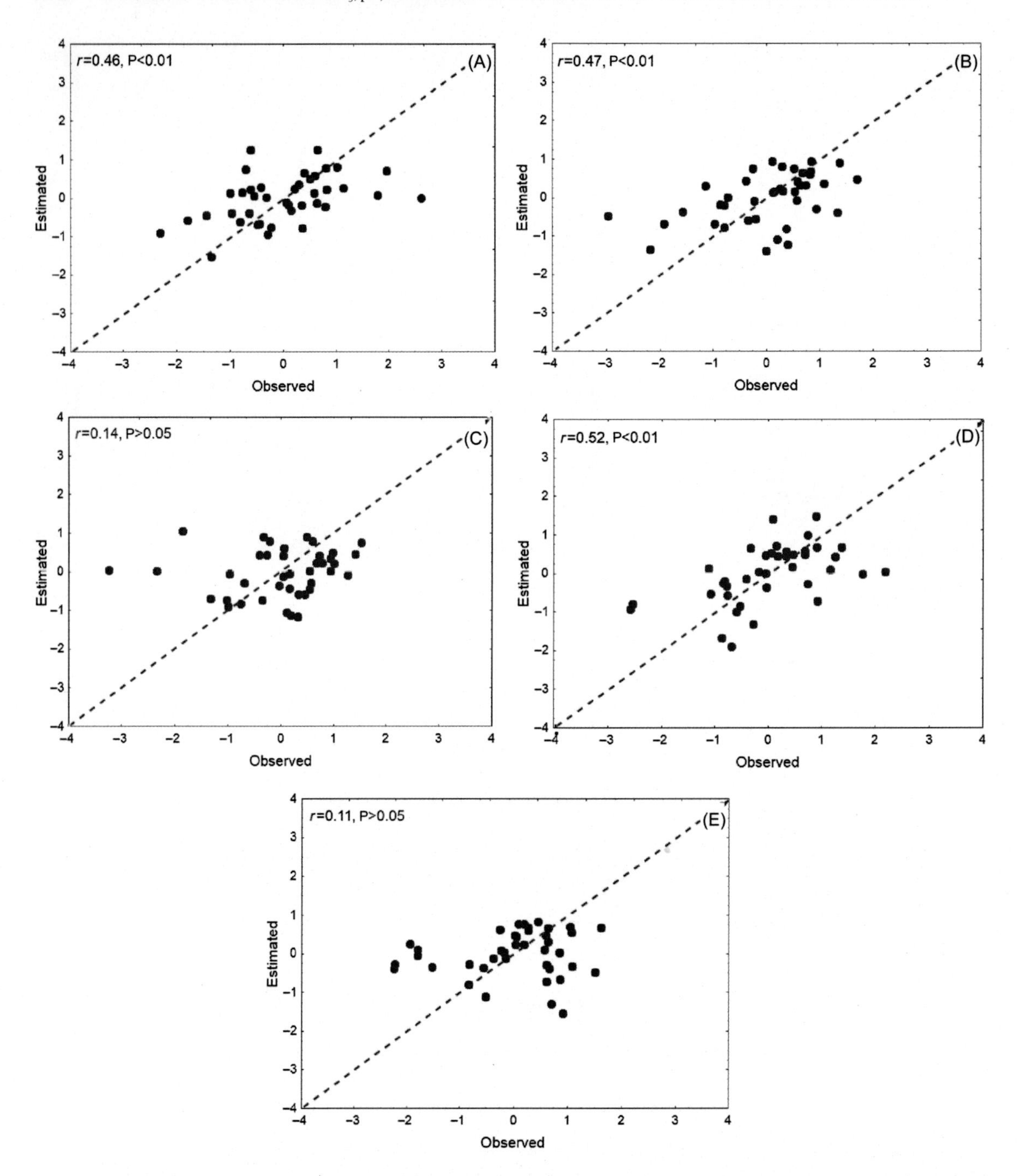

FIGURE 10.3 Scaterplots of the observed versus estimated values: (A) factor 1, (B) factor 2, (C) factor 3, (D) factor 4, (E) factor 5.

IMPLICATIONS FOR SUSTAINABLE MANAGEMENT OF FIRE-AFFECTED AREAS

The pilot study carried out was not able to represent all of the burned area, and there is a lack of larger scales studies about mapping ash properties. The studies available about the spatial distribution of ash nutrients dynamics were carried out on small plots (Pereira et al., 2010). In the immediate period after a fire, it is important to know the type and amount of nutrients available in ash since it is one of the most important nutrient sources for plant recuperation. In this context, it is important to invest in studies at larger scales (e.g., entire burned areas, catchments) that could include more covariates. Nevertheless this can be limited by two aspects. The first is the high number of samples required for detailed mapping analyses that normally are time and cost-consuming (Brevik et al., 2016). Some studies have used pre- and postfire satellite images to predict ash loads (Chafer et al., 2016) and identify fire severity based on ash reflectance (Smith et al., 2005); however, no attempt was made to estimate ash nutrients using fire severity information from remote sensing or proximal sensing. These techniques were applied only to estimate soil properties and vegetation recuperation. It is well known that ash $CaCO_3$ (a proxy of fire severity), pH, and extractable elements depend on fire severity (Pereira et al., 2012). This may be promising in the future for a rapid and quick assessment of the ash nutrients availability in fire-affected areas. The second limitation is the high mobility of ash, especially in areas affected by high severity fires, that can be easily eroded by the wind and (re)distributed inside and outside the burned area. This means that in the space of seconds, minutes, or hours, the ash spatial pattern and the nutrient content can change their spatial disposition; this can also create a problem in the analyses of satellite images in the postfire period. The analysis of a large number of images is required to tackle this problem and to understand the spatiotemporal pattern of ash nutrients.

From the management point of view, it is extremely important not to intervene in burned areas immediately after the fire, when the soils are most vulnerable. It is important to maintain ash cover and minimize foot and mechanical traffic because they increase the vulnerability of ash to erosion. Ash is an extremely valuable, low-cost protection against soil degradation after a fire as demonstrated in previous works (Cerdà and Doerr, 2008). In this context, mapping ash cover and nutrient contents is important to know their capacity to protect soil and supply nutrients in the period immediately after a fire and identify whether or not there is a need for management interventions. However, the option of intervening may be negative. The management of burned areas using heavy machinery to extract burned wood or plant trees affects ash distribution and increases soil vulnerability to erosion, which is considered one of the most important causes of land degradation in fire-affected areas (Shakesby, 2011; Tessler et al., 2016). In some cases it is better to allow the area to recover without any intervention, since vegetation is strongly resilient to fire and recovers very fast, especially in Mediterranean environments (Gonzalez-De Vega et al., 2016; Lopez-Serrano et al., 2016; Lucas-Borja et al., 2016).

Ash cover is very important for the sustainable management of fire-affected areas since it provides the first and most important form of soil protection after a fire and the nutrients necessary for plant recuperation. Maintaining ash cover and avoiding disturbance of burned areas is the first step for sustainable management of these areas and to facilitate the natural recovery of fire-affected areas.

CONCLUSIONS

PCA was used to group ash properties and facilitate data interpretation. Five factors were retained. Factor 1 had high positive loadings for EC, Ca, and Mg and negative loadings for Al and Fe, factor 2 high positive loadings for TP and Si, factor 3 for Mn and Fe, and factor 5 for Na and K. Factor 4 showed high positive loadings for $CaCO_3$ and pH. The Gaussian model was the best fit for the factor 1 variogram and wave (hole effect) was the most accurate model for the variograms produced by factors 2, 3, and 5. The linear model was the best adjusted to factor 4. The extracted factors had a different spatial structure which was observed in the maps produced. The spatial distribution of the loadings for factor 1 had a specific pattern where the content of EC, Ca, and Mg was high and the content of Al and Fe were low. The variables explained by factor 4 had a major concentration in the eastern part of the area of interest and were low in the central and west. Finally the maps produced by factors 2, 3, and 5 showed a cyclic pattern (the factors 3 and 5 maps also had a high small-scale variance) common in burned areas due the different fuel conditions that existed previous to the fire. The ordinary kriging method was better at estimating the loadings from the factors 1, 2, and 4. The loadings of factors 3 and 5 were poorly estimated due to the high short-scale variability.

Acknowledgments

The authors gratefully acknowledge the support of the projects: (1) soil quality, erosion control, and plant cover recovery under different postfire management scenarios (POSTFIRE), funded by the Spanish Ministry of Economy and Competitiveness (CGL2013-47862-C2-1-R), (2) Preventing and Remediating Degradation of Soils in Europe Through Land Care (RECARE) funded by the European Commission (FP7-ENV-2013-TWO STAGE), and (3) COST action ES1306 (connecting European connectivity research). Paulo Pereira and Artemi Cerdà would like to gratefully acknowledge the support of COST action ES1306 (connecting European connectivity research). E.C. Brevik was partially supported by the National Science Foundation under Grant Number IIA-1355466 during this project.

References

Abdi, H., 2003. Factor rotations in factor analysis In: Lewis-Beck, M. Bryman, A. Futing, T. (Eds.), Encyclopedia of Social Sciences Research Methods, 2003. Sage, Thousand Oaks, CA, pp. 1–8.

Abson, D.J., Dougill, A.J., Stringer, L.C., 2012. Using principal component analysis for information-rich socio-ecological vulnerability mapping in Southern Africa. Appl. Geogr. 35, 515–524.

Balaguer-Beser, A., Ruiz, L.A., Hermosilla, T., Recio, J.A., 2013. Using semivariogram indices to analyse heterogeneity in spatial patterns in remotely sensed data. Comput. Geosci. 50, 115–127.

Bodi, M., Martin, D.A., Santin, C., Balfour, V., Doerr, S.H., Pereira, P., et al., 2014. Wildland fire ash: production, composition, and eco-hydro geomorphic effects. Earth-Sci. Rev. 130, 103–127.

Brevik, E.C., Calzolari, C., Miller, B.A., Pereira, P., Kabala, C., Baumgarten, A., et al., 2016. Soil mapping, classification, and modeling: history and future directions. Geoderma 264, 256–274.

Bro, R., Smilde, A.K., 2014. Principal component analysis. Anal. Methods 6, 2812–2831.

Brook, A., Wittenberg, L., 2016. Ash-mineralogical composition and physical structure. Sci. Total Environ. http://www.sciencedirect.com/science/article/pii/S0048969716303333.

Campos, I., Abrantes, N., Keiser, J.J., Vale, C., Pereira, P., 2016. Major and trace elements in soils and ashes of eucalypt and pine forest plantations in Portugal following a wildfire. Sci. Total Environ. http://www.sciencedirect.com/science/article/pii/S0048969716301863.

Cerdà, A., Doerr, S.H., 2008. The effect of ash and needle cover on surface runoff and erosion In the immediate post-fire period. Catena 74, 256–263.

Chafer, C.J., Santin, C., Doerr, S.H., 2016. Modelling and quantifying the spatial distribution of post-wildfire ash loads. Int. J. Wildland Fire 25, 249–255.

Chien, Y.J., Lee, D.Y., Guo, H.Y., 1997. Geostatistical analysis of soil properties of mid-west Taiwan soils. Soil Sci. 162, 291–298.

Chihi, H., Bedir, M., Belayouni, H., 2013. Variogram identification aided by a structural framework for improved geometric modelling of faulted reservoirs: Jaffara basin, Southern Tunisia. Nat. Resourc. Res. 22, 139–161.

Chung, S.O., Sudduth, K.A., Drummond, S.T., Kitchen, N.R., 2014. Spatial variability of soil properties using nested variograms at multiple scales. J. Biosyst. Eng. 39, 377–388.

Demsar, U., Harris, P., Brundson, C., Fotheringham, A.S., McLeone, S., 2013. Principal component analysis on spatial data: an overview. Ann. Assoc. Am. Geogr. 103, 106–128.

Dimitrakopoulos, A.P., Papaioannou, K.K., 2001. Flammability assessment of Mediterranean forest fuels. Fire Technol. 37, 143–152.

Diodato, N., Bellocchi, G., 2016. Enhanced propagation of rainfall kinetic energy in UK. Theor. Appl. Climatol. http://dx.doi.org/10.1007/s00704-016-1863-1.

FAO, 2006. http://www.fao.org/3/a-a0510e.pdf.

Ganteaume, A., Lampin-Maillet, C., Guijarro, M., Hernando, C., Jappiot, M., Fonturbel, T., et al., 2009. Spot fires: fuel bed flammability and capability of firebrands to ignite fuel beds. Int. J. Wildland Fire 18, 951–969.

Gong, G., Mattevada, S., O'Bryant, S.E., 2014. Comparision of the accuracy of kriging and IDW interpolations in estimating groundwater arsenic concentrations in Texas. Environ. Res. 130, 59–69.

Gonzalez-De Vega, S., De las Heras, J., Moya, D., 2016. Resilience of Mediterranean terrestrial ecosystems and fire severity in semi-arid areas: responses of Aleppo pine forests in short, mid and long term. Sci. Total Environ. http://www.sciencedirect.com/science/article/pii/S004896971630540X.

Hengl, T., Mendes de Jesus, J., MacMillan, R.A., Batjes, N.H., Heuvelink, G.B.M., Ribeiro, E., et al., 2014. SoilGrids1km—global soil information based on automated mapping. PLoS One 9, e105992.

Henriksson, S., Hagberg, J., Backstrom, M., Persson, I., Lindstrom, G., 2013. Assessment of PCDD/Fs levels in soils at a contaminated sawmill site in Sweden—a GIS and PCA approach to interpret the contamination pattern and distribution. Environ. Pollut. 180, 19–26.

Hosseini, M., Keizer, J.J., Pelayo, O.G., Prats, S.A., Ritsema, C., Geissen, V., 2016. Effect of fire frequency on runoff, soil erosion, and loss of organic matter at the micro-plot scale in north-central Portugal. Geoderma 269, 126–137.

Jemo, M., Jayeoba, O.J., Alabi, T., Montes, A.J., 2014. Geostatistical mapping of soil fertility constrains for yam based cropping systems of North-central. Geoderma Regional 2-3, 102–109.

Leon, J., Bodi, M.B., Cerdà, A., Badia, D., 2013. The contrasted reponse of ash to wetting: the effects of ash type, thickness and rainfall events. Geoderma 209-210, 143–152.

Li, X., Liu, L., Wang, Y., Luo, G., Chen, X., Yang, X., et al., 2013. Heavy metal contamination of urban soil and in an old industrial city (Shenyang) in Northeast China. Geoderma 192, 50–58.

Lopez-Serrano, F.R., Rubio, E., Dadi, T., Moya, D., Andres-Abellan, M., Garcia-Morote, F.A., et al., 2016. Influences of recovery from wildfire and thining on soil respiration of a Mediterranean mixed forest. Sci. Total Environ. http://www.sciencedirect.com/science/article/pii/S0048969716306787.

Lu, G.Y., Wong, D.W., 2008. An adaptive inverse-distance weighting spatial interpolation technique. Comput. Geosci. 35, 1044–1055.

Lucas-Borja, M.E., Ahrazem, O., Candel-Perez, D., Moya, D., Fonseca, T., Hernandez Tecles, E., et al., 2016. Evaluation of fire recurrence effect on genetic diversity in maritime pine (*PinuspinasterAit.*) stands using inter-simple sequence repeated profiles. Sci. Total Environ. http://www.sciencedirect.com/science/article/pii/S0048969716301061.

Martin, M.P., Orton, T.G., Lacarce, E., Meersmans, J., Saby, N.P.A., Paroissen, J.B., et al., 2014. Evaluation of modelling approaches for predicting the spatial distribution of soil organic carbon stocks at the national scale. Geoderma 223-225, 97–107.

Martinez-Murillo, J.F., Hueso-Gonzalez, P., Ruiz-Sinoga, J.D., Lavee, H., 2016. Short-term experimental fire effects in soil and water losses in Southern Europe. Land Degrad. Dev. 27, 1513–1522.

Mosammam, A.M., 2015. The reverse dimple in potential negative-value sdpace-time covariance models. Stoch. Environ. Res. Risk Assess. 29, 599–607.

Olsen, R.L., Chappell, R.W., Loftis, J.C., 2012. Water quality sample collection, data measurement and results presentation for principal component analysis—literature review and Illinois watershed case. Water Res. 46, 3110–3122.

Pausas, J.G., Alessio, G.A., Moreira, B., Segarra-Moragues, J.G., 2016. Secondary compounds enhance flammability in a Mediterranean plant. Oecologia 180, 103–110.

Pereira, P., Jordán, A., Cerdà, A., Martin, D., 2015. Editorial: the role of ash in fire-affected ecosystems. Catena 135, 337–339.

Pereira, P., Úbeda, X., Baltrenaite, E., 2010. Mapping total nitrogen in ash after a wildfire. A microplot analysis. Ekologija 56, 144–152.

Pereira, P., Úbeda, X., Martin, D., 2012. Fire severity effects on ash chemical composition and water-extractable elements. Geoderma 191, 105–114.

Pereira, P., Úbeda, X., Outeiro, L., Martin, D.A., 2009. Factor analysis applied to fire temperature effects on water quality. In: Gomez, E., Alvarez, K. (Eds.), Forest Fires: Detection, Suppression and Prevention, Series Natural Disaster Research, Prediction and Mitigation. Nova Science Publishers, Inc, Hauppauge, NY, pp. 273–285. Chapter 9.

Pernet, C.R., Wilcox, R., Rousselet, G.A., 2013. Robust correlation analysis: false positive and power validation using a new open Matlab toolbox. Front. Psychol. 3 http://dx.doi.org/10.3389/fpsyg.2012.00606.

Rodriguez-Martin, J.A., Ramos-Miras, J.J., Boluda, R., Gil, C., 2013. Spatial relations of heavy metals in arable and greenhouse soils of a Mediterranean environment region (Spain). Geoderma 200-201, 180–188.

Shahbazi, F., Aliasgharzad, N., Ebrahimzad, S.A., Najafi, N., 2013. Geostatistical analysis for predicting soil biological maps under different scenarios of land use. Eur. J. Soil Biol. 55, 20–27.

Shakesby, R.A., 2011. Post-fire soil erosion in the Mediterranean: review and future research directions. Earth-Sci. Rev. 105, 71–100.

Smith, A.M.S., Wooster, M.J., Drake, N.A., Dipotso, F.M., Falkowski, M.J., Hudak, A.T., 2005. Testing the potential of multi-spectral remote sensing for retrospectively estimating fire severity in African Savannahs. Remote Sens. Environ. 97, 92–115.

StatSoft Inc. 2006. STATISTICA, Version 6.0., Tulsa, OK.

Stevens, A., Miralles, I., van Wesemael, B., 2012. Soil organic carbon predictions by airborne imaging spectroscopy: comparing cross validation and validation. Soil Sci. Soc. Am. J. 76, 2174–2183.

Tessler, N., Wittenberg, L., Greenbaum, N., 2016. Vegetation cover and species richness after recurrent forest fires in Mount Carmel, Israel. Sci. Total Environ. http://www.sciencedirect.com/science/article/pii/S0048969716303230.

Triki, I., Trabelsi, N., Hentati, I., Zairi, M., 2014. Groudwater levels time series sensivity to pluviometry and air temperature: a geostatistical approach to Sfax region, Tunisia. Environ. Monit. Assess. 186, 1593–1608.

Yao, X., Fu, B., Lu, Y., Sun, F., Wang, S., Liu, M., 2013. Comparison of four spatial interpolation methods for estimating soil moisture in a complex terrain catchment. PLoS One 8, e54660.

Yuan, G.L., Sun, T.H., Han, P., Li, J., Lang, X.X., 2014. Source identification and ecological risk assessment of heavy metals in topsoil using environmental geochemical mapping: typical urban renewal area in Beijing, China. J. Geochem. Explor. 136, 40–47.

CHAPTER

11

Human-Impacted Catenas in North-Central Iowa, United States: Ramifications for Soil Mapping

Jenny L. Richter and C. Lee Burras

Iowa State University, Ames, IA, United States

CATENAS AND SOIL MAPPING

One of the things that makes soils so interesting is their variability. The challenge for a pedologist is to find coherence and patterns within the variability. This is especially true when mapping soils as it is not feasible to sample each pedon individually, and therefore an understanding of soil variability across the landscape is necessary in order to make predictions about the underlying distribution of properties at a given location (Soil Survey Division Staff, 1993). According to Schaetzl and Anderson (2005), V.V. Dokuchaev was one of the first modern pedologists to create a model explaining the factors that affect variability of soils across landscapes. He concluded that a soil's properties were a function of five environmental factors: climate, organisms, parent material, age, and relief (Schaetzl and Anderson, 2005). Jenny (1941) reemphasized the model and solidified its importance as an influential soil genesis model. This five-factor model is useful for explaining similarities between soils at different locations, because a similar set of soil-forming factors will create like soils, but it tends to focus on individual pedons without regard to the hillslope processes that connect them. The catena concept, however, incorporates the idea of the five soil-forming factors along with the concept that soils form a continuous system via interlinked hillslope processes, making it more practical for soil mapping (Milne, 1935; Milne, 1936; Soil Survey Division Staff, 1993). The interactions between soils across a hillslope have been repeatedly validated, especially for sediment and hydrologic connectivity across landscapes through examination of sediment/water transfer from one location to another (e.g., see Bracken et al., 2015; Detty and McGuire, 2010; Logsdon, 2007; Logsdon et al., 2009; Masselink et al., 2016; McGuire and McDonnell, 2010).

Soil Mapping and Process Modeling for Sustainable Land Use Management.
DOI: http://dx.doi.org/10.1016/B978-0-12-805200-6.00011-6

The catena, as defined by Milne (1935), is "a unit of mapping convenience..., a grouping of soils which while they fall wide apart in a natural system of classification on account of fundamental and morphological differences, are yet linked in their occurrence by conditions of topography and are repeated in the same relationships to each other wherever the same conditions are met with." Soils along a hillslope are linked pedologically through geomorphic erosion and sedimentation processes as well as hydrologic water and solute movement (Schaetzl and Anderson, 2005). Milne likened the sequence of soils across a hillslope to a horizontal soil profile, with the hilltop analogous to the surface horizon and downslope soils analogous to subsurface horizons. Components of the upper hill are eroded and/or eluviated downhill and cause subsequent changes in soil characteristics at lower landscape positions. Lowland soils can also influence upland soils. For example, solutes can be moved laterally or upward within the hillslope through hydraulic potential gradients (Hillel, 2004; Ibrahim and Burras, 2012; Ibrahim et al., 2015; Logsdon et al., 2009). These hillslope processes create a characteristic sequence of soils repeated across the landscape wherever soil-forming factors and hillslope linkages remain constant.

The recurring nature of the catena makes it a powerful tool for mapping large swaths of land without extensive sampling, which is why it remains widely used in soil survey. Experienced soil mappers use a mental model of catenas to predict soil distribution across the landscape and select sampling sites for validation of soil properties in the field. Another example of the role catenas play in US soil survey is the use of soil associations, which are comprised of soils adjacent to one another that occur in specific patterns across the landscape (Soil Survey Staff, 2007). While they do not always constitute a catena—for example, alluvial soils on a floodplain are operating under different dynamics than nearby upland soils—it is common for soil associations in the Midwestern US to be catenas. Small-scale soil association maps are thereby embedded with information that can be interpreted at the hillslope scale by those with an understanding of catenas.

IOWA CATENAS: THE DES MOINES LOBE

Clarion-Nicollet-Webster Catena

The Clarion-Nicollet-Webster (CNW) soil association is a catena that dominates the Des Moines Lobe in north-central Iowa (Fig. 11.1). Soils within this catena are classified as Mollisols according to US Soil Taxonomy and have similar parent material, climate, historic vegetation, and age. However, these factors vary somewhat across the landscape due to the hummocky, closed-basin topography, resulting in a systematic variation of soil properties from summit to toeslope. By definition the topography changes across a hillslope. These topographic changes lead to differences in infiltration and solute movement as well as erosion-sedimentation processes that affect pedogenesis and soil parent material distribution (Fig. 11.2). Soils on the upper hillslope are formed in loamy, calcareous till deposited roughly 14,000 years before present (BP) (Ruhe and Scholtes, 1959), while soils lower on the landscape are formed in the more recently deposited finer hillslope sediment (Walker and Ruhe, 1968; Burras and Scholtes, 1987). The microclimate also varies across the hill, with wetter soils lower on the landscape where the water table is closer to the surface. Differences in wetness, in turn, cause differences in native vegetation. A transition from tallgrass prairie to wet prairie to marsh historically occurred as one moved from the summit to basin center (Soil Survey

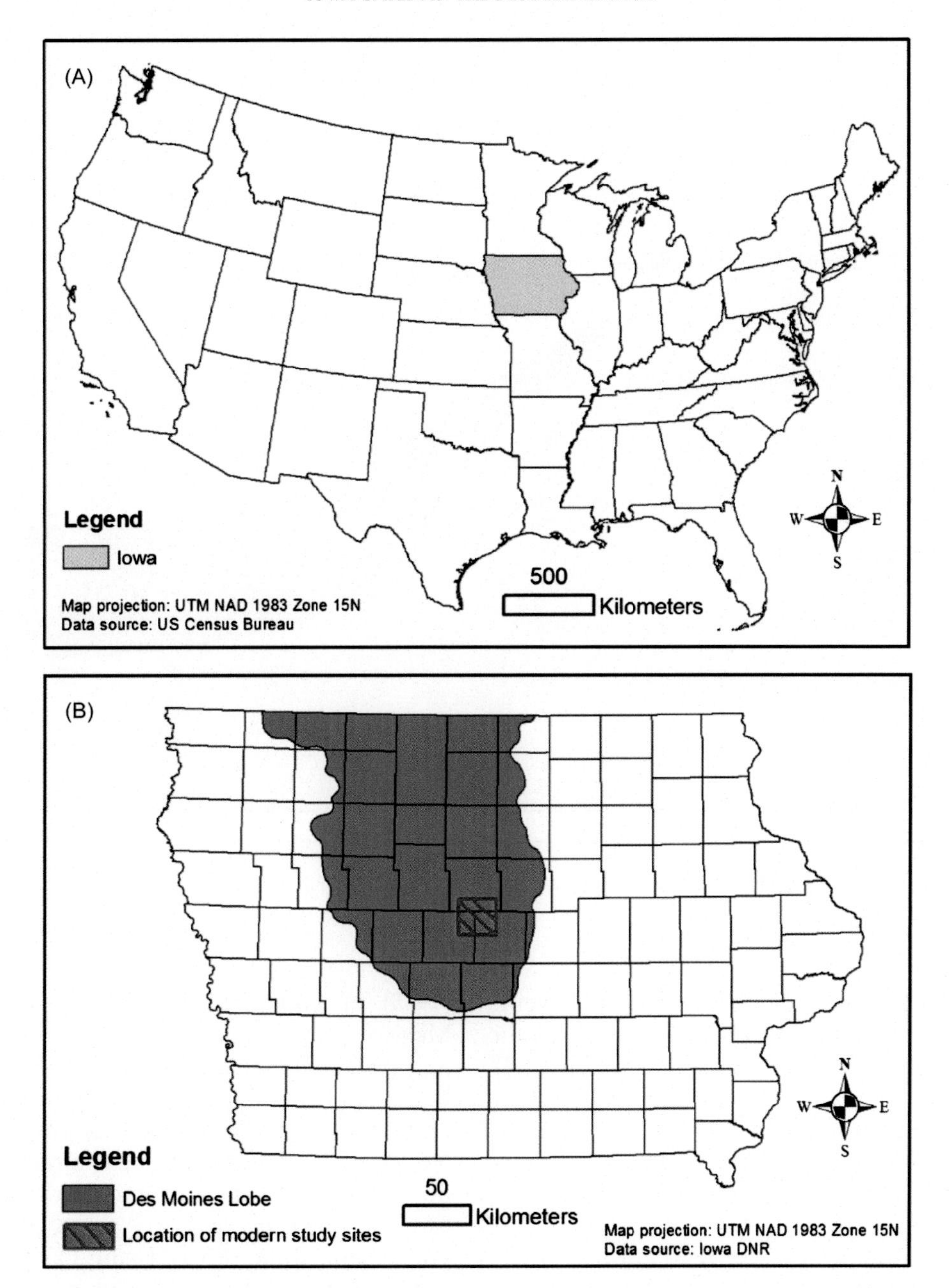

FIGURE 11.1 (A) Location of Iowa within the continental United States. (B) Location of the Des Moines Lobe within Iowa; location of modern catena study sites on the Des Moines Lobe.

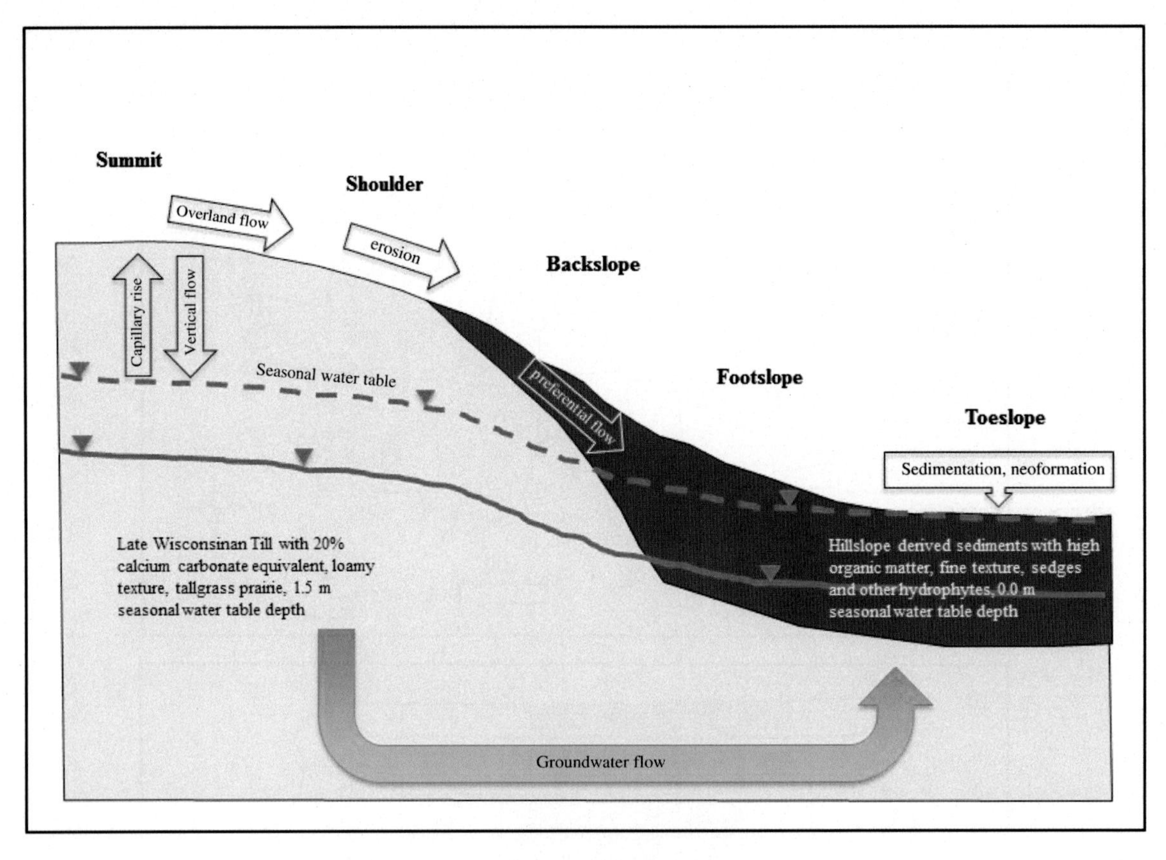

FIGURE 11.2 Classic catena model in closed-basin system of north-central Iowa.

Staff, 2016). This dynamic hillslope unit recurs across the Des Moines Lobe and comprises the CNW catena (Tables 11.1 and 11.2 and Fig. 11.3). The CNW catena is the most extensive catena in Iowa (Simonson et al., 1953) and accounts for roughly 76% of land area on the Des Moines Lobe and 16% of land area in Iowa (Table 11.3) (Miller et al., 2016).

Historic Catena Research

Our primary understanding of catena dynamics arises from studies more than 40 years old, such as the work done by Milne (1935) in East Africa, Ruhe (1967) in New Mexico, and the twin studies in Iowa by Ruhe and Walker (1968) and Walker and Ruhe (1968). Our understanding of the soil–landscape system on the Des Moines Lobe is predominantly based on Walker and Ruhes' research studying five closed basins on moraines of the Des Moines Lobe in Iowa (Walker, 1965, 1966; Walker and Ruhe, 1968). Using basin stratigraphy as well as fossil and pollen records, they determined that there were four major episodes of accumulation within the basins after deglaciation that were influenced by landscape stability (Fig. 11.4) (Walker, 1965, 1966). Two periods

TABLE 11.1 Characteristics of the Soil Series Commonly Found Along the CNW Catena in North-Central Iowa

Soil Series	Hillslope Position	Drainage	Native Vegetation	Parent Material	Stratigraphic Unit
Clarion	Summit	MW	Tallgrass Prairie	Glacial till	Morgan Member, Dows Formation
Omsrud	Shoulder	W	Tallgrass Prairie	Glacial till	Morgan Member, Dows Formation
Nicollet	Backslope	SWP	Tallgrass Prairie	Glacial till	Morgan Member, Dows Formation
Terril	Footslope	MW	Tallgrass Prairie	Colluvium	Flack Member, Deforest Formation
Webster	Footslope	P	Wet Prairie	Hillslope sediment over till	Woden Member, Deforest Formation
Canisteo	Footslope	P	Wet Prairie	Hillslope sediment over till	Woden Member, Deforest Formation
Harps	Toeslope	P	Calcareous Rim Prairie	Hillslope sediment	Woden Member, Deforest Formation
Okoboji	Toeslope	VP	Pothole Marsh	Hillslope sediment	Woden Member, Deforest Formation
Klossner	Toeslope	VP	Organic Pothole Marsh	Organic material over hillslope sediment	Woden Member, Deforest Formation

Drainage classes: *W*, well drained; *MW*, moderately well drained; *SWP*, somewhat poorly drained; *P*, poorly drained; *VP*, very poorly drained.

TABLE 11.2 Taxonomic Classification for Soils Typically Found Along the CNW Catena (Soil Survey Staff, 2016)

Soil Series	Soil Order	Soil Taxonomic Classification
Clarion	Mollisol	Fine-loamy, mixed, superactive, mesic Typic Hapludoll
Omsrud	Mollisol	Fine-loamy, mixed, superactive, mesic Typic Hapludoll
Nicollet	Mollisol	Fine-loamy, mixed, superactive, mesic Aquic Hapludoll
Terril	Mollisol	Fine-loamy, mixed, superactive, mesic Cumulic Hapludoll
Webster	Mollisol	Fine-loamy, mixed, superactive, mesic Typic Endoaquoll
Canisteo	Mollisol	Fine-loamy, mixed, superactive, calcareous, mesic Typic Endoaquoll
Harps	Mollisol	Fine-loamy, mixed, superactive, mesic Typic Calciaquoll
Okoboji	Mollisol	Fine, smectitic, mesic Cumulic Vertic Endoaquoll
Klossner	Histosol	Loamy, mixed, euic, mesic Terric Haplosaprist

characterized by high erosion and sedimentation rates occurred from 13,000 to 10,500 BP and 8000 to 3000 BP when the landscape was unstable due to shifting vegetation, which led to higher erosion rates and subsequent deposition of silty mineral sediment in the basins. During the two more stable periods from 10,500 to 8000 BP and 3000 to 150 BP, perennial vegetation

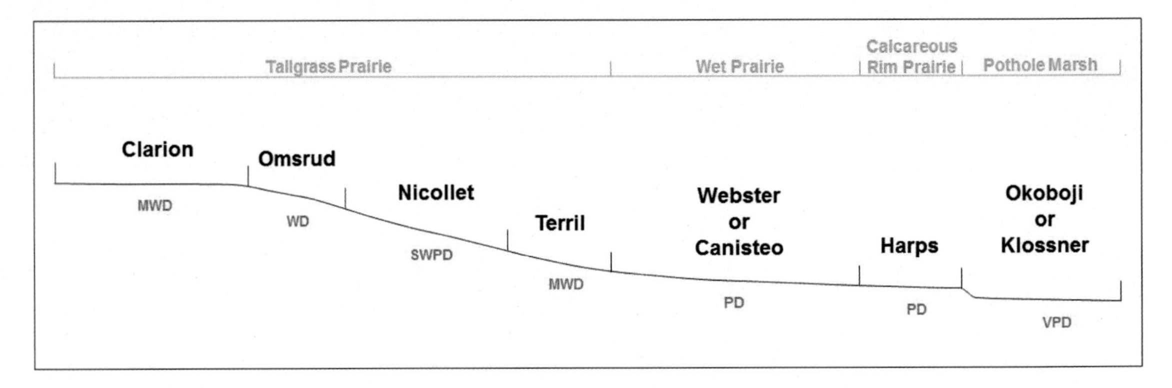

FIGURE 11.3 Hillslope cross-section showing soil series, drainage class, and native vegetation with landscape position for the CNW catena. (Drainage classes: *WD*, well drained; *MWD*, moderately well drained; *SWPD*, somewhat poorly drained; *PD*, poorly drained; *VPD*, very poorly drained).

TABLE 11.3 Distribution of the CNW Catena by Land Area

Soil Series	Total Hectares Mapped (USA)	Total Hectares Mapped (IA)	%Land Area (IA)	%Land Area (IA DM Lobe)
Clarion	930,468	731,183	5	24
Omsrud	7399	4768	<1	<1
Nicollet	632,917	446,482	3	14
Terril	73,477	31,019	<1	1
Webster	712,349	412,911	3	13
Canisteo	741,108	474,558	3	15
Harps	213,953	110,313	1	4
Okoboji	382,984	133,456	1	4
Klossner	94,319	15,195	<1	1
Total	**3,788,973**	**2,359,885**	**16**	**76**

Source: Adapted from Miller, G.A., Sassman, A.M., Burras, C.L., 2016. Iowa soil properties and interpretations database, ISPAID Version 8.1. Iowa State University, Ames, IA. Available from: http://www.extension.iastate.edu/soils/ispaid (accessed 04.04.16); California Soil Resource Lab, 2016. Series Extent Explorer. University of California, Davis, CA. Available from: http://casoilresource.lawr.ucdavis.edu/ (accessed 04.01.16).

minimized erosion and a thick strata of organic peat/muck accumulated in the basins. Erosion and sedimentation rates for each of these periods are shown in Table 11.4.

Walker (1965, 1966) and Walker and Ruhe (1968) also found several soil trends based on their catena studies (Figs. 11.5 and 11.6). For example, the weighted average soil particle size, or geometric mean, across a hillslope tends to decrease as one moves from the summit to the toeslope (Figs. 11.5 and 11.6A) (Walker, 1966; Walker and Ruhe, 1968). This is because fine sediment (e.g., fine sand, silt, and clay) is preferentially eroded from the upper hillslope, leaving behind the coarse sediment. The eroded sediment is then sorted as it is transported downslope such that the coarsest of this material is deposited closer to the source and finer material is transported further out into the basin (Walker and Ruhe, 1968). A low geometric mean is associated with higher clay content and lower sand content, so it follows that clay content increases from the summit to toeslope while sand content decreases. Walker and Ruhe (1968) also found a systematic

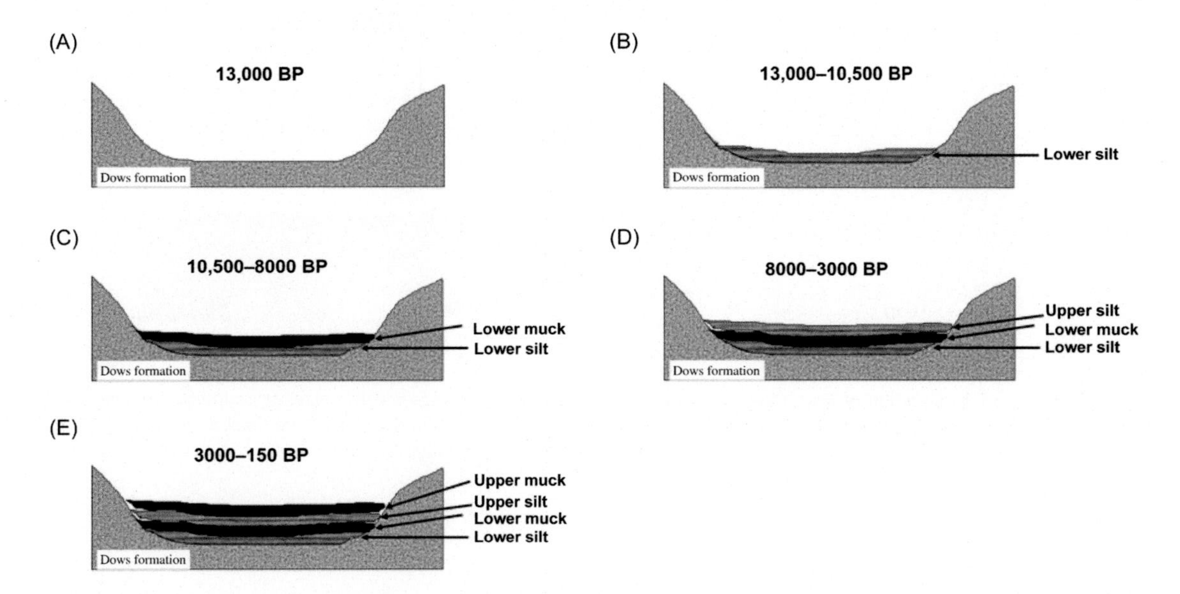

FIGURE 11.4 Cross-section of stratigraphy at basin center on the Des Moines Lobe. (A) Glacial drift exposed after ice retreat. (B) Erosion and deposition of lower silt in basin center. (C) Accumulation of lower muck with little upland erosion. (D) Erosion and deposition of upper silt in basin center. (E) Accumulation of upper muck with little upland erosion. *Source: Adapted from Walker, P.H., 1965. Soil and Geomorphic History in Selected Areas of the Cary Till, Iowa. Ph.D. Diss., Iowa State University, Ames, IA.*

TABLE 11.4 Average Erosion and Sedimentation Rates for Closed Basins on End Moraines of the Des Moines Lobe, Iowa

Unit	Time (BP)	Erosion Rate ($Mg\,ha^{-1}\,year^{-1}$)	Total Erosion (m)	Accumulation Rate (cm per 100 years)	Total Accumulation (m)
Lower silts	13,000–10,500	2.4	0.40	3.3	0.83
Lower muck	10,500–8000	1.1	0.12	15.2	2.28
Upper silts	8000–3000	2.5	0.80	5.3	2.65
Upper muck	3000–150	0.3	0.06	2.7	0.77

Source: Adapted from Walker, P.H., 1965. Soil and Geomorphic History in Selected Areas of the Cary Till, Iowa. Ph.D. Diss., Iowa State University, Ames, IA.

distribution of soil organic carbon (SOC) across a hillslope (Fig. 11.6B). Some factors that lead to an increase in SOC toward the toeslope include erosion and sedimentation, differences in productivity and oxidation rates, and greater physical protection of SOC within microaggregates (Doetterl et al., 2012; Ritchie et al., 2007).

AGRICULTURAL LAND USE: AGENT OF SOIL CHANGE

The modern relevance of predictable soil distribution across catenas is increasingly uncertain due to mounting evidence that humans have aggressively, albeit often unintentionally,

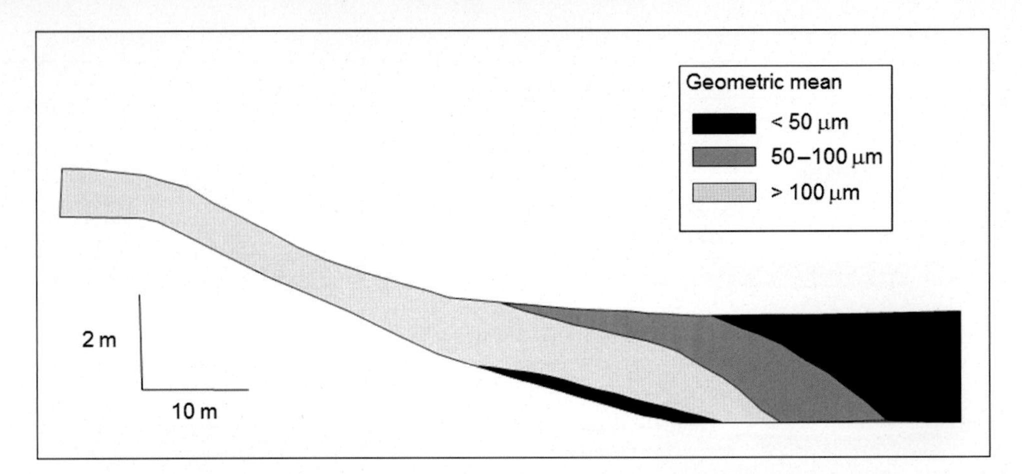

FIGURE 11.5 Cross-section of Jewell Bog in Iowa showing weighted average particle size, or geometric mean, distribution. *Source: Adapted from Walker, P.H., 1966. Postglacial Environments in Relation to Landscape and Soils on the Cary Drift, Iowa (No. 549). Agriculture and Home Economics Experiment Station, Iowa State University, Ames, IA.*

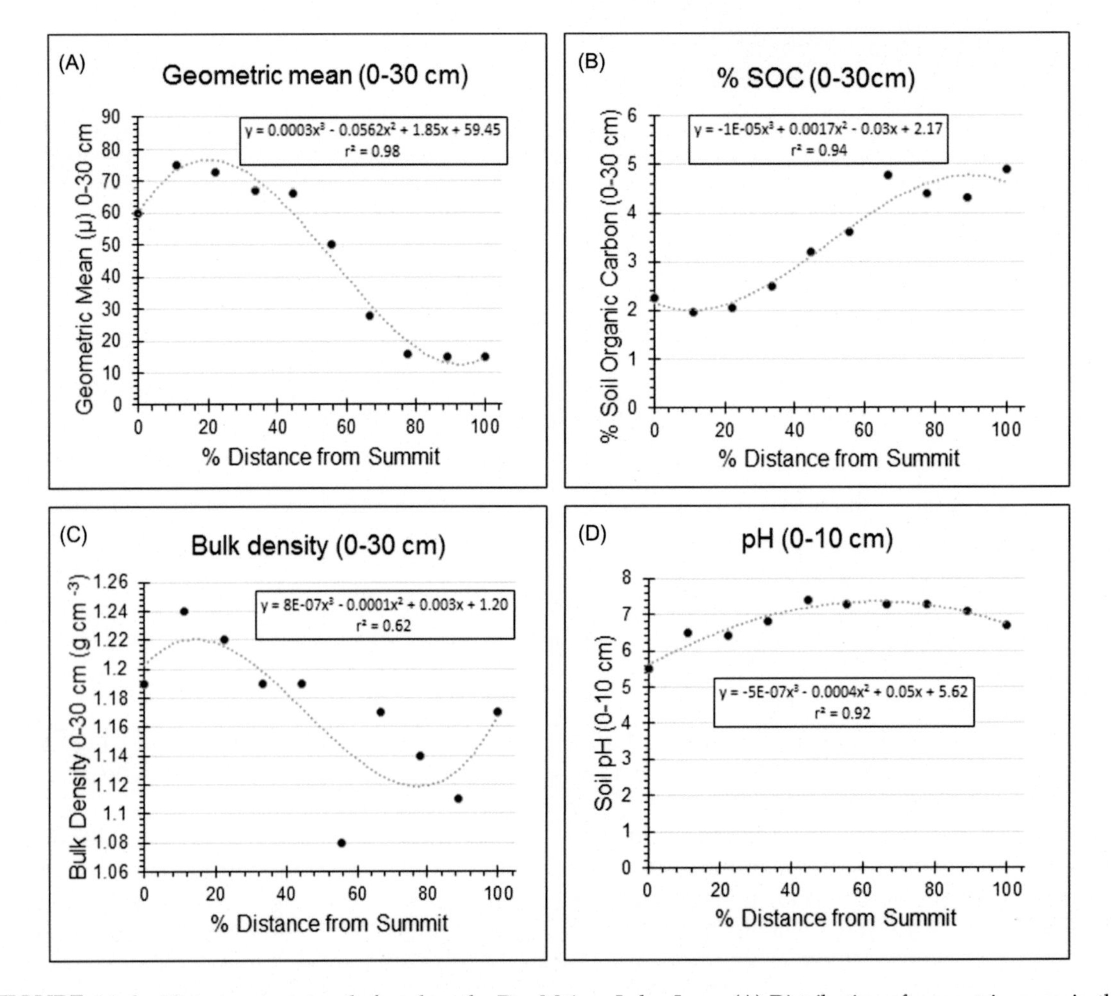

FIGURE 11.6 Historic catena trends found on the Des Moines Lobe, Iowa. (A) Distribution of geometric mean in the upper 30 cm across a hillslope. (B) Distribution of SOC in the upper 30 cm across a hillslope. (C) Distribution of bulk density in the upper 30 cm across a hillslope. (D) Distribution of soil pH in the upper 10 cm across a hillslope. *Source: Modified from Walker, P.H., Ruhe, R.V., 1968. Hillslope models and soil formation II. Closed systems. Int. Soc. Soil Sci. Trans. 4, 561–568; Walker, P.H., 1965. Soil and Geomorphic History in Selected Areas of the Cary Till, Iowa. Ph.D. Diss., Iowa State University, Ames, IA.*

changed the soil–landscape system due to agricultural practices such as tile drainage and drainage ditches, tillage-induced SOC losses, erosion and sedimentation, use of fertilizers, etc. Collectively these human impacts are referred to as "land use and management" (LUM). The literature is replete with examples of LUM during the 19th to 21st centuries causing minor to dramatic changes within soil profiles or across landscapes. Documented property changes from LUM include bulk density, porosity, SOC, N, P, K, cation exchange capacity, pH, soil structure, stable aggregates, epipedon thickness, and solum thickness (Barak et al., 1997; Blanco-Canqui et al., 2010; Bouma and Hole, 1971; Bouman et al., 1995; Bowman et al., 1990; Bowman and Halvorson, 1998; Chendev et al., 2012; Coote and Ramsey, 1983; Davidson and Ackerman, 1993; Grandy and Robertson, 2006; Gregorich and Anderson, 1985; Hayes and Vepraskas, 2000; Indorante et al., 2014; Konen, 1999; Mann, 1985; Papiernik et al., 2007; Pennock, 2003; Saviozzi et al., 2001; Veenstra and Burras, 2012, 2015; Voorhees and Lindstrom, 1984).

At times, LUM increases landscape connectivity and/or the timescale at which connectivity occurs, while in other instances these factors are decreased (Fryirs et al., 2007; Warner, 2006). For example, high density perennial vegetation decreases rates of erosion and sedimentation, thereby decreasing sediment connectivity across the landscape, while artificial drainage networks increase both spatial and temporal hydrologic connectivity (Sandercock and Hooke, 2011; Warner, 2006). Variations in LUM can therefore have enormous influence on hillslope connectivity and catena functioning that ultimately lead to different pedogenic responses and soil change. Given the assortment of LUM, it is plausible that multiple pathways of catena development are emerging as opposed to a single trajectory based on initial soil-forming factors.

Under some circumstances, LUM can negate the relationship between soil properties and topography. For instance, Houben (2008) found that over 1300 years of LUM in Germany caused near-randomness with regard to soil truncation and burial that has no correlation to modern topography. He cited tillage erosion and field borders as important factors that influence the occurrence of erosion and sedimentation outside natural topographic locales and concluded that the catena concept fails under these fragmented hillslope conditions. Following are but a few examples of LUM that have the potential to alter catena functioning and lead to divergent soil properties.

Subsurface Drainage

Subsurface drainage is common throughout parts of the United States, particularly in the Midwestern corn-belt region, and usually coincides with intense agricultural production on glaciated landscapes. These artificial drainage networks have altered the hydrologic environment by creating a rapid escape of soil water and its constituents from the soil system. In Iowa, it is estimated that about 3.2 million hectares (22%) of land has subsurface drainage (Schilling and Helmers, 2008), the majority of which is found on the Des Moines Lobe. A typical square kilometer of land in this region now has more than 10 linear km of tile drainage systems that were installed to lower the water table in closed basins and broad, poorly drained upland flats. The drainage tiles are connected to constructed drainage ditches that feed into streams and rivers, many of which have been straightened. While eroded sediment continues to be captured and stored within the basins, colloids, and solutes such as 2:1 phyllosilicates, iron, pesticides, phosphorus, and nitrate are leached from the soil and rapidly transported away from the basins in these constructed drainage networks (Kladivko et al., 1991;

Mercier et al., 2000; Smith et al., 2015). Thus the drained catena has hydrologically become like that of an open basin while maintaining its closed-basin nature of sediment capture.

The goal of subsurface drainage is to manipulate the microclimate through lowering of the water table to make the soil drier and warmer, but because the soil is a complex system heavily influenced by hydrology, other pedogenic changes follow. Helmers et al. (2012) found that the average depth to water table in drained soils is 40–70 cm lower than undrained soils in Iowa, while Eidem et al. (1999) state that the water table may be 1.5–2 m lower in drained versus undrained watersheds. Drained soils have a higher oxygen level and redox potential deeper in the profile than they did prior to drainage, which directly affects soil redoximorphic features. Veenstra and Burras (2015) found that gray depletions occur lower in the profile after 50 years of agriculture across all hillslope positions. The largest change was found at the summit, where gray depletions are up to 39 cm deeper, while the smallest change is a deepening of 17 cm at the backslope. They attribute the deepening of gray depletions to a lower water table from tile drainage. Tile drainage can also introduce more heterogeneity in soil properties across the landscape based on proximity to tile lines. Montagne et al. (2008) found that Mn oxides increase near tile lines and loss of smectites is higher due to greater eluviation near the drain. Likewise, Hayes and Vepraskas (2000) found more Fe masses with proximity to tile lines. All of these pedogenic changes indicate that the soil system is functioning differently in areas with tile drainage versus those that are undrained.

Erosion and Sedimentation

One of the most evident impacts of LUM is accelerated soil erosion. The magnitude and frequency of sediment movement has been affected by various tillage operations and annual versus perennial vegetative cover that leaves the soil bare and susceptible to erosion for much of the year. Natural Holocene erosion rates under perennial vegetation tend to be less than 3 $Mg\,ha^{-1}\,year^{-1}$ (Burras and Scholtes, 1987; Ruhe and Daniels, 1965; Turnage et al., 1997; Walker, 1966). In contrast, average erosion rates on cultivated land can be much greater, on the order of 11–36 $Mg\,ha^{-1}\,year^{-1}$ (Afshar et al., 2010; Konen, 1999; Kreznor et al., 1990; Norton, 1986; Papiernik et al., 2007; Ruhe and Daniels, 1965; Theocharopoulos et al., 2003; Turnage et al., 1997).

The role of tillage and water erosion in soil redistribution across catenas has been highlighted by Freilinghaus et al. (1998) as well as De Alba et al. (2004). Rates of erosion/sedimentation depend on many factors including LUM, topography, parent material, and soil hydrologic properties (Freilinghaus et al., 1998; Papiernik et al., 2005). The morphologic changes that can occur from erosion and sedimentation are so great that De Alba et al. (2004) proposed a new catena model to account for soil redistribution by tillage. Under perennial vegetation with only slight erosion, pedogenesis on the upper hillslope is occurring at an equal or faster rate than erosion, leading to a thicker solum over time, while sedimentation is minor on the lower landscape. As erosion rates increase, sediment removal begins to outpace pedogenic weathering and the epipedon can be truncated or even completely removed (De Alba et al., 2004; Phillips et al., 1999). With moderate erosion, this leads to thickening of soils in depositional areas, while in cases of severe erosion the sedimentation rate is so high that rapid and deep burial of the original surface occurs (De Alba et al., 2004), often with coarser sediment that is lighter in color and lower in SOC than the buried soil. Areas of erosion such as the shoulder and backslope often have shallower soils with thinner A-horizons and sola, lower SOC, N and P, along with higher inorganic carbon content, pH, and bulk densities (Gregorich and Anderson, 1985; Papiernik et al., 2007).

Over the past 50 years the farm sector has responded aggressively to manage erosion (Helms, 1990; Soil Conservation Staff, 2015). Some farmers have used terraces; others have used no-till, reduced tillage, contour tillage, or crop rotations, to name just a few approaches (Karlen et al., 2010; Soil Conservation Staff, 2016). However, these practices and their impacts on soils vary tremendously from farm to farm, which means any given landscape likely has undergone a different set of detailed management than its comparable analogs in the region. This supports the idea that today's human-impacted catenas will have greater variability and ambiguity spatially than their native counterparts.

Soil Organic Carbon

Cultivation leads to decreases in SOC (Davidson and Ackerman, 1993; Guo and Gifford, 2002; Saviozzi et al., 2001), especially related to soils with naturally high SOC content. The primary *in situ* agent causing reduced SOC content is increased aerobic decomposition resulting from increased oxygen due to tillage mixing and aggregate breakdown, while the primary *ex situ* agent is erosion. We speculate that a lower water table along with increased nutrients from fertilization are less well documented components of SOC loss. Variations in the vegetation and type of tillage used cause differences in SOC over time (Guo and Gifford, 2002; Moorman et al., 2004). Given that SOC affects many important soil properties including anion and cation exchange capacity, soil structure, stable aggregate content, and water holding capacity, it is clear that changes in SOC distribution from LUM influence soil functioning at multiple scales.

Soil Amendments

In order to maintain high plant productivity, large quantities of soil amendments are applied each year such as fertilizers, ag lime, and pesticides. Although these amendments provide benefits to crops, they have unintended consequences. Anhydrous ammonia and urea are two of the most commonly used N fertilizers in the United States, and while they are not acidic themselves, they become acid forming with microbial oxidation (Barak et al., 1997). Shortly after application, inorganic N fertilizers cause rapid fluctuations in pH that facilitate dissolution and translocation of organic matter and carbonate minerals (Veenstra and Burras, 2015). Over time, application of excess ammoniacal N fertilizers along with loss of alkalinity through crop uptake and removal lead to soil acidification (Barak et al., 1997; Bouman et al., 1995; Guo et al., 2010). Increased acidity, in turn, leads to changes in other soil properties such as a decrease in cation exchange capacity, eluviation of base cations lower in the profile, increased mobilization of Mn and Al, and more intense clay weathering (Barak et al., 1997; Blake et al., 1999; Blevins et al., 1977; Fortner et al., 2012; Schwab et al., 1989). Liming agents are often applied to counteract the acidifying effects of N fertilizers, but application varies by farm, leading to greater spatial heterogeneity in pH based on management history.

Soil Taxonomy

US Soil Taxonomy relies heavily on chemical and morphological properties to classify soils, but these properties are vulnerable to changes from LUM, which can result in corresponding changes to a soil's classification. Because Mollisols are classified based on their epipedons, they are particularly susceptible to morphing into Alfisols, Inceptisols, or Entisols due to acidification, erosion, or compaction of the surface through LUM. Likewise, Histosols can rapidly be converted to other soil orders due to wind erosion and increased microbial decomposition from tillage and subsurface drainage that decrease the thickness and SOC content of the epipedon. These processes can cause rapid subsidence in Histosols (Millette et al., 1982). It is also possible for Alfisols to become Mollisols with tillage and cropping that cause deepening of SOC (Chendev et al., 2012).

Ultimately, soils can and often do change classification. In a reexamination of reference pedons in Iowa, Veenstra and Burras (2012) found that 60% of benchmark pedons they sampled have changed classification within the last 50 years, with 19% of those placed in a completely different soil order. Veenstra and Burras studied pedons on landscapes with an average slope greater than 2%. Likewise, Mokma et al. (1996) found that 78% of both moderately and severely eroded Mollisols they sampled have changed classification compared to slightly eroded counterparts, while 20% of moderately eroded and 40% of severely eroded Alfisols have changed classification. Thus it is possible for soils to change classification over short periods of time as a result of LUM.

CASE STUDY: MODERN CATENAS ON THE DES MOINES LOBE

Individually and collectively, the preceding studies indicate that noteworthy changes in soil–landscape relationships occur at the decadal timescale for farmed settings in a manner more or less proportional to the intensity of land use. Therefore soil maps are increasingly at risk of being based on outdated catenic models because those models neglect to take into account the effects of LUM. The objective of this study is to evaluate current soil distribution and processes within catenas on the Des Moines Lobe, Iowa.

Materials and Methods

Four benchmark catenas that have been in continuous agricultural production were sampled on the Des Moines Lobe in north-central Iowa (Fig. 11.1B and Tables 11.5 and 11.6). Each location is the site of a historic catena study. The original studies were Walker and Ruhe (1968) in Hamilton County, Burras and Scholtes (1987) in Story County, Steinwand and Fenton (1995) in Boone County, and Konen (1999) in Boone County. For simplicity sake, each site will now be referred to by only the original lead author's name. All of the sites have been in continuous agricultural production during the intervening years. The Burras and Steinwand sites have been in continuous row crop agriculture, while the Walker and Konen sites have had row crops as well as pasture or hay. All four sites have subsurface tile drainage systems, which is very common for this region.

At each site, five to seven soil cores were taken from one hillslope transect using a truck-mounted hydraulic soil sampler (Giddings Machine Company, Windsor, CO). The cores

TABLE 11.5 Characteristics of the Four Catena Study Sites in North-Central Iowa

Site	GPS Coordinates of Site Center (UTM Zone 15N: mE, mN)	Land Use History	Modern Relief (m)	Average %Slope (rise/run)	Transect Length (m)	Sample Spacing (m)	Cores Described
Walker	442598, 4678297	Variable	6	3	180	30	7
Burras	449035, 4661717	Row crops	1	1	80	20	5
Steinwand	436750, 4659041	Row crops	4	3	135	27	6
Konen	421774, 4672708	Variable	5	6	88	22	5

TABLE 11.6 Current Soil Taxonomic Classification and Drainage Class Across Catenas at Four Study Sites in North-Central Iowa

Site	Hillslope Position	Soil Series	Soil Order	Soil Taxonomic Classification	Drainage
Walker	Summit	Lester taxadjunct	Alfisol	Fine, mixed, superactive, mesic Mollic Hapludalf	SWP
	Shoulder	Clarion	Mollisol	Fine-loamy, mixed, superactive, mesic Typic Hapludoll	MW
	Backslope	Terril	Mollisol	Fine-loamy, mixed, superactive, mesic Cumulic Hapludoll	MW
	Footslope	(No match)	Entisol	Fine-loamy, mixed, superactive, mesic Aquic Udorthent	SWP
	Toeslope	Klossner	Histosol	Loamy, mixed, euic, mesic Terric Haplosaprist	VP
	Toeslope	Kimvar	Mollisol	Fine-loamy, mixed, superactive, mesic Histic Endoaquoll	VP
	Toeslope	Kimvar	Mollisol	Fine-loamy, mixed, superactive, mesic Histic Endoaquoll	VP
Burras	Summit	Nicollet taxadjunct	Mollisol	Coarse-loamy, mixed, superactive, mesic Aquic Hapludoll	MW
	Shoulder	Nicollet	Mollisol	Fine-loamy, mixed, superactive, mesic Aquic Hapludoll	MW
	Backslope	Nicollet	Mollisol	Fine-loamy, mixed, superactive, mesic Aquic Hapludoll	SWP
	Footslope	Floyd	Mollisol	Fine-loamy, mixed, superactive, mesic Aquic Pachic Hapludoll	SWP
	Toeslope	Nevin taxadjunct	Mollisol	Fine, mixed, superactive, mesic Aquic Pachic Argiudoll	P
Steinwand	Summit	Clarion	Mollisol	Fine-loamy, mixed, superactive, mesic Typic Hapludoll	MW
	Shoulder	Storden taxadjunct	Inceptisol	Coarse-loamy, mixed, superactive, mesic Typic Eutrudept	W
	Backslope	Terril taxadjunct	Mollisol	Coarse-loamy, mixed, superactive, mesic Cumulic Hapludoll	SWP
	Footslope	Terril	Mollisol	Fine-loamy, mixed, superactive, mesic Cumulic Hapludoll	P
	Toeslope	Colo	Mollisol	Fine-silty, mixed, superactive, mesic Cumulic Endoaquoll	P
	Toeslope	Calco	Mollisol	Fine-silty, mixed, superactive, calcareous, mesic Cumulic Endoaquoll	P
Konen	Summit	Lindley	Alfisol	Fine-loamy, mixed, superactive, mesic Typic Hapludalf	MW
	Shoulder	Storden	Inceptisol	Fine-loamy, mixed, superactive, mesic Typic Eutrudept	MW
	Backslope	Nevin taxadjunct	Mollisol	Fine-loamy, mixed, superactive, mesic Aquic Pachic Argiudoll	SWP
	Footslope	Nevin taxadjunct	Mollisol	Fine-loamy, mixed, superactive, mesic Aquic Pachic Argiudoll	SWP
	Toeslope	Jameston	Mollisol	Fine-loamy, mixed, superactive mesic Typic Argiaquoll	P

Drainage classes: *W*, well drained; *MW*, moderately well drained; *SWP*, somewhat poorly drained; *P*, poorly drained; *VP*, very poorly drained.

were sampled using a 5-cm diameter tube and were spaced at 20–30 m intervals along the hillslope. Sampling depth varied depending on stoniness and ability of the sampler to penetrate the soil, but was generally between 1 and 3 m. The soil cores were described and the following information was recorded based on Schoeneberger et al. (2002): horizon depth, color, texture, rock fragments, structure, consistence, coatings, roots, pores, effervescence, and redoximorphic features. Each horizon sample was oven dried, weighed, and ground to pass through a 2 mm sieve. Coarse fragments larger than 2-mm were collected and weighed separately.

Soil pH was measured in a mixture of one part soil to one part water. SOC content was determined by loss on ignition (Konen et al., 2002). Particle size analysis was completed using the pipette method described by Konen (1999). Sand was fractionated using a nest of sieves (1.0, 0.5, 0.25, 0.1, and 0.047 mm) on a shaker. After clay and sand fractions were determined, silt was determined by difference. The geometric mean herein is the weighted average particle size of the very coarse, coarse, medium, fine, and very fine sand fractions along with the silt and clay fractions for that horizon, using the midpoint of each size range. For instance, the percentage of very coarse sand, which is the 1000–2000 μm size fraction, is multiplied by 1500 μm. The percentage of coarse sand (500–1000 μm diameter) is multiplied by 750 μm, etc., and the sum of all seven weighted fractions is the geometric mean for that horizon. Bulk density was calculated for each horizon using the core method (Method 4A3, National Soil Survey Center) (Soil Survey Staff, 1996).

Results and Discussion

Soil pH

The pH at the surface (0–10 cm) varies across the different catenas, with an average trendline for the four sites indicating that the pH at the summit and toeslope are similar, with a slight decrease in pH at the foot of the hill (Fig. 11.7A). However, this relationship between pH (0–10 cm) and distance from the summit when data are combined for all four sites is weak ($r^2 = 0.08$). When each catena is evaluated individually, there is a strong relationship at the Burras (Fig. 11.7B) ($r^2 = 0.94$), Konen (Fig. 11.7C) ($r^2 = 0.93$), and Steinwand (Fig. 11.7E) ($r^2 = 0.98$) sites and moderately at the Walker site (Fig. 11.7D) ($r^2 = 0.50$). Two reasons for high pH in the epipedon of the summit, shoulder, and backslope are addition of liming agents and exposure of unleached, calcareous till through tillage and erosion. On the lower landscape, pH is affected by liming agents and pedogenic addition of carbonates through hydrologic processes. Salts such as calcium carbonate can be leached from the upper hillslope and moved to the bottom of the hill through groundwater flow, where they are later pulled upward in the profile by capillary suction as the soil becomes dry at the surface through evaporative processes.

At the Burras site (Fig. 11.7B), pH increases from the summit to the footslope and then drops slightly at the toeslope. Redmond and McClelland (1959) found that accumulation of salts occurs in closed basins at the edge of seasonally ponded areas (i.e., at the footslope) where evaporation and precipitation outweigh leaching, while the interior of ponded areas (i.e., at the toeslope) have less salt accumulation due to greater leaching from pooled water. This could explain the distribution of pH at the Burras site. Bases are weathered and leached from the upper hillslope while the highest evaporation, and therefore addition of pedogenic carbonates that increase pH, is likely at the footslope of this site along the edge of wetter areas and where carbonate-rich soil water from upslope collects. At the toeslope, carbonates are leached and exported through tile lines and through coarse outwash strata that allow natural export between linked basins,

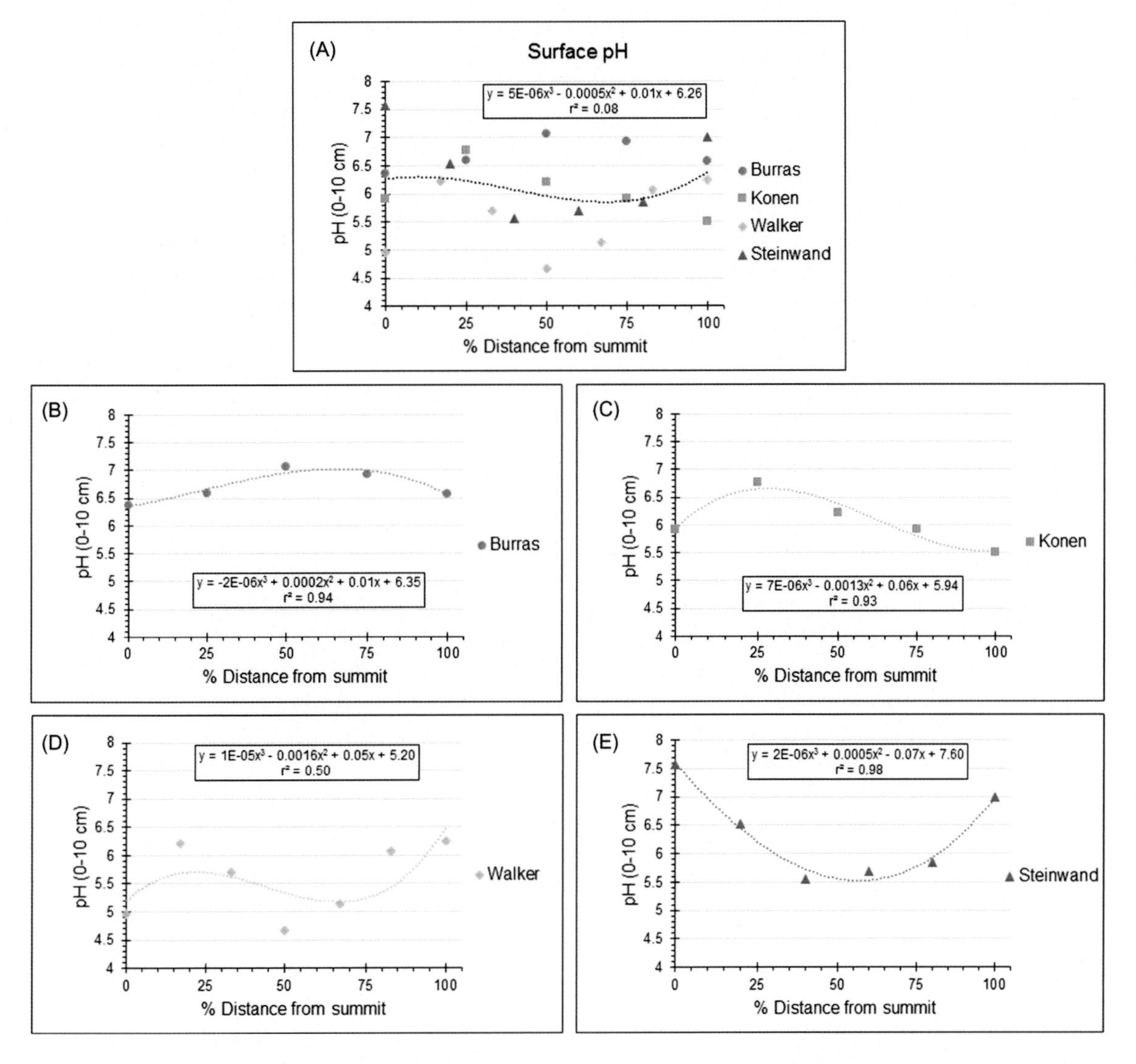

FIGURE 11.7 Current catena pH trends found on the Des Moines Lobe, Iowa. (A) Composite trendline for soil pH in the upper 10 cm across all four basins. (B) Trendline for soil pH in the upper 10 cm across Burras catena. (C) Trendline for soil pH in the upper 10 cm across Konen catena. (D) Trendline for soil pH in the upper 10 cm across Walker catena. (E) Trendline for soil pH in the upper 10 cm across Steinwand catena.

leading to a decreased pH in comparison to the footslope.

Papiernik et al. (2005) found that pH decreases across a hillslope in eroded areas with calcareous parent material, which they attribute to exposure and mixing of calcareous subsoil in upslope convex areas and deposition of soils lower in inorganic carbon downslope as a result of intensive tillage. Likewise, Sherrod et al. (2015) found that calcium carbonate is highest at the summit and lowest at the toeslope. At the Konen site (Fig. 11.7C), this trend of decreasing pH downslope is apparent with the exception of the summit, where pH is lower than the shoulder and backslope. Weathering, along with addition of nitrogen fertilizers and crop removal, are likely responsible for the lower pH at the summit. The highest pH occurs at the shoulder where there has been heavy erosion and exposure of calcareous till, followed by a steady decrease in pH toward the foot of the hill. It is probable that the entire hillslope had a lower pH due to LUM until erosion became severe enough upslope to expose the underlying calcareous till. The higher pH till was then moved downslope, creating the steady decline in pH with declining deposition of the calcareous sediment. At the Walker site (Fig. 11.7D) a process may be occurring similar to that at the Konen site, except that pedogenic accumulation of carbonates is occurring at the toeslope. The Steinwand site (Fig. 11.7E), in turn, appears to have a trend similar to the Walker site except that the highest pH occurs at the summit as opposed to the shoulder, which is most likely due to a higher erosion rate on the summit that exposes the calcareous till, along with addition of liming agents. Each of the catenas in this study exhibits a unique hillslope distribution of pH which appears to be a result of natural factors as well as distinct anthropogenic factors.

Thickness of Mollic Colors

The upper part of the solum in most of north-central Iowa is historically comprised of a dark, organic-matter-rich epipedon with mollic colors, which are defined as having a moist Munsell color value and chroma of 3 or less. The maximum depth to mollic colors can become thinner in a soil profile over time due to erosion, compaction, decomposition of organic matter, or gleying. Alternatively, it can thicken from deposits of dark sediment, tillage mixing of dark surface horizons with lighter subsurface zones, translocation of organic matter, or addition of organic matter by plant roots or other organic amendments. Because these factors vary across management practices and across hillslopes, mollic color thickness differs across the landscape.

When examining the maximum depth of mollic colors across all four catenas (Fig. 11.8A), distance along the hillslope and depth of mollic colors are only somewhat related ($r^2 = 0.60$). The trendline indicates thickness of mollic colors is similar at the summit and shoulder followed by a steady increase before leveling out at the toeslope. Many studies have found that the thickness of the A-horizon increases downslope due to greater leaching and biomass production, lower erosion rates, and higher sedimentation rates compared to the upper hillslope (De Alba et al., 2004; Gregorich and Anderson, 1985; Papiernik et al., 2007; Veenstra and Burras, 2015). Given that mollic colors are associated with the A-horizon, these factors likely explain the increase in mollic thickness we see downslope.

Analyzing the catenas individually, the Burras (Fig. 11.8B) ($r^2 = 0.99$), Konen (Fig. 11.8C) ($r^2 = 0.99$), and Steinwand (Fig. 11.8E) ($r^2 = 0.93$) sites each have a strong relationship between distance along the hillslope and depth of mollic colors, while the Walker site (Fig. 11.8D) ($r^2 = 0.47$) has a more moderate relationship. The shallowest depth to mollic colors at the Burras site (Fig. 11.8B) occurs at the summit and gets progressively deeper toward the toeslope, with a more or less linear increase in mollic color thickness until the basin center, at which point it thickens drastically. The thick, dark deposit in the basin center is a reflection of the original moderate relief of 3 m after deglaciation, while the present lower relief landscape is a result of backwearing and infilling of the basin with organic-rich sediment from upslope (Burras and Scholtes, 1987). At the Konen site (Fig. 11.8C) the mollic epipedon on the summit and shoulder has been completely removed by recent erosion, but mollic thickness increases

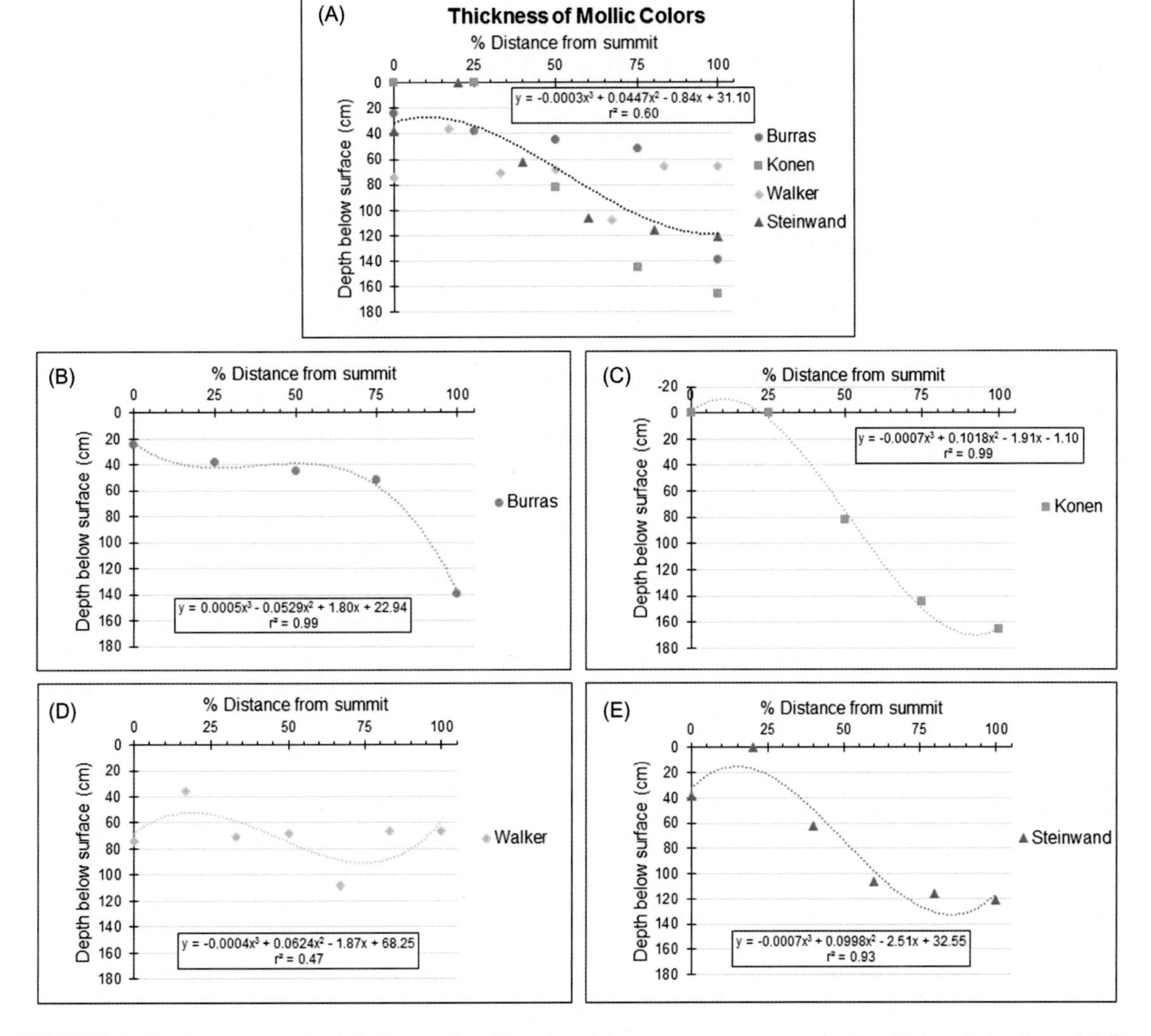

FIGURE 11.8 Current trends in thickness of mollic colors found across catenas on the Des Moines Lobe, Iowa. Mollic colors have a moist Munsell value and chroma ≤ 3. (A) Composite trendline for depth of mollic colors across all four basins. (B) Trendline for depth of mollic colors across Burras catena. (C) Trendline for depth of mollic colors across Konen catena. (D) Trendline for depth of mollic colors across Walker catena. (E) Trendline for depth of mollic colors across Steinwand catena.

rapidly as one moves downslope due to the thick deposits of dark, eroded sediment. The shoulder at the Steinwand site (Fig. 11.8E) no longer has mollic colors because of removal by erosion, but there is an increased thickness as one moves downslope.

At the Walker site (Fig. 11.8D) the shoulder has the thinnest layer of mollic colors due to erosion. The toeslope nearest the summit has the thickest layer, presumably due to deposition of darker sediment from upslope. The other hillslope positions at the Walker site (summit, backslope, footslope, and toeslope further from the summit) all have similar thickness of mollic colors. We attribute the thickness at the summit to periods of perennial vegetation along with occasional aeolian deposition of organic-rich sediment. It is also possible that SOC in the epipedon has been dissolved, translocated, and precipitated deeper in the profile, as Veenstra and Burras (2015) found in cultivated soils, which could lead to deepening of mollic colors. The toeslope near the basin center does not have an increased thickness of mollic colors like the other sites. Previous research by Millette et al. (1982) found that drainage and cultivation of organic soils can cause subsidence of up to 7 cm per year, so we presume that subsurface drainage has enabled rapid microbial decomposition of the histic epipedon, causing thinning of the darker horizons.

Soil Organic Carbon

Soil color is highly correlated with SOC, where darker colors indicate higher SOC content (Konen et al., 2003; Wills et al., 2007). Given this relationship, there are similarities between the maximum depth of mollic colors and the depth in the soil to <1% SOC across the catenas. For all four catenas, there is a general trend of increased depth to <1% SOC as one moves further from the summit (Fig. 11.9A), but the relationship is weak ($r^2 = 0.18$). Many other studies have found similar results of increasing SOC downslope (Ellerbrock et al., 2016; Khan et al., 2013; Kruczkowska, 2015; Moorman et al., 2004; Papiernik et al., 2005). Organic carbon content increases toward the toeslope for a number of reasons including removal of organic-rich sediment from upslope and subsequent deposition downslope, differences in plant productivity, increasing clay and microaggregate content that protects SOC from decomposition via fixation, a higher water table that inhibits aerobic microbial decomposition, and a shorter period of microbial activity throughout the year due to delayed soil warming in wetter areas (Doetterl et al., 2012; Moorman et al., 2004; Ritchie et al., 2007; Sahrawat, 2003; Song and Wang, 2014; Thompson and Bell, 1998; Van Vleck, 2011).

The Burras (Fig. 11.9B) ($r^2 = 0.98$), Walker (Fig. 11.9D) ($r^2 = 0.49$), and Steinwand (Fig. 11.9E) ($r^2 = 0.95$) sites individually have similar trends for depth to <1% SOC as they do for depth to maximum mollic colors across the hillslope due to the high correlation between soil color and SOC. However, depth to <1% SOC at the Konen site (Fig. 11.9C) ($r^2 = 0.56$) is much different than depth to maximum mollic colors. There has been considerable modern erosion at this site (Konen, 1999) which has buried the original soils on the lower landscape with sediment that contains lower amounts of SOC but still retains mollic colors. Therefore when examining the first depth with <1% SOC, the layer is very shallow across the whole Konen catena despite a buried soil with higher SOC.

The majority of the SOC is near the soil surface where plant roots grow, making it vulnerable to change through land use, so we also examined the weighted average percent SOC in the upper 30 cm. The combined, though weak, trend across all four catenas is an increase in percent SOC from the summit to the toeslope in the upper 30 cm (Fig. 11.10A) ($r^2 = 0.23$). The SOC in the summit ranges from 0.6% at the Steinwand site to 4% at the Walker site, which we attribute to differences in management such as crop rotation and tillage, along with associated erosion processes that affect SOC distribution. The larger catenas (Walker and Steinwand) have the highest SOC content in the basin center because of their almost lake-like conditions when the water table is high in wet years.

For each individual catena the relationship between the distance from the summit and weighted average percent SOC in the upper 30 cm is strong. At the Burras site (Fig. 11.10B) ($r^2 = 0.94$), percent SOC increases linearly from summit to toeslope. We speculate that LUM impacts are muted on the low relief landscape, leading to the linear trend. At the Konen site

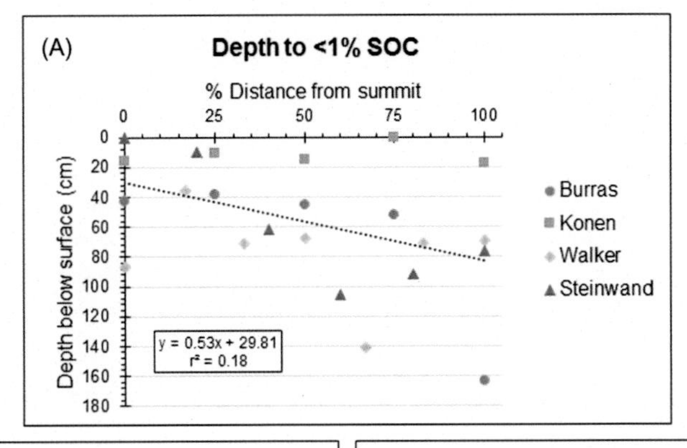

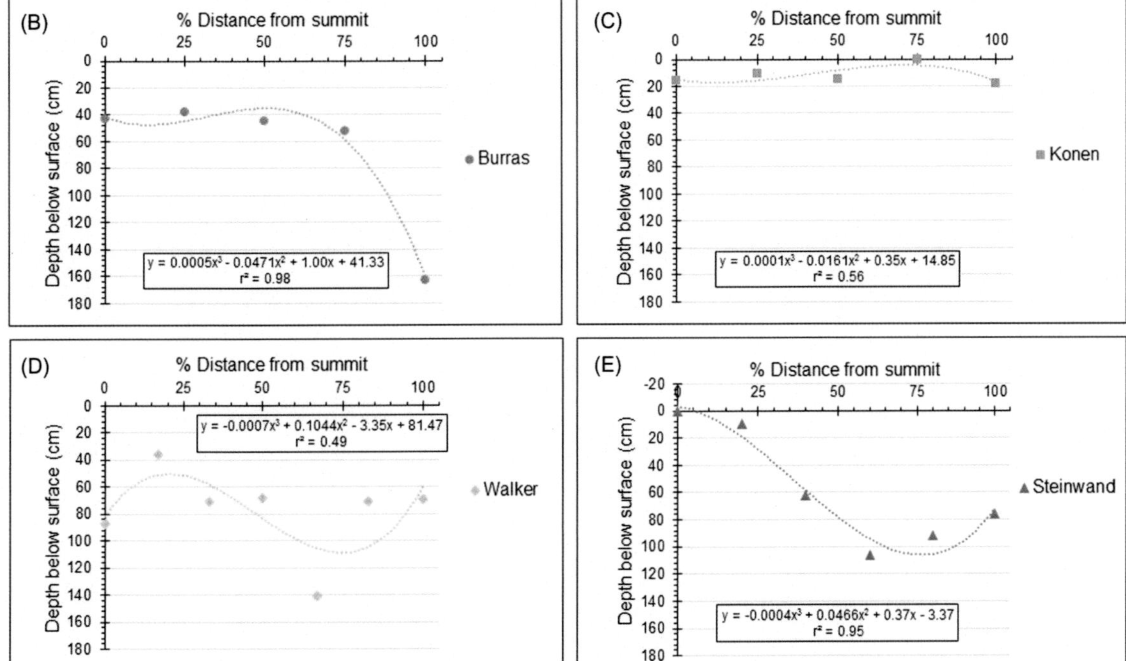

FIGURE 11.9 Current trends in SOC found across catenas on the Des Moines Lobe, Iowa. (A) Composite trendline for depth to less than 1% SOC across all four basins. (B) Trendline for depth to less than 1% SOC across Burras catena. (C) Trendline for depth to less than 1% SOC across Konen catena. (D) Trendline for depth to less than 1% SOC across Walker catena. (E) Trendline for depth to less than 1% SOC across Steinwand catena.

(Fig. 11.10C) ($r^2 = 0.97$), there is little change across the hillslope, due to the previously mentioned erosion and sedimentation that homogenized the distribution of SOC. At the Walker site (Fig. 11.10D) ($r^2 = 0.96$), there is a dramatic increase of SOC for the upper 30cm in the basin center, which is a relic of the bog history that facilitated formation of a Histosol with very poor drainage. The Steinwand site (Fig. 11.10E) ($r^2 = 0.99$) shows a trend in SOC that is a hybrid of the Burras and Walker sites, which was expected because its size and steepness fall between these basins.

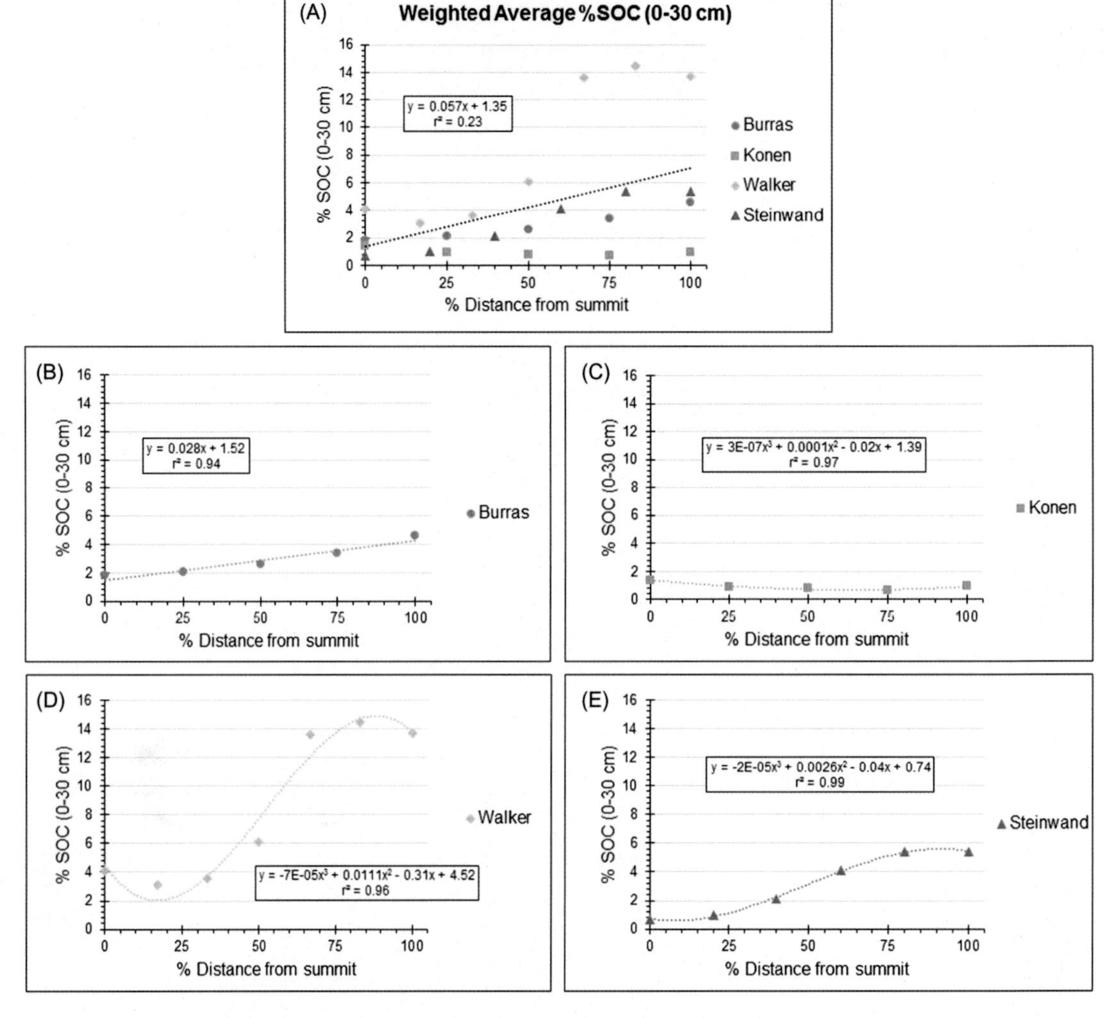

FIGURE 11.10 Current trends in SOC for upper 30 cm found across catenas on the Des Moines Lobe, Iowa. (A) Composite trendline for weighted average percent SOC in the upper 30 cm across all four basins. (B) Trendline for percent SOC in the upper 30 cm across Burras catena. (C) Trendline for percent SOC in the upper 30 cm across Konen catena. (D) Trendline for percent SOC in the upper 30 cm across Walker catena. (E) Trendline for percent SOC in the upper 30 cm across Steinwand catena.

Geometric Mean

Geometric mean increases as the amount of silt and sand increase, especially coarser sand fractions. Walker (1966) found that geometric mean decreases from the summit to the toeslope because of sediment sorting during erosion and sedimentation (Fig. 11.5). Cross-sections of the geometric mean particle size distribution for the four modern catenas are shown in Fig. 11.11. The upper hillslopes at the Burras and Steinwand sites have high geometric mean particle size (>100 μm) like that found

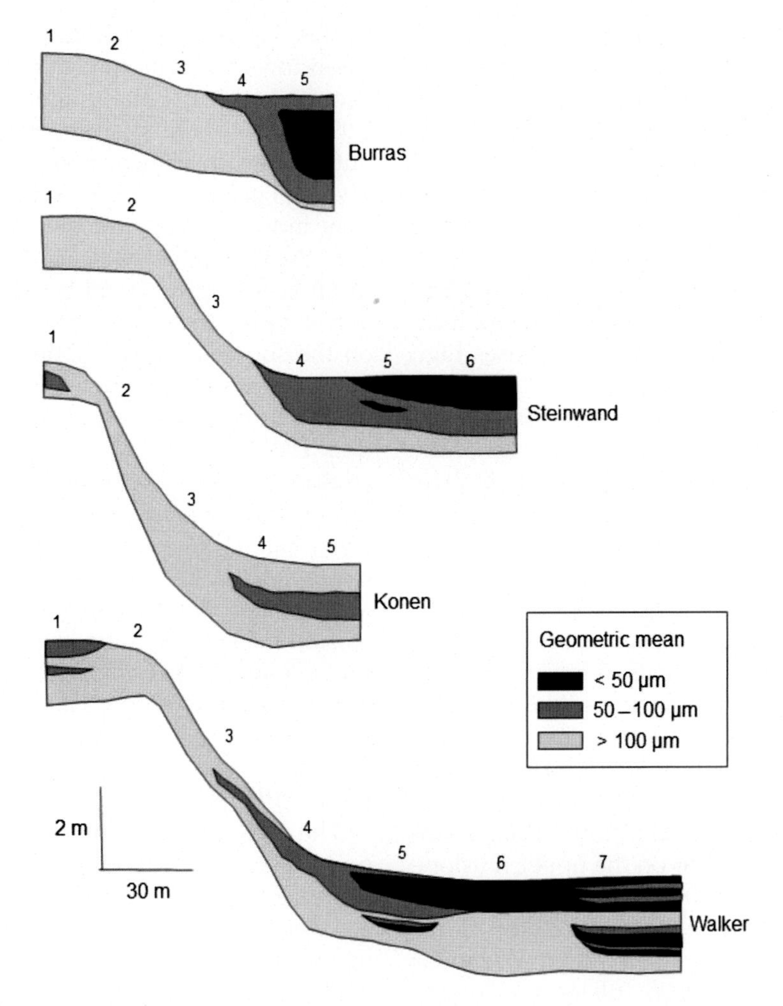

FIGURE 11.11 Cross-section of each catena showing geometric mean distribution. The numbers above the catenas are sampling locations along each transect.

in unsorted glacial till. The Konen and Walker sites have a high geometric mean at the summit as well, but they also have subsurface layers with a lower geometric mean (50–100 μm). These areas of lower geometric mean at the summit coincide with weathering and formation of clay-enriched zones. While geometric mean distribution is commonly associated with geomorphic processes, pedogenic processes such as those that lead to argillic horizon formation at the summit can also affect geometric mean. A summit with lower geometric mean can also occur with loess deposition. In the case of the Walker site, all of the factors are likely.

Basin size appears to have an effect on geometric mean distribution across the catenas. At the Burras and Konen sites, which are smaller basins, coarser sediment has been deposited over the finer sediment in the basin centers. The Konen site, which has higher relief with slightly sandier parent material, has much coarser basin sediments compared to the Burras site as illustrated by the higher geometric mean. Bands of coarser deposits also occur in the basin of the Walker site intermittently.

We attribute higher geometric mean in basin centers to episodic high erosion rates induced by agriculture. The high variability both inter and intrabasin is likely caused by differences in sandy deposits from sheet and rill erosion. In particular, rill erosion causes development of small alluvial fans that vary in size and location with each rainfall event. This causes high variability in both location and extent of coarse textured deposits, as illustrated by the Walker geometric mean cross section at the base of the hill. Heavy rainfall events coupled with bare soil increase overland flow for a given event and subsequently increase the likelihood of rill erosion and alluvial fan development.

Another factor in particle size distribution is the type of causative agent contributing to erosion. Two primary agents of erosion in agriculture are water and tillage. Sediment moved by water is sorted such that finer particles (e.g., clay and organic matter) travel further downslope, but sediment moved via tillage is unsorted (Zheng-An et al., 2010). Tillage erosion is usually dominant on the upper hillslope (Zheng-An et al., 2010) and facilitates subsequent water erosion (Wang et al., 2016). While both types of erosion often occur at similar rates (Van Oost et al., 2006), tillage erosion can easily outpace water erosion and cause the majority of soil redistribution in a field (Chartin et al., 2013).

In the upper 30 cm of the catenas, there is a weak trend of decreasing geometric mean as one moves further from the summit when using data from all four sites (Fig. 11.12A) (r^2 = 0.16). This is supported by Boling et al. (2008) as well as Khan et al. (2013), who found that clay content increases downslope due to erosion and sedimentation processes, which would lead to a decrease in geometric mean. The individual sites exhibit strong trends between distance from the summit and geometric mean. The Burras site has a linear decrease in geometric mean toward the basin center, with the highest geometric mean at the summit (Fig. 11.12B) (r^2 = 0.99). The Steinwand site shows a similar trend but is more curvilinear (Fig. 11.12E) (r^2 = 0.99). The high geometric mean at the summit of these sites indicates that extensive erosion is occurring. At the Walker site, the highest geometric mean in the upper 30 cm is found at the shoulder as opposed to the summit, due to erosion on the shoulder exposing till with higher geometric mean and pedogenesis/loess deposits on the summit decreasing the geometric mean (Fig. 11.12D) (r^2 = 0.98). The Konen site has an increase in geometric mean from the summit to the toeslope, which is the opposite of the other sites (Fig. 11.12C) (r^2 = 0.96). There has been weathering at the summit, as indicated by the formation of an argillic horizon, along with high erosion rates in the basin. The erosion has been severe enough that coarser sediment has been transported downslope, leading to a higher geometric mean.

Bulk Density

The weighted average bulk density for the upper 30 cm across multiple sites tends to decrease with distance from the summit (Fig. 11.13A) (r^2 = 0.10). When looking at the individual basins, the Burras (Fig. 11.13B) (r^2 = 0.94), Walker (Fig. 11.13D) (r^2 = 0.98), and Steinwand (Fig. 11.13E) (r^2 = 0.99) sites exhibit a strong trend of decreasing bulk density with distance from the summit, with linear to curvilinear distribution. Gregorich and Anderson (1985) found similar trends of decreasing bulk density across cultivated catenas, as did Malo et al. (1974) and Khan et al. (2013). The decrease in bulk density across catenas is attributed to higher amounts of SOC, which is low in density, along with increases in clay, both of which aid in the formation of stable aggregates that increase resiliency of soil structure and preserve soil porosity, thereby decreasing bulk density. Additionally the effects of wetting and drying

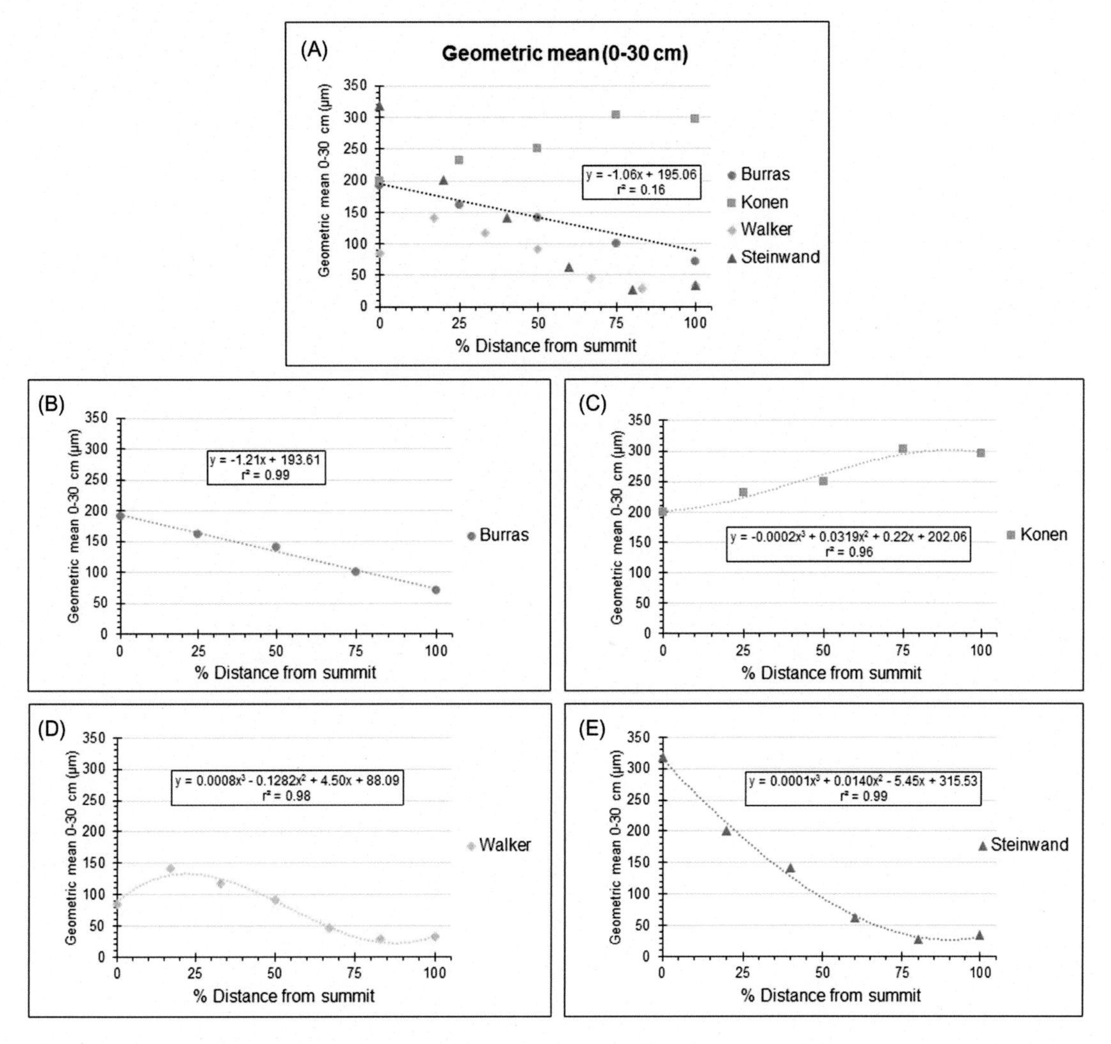

FIGURE 11.12 Current trends in geometric mean (0–30 cm) found across catenas on the Des Moines Lobe, Iowa. (A) Composite trendline for geometric mean in upper 30 cm across all four basins. (B) Trendline for geometric mean in upper 30 cm across Burras catena. (C) Trendline for geometric mean in upper 30 cm across Konen catena. (D) Trendline for geometric mean in upper 30 cm across Walker catena. (E) Trendline for geometric mean in upper 30 cm across Steinwand catena.

cycles are amplified with increasing amounts of shrink-swell clays, which contributes to lower bulk density. At the Konen site (Fig. 11.13C) (r^2 = 0.95), bulk densities increase toward the basin center. There is an increase in geometric mean (i.e., more sand and less clay) and no increase in SOC within the upper 30 cm across the catena, which is the opposite of that found at the other three sites. These factors could contribute to low aggregate stability and make the area vulnerable to compaction. The sandy loam texture of soils on the lower hillslope at the Konen site is also naturally more prone to compaction because of the possibility of tighter

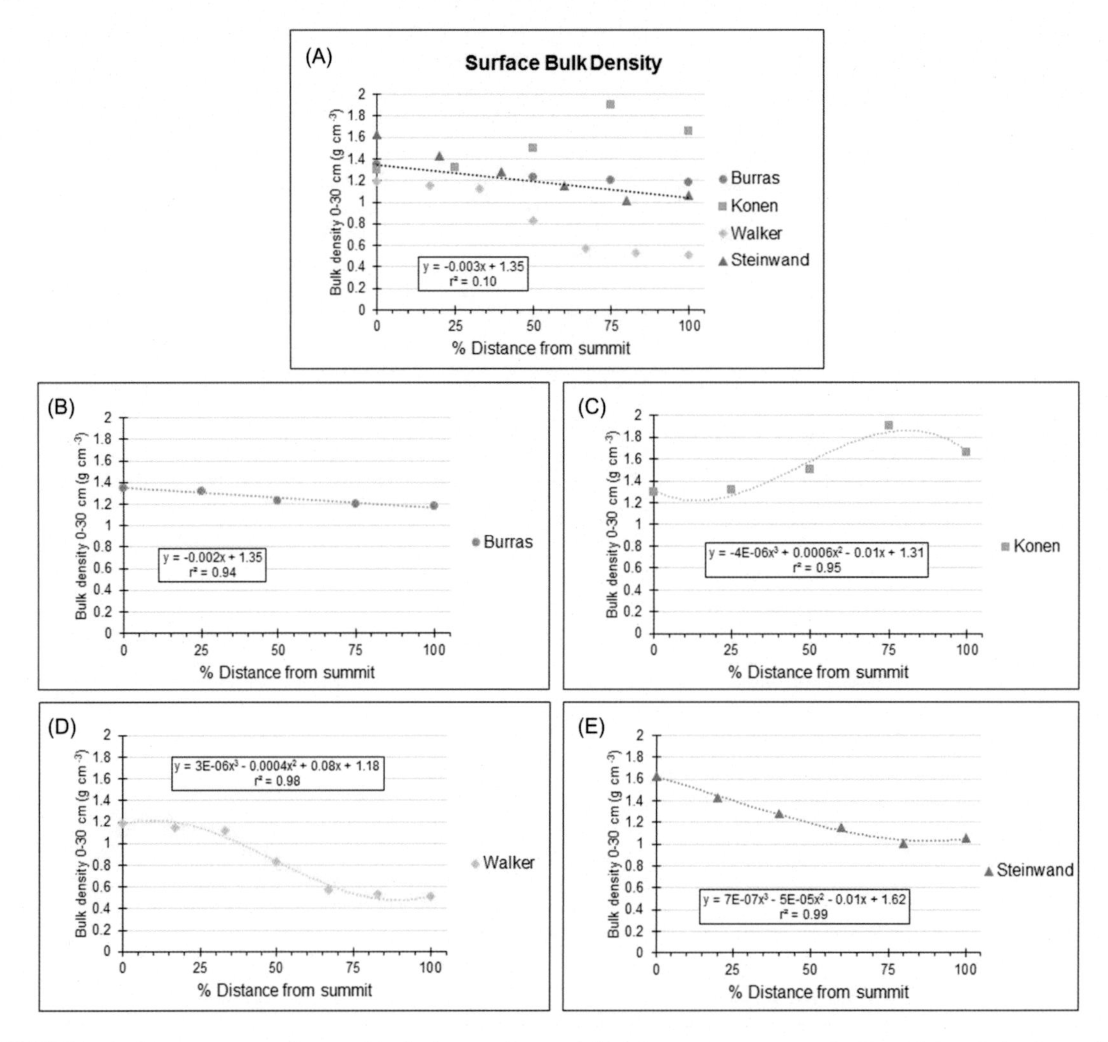

FIGURE 11.13 Current trends in soil bulk density (0–30 cm) found across catenas on the Des Moines Lobe, Iowa. (A) Composite trendline for bulk density in upper 30 cm across all four basins. (B) Trendline for bulk density in upper 30 cm across Burras catena. (C) Trendline for bulk density in upper 30 cm across Konen catena. (D) Trendline for bulk density in upper 30 cm across Walker catena. (E) Trendline for bulk density in upper 30 cm across Steinwand catena.

packing given the size of the particles involved (Schaetzl and Anderson, 2005).

Ramifications for Soil Mapping

To reevaluate the preceding tables and figures in contemporary geospatially explicit mapping language, the results herein demonstrate that covariate properties are distributed in vertical and horizontal patterns specific to their individual catenas. As a result, validating a new map via the integration of legacy maps with digital elevation models, light detection and ranging, and other geospatially intense modern data will be difficult and require perhaps unexpectedly greater groundtruthing than normal for such a well-studied and mapped soilscape. This is demonstrated by qualitatively examining any and all of the covariate (i.e., individual property) distributions across our catenas (Figs. 11.7–11.13) and reflecting upon how those properties created the soil series and classification challenges we presented (Table 11.6). The US Department of Agriculture Natural Resources Conservation Service (USDA-NRCS) maps and the original historic scientific findings are geospatially different from our findings, with the degree of difference varying basin by basin. This means a next generation map will need to uniquely weight its reliance on legacy data and accept less consistent abilities to map and verify modern soil series and individual property distributions.

More explicitly, our results demonstrate a decoupling of covariate relationships in each of our basins. This is important given that mappers (past and modern) on the Des Moines Lobe of Iowa, Minnesota, and South Dakota have relied on a consistency between closed-basin landscape position, slope steepness, SOC content, water table depth (or at least depth to redoximorphic features), clay content, and type of strata as they identified soil series and calculated masses of SOC per unit area, etc. Those consistencies no longer exist, which means legacy maps are becoming less accurate. On a positive note, we were able to explain property (covariate) distributions within each individual basin. As a result, modern mapping techniques will be developed that are able to remap all the soilscapes of the Des Moines Lobe. We are just not sure what relationships will have to be developed.

CONCLUSIONS

The catena model is used to predict the distribution of soil properties across landscapes and is therefore central to soil mapping. It has been historically validated in many locations. As a result, countries worldwide rely on the catena for creating initial soil maps and map updates. However, we found soil distribution in Iowa is hillslope-by-LUM dependent. Combining data from multiple hillslopes results in a relatively weak soil–landscape relationship, which means that extrapolating from one hillslope to the next is problematic. This appears to be especially true in areas of intensive cropping that include tile drainage, heavy use of fertilizers, and significant tillage and erosion. In other words, differences in management history from one location to the next make soils more unpredictable across the landscape as catena functioning diverges. This indicates ongoing field work is especially important to the creation of precision soil maps. It further indicates those precision maps will need to rely upon a new catena model that integrates human impacts vis-à-vis pedology, hydrology, and geomorphology.

References

Afshar, F.A., Ayoubi, S., Jalalian, A., 2010. Soil redistribution rate and its relationship with soil organic carbon and total nitrogen using 137 Cs technique in a cultivated complex hillslope in western Iran. J. Environ. Radioactiv. 101 (8), 606–614. http://dx.doi.org/10.1016/j.jenvrad.2010.03.008.

Barak, P., Jobe, B.O., Krueger, A.R., Peterson, L.A., Laird, D.A., 1997. Effects of long-term soil acidification due to nitrogen fertilizer inputs in Wisconsin. Plant Soil 197 (1), 61–69. http://dx.doi.org/10.1023/A:1004297607070.

Blake, L., Goulding, K.W.T., Mott, C.J.B., Johnston, A.E., 1999. Changes in soil chemistry accompanying acidification over more than 100 years under woodland and grass at Rothamsted Experimental Station, UK. Eur. J. Soil Sci. 50 (3), 401–412. http://dx.doi.org/10.1046/j.1365-2389.1999.00253.x.

Blanco-Canqui, H., Stone, L.R., Stahlman, P.W., 2010. Soil response to long-term cropping systems on an Argiustoll in the central Great Plains. Soil Sci. Soc. Am. J. 74 (2), 602–611. http://dx.doi.org/10.2136/sssaj2009.0214.

Blevins, R.L., Thomas, G.W., Cornelius, P.L., 1977. Influence of no-tillage and nitrogen fertilization on certain soil properties after 5 years of continuous corn. Agron. J. 69

(3), 383–386. http://dx.doi.org/10.2134/agronj1977.00021962006900030013x.

Boling, A.A., Tuong, T.P., Suganda, H., Konboon, Y., Harnpichitvitaya, D., Bouman, B.A.M., et al., 2008. The effect of toposequence position on soil properties, hydrology, and yield of rainfed lowland rice in Southeast Asia. Field Crop. Res. 106 (1), 22–33. http://dx.doi.org/10.1016/j.fcr.2007.10.013.

Bouma, J., Hole, F.D., 1971. Soil structure and hydraulic conductivity of adjacent virgin and cultivated pedons at two sites: a Typic Argiudoll (silt loam) and a Typic Eutrochrept (clay). Soil Sci. Soc. Am. J. 35 (2), 316–319. http://dx.doi.org/10.2136/sssaj1971.03615995003500020039x.

Bouman, O.T., Curtin, D., Campbell, C.A., Biederbeck, V.O., Ukrainetz, H., 1995. Soil acidification from long-term use of anhydrous ammonia and urea. Soil Sci. Soc. Am. J. 59 (5), 1488–1494. http://dx.doi.org/10.2136/sssaj1995.03615995005900050039x.

Bowman, R.A., Halvorson, A.D., 1998. Soil chemical changes after nine years of differential N fertilization in a no-till dryland wheat-corn-fallow rotation. Soil Sci. 163 (3), 241–247. http://dx.doi.org/10.1097/00010694-199803000-00009.

Bowman, R.A., Reeder, J.D., Lober, R.W., 1990. Changes in soil properties in a central plains rangeland soil after 3, 20, and 60 years of cultivation. Soil Sci. 150 (6), 851–857.

Bracken, L.J., Turnbull, L., Wainwright, J., Bogaart, P., 2015. Sediment connectivity: a framework for understanding sediment transfer at multiple scales. Earth Surf. Proc. Land. 40 (2), 177–188. http://dx.doi.org/10.1002/esp.3635.

Burras, C.L., Scholtes, W.H., 1987. Basin properties and postglacial erosion rates of minor moraines in Iowa. Soil Sci. Soc. Am. J. 51 (6), 1541–1547. http://dx.doi.org/10.2136/sssaj1987.03615995005100060025x.

California Soil Resource Lab, 2016. Series Extent Explorer. University of California, Davis, CA. Available from: http://casoilresource.lawr.ucdavis.edu/. (accessed 04.01.16).

Chartin, C., Evrard, O., Salvador-Blanes, S., Hinschberger, F., Van Oost, K., Lefèvre, I., et al., 2013. Quantifying and modelling the impact of land consolidation and field borders on soil redistribution in agricultural landscapes (1954–2009). Catena 110, 184–195. http://dx.doi.org/10.1016/j.catena.2013.06.006.

Chendev, Y.G., Burras, C.L., Sauer, T.J., 2012. Transformation of forest soils in Iowa (United States) under the impact of long-term agricultural development. Eurasian Soil Sci. 45 (4), 357–367. http://dx.doi.org/10.1134/S1064229312040035.

Coote, D.R., Ramsey, J.F., 1983. Quantification of the effects of over 35 years of intensive cultivation on four soils. Can. J. Soil Sci. 63 (1), 1–14. http://dx.doi.org/10.4141/cjss83-001.

Davidson, E.A., Ackerman, I.L., 1993. Changes in soil carbon inventories following cultivation of previously untilled soils. Biogeochemistry 20 (3), 161–193. http://dx.doi.org/10.1007/BF00000786.

De Alba, S., Lindstrom, M., Schumacher, T.E., Malo, D.D., 2004. Soil landscape evolution due to soil redistribution by tillage: a new conceptual model of soil catena evolution in agricultural landscapes. Catena 58 (1), 77–100. http://dx.doi.org/10.1016/j.catena.2003.12.004.

Detty, J.M., McGuire, K.J., 2010. Topographic controls on shallow groundwater dynamics: implications of hydrologic connectivity between hillslopes and riparian zones in a till mantled catchment. Hydrol. Proc 24 (16), 2222–2236. http://dx.doi.org/10.1002/hyp.7656.

Doetterl, S., Six, J., Van Wesemael, B., Van Oost, K., 2012. Carbon cycling in eroding landscapes: geomorphic controls on soil organic C pool composition and C stabilization. Glob. Change Biol. 18 (7), 2218–2232. http://dx.doi.org/10.1111/j.1365-2486.2012.02680.x.

Eidem, J.M., Simpkins, W.W., Burkart, M.R., 1999. Geology, groundwater flow, and water quality in the Walnut Creek watershed. J. Environ. Qual. 28 (1), 60–69. http://dx.doi.org/10.2134/jeq1999.00472425002800010006x.

Ellerbrock, R.H., Gerke, H.H., Deumlich, D., 2016. Soil organic matter composition along a slope in an erosion-affected arable landscape in North East Germany. Soil Till. Res. 156, 209–218. http://dx.doi.org/10.1016/j.still.2015.08.014.

Fortner, S.K., Lyons, W.B., Carey, A.E., Shipitalo, M.J., Welch, S.A., Welch, K.A., 2012. Silicate weathering and CO_2 consumption within agricultural landscapes, the Ohio-Tennessee River Basin, USA. Biogeosciences 9 (3), 941–955. http://dx.doi.org/10.5194/bgd-8-9431-2011.

Frielinghaus, M., Bork, H.R., Kocmit, A., Schmidt, R., 1998. Characterization of soil change by erosion in the German-Polish moraine regions. In: Ruellan, A., Jarnagne, M. (Eds.), Proc. 16th World Conf. Soil Sci., Montpellier, France, pp. 20–25.

Fryirs, K.A., Brierley, G.J., Preston, N.J., Kasai, M., 2007. Buffers, barriers and blankets: the (dis) connectivity of catchment-scale sediment cascades. Catena 70 (1), 49–67. http://dx.doi.org/10.1016/j.catena.2006.07.007.

Grandy, A.S., Robertson, G.P., 2006. Aggregation and organic matter protection following tillage of a previously uncultivated soil. Soil Sci. Soc. Am. J. 70 (4), 1398–1406. http://dx.doi.org/10.2136/sssaj2005.0313.

Gregorich, E.G., Anderson, D.W., 1985. Effects of cultivation and erosion on soils of four toposequences in the Canadian prairies. Geoderma 36 (3), 343–354. http://dx.doi.org/10.1016/0016-7061(85)90012-6.

Guo, J.H., Liu, X.J., Zhang, Y., Shen, J.L., Han, W.X., Zhang, W.F., et al., 2010. Significant acidification in major Chinese croplands. Science 327 (5968), 1008–1010. http://dx.doi.org/10.1126/science.1182570.

Guo, L.B., Gifford, R.M., 2002. Soil carbon stocks and land use change: a meta analysis. Glob. Change Biol. 8 (4), 345–360. http://dx.doi.org/10.1046/j.1354-1013.2002.00486.x.

Hayes, W.A., Vepraskas, M.J., 2000. Morphological changes in soils produced when hydrology is altered by ditching. Soil Sci. Soc. Am. J. 64 (5), 1893–1904. http://dx.doi.org/10.2136/sssaj2000.6451893x.

Helmers, M., Christianson, R., Brenneman, G., Lockett, D., Pederson, C., 2012. Water table, drainage, and yield response to drainage water management in southeast Iowa. J. Soil Water Conserv. 67 (6), 495–501. http://dx.doi.org/10.2489/jswc.67.6.495.

Helms, D., 1990. Conserving the plains: the soil conservation service in the Great Plains. Agric. Hist. 64 (2), 58–73. Stable URL: http://www.jstor.org/stable/3743797.

Hillel, D., 2004. Introduction to Environmental Soil Physics. Academic Press, USA.

Houben, P., 2008. Scale linkage and contingency effects of field-scale and hillslope-scale controls of long-term soil erosion: anthropogeomorphic sediment flux in agricultural loess watersheds of Southern Germany. Geomorphology 101 (1), 172–191. http://dx.doi.org/10.1016/j.geomorph.2008.06.007.

Ibrahim, M.A., Burras, C.L., 2012. Clay movement in sand columns and its pedological ramifications. Soil Horizons 53 (2), 27–31. http://dx.doi.org/10.2136/sh12-01-0004.

Ibrahim, M.A., Elnaka, E.A., Burras, C.L., 2015. Clay upward movement in sand columns under partially saturated conditions. Soil Sci. Soc. Am. J. 79 (3), 896–902. http://dx.doi.org/10.2136/sssaj2015.01.0005.

Indorante, S.J., Kabrick, J.M., Lee, B.D., Maatta, J.M., 2014. Quantifying soil profile change caused by land use in central Missouri loess hillslopes. Soil Sci. Soc. Am. J. 78 (1), 225–237. http://dx.doi.org/10.2136/sssaj2013.07.0285.

Jenny, H., 1941. Factors of Soil Formation: A System of Quantitative Pedology. Macgraw Hill, New York.

Karlen, D.L., Dinnes, D.L., Singer, J.W., 2010. Midwest soil and water conservation: past, present, and future. In: Soil and Water Conservation Advances in the US: Past Efforts—Future Outlook. Soil Sci. Soc. Am., Inc., Madison, WI.

Khan, F., Hayat, Z., Ahmad, W., Ramzan, M., Shah, Z., Sharif, M., et al., 2013. Effect of slope position on physico-chemical properties of eroded soil. Soil Environ. 32 (1), 22–28.

Kladivko, E.J., Van Scoyoc, G.E., Monke, E.J., Oates, K.M., Pask, W., 1991. Pesticide and nutrient movement into subsurface tile drains on a silt loam soil in Indiana. J. Environ. Qual. 20 (1), 264–270. http://dx.doi.org/10.2134/jeq1991.00472425002000010043x.

Konen, M.E., 1999. Human Impacts on Soils and Geomorphic Processes on the Des Moines Lobe, Iowa. PhD Diss., Iowa State University, Ames, IA. < http://lib.dr.iastate.edu/rtd/12143> .

Konen, M.E., Burras, C.L., Sandor, J.A., 2003. Organic carbon, texture, and quantitative color measurement relationships for cultivated soils in north central Iowa. Soil Sci. Soc. Am. J. 67 (6), 1823–1830. http://dx.doi.org/10.2136/sssaj2003.1823.

Konen, M.E., Jacobs, P.M., Burras, C.L., Talaga, B.J., Mason, J.A., 2002. Equations for predicting soil organic carbon using loss-on-ignition for north central US soils. Soil Sci. Soc. Am. J. 66 (6), 1878–1881. http://dx.doi.org/10.2136/sssaj2002.1878.

Kreznor, W.R., Olson, K.R., Jones, R.L., Johnson, D.L., 1990. Quantification of postsettlement deposition in a northwestern Illinois sediment basin. Soil Sci. Soc. Am. J. 54 (5), 1393–1401. http://dx.doi.org/10.2136/sssaj1990.03615995005400050031x.

Kruczkowska, B., 2015. The use of kettle holes for reconstructing former soil cover in different types of land use. Geogr. Pol. 89 (3) http://dx.doi.org/10.7163/GPol.0060.

Logsdon, S.D., 2007. Subsurface lateral transport in glacial till soils. T. ASABE 50 (3), 875–883. http://dx.doi.org/10.13031/2013.23152.

Logsdon, S.D., Hernandez-Ramirez, G., Hatfield, J.L., Sauer, T.J., Prueger, J.H., Schilling, K.E., 2009. Soil water and shallow groundwater relations in an agricultural hillslope. Soil Sci. Soc. Am. J. 73 (5), 1461–1468. http://dx.doi.org/10.2136/sssaj2008.0385.

Malo, D.D., Worcester, B.K., Cassel, D.K., Matzdorf, K.D., 1974. Soil–landscape relationships in a closed drainage system. Soil Sci. Soc. Am. J. 38 (5), 813–818. http://dx.doi.org/10.2136/sssaj1974.03615995003800050034x.

Mann, L.K., 1985. A regional comparison of carbon in cultivated and uncultivated Alfisols and Mollisols in the central United States. Geoderma 36 (3–4), 241–253. http://dx.doi.org/10.1016/0016-7061(85)90005-9.

Masselink, R.J., Keesstra, S.D., Temme, A.J., Seeger, M., Giménez, R., Casalí, J., 2016. Modelling discharge and sediment yield at catchment scale using connectivity components. Land Degrad. Dev. 27 (4), 933–945. http://dx.doi.org/10.1002/ldr.2512.

McGuire, K.J., McDonnell, J.J., 2010. Hydrological connectivity of hillslopes and streams: characteristic time scales and nonlinearities. Water Resour. Res. 46 (10) http://dx.doi.org/10.1029/2010WR009341.

Mercier, P., Denaix, L., Robert, M., de Marsily, G., 2000. Colloid transfer by subsurface drainage. Comptes

Rendus de l'Academie des Sciences Series IIA Earth Planet. Sci. 3 (331), 195–202.

Miller, G.A., Sassman, A.,M., Burras, C.L., 2016. Iowa soil properties and interpretations database, ISPAID Version 8.1. Iowa State University, Ames, IA, Available from: http://www.extension.iastate.edu/soils/ispaid. (accessed 04.04.16).

Millette, J.A., Vigier, B., Broughton, R.S., 1982. An evaluation of the drainage and subsidence of some organic soils in Quebec. Can. Agric. Eng. 24, 5–10.

Milne, G., 1935. Some suggested units of classification and mapping particularly for East African soils. Soils Res 4 (3), 183–198.

Milne, G., 1936. Normal erosion as a factor in soil profile development. Nature 138, 548–549. http://dx.doi.org/10.1038/138548c0.

Mokma, D.L., Fenton, T.E., Olson, K.R., 1996. Effect of erosion on morphology and classification of soils in the north central United States. J. Soil Water Conserv. 51 (2), 171–175.

Montagne, D., Cornu, S., Le Forestier, L., Hardy, M., Josière, O., Caner, L., et al., 2008. Impact of drainage on soil-forming mechanisms in a French Albeluvisol: input of mineralogical data in mass-balance modelling. Geoderma 145 (3), 426–438. http://dx.doi.org/10.1016/j.geoderma.2008.02.005.

Moorman, T.B., Cambardella, C.A., James, D.E., Karlen, D.L., Kramer, L.A., 2004. Quantification of tillage and landscape effects on soil carbon in small Iowa watersheds. Soil Till. Res. 78 (2), 225–236. http://dx.doi.org/10.1016/j.still.2004.02.014.

Norton, L.D., 1986. Erosion-sedimentation in a closed drainage basin in northwest Indiana. Soil Sci. Soc. Am. J. 50 (1), 209–213. http://dx.doi.org/10.2136/sssaj1986.03615995005000010040x.

Papiernik, S.K., Lindstrom, M.J., Schumacher, J.A., Farenhorst, A., Stephens, K.D., Schumacher, T.E., et al., 2005. Variation in soil properties and crop yield across an eroded prairie landscape. J. Soil Water Conserv. 60 (6), 388–395.

Papiernik, S.K., Lindstrom, M.J., Schumacher, T.E., Schumacher, J.A., Malo, D.D., Lobb, D.A., 2007. Characterization of soil profiles in a landscape affected by long-term tillage. Soil Till. Res. 93 (2), 335–345. http://dx.doi.org/10.1016/j.still.2006.05.007.

Pennock, D.J., 2003. Multi-site assessment of cultivation-induced soil change using revised landform segmentation procedures. Can. J. Soil Sci. 83 (5), 565–580. http://dx.doi.org/10.4141/S03-010.

Phillips, J.D., Gares, P.A., Slattery, M.C., 1999. Agricultural soil redistribution and landscape complexity. Landsc. Ecol. 14 (2), 197–211. http://dx.doi.org/10.1023/A:1008024213440.

Redmond, C.E., McClelland, J.E., 1959. The occurrence and distribution of lime in calcium carbonate Solonchak and associated soils of eastern North Dakota. Soil Sci. Soc. Am. J. 23 (1), 61–65. http://dx.doi.org/10.2136/sssaj1959.03615995002300010024x.

Ritchie, J.C., McCarty, G.W., Venteris, E.R., Kaspar, T.C., 2007. Soil and soil organic carbon redistribution on the landscape. Geomorphology 89 (1), 163–171. http://dx.doi.org/10.1016/j.geomorph.2006.07.021.

Ruhe, R.V., Daniels, R.B., 1965. Landscape erosion-geologic and historic. J. Soil Water Cons. 20, 52–57.

Ruhe, R.V., Walker, P.H., 1968. Hillslope models and soil formation. I. Open systems. Int. Soc. Soil Sci. Trans. 4, 551–560.

Ruhe, R.V., 1967. Geomorphic Surfaces and Surficial Deposits in Southern New Mexico. Memoir 18. New Mexico Bureau of Mines and Mineral Resources, Socorro, NM, USA.

Ruhe, R.V., Scholtes, W.H., 1959. Important elements in the classification of the Wisconsin glacial stage: A discussion. J. Geol. 67 (5), 585–593. Stable URL: http://www.jstor.org/stable/30056117.

Sahrawat, K.L., 2003. Organic matter accumulation in submerged soils. Adv. Agron. 81, 169–201. http://dx.doi.org/10.1016/S0065-2113(03)81004-0.

Sandercock, P.J., Hooke, J.M., 2011. Vegetation effects on sediment connectivity and processes in an ephemeral channel in SE Spain. J. Arid Environ. 75 (3), 239–254. http://dx.doi.org/10.1016/j.jaridenv.2010.10.005.

Saviozzi, A., Levi-Minzi, R., Cardelli, R., Riffaldi, R., 2001. A comparison of soil quality in adjacent cultivated, forest and native grassland soils. Plant Soil 233 (2), 251–259. http://dx.doi.org/10.1023/A:1010526209076.

Schaetzl, R.J., Anderson, S., 2005. Soils: Genesis and Geomorphology. Cambridge University Press, New York.

Schilling, K.E., Helmers, M., 2008. Effects of subsurface drainage tiles on streamflow in Iowa agricultural watersheds: exploratory hydrograph analysis. Hydrol. Process. 22 (23). http://dx.doi.org/4497-4506.10.1002/hyp.7052.

Schoeneberger, P.J., Wysocki, D.A., Benham, E.C., Broderson, W.D., 2002. Field Book for Describing and Sampling Soils, Version 2.0. Natural Resources Conservation Service, National Soil Survey Center, Lincoln, NE.

Schwab, A.P., Ransom, M.D., Owensby, C.E., 1989. Exchange properties of an Argiustoll: effects of long-term ammonium nitrate fertilization. Soil Sci. Soc. Am. J. 53 (5), 1412–1417. http://dx.doi.org/10.2136/sssaj1989.03615995005300050018x.

Sherrod, L.A., Erskine, R.H., Green, T.R., 2015. Spatial patterns and cross-correlations of temporal changes in soil carbonates and surface elevation in a winter wheat–fallow cropping system. Soil Sci. Soc. Am. J. 79 (2), 417–427. http://dx.doi.org/10.2136/sssaj2014.05.0222.

Simonson, R.W., Riecken, F.F., Smith, G.D., 1953. Understanding Iowa Soils. Wm. C. Brown Co., Dubuque, IA.

Smith, D.R., King, K.W., Johnson, L., Francesconi, W., Richards, P., Baker, D., et al., 2015. Surface runoff and tile drainage transport of phosphorus in the Midwestern United States. J. Environ. Qual. 44 (2), 495–502. http://dx.doi.org/10.2134/jeq2014.04.0176.

Soil Conservation Staff, 2015. 80 years helping people help the land: a brief history of NRCS. Natural Resources Conservation Service, United States Department of Agriculture. Available from: http://www.nrcs.usda.gov/wps/portal/nrcs/detail/national/about/history/?cid=nrcs143_021392 (accessed 30.08.16).

Soil Conservation Staff, 2016. Conservation practices. National conservation practice standards. Natural Resources Conservation Service, United States Department of Agriculture. Available from: http://www.nrcs.usda.gov/wps/portal/nrcs/detailfull/national/technical/cp/ncps/?cid=nrcs143_026849 (accessed 30.08.16).

Soil Survey Division Staff, 1993. Soil Survey Manual. Handbook no. 18. Natural Resources Conservation Service, United States Department of Agriculture.

Soil Survey Staff, 1996. Soil Survey Laboratory Methods Manual. Soil Survey Investigations Report No. 42 version 3.0. Natural Resources Conservation Service, United States Department of Agriculture.

Soil Survey Staff, 2007. From the Ground Down: An Introduction to Iowa Soil Surveys. Natural Resources Conservation Service, United States Department of Agriculture. Available from: http://www.nrcs.usda.gov/Internet/FSE_DOCUMENTS/nrcs142p2_006038.pdf. (accessed 30.08.16).

Soil Survey Staff, 2016. Natural Resources Conservation Service, United States Department of Agriculture. Web Soil Survey, Official Soil Series Descriptions. Available from: http://www.nrcs.usda.gov/wps/portal/nrcs/detail/soils/survey/geo/?cid=nrcs142p2_053587. (accessed 03.03.16).

Song, Z., Müller, K., Wang, H., 2014. Biogeochemical silicon cycle and carbon sequestration in agricultural ecosystems. Earth-Sci. Rev. 139, 268–278. http://dx.doi.org/10.1016/j.earscirev.2014.09.009.

Steinwand, A.L., Fenton, T.E., 1995. Landscape evolution and shallow groundwater hydrology of a till landscape in central Iowa. Soil Sci. Soc. Am. J. 59, 1370–1377. http://dx.doi.org/10.2136/sssaj1995.03615995005900050025x.

Theocharopoulos, S.P., Florou, H., Walling, D.E., Kalantzakos, H., Christou, M., Tountas, P., et al., 2003. Soil erosion and deposition rates in a cultivated catchment area in central Greece, estimated using the 137Cs technique. Soil Till. Res. 69 (1), 153–162. http://dx.doi.org/10.1016/S0167-1987(02)00136-8.

Thompson, J.A., Bell, J.C., 1998. Hydric conditions and hydromorphic properties within a Mollisol catena in southeastern Minnesota. Soil Sci. Soc. Am. J. 62 (4), 1116–1125. http://dx.doi.org/10.2136/sssaj1998.03615995006200040037x.

Turnage, K.M., Lee, S.Y., Foss, J.E., Kim, K.H., Larsen, I.L., 1997. Comparison of soil erosion and deposition rates using radiocesium, RUSLE, and buried soils in dolines in East Tennessee. Environ. Geol. 29 (1-2), 1–10. http://dx.doi.org/10.1007/s002540050097.

Van Oost, K., Govers, G., De Alba, S., Quine, T.A., 2006. Tillage erosion: a review of controlling factors and implications for soil quality. Prog. Phys. Geog. 30 (4), 443–466. http://dx.doi.org/10.1191/0309133306pp487ra.

Van Vleck, H.E., 2011. Impacts of Agricultural Management and Landscape Factors on Soil Carbon and Nitrogen Cycling. Ph.D. Diss., University of Minnesota. Stable URL: http://hdl.handle.net/11299/162273.

Veenstra, J.J., Burras, C.L., 2012. Effects of agriculture on the classification of Black soils in the Midwestern United States. Can. J. Soil Sci. 92 (3), 403–411. http://dx.doi.org/10.4141/cjss2010-018.

Veenstra, J.J., Burras, C.L., 2015. Soil profile transformation after 50 years of agricultural land use. Soil Sci. Soc. Am. J. 79 (4), 1154–1162. http://dx.doi.org/10.2136/sssaj2015.01.0027.

Voorhees, W.B., Lindstrom, M.J., 1984. Long-term effects of tillage method on soil tilth independent of wheel traffic compaction. Soil Sci. Soc. Am. J. 48 (1), 152–156. http://dx.doi.org/10.2136/sssaj1984.03615995004800010028x.

Walker, P.H., 1965. Soil and Geomorphic History in Selected Areas of the Cary Till, Iowa. Ph.D. Diss., Iowa State University, Ames, IA.

Walker, P.H., 1966. Postglacial environments in relation to landscape and soils on the Cary Drift, Iowa (No. 549). Agriculture and Home Economics Experiment Station, Iowa State University, Ames, IA.

Walker, P.H., Ruhe, R.V., 1968. Hillslope models and soil formation II. Closed systems. Int. Soc. Soil Sci. Trans. 4, 561–568.

Wang, Y., Zhang, J.H., Zhang, Z.H., Jia, L.Z., 2016. Impact of tillage erosion on water erosion in a hilly landscape. Sci. Total Environ. 551, 522–532. http://dx.doi.org/10.1016/j.scitotenv.2016.02.045.

Warner, R.F., 2006. Natural and artificial linkages and discontinuities in a Mediterranean landscape: some case studies from the Durance Valley, France. Catena 66 (3), 236–250. http://dx.doi.org/10.1016/j.catena.2006.02.004.

Wills, S.A., Burras, C.L., Sandor, J.A., 2007. Prediction of soil organic carbon content using field and laboratory measurements of soil color. Soil Sci. Soc. Am. J. 71 (2), 380–388. http://dx.doi.org/10.2136/sssaj2005.0384.

Zheng-An, S.U., Zhang, J.H., Xiao-Jun, N.I.E., 2010. Effect of soil erosion on soil properties and crop yields on slopes in the Sichuan Basin, China. Pedosphere 20 (6), 736–746. http://dx.doi.org/10.1016/S1002-0160(10)60064-1.

CHAPTER

12

Mapping Soil Vulnerability to Floods Under Varying Land Use and Climate

Abdallah Alaoui
University of Bern, Bern, Switzerland

INTRODUCTION

Flooding is a natural process that is difficult to control and avoid. But the processes and conditions that result in floods are often predictable and usually occur in the same areas, known as floodplains. Most floodplains are flat and fertile, which is why they are often inhabited and developed by humans despite the risk of flooding (Huppert and Sparks, 2006).

Floods can cause considerable damage to property and infrastructure, threaten human lives and cost millions in emergency assistance, clean-up, and remediation. Various regions in Europe have suffered from severe flooding over the last four decades. Examples in Germany are the Rhine floods in 1993, which caused 530 million euros of damage; the Rhine floods in 1995, where damage amounted to 280 million euros; and the Oder flood in 1997, which caused damage worth 330 million euros. Floods in the Elbe and Danube catchments in August 2002 resulted in 38 fatalities and financial damage of more than 18,000 million Euros (Kron, 2004; Büchele et al., 2006; Hilker et al., 2009). In Switzerland, floods have mainly affected regions in the Alpine foothills and in the Central Alps. The bulk of the total cost of damage is due to a small number of major events which, taken together, caused financial damage of nearly 8000 million euros, including damage from debris flows, landslides, and rockfalls. These disasters may be exacerbated by climate change, which may lead to changes in the frequency, intensity, spatial extent, duration, and timing of rainfall events. The precipitation scenario developed by Frei (2007) predicts a decrease in rainfall of about 19% in summer and an increase of about 11% in winter in southern Switzerland by 2050. Extreme events (droughts, heavy precipitation) are expected to become more frequent. Heavy precipitation might trigger natural hazards (floods, landslides, mudflows, soil erosion) and lead us to believe that major events in the past are not unique and that similar events will occur in the future.

Environmental planners need solutions that address daily ability to effectively predict and respond to repetitive urban problems and future market fluctuations. The success of planners in combating chronic urban problems

Soil Mapping and Process Modeling for Sustainable Land Use Management.
DOI: http://dx.doi.org/10.1016/B978-0-12-805200-6.00012-8

is largely determined by their ability to use effective tools and planning support systems that allow them to make informed decisions, based on actionable intelligence (Steiniger and Geoffrey, 2009).

Earth observation techniques can contribute to more accurate flood hazard modeling, and they can be used to assess damage to residential properties, infrastructure, and agricultural crops. Today, planners around the world use geographic information systems (GISs) with a variety of applications. Floodplain maps can be the most valuable tools for avoiding severe social and economic losses from floods. If they are accurate and up to date, floodplain maps can also improve public safety. Accurate floodplain maps are the keys to better floodplain management (Simonovic, 1993).

Flood prediction requires quantitative knowledge on infiltration- and runoff dynamics. A large amount of research at small spatial scales with the aim to understand these dynamics has been carried out (e.g., Baigorria and Romero, 2007; Buttle and McDonald, 2002; McDonnell, 2003; Weiler, 2005). When up-scaling such small-scale investigations to the catchment scale, the organization of the catchment (connectivity and patchiness) need to be taken into account (Fiener et al., 2011). Conceptual models typically lump numerous surface properties and do not consider subsurface processes explicitly. They are efficient in computational runtimes and can be applied to large spatial scales. They are therefore widely used. One example of a commonly used conceptual approach is that of terrain analysis using digital elevation models (TauDEMs) (Tarboton, 1997). This simple approach makes it possible to generate high-resolution maps of flow networks at any spatial scale that are suited for practical use. Digital data generated by this approach also have the advantage that they can be readily imported and analyzed in GIS. Advances in GIS technology and the increasing availability and quality of digital elevation models (DEMs) have greatly expanded the potential application of TauDEMs to many hydrological, hydraulic, water resource, and environmental investigations (Steiniger and Geoffrey, 2009). The continuity of the DEM is an important contributor to interpolated gradient values, potentially affecting energy estimates as well as flow directions (Tarboton, 1997).

The main objective of this work is to introduce a new method for mapping soil vulnerability to floods. The specific objectives of this study are to (1) to identify and characterize the flow processes at the plot scale, and (2) to up-scale these flow processes at the catchment scale, by combining maps of flow directions according to Tarboton (1997), maps of zones of predisposition to surface runoff and in situ sprinkling experiments.

MATERIAL AND METHODS

Site Description

Our concept for mapping soil vulnerability to floods was applied to the experimental area called Innerrüteni/Hälfis in Kandergrund located 1110 m above sea level (masl) at the east side of the Kander Valley, which spreads out in the north-south direction in the Bernese Oberland in Switzerland (Furrer et al., 1993). On the west side of the investigated area, the Kander River is located at 793 masl and can have streamflow discharge as high as $7.7\,m^3\,s^{-1}$, as observed during the period between 1961 and 1980 (Schädler and Weingartner, 1992). On the eastern part of the investigated area the glacier deposit of the Kandergrund is divided into two main compartments (Furrer et al., 1993): a moraine till deposit at the west side and slope debris on the east side. The mean annual temperature is 5.9°C and the total annual precipitation is 1274 mm. In addition to grassland soil (G1–G7), two types of forest were considered in this study: Fa (Fa1, Fa2, and Fa3) and Fb (Fb1,

Fb2, and Fb3) corresponding to Spruce forest (*Calamagrostio variae-Pieceetum*) and Fir-Beech forest (*Adenostylo alliariae-Abieti-Fagetum*), respectively. The soil is described as Dystric Cambisol (WRB/FAO) in the grassland and in forest soil Fb, and as Rendzic Leptosol (WRB) (Rendzina (FAO)) in forest soil Fa (USDA, Soil Survey Staff, 2003). Its texture consists of clay loam to a depth of 0.60 m in grassland and silt loam in forest soil to a depth of 0.75 m. Its organic carbon content varies from 9 to 9.6% in grassland and from 8.3 to 12.5% in forest soil. A pH varying from 5 to 8 was measured in topsoil and subsoil, respectively, in both grassland and forest soils. At both locations the soil is relatively shallow and the weathered bedrock starts at depths below about 0.30 m.

In Situ Field Rainfall Simulations

The experiments were carried out during summer 2008 (April–October). On the 1 m^2 plots, rainfall simulation was supplied by a rainfall simulator: a metallic disc with a surface of 1 m^2 which is perforated with 100 holes attached to small tubes that lead into a reservoir. Rainfall simulation was applied from a height of 0.50 m from the ground. The metallic disc is moved by an electric motor, and the rainfall intensity is controlled by a flowmeter (Alaoui and Helbling, 2006). The duration of each rainfall simulation was 1 hour. The intensities of rainfall simulations were 24, 36, and 48 mm h^{-1}. In this area the 100-year return period of rainfall is 60 mm h^{-1} (Alaoui et al., 2012). In total, 57 sprinkling experiments carried out on grassland and forest soils with three different rainfall-simulation intensities were used to up-scale surface runoff in the small catchment under consideration. Surface runoff out of each plot of 1 m^2 surface area was measured during the rainfall simulation, using a metallic sheet (1 × 0.50 m) inserted in a soil profile at 0.05–0.10 m depth to collect surface runoff along a width of 1 m. The volume of collected water was measured with a flowmeter (Alaoui and Helbling, 2006) and stored automatically in a datalogger (CR10X, Campbell Scientific Inc.). In this study, runoff coefficient (RC) specific to an individual rainstorm (i.e., rainfall simulation) is defined as surface runoff divided by the corresponding rainfall, both expressed as depth over plot area (mm).

Principle of the Extrapolation of Runoff Process

The hydrological response units (HRUs) were first delineated according to soil type, geology, vegetation, and field slope. Sprinkling experiments of three different intensities were carried out on each plot of a given slope to attribute a specific value of runoff to a class category.

Predisposition to surface runoff was then mapped using the classification of Markart et al. (2004), who attributed a value of RC and slope to each of six classes corresponding to an increasing risk of surface runoff from 1 (no risk) to 6 (very high risk); after attributing the RC to each flow direction of a pixel of 2 m × 2 m resolution, flow directions were then determined at the catchment scale according to the principle of Tarboton (1997) using ArcGIS, version 9.3 (Fig. 12.1).

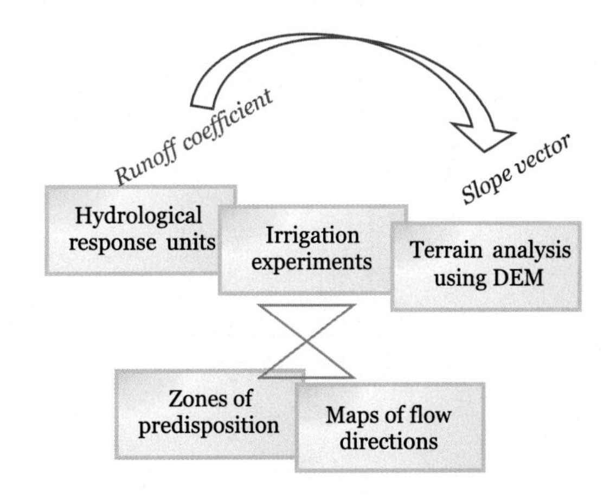

FIGURE 12.1 Steps to realize maps of cumulated runoff in Innerrrüteni in Kandergrund.

RESULTS AND DISCUSSIONS

Compacted grassland soil promotes surface runoff and increases the risk to surface runoff excess with increasing rainfall intensities (Alaoui et al., 2012). Furthermore, dry antecedent conditions and related water repellence of the forested catchments prevent surface runoff generation (e.g., Badoux et al., 2006; Schwarz, 1986; Kohl et al., 1997).

The storage capacity of soil exerts dominant control on surface runoff at a local scale which in turn is controlled by soil vegetation type. By applying a statistic t-test ($\alpha = 5\%$), one can say that the values obtained for forest soils were significantly higher ($\alpha = 5\%$) than those for grassland (Fig. 12.2A). This is due, on the one hand, to more intense root water uptake capacity by trees, and on the other, by the larger unsaturated hydraulic conductivity of forest

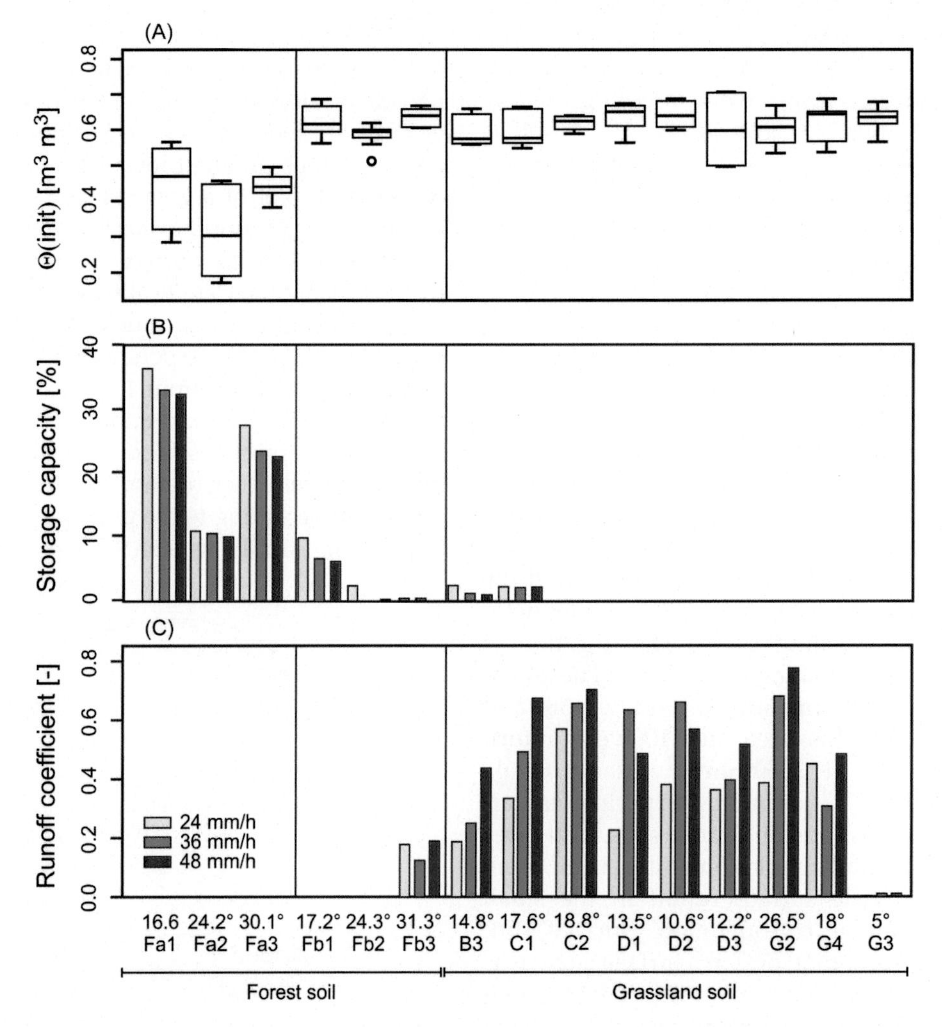

FIGURE 12.2 Initial soil water content (A), storage capacity (B), and Runoff coefficient, RC (C); storage capacity is obtained from the difference between total porosity and maximum water content measured during the infiltration phase (during sprinkling experiment).

soil as compared to grassland. These results showed that the two types of vegetation have distinct effects on soil structure principally leading to two major flow processes, vertical infiltration in forest soil and surface runoff in grassland soil (Alaoui et al., 2012). In this study, we have also shown, in addition to the other parameters that field slope plays a major role in surface runoff generation as shown in this figure: forest soil does not generate any surface runoff except at the highest slope angle (31.3°) defined as a threshold belong which surface runoff occurs in this case study (Fig. 12.2C). Similarly, Badoux et al. (2006) reported that sprinkling experiments on dry to humid Cambisols in forested catchments result in no or only low RC values varying between 0.01 and 0.16. They showed that RC was considerably higher after dry antecedent conditions than after wet antecedent conditions, which is probably due to water repellency. In contrast, artificial high-intensity precipitation on plots of 1 m^2 plots in forest soil leads to high RC values (from 0.39 to 0.94) on humid to wet gleysols (Badoux et al., 2006).

Examination of the storage capacity (Δsat) shows that the values obtained for forest soil are significantly higher (α = 5%) than those for grassland soil (Fig. 12.2B). Any influence of external weather factors can be ruled out because the infiltration experiments were carried out during the same periods in the summer season, excluding any precipitation events during at least 1 week. The higher storage capacity of forest soil is probably due, on the one hand, to more intense root water uptake capacity by trees (Fig. 12.2B), and on the other, by the larger unsaturated hydraulic conductivity of forest soil. In fact, Fig. 12.2 exhibits significantly higher (α = 5%) values of $\Delta\theta$ especially in forest soil Fa (Group 1) compared to grassland soil (Group 3). In this study, we have also shown, in addition to the other parameters that field slope plays a major role in surface runoff generation as shown in this figure: forest soil does not generate any surface runoff except at the highest slope angle (31.3°) defined as a threshold belong which surface runoff occurs in this case study.

The maps of the zones with a predisposition to surface runoff classified according to Markart et al. (2004) showed areas in the low predisposition classes located principally in the eastern and central parts of the catchment under consideration, corresponding to limestone outcrop and forest soils, respectively (Fig. 12.3). In addition, it appears that increasing the rainfall-simulation intensity from 24 to 48 $mm\,h^{-1}$ increases the risk of the predisposition to surface runoff from 3 to 5, respectively, in the major parts of the catchment on both sides of the forest. Additional insights can be drawn from these maps: grassland areas are more subject to surface runoff with variable values, depending especially on field slopes. In contrast, forest soils generate vertical percolation in all cases except on the plot of a slope greater than 31.3°.

The maps of the flow directions obtained using the TauDEM (Tarboton, 1997) shown in Fig. 12.4 highlight two main observations:

1. Surface runoff converges in the channels which constitute the main lateral flow pathways and the cumulated runoff increases, generally in the direction of flow;
2. Two main areas composing the catchment under consideration are distinguished: A (2.3 km^2) and B (1.5 km^2) which contain two independent flow networks resulting in final outlets O-A7 and O-B6, respectively (Fig. 12.3).

For a rainfall intensity of 48 $mm\,h^{-1}$, estimated volume was 40,270 m^3 in O-A7 and 18,030 m^3 in O-B6, showing the great difference between the subcatchments (Table 12.1). This is also the case for the RC which is twice as high in A as in B. Increasing rainfall intensity, from 24 to 36 $mm\,h^{-1}$, increases the risk of the predisposition to surface runoff from weak to

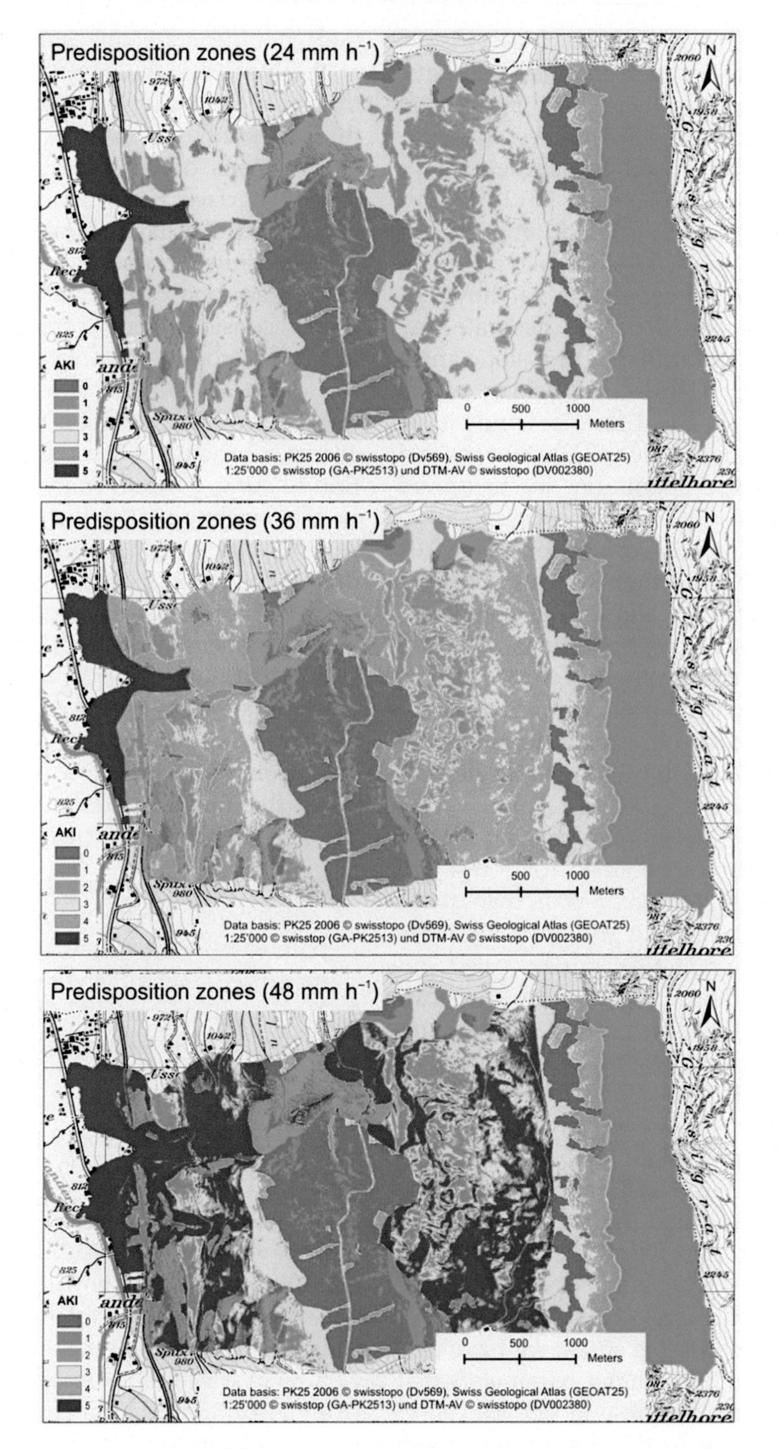

FIGURE 12.3 Maps of the predisposition zones (according to Markart et al., 2004) in the Kandergrund valley, produced with sprinkling experiments and delineation of HRUs, defined from the soil data, geology, topography, and vegetation.

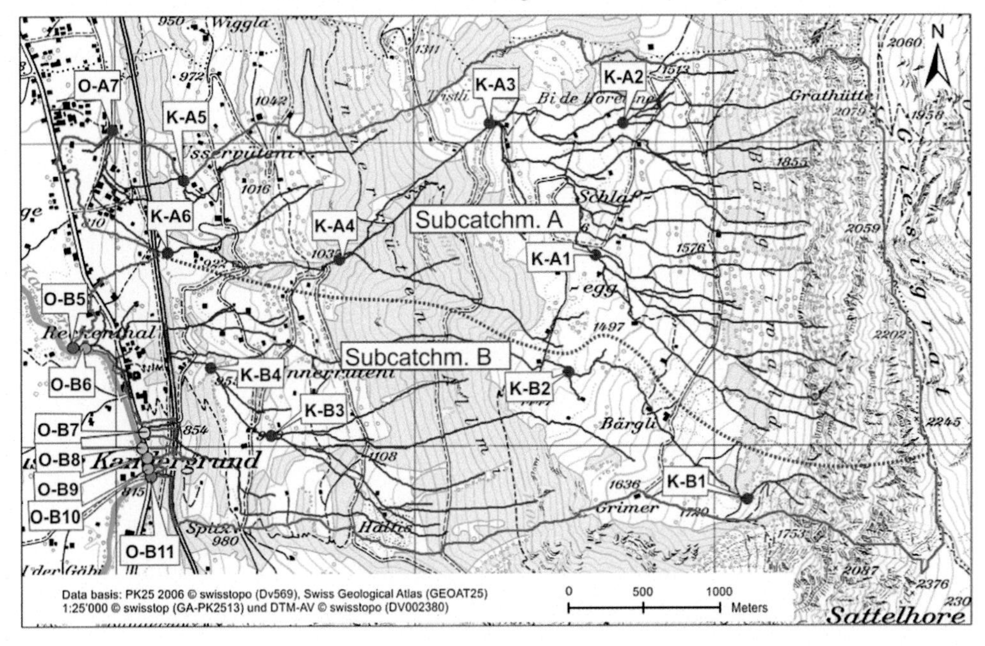

FIGURE 12.4 Map of the flow network defined by the terrain analysis using digital elevation model according to Tarboton (1997).

medium when considering the entire catchment (Fig. 12.4).

The flood event that occurred on the October 10, 2011 (with about $100\,m^3\,s^{-1}$) caused severe damage in the cantons of Valais (e.g., Lötschental) and Berne (e.g., Kandertal) as well as in some parts of Central Switzerland. Although the Federal Office for the Environment (FOEN) operationally forecasts the discharge of several river systems through a dense network of discharge gauging stations in Switzerland, flood peak of the river Kander (Bernese Oberland) was strongly underestimated—below the warning level. As our study region belongs this affected area, it aims first to demonstrate that a better planning of the land use in the region can significantly reduce peak discharge if their succession and connectivity are taken into account.

CONCLUSIONS

The main aim of this study was to introduce a new methodology to delineate the zones of predisposition to floods based on TauDEM and sprinkling experiments under various conditions of land use.

The marked differences in the textural and structural porosities between forest and grassland plots appear to control runoff processes. On the one hand, forest soil has a higher storage capacity than grassland soil, probably caused by large unsaturated hydraulic conductivity and root water uptake resulting in lower surface runoff. On the other hand, fine material present in the topmost 10 cm helps to generate a structure that is probably unfavorable to vertically downward percolation and thus enhances surface runoff as observed on the grassland

TABLE 12.1 Calculated Water Volumes and Runoff Coefficients in the Catchment Under Consideration According to Tarboton (1997)

Outlet	**Volume (m^3)**			**Runoff Coefficient**		
See Fig.	$24\,mm\,h^{-1}$	$36\,mm\,h^{-1}$	$48\,mm\,h^{-1}$	$24\,mm\,h^{-1}$	$36\,mm\,h^{-1}$	$48\,mm\,h^{-1}$
K-A1	3070	5543	7890	0.0338	0.0407	0.0434
K-A2	300	577	841	0.0033	0.0042	0.0046
K-A3	3906	7490	10,746	0.0430	0.0550	0.0592
K-A4	4944	9621	13,972	0.0545	0.0706	0.0769
K-A5	1108	2486	3924	0.0122	0.0183	0.0216
K-A6	3379	6606	9674	0.0372	0.0485	0.0533
O-A7 (A)	14,083	27,358	40,273	0.16	0.20	0.22
K-B1	321	474	653	0.0035	0.0035	0.0036
K-B2	1698	3580	5521	0.0187	0.0263	0.0304
K-B3	1038	1844	2785	0.0114	0.0135	0.0153
K-B4	1018	2160	3263	0.0112	0.0159	0.0180
O-B5	24	36	48	0.0003	0.0003	0.0003
O-B6	6027	11,993	18,029	0.0664	0.0881	0.0993
O-B7	1121	2371	3554	0.0123	0.0174	0.0196
O-B8	61	119	182	0.0007	0.0009	0.0010
O-B9	10	19	29	0.0001	0.0001	0.0002
O-B10	11	21	31	0.0001	0.0002	0.0002
O-B11	90	196	309	0.0010	0.0014	0.0017
Subcatchm. B	7344	14,755	22,182	0.08	0.11	0.12
Whole atchm.	21,427	42,113	62,455	0.24	0.31	0.34

plots. However, within each soil category, slope plays an important role in generating surface runoff.

Up-scaling runoff processes using TauDEM based on sprinkling experiments gave more quantitative insight into flow processes such as flow directions and runoff quantification and traced the hydrological connectivity between the zones of predisposition. Our case study showed that a better planning of land use can significantly reduce peak discharge if their succession and their connectivity are taken into account.

References

Alaoui, A., Helbling, A., 2006. Evaluation of soil compaction using hydrodynamic water content variation: comparison between compacted and non-compacted soil. Geoderma 134, 97–108.

Alaoui, A., Spiess, P., Beyeler, M., Weingartner, R., 2012. Up-scaling surface runoff from plot to catchment scale. Hydrol. Res. 43 (4), 531–546.

Badoux, A., Witzig, J., Germann, P., Kienholz, H., Lüscher, P., Weingartner, R., et al., 2006. Investigations on the runoff generation at the profile and plot scales, Swiss Emmental. Hydrol. Proc 20, 377–394.

Baigorria, G.A., Romero, C.C., 2007. Assessment of erosion hotspots in a watershed: integrating the WEPP model and GIS in a case study in the Peruvian Andes. Environ. Model. Softw 22, 1175–1183.

Büchele, B., Kreibich, H., Kron, A., Thieken, A., Ihringer, J., Oberle, P., et al., 2006. Flood-risk mapping: contributions towards an enhanced assessment of extreme events and associated risks. Nat. Hazards Earth Syst. Sci. 6, 485–503.

Buttle, J.M., McDonald, D.J., 2002. Coupled vertical and lateral preferential flow on a forest slope. Water Resour. Res. 38, 1060.

Fiener, P., Auerswald, K., Van Oost, K., 2011. Spatio-temporal patterns in land use and management affecting surface runoff response of agricultural catchments—a review. Earth-Sci. Rev. 106, 92–104.

Frei, C., 2007. Climate Change and Switzerland 2050—Impacts on Environment, Society and Economy. Advisory Body on Climate Change (OcCC).12–16, Available from: http://www.occc.ch.

Furrer, H., Huber, K., Adrian, H., Baud, A., Flück, W., Preiswerk, C., et al., 1993. Geologischer Atlas der Schweiz, 1:250000, Adelboden (LK 1247), Blatt 87 [Swiss Geological Atlas, 1:25,000, Adelboden (LK 1247), Sheet no. 87]. Landshydrologie und-Geologie, Bern, Switzerland.

Hilker, N., Badoux, A., Hegg, C., 2009. The Swiss flood and landslide damage database 1972–2007. Nat. Hazards Earth Syst. Sci. 9, 913–925.

Huppert, H.E., Sparks, S.J., 2006. Extreme natural hazards: population growth, globalization and environmental change. Philos. Trans. R. Soc. A 364, 1875–1888.

Kohl, B., Markart, G., Stary, U., Proske, H., Trinkaus, P., 1997. Abfluss-und Infiltrationsverhalten von Böden unter Fichtenalbeständen in der Gleinalm (Stmk.) [Drainage and infiltration characteristics of spruce forest soils in the 'Gleinalm' alp (Steiermark)]. Berichte der Forstlichen Bundesversuchsanstalt Wien 96, 27–32.

Kron, W., 2004. Zunehmende Überschwemmungsschäden: Eine Gefahr für die Versicherungswirtschaft? (Increasing flood damage: a hazard for the insurance industry?), ATV-DVWK: Bundestagung 15–16 September 2004 in Würzburg, DCM, Meckenheim, pp. 47–63.

Markart, G., Kohl, B., Sotier, B., Schauer, T., Bunza, G., Stern, R., 2004. Provisorische Geländeanleitung zur Abschätzung des Oberflächenabflussbeiwertes auf alpinen Boden-/Vegetationseinheiten bei konvektiven Starkregen [A simple Code of Practice for Assessment of Surface Runoff Coefficients for Alpine Soil-/Vegetation Units in Torrential Rain] (Version 1.0). BFW Dokumentationen. Bundesamt und Forschungszentrum für Wald, Wien.

McDonnell, J.J., 2003. Where does water go when it rains? Moving beyond the variable source area concept of rainfall–runoff response. Hydrol. Proc. 17, 1869–1875.

Schädler, B., Weingartner, R., 1992. Natürliche Abflüsse 1961–1980 [Natural discharge 1961–1980]. In: Hydrologischer Atlas der Schweiz, Tafel 5.4. Bundesamt für Landestopographie, Wabern, Switzerland.

Schwarz, O., 1986. Zum Abflussverhalten von Waldböden bei künstlicher Beregnung [Runoff characteristics of forest soils as affected by artificial irrigation]. In: Einselem, G. (Ed.), DFG–Research Report on the Landscape Ecological Project 'Schönuch'. VCH Publisher, Weinheim, pp. 161–179.

Simonovic, S.P., 1993. Flood control management by integrating GIS with expert systems: Innipeg city case study. A Paper Presentation in HrydroGIS Conference, Vienna.

Soil Survey Staff, 2003. Keys to Soil Taxonomy. Soil Survey Staff, 0120. U.S. Dept. of Agriculture, Natural Resources Conservation Service, Washington, D.C.

Steiniger, S., Geoffrey, J.H., 2009. Free and open source geographic information tools for landscape ecology. Ecol. Infor. 4, 183–195. http://dx.doi.org/10.1016/j.ecoinf.2009.07.004.

Tarboton, D.G., 1997. A new method for the determination of flow directions and upslope areas in grid digital elevation models. Water Resour. Res. 33, 309–319.

Weiler, M., 2005. An infiltration model based on flow variability in macropores: development, sensitivity analysis and applications. J. Hydrol. 310, 294–315.

Further Reading

Moore, I.D., Grayson, R.B., Ladson, A.R., 1991. Digital terrain modelling: a review of hydrological, geomorphological, and biological applications. Hydrol. Proc. 5, 3–30.

Index

Note: Page numbers followed by "*f*" and "*t*" refer to figures and tables, respectively.

L

M

T

U

Printed in the United States
By Bookmasters